AIR POLLUTION AND CONTROL

(Text book for Undergraduate and Graduate courses in Environmental, Chemical and Mechanical Enginnering, Competitive Examinations of IES, IAS, GATE, IFS, PCS, AMIE (Section B) and a comprehensive reference for Technologists, Designers and Professional Engineers)

ER. RAJNI KANT
Retired Chief Engineer,
UP Electricity Board,
and
Former Consultant and Adviser to
Various Design and Engineering Organizations in India

DR. KESHAV KANT
Retired Professor of Mechanical Engineering
Indian Institute of Technology, Kanpur, India

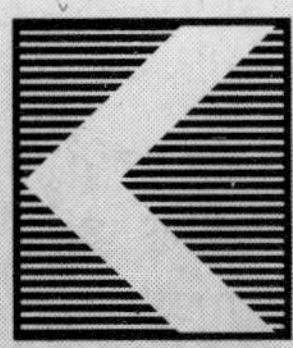

KHANNA BOOK PUBLISHING CO. (P) LTD.
Publisher of Engineering and Computer Books
4C/4344, Ansari Road, Darya Ganj, New Delhi-110002
Phone : 011-23244447-48 **Mobile:** +91-9910909320
E-mail : contact@khannabooks.com
Website : www.khannabooks.com

Dear Readers,

To prevent the piracy, this book is secured with HIGH SECURITY HOLOGRAM on the front title cover. In case you don't find the hologram on the front cover title, please write us to at contact@khannabooks.com or whatsapp us at +91-99109 09320 and avail special gift voucher for yourself.

Specimen of Hologram on front Cover title:

Moreover, there is a SPECIAL DISCOUNT COUPON for you with EVERY HOLOGRAM.

How to avail this SPECIAL DISCOUNT:

Step 1: Scratch the hologram
Step 2: Under the scratch area, your "coupon code" is available
Step 3: Logon to www.khannabooks.com
Step 4: Use your "coupon code" in the shopping cart and get your copy at a special discount
Step 5: Enjoy your reading!

Price: ₹ 450.00

ISBN: 978-93-86173-30-0

Air Pollution and Control
by Rajni Kant, Keshav Kant

First Edition: 2018
Reprint Edition: 2023

To view complete list of books, Please scan the QR Code:

Published by:
Khanna Book Publishing Co. (P) Ltd.
CIN: U22110DL1998PTC095547

Visit us at: **www.khannabooks.com**
Write us at: **contact@khannabooks.com**

Printed in India by: Print Point, Delhi.

Dedicated to the Memory of

Our Beloved Mother (Late)

Mrs. Ram Mohini

and

Our Beloved Father (Late)

Mr. Mathura Prasad

a Versatile Scholar, Teacher, Educationist and Guide

Preface

Environmental issues have gained great importance in the last two decades, but recently the 'Environmental Pollution and Global warming' have assumed menacing proportions which is very adversely affecting the life on earth particularly in developing countries and may become even worse in future. The need of the hour is to control environmental pollution and reduce other undesirable emissions and effluents. Little can be done to control natural causes, but much can done to reduce and remedy the anthropogenic causes of these problems. It is essential to ensure that the threat of environmental pollution is drastically reduced while the development works to fulfill the human needs can be carried out without adverse environmental effects.

This book is devoted to the subject of Air Pollution and Control. It may not be out of place to mention that the same authors have earlier published a book on '***Water pollution- Management, Control and Treatment***' in 2010.

The goal of the authors is to provide rigorous and refreshing treatment of the subject covering present day technology and developments. This book has been divided in various chapters systematically as given in the list of contents. The book presents enormous information and covers all aspects of air pollution in various sectors like power including nuclear power, industrial, domestic, transport (including surface transport, aviation and marine transport), agriculture, forestry, waste handling and disposal, etc. The monitoring, measurements of pollutants, ill-effects, and emission limits of pollutants, safety and control of radioactivity, structure of environmental impact reports, etc. are also covered along with control measures. The authors have attempted to reach every section of readers including:

Students of particularly Environmental and Mechanical engineering and also other branches of Engineering (ii) Technologists, designers, operation and maintenance engineers of industries, electrical power plants, heat and power utilities; (iii) Aspirants for competitive examinations for IAS, IES, IFS, PCS, and aspirants for various state and private technical services, etc. (iv) General readers who are interested in the field for better understanding and knowledge. Each Chapter is followed by a list of questions based upon the text.

The contents of the book have been evolved over a long period of time and careful considerations of the requirements of various sections of readers have been given. The first author has been a practicing engineer, consultant and a designer who has been actively involved in the field of thermal power plants and control of environmental pollution for almost 50 years and has widely published in the area. The second author has devoted his entire life as an academician pursuing both teaching and research at I.I.T. Kanpur and has made significant contributions.

June 2, 2018 **Er. Rajni Kant and Dr. Keshav Kant**

About the Authors

Er. Rajni Kant retired from UP State Electricity Board as Chief Engineer, Thermal Design and Engineering following which he remained Consultant / Advisor to a number of Design and Enginnering Organizations namely Premier Engineering Technologies-Engineering and Management Consultants, New Delhi, M/S Premier Mott Macdonald (Consulting Engineers) New Delhi and other organizations.

Throughout his career both in UPSEB as well as a Consultant / Adviser (after superannuation), he has actively participated in Planning, Design and Engineering, Development, Construction & Erection, Testing & Commissioning and Operation & Maintenance of Thermal Power Plants of various capacities. He was also deeply involved in Environmental studies, Planning and Environmental design of various power plants.

Er. Rajni Kant, an Engineering graduate from BHU in 1954 (now IIT BHU Varanasi), received an extensive training in the works of M/S Hitachi Ltd. (Japan) and also at large capacity Power Stations and Projects of Georgia Power Company (USA). He has more than 50 years of experience to his credit in the aforesaid fields and has authored about 40 papers in various fields including Power Plants and Environmental Engineering which were published in National and International Journals and Proceedings of National and International Conferences. He is a member of several professional Organizations namely; (1) Association of Technical Scholarship (Japan), (2) Association on Electrical Generation, Transmission & Distribution (Afro-Asian Region), (3) Central Board of Irrigation & Power (India) and (4) International Society of Environmental Botanists. Er. Rajni Kant along with Dr. Keshav Kant have also authored a book entitled, 'Water Pollution – Management, Control and Treatment' in 2010, and its revised edition is in press.

Dr. Keshav Kant is a former Professor of Mechanical Engineering, I.I.T. Kanpur, who after superannuation from I.I.T. Kanpur in June 2005, served as Professor and Head of Mechanical Engineering Department and later as the Director in a number of private Engineering Colleges/ Institutes. He has 3 Monographs, 5 Technical reports, 90 publications to his credit in National and International Journals and Proc. of National and International Conferences.

A graduate from Birla Institute of Technology Mesra, Ranchi in 1964 and Master of Engineering with Honours from the University of Roorkee (now IIT,Roorkee) in 1966, he served as a faculty of Mechanical Engineering mostly at IIT Kanpur since 1967. He obtained his Ph.D. Degree in 1983 from McGill University Montreal, Canada. He has been the recipient of several awards from his Alma-maters. He is Included in 'Who's Who in the World', 2010 edition. He has significantly contributed in developing Undergraduate and Graduate Courses in the Department and Thermal Science Laboratories at I.I.T. Kanpur. He has guided 7 Ph.Ds, 45 M.Techs, 30 senior UG Design Projects and 35 AMIE Students. He is the author of 3 books.

Professor Keshav Kant has been involved with Technology Development related to Surfactants, Simulated Moving bed Heat Regenerators and Retrofitting Small Capacity Refrigeration Systems using Eco-friendly Refrigerants. He has also been involved with Software development related to heat exchangers and Air Conditioning System Design. He has also been involved with several sponsored research projects and consultancy projects. He has been Visiting Professor to several U.S. Universities. He has been an invited speaker to many universities abroad and Keynote speaker in many National and International conferences, and many Institutions in India. He has been a reviewer for the J. of IIR Paris and J. of IE (India). He has been accredited as one of the most distinguished teachers at IIT Kanpur and U.S. Universities. As a Chairman of the Kanpur Centre, (IE) India for 2 years and member of its executive for 17 years, he has organized several technical events related to 'Water Resources' and 'environment' for the awareness of the people in the society. He is member of several Professional Organizations namely: (1) Fellow of the Institution of Engineers (India), (2) Member of Indian Society for Technical Education, (3) Member of the Indian Society for Heat and Mass Transfer and (4) Member of Indian Science Congress.

Acknowledgements

The authors very gratefully acknowledge the direct and indirect contributions by various Technologists, Scientists, Environmentalists, Private and State authorities for their help and suggestions in writing this book on **'Air Pollution and Control'.**

The authors would like to express their gratitude for many useful suggestions to Professor Arun S. Mujumdar – Former Professor of Chemical Engineerinig, McGill University, Montreal, Canada and to Professor Ashok Mulchandani, Distinguished Professor of Chemical and Environmental Engineering at the University of California, Riverside, California, USA.

The authors would also like to express their sincere thanks and appreciation to their sisters Ms. Kanak Saxena and Dr. (Ms.) Padma Vati for their painstaking efforts in typing the manuscript and making diagrams.

Er. Rajni Kant
Dr. Keshav Kant

Foreword

It is my privilege to write this forwarding piece for the book entitled, "**Air Pollution and Control**" written by **Er. Rajni Kant** and **Prof. Keshav Kant**. I found this book to be one of the most relevant and detailed texts written in India on this topic. The book takes off with topics describing the extent of air pollution and its effect on ecology as well as human beings. Then it describes various sources of gaseous and particulate air pollutants with inclusion of several case studies including the Bhopal gas tragedy. It covers various new topics like bioaerosols or aeroallergens and hazardous air pollutants including diesel exhaust and dioxins.

This book overall consists of twenty chapters and is apt to be employed as the most comprehensive text for graduate students, practising engineers and teachers alike. It is worth to highlight a few chapters that truly stand out. Chapter 3 describes the meteorological conditions and their role in dispersion and transport of air pollutants. It also discusses various advance source models based on a set of mathematical equations effectively used by scientists to make chimney plume prediction based on source strength and meteorological information. Chapter 8 gives ample details of the various industrial processes that subsequently result in the formation of ambient air pollutants. It also describes state of the art industrial control technologies to curb these pollutants at the source itself using technologies like electrostatic precipitators and dry and wet type of scrubbers. Chapter 10 exclusively talks about an important topic of mercury emissions largely from the coal based thermal power plants and engineering control mechanisms that may help in appropriate capturing of mercury fumes from these sources.

This book takes a holistic view of the air pollution problem and includes topics like effect of air pollution on semiconductor industries and also talks about emissions from solid waste and sewage treatment plants, nuclear waste, and evolution of air pollutant standards and development of air quality indices. I think it will be a popular choice among students and will bridge the long felt gap for a comprehensive reference on Air Pollution and Control relevant for Indian conditions.

Professor Tarun Gupta
Ph.D. (Harvard University), P.K. Kelkar Fellow
Department of Civil Engineering,
Indian Institute of Technology, Kanpur, India

Contents

CHAPTER 1

INTRODUCTION

1.1 AIR POLLUTION

Air pollution can be defined as the presence of contaminants in atmospheric air in concentrations which are harmful to humans, animals, plants, metals and materials, and environment in general. The air contaminants may be gases, vapours or suspended particulate matter, bacteria, virus, fungi spores, pollens, aerosols, excessive sound, nuclides / radiation, excessive heat (thermal pollution), etc. These may come from natural causes and / or anthropogenic causes, but the level of pollution due to anthropogenic causes is growing very fast.

According to 'Global Burden Disease (GBD) report of the US based Health Effect Institute, air pollution and related diseases are the fifth largest killer in india and are responsible for six fold premature deaths (2010) as compared to that in 2000 (when the figure was 100,000). On release of 'GBD Report', the Centre for Sciences and Environment—a New Delhi based non-profit Organization, analysed the data of 2010 of Central Pollution Board, India. The findings of analyses were disturbing. According to findings, about half of Indian cities were reaching severe particulate pollution along with oxides of nitrogen, ozone, air toxins, etc. In fact most of the 400 locations monitored in 2010 reported high or critical levels of particulate matter. The rise of particulate matter in smaller cities is faster than in Delhi. Pollution in rural areas in India is also getting as bad as in cities.

The data of the Central Pollution Control Board (2010) shows that smaller cities such as Solapur in Maharashtra, Rajkot in Gujarat, Yamnanagar and Faridabad in Haryana, Ghaziabad and Meerut in UP, Paontasahib in Himanchal, Vijaiwada in Andhra, Nagaon in Assam, Gwalior in MP, Howrah in West Bengal, etc. have footprints of bigger cities as regards air pollution. In the last two years, Gwalior ranked as most polluted with regard to 'Particulate Matter', and Howrah topped the list of most polluted with regard to 'Oxides of Nitrogen'.

At present the industrial growth of India is much below that of China, but if this growth rate increases, the pollution would be much greater.

The Control of pollutants from anthropogenic sources should be focussed upon, as the natural sources of pollution, in general, cannot be controlled. If the pollution from anthropogenic sources is controlled, most of the problems due to pollution can be surmounted. This is a big challenge in view of increase in population and consequent pressing need for development of power, industries, surface transport, shipping, electronic communication, mining of coal and minerals, mechanised agriculture and food security, defence systems, etc. The forest cover has been decreasing on account of use of more and more forest land for industries, mining, farming, power stations, housing, and other projects, besides cutting of trees for timber or fuel wood and unwanted destructions of forest wealth. This is, also, one of the causes of far exceeding the limit or capacity of our natural bio-geo-chemical environment to clean itself. What is true of India, is true for all developing and underdeveloped countries.

Natural Sources of Air Pollutants

The natural sources of pollutants may include:

- Volcanic eruptions and fires [which may emit large amount of dust, oxides of sulphur, hydrogen sulphide, carbon monoxide, chlorine, oxides of nitrogen, etc.];
- Sand/ dust storms and shifting/ deflation of sand;
- Forest fires, other accidental fires, petroleum well fires, coal/ lignite mine fires spontaneous combustion of coal/ lignite;
- Release of methane by cudding and digestion of food by animals;
- Release of gases from oceans on natural rise of temperature mainly methane and carbon dioxide;
- Salt particle spray from seas;
- Pollens from plant flowers;
- Fungal spores, bacteria, and viruses;
- Nuclides / radiation from natural sources including radioactive decay within earth crust and emission of radon gas.
- Decay of organic matter
- Bacterial action on soil and natural environment.

Anthropogenic Sources of Air Pollutants

The anthropogenic sources of pollutants may include:

- Various industries;
- Fossil fuel burning power stations;
- Surface transports;
- Aviation;
- Shipping/ water way transports;
- Nuclear power stations;
- Petroleum exploration, production, transport and refining,
- Transport, storage, and distribution of petroleum products
- Mining of various ores and materials like coal, lignite, etc.
- Transport of mined products, their refining/ benefaction, and use;
- Metallurgical operations
- Services like painting, varnishing, solvent use, spraying of insecticides, welding, washing and dyeing, etc.
- Chemical and petrochemicals industries, and other industries like leather, sugar, distilleries, cement, textile, pulp and paper, rubber, plastics, etc.
- Fires for cooking and heating, backyard fires, wilful burning of wastes, etc.;
- Agricultural activities including use of pesticides, 'Jhum' cultivation, burning of biomass, emission from machineries, etc.;
- Municipal sewage and solid wastes and landfills, hospital wastes, hazardous waste incinerators, electronic wastes (e-wastes);
- Military/ defence weapons including nuclear, chemical and germ warfare;
- Space exploration and rocketry; etc.

Ground Level Air Pollution

It may be mentioned that the increase in ground level concentration of pollutants over the background concentration in immediate vicinity of the source of emission is related to mass rate of emission of volume of pollutant gases discharged per unit time, height at which the contaminant gases are discharged (i.e. stack height plus plume rise), difference in temperature of stack gases and surrounding air, wind velocity, horizontal and vertical mixing of pollutants in air, settling velocity of polluting substances, etc. The ground level pollution is determined by actual measurements at different times of the day and in different seasons and then eight hourly average, twenty-four hourly average, and annual average are determined as per guide lines in Indian 'Environment Protection Rules'. However, the 'increase in emission of pollutants' directly affects the increase in ground level pollution and therefore there is utmost need of reduction of emissions.

Major Air Pollutants

Major primary pollutants of air include the following:

- Carbon monoxide
- Oxides of nitrogen represented by NOx – mostly nitric oxide (NO) and nitrogen dioxide ((NO_2)
- Nitrous oxide ((N_2O) – also a greenhouse gas
- Oxides of sulphur represented by SOx (i.e. SO_2 and SO_3)
- Hydrogen sulphide, mercaptans,
- Fluorine, hydrogen fluoride gas and other gaseous fluorine compounds
- Chlorine, hydrogen chloride gas or vapours, and other gaseous chlorine compounds
- Methane – also a greenhouse gas
- Hydrocarbons – unburnt gaseous hydrocarbon fuel or liquid fuel vapours and Volatile Organic Compounds (VOC) including Polycyclic Aromatic Hydrocarbons (PAH) and dioxins
- Ammonia (NH_3)
- Ozone–also a greenhouse gas
- Ozone depleting substances such as CFCs (Chlorofluorocarbons), HCFCs (Hydro chlorofluorocarbons), etc. in atmosphere. [These also create greenhouse effect]
- Suspended particulate Matter (SPM), Respirable Suspended Particulate matter (RSP) / Aerosols
- Micro particles of toxic metals or their compounds such as arsenic, lead, mercury, barium, beryllium, boron, cadmium, chromium, vanadium, manganese, nickel, phosphorous, selenium, zinc, iron, etc.
- Biological pollutants such as aero-allergens, bacteria, fungus, virus, etc.
- Airborne fumes, vapours and dust of 'Pesticides, insecticides, herbicides'
- Nuclides / radioactive gases

It may be mentioned here that the *carbon dioxide* is a normal constituent of atmosphere and is not rated as a pollutant, but it is a *greenhouse gas* and its rise of content in air is unwelcome because of its *global warming* effect.*

During the combustion of fuels, some of the pollutants may be formed, but others may already be present in fuels and in the oxidizing agent (usually air/ oxygen) used in the combustion process.

***Note: Greenhouse gases, greenhouse effect, global warming**

The solar energy in the form of heat is retained in atmosphere mainly in some gases present in atmosphere which are called 'Greenhouse gases' and the effect of heat retention by such gases is termed as

'greenhouse effect'. The increase in content of greenhouse gases in atmosphere raises the average atmospheric temperatures which is called 'Global warming'.
It may be mentioned that apart from carbon dioxide, methane, nitrous oxide, ozone in troposphere and CFCs, HCFC, and HFCs are important greenhouse gases which are causing global warming. As water vapour present in atmosphere also retains lot of heat, it is also included in the category of greenhouse gases, but the amount of water vapour in atmosphere varies from season to season and is not cumulative while other gases have a cumulative effect. (Man has no control over water vapour.)

Secondary Pollutants

Apart from primary pollutants, there are some secondary pollutants (which are formed in atmosphere even though not emitted from primary sources). These may include:

- Acid vapours,
- Aldehydes,
- Photo chemical smog,
- Peroxy Acetyl Nitrate (PAN),
- Ozone

1.2 IDENTIFICATION OF AIR POLLUTION IN AN AREA

The environmental air pollution in an area may be identified by many ways as described below:

1.2.1 Sensory Recognition of Air Pollution

Air pollution may be recognized by unusual odours, fall in visibility, irritation in eyes, skin irritation, breathing difficulty, change in mouth taste, etc. The sensory recognition is immediate and is particularly sensitive for air pollution.

1.2.2 Bio-indication of Pollution by Plants/'Effects of Pollution on Plants'

The environmental pollution stress (including pollution of air, water, or land) on plants is evident by:

- Loss or reduction of chlorophyll/green pigment of leaves, turning of leaves pale green or yellowish, bleaching of leaves termed as 'Chlorosis'. (This may be result of environmental pollution or deficiency of some nutrient.)
- Downward curvature of leaf due to higher growth rate on its upper surface than that of its lower surface termed as 'Epinasty'. (This may be due to environmental pollution or attack by pest or virus.)
- Dropping of leaves prematurely and un-seasonally termed as 'Abscission'. (This may be due to environmental pollution or malnutrition.)
- Collapsing of leave tissue or death of some living cells of leaves termed as 'Necrosis' of leaves. (This may be due to many reasons including environmental pollution, attack by pest/virus, etc.)
- Flowering and fruitarian changes, damages, premature droppings, lack of formation of fruit, etc. (This may be due to pollution or lack of nutrition or other stresses in the area. It may be also noted that the yield of a crop is related to the destruction of leaves by a pollutant.)
- Suppressed growth of plant and premature aging of plants in the area. (This may be due to many factors including pollution.)

Note: *Total dose of air pollutant is the product of concentration of pollutant in air and time of exposure to the pollutant. It may be noted that a high concentration of a pollutant for even short time may produce* *acute injury to plant species, while the same total dose of air pollutant in low concentration for a longer time may not produce perceptible injury as plants would defend themselves by their own natural defence mechanisms.*

Bio-Indication of Some Specific Air Pollutants by Plants

(i) ***Sulphur dioxide***:

Indicator plants: Wheat, Barley, Cotton, Apple, Alfalfa, etc. [Sensitivity concentration more than 0.1 ppm]

Effects: In broad leaf plants, SO_2 causes marginal and inter-vein white to straw blotched areas distributed over leaf. In grass like leaf plants it causes irregular necrotic streaks either side of mid-vein, while in conifers, it causes browning of necrotic tips of needles. In general, the sulphur dioxide causes bleaching and even necrosis in severe cases. However, such symptoms may appear on plants due to other causes like drought /winter or attack by fungus/ bacteria / pest.

(ii) ***Fluorides:***

Indicator plants: Gladiolus, Italian prune, Chinese apricot, etc. (Even a concentration of 0.1 ppb for 5 weeks affects gladiolus).

Effects: In broad leaf plants, fluorides cause accentuated dark red-brown band between green leaf and necrotic border around the entire leaf. In grass like leaf plants and conifers, fluorides create tip burns, and accentuated line between live and necrotic tissue. In general, Fluoride causes browning and necrosis of leaf tips in the long run. However, similar damages may be seen due to other causes like high temperatures, drought, severe winter, bacterial/ pest/ fungal attack, SO_2 injury, etc.

(iii) ***Ozone:***

Indicator plants: Potato, Tomato, Tobacco, Spinach, Beans, Begonia, etc. [Sensitivity concentration even 0.02 ppm for 4 to 8 hours or 0.05 ppm for 1 to 2 hours]

Effects: In broad leaf plants, it causes white flecks uniformly distributed over leaf surface, while in conifers it causes brown necrotic needle tips called tip burn. In grass like leaf plants, it causes scattered markings distributed over the surface. In general, ozone causes flecks spots or streaks on upper surface of leaves, but in severe cases bleaching, collapse of leaves, and necrosis. Such symptoms may also appear due to other causes like some types of rust, damage by red spider, mite, sulphur dioxide, etc.

In Europe, experiments with Ozone-sensitive tobacco plant (Nicotiana tabacum variety Bel-W30 demonstrated a very clear gradient of ozone impact with increasing of plant injury levels from North and North west of Europe to Central and Southern regions.

(iv) ***Peroxyacetyl nitrate (PAN):***

Indicator plants: Petunia, Romaine lettuce, Pinto bean, Blue grass, etc. [Sensitivity more than 0.01 ppm]

PAN Effects: Broad leaf plants–underside collapses, leaves become glazed/ silvered/ bronzed. Grass like leaved plants–bleaching and collapsing of leaves. Conifers show non-specific blight / withering of leaves. In general, PAN causes bronzing or silvering spots on the lower surface of some leaves and suppressed growth.

(v) ***Nitrogen dioxide:***

Indicator plants: Tomato, Bean, Tobacco, etc. (Acute injury at concentration of 2 to 5 ppm)

Effects: Nitrogen dioxide in severe cases of pollution causes brown margins, brownish black spots on leaves or bleaching.

(vi) ***Polycyclic Aromatic Hydrocarbons (PAHs):*** Lichens act as bio-monitors for polycyclic aromatic hydrocarbons. Lichens have no roots and take all nutrition (including water) from air. In the process they absorb/adsorb various substances from atmosphere including air pollutants. They accumulate PAHs and can be used for their bio-monitoring. *[For details about Lichens, see Chapter 20, Para 20.5].* Curle kale (Brassica oleracea) also accumulates organic air pollutants such as PAHs in its waxy leaves. These can act as bio-monitors for PAHs.

Resistant Plants for Sulphur dioxide, Fluorides, Ozone, and PAN

Sulphur dioxide: Potato, onion, corn, maple, etc.

Fluorides: Rose, tobacco, cotton, tomato, alfalfa, etc.

Ozone: Mint, geranium, gladiolus, pepper, maple, etc.

PAN: Cabbage, corn, wheat, pansy, etc.

Notes:

1. *For monitoring of air pollution especially heavy metals ' Moss' has been used as bio-indicator in the high traffic zone of western express highway of Mumbai.*
2. *Lichen Genus Usnea is suitable organism for term air pollution monitoring in the world. The density and abundance of Usnea species can be correlated with changes in the forest ecosystem. The species has a preference for temperate regions and therefore it helps to evaluate the shifts due to climate changes.*
3. *Lichen diversity and distribution pattern is an important biometric tool to assess the prevailing environmental condition in the area. There is a distinct increase in number species and their density when we move away from air -polluted areas.*

1.2.3 Effects of Air Pollution on Animals and Human Beings in the Area

Pollution affects animals and humans through breathing, contact with air, contaminated water and food. The polluted air can pollute water, grains, fruits, vegetables and many substances used for food by animals and humans. Radiation from radioactive substances affects all body parts and organs. The environmental pollution particularly of air and water in an area may be the cause of many diseases of lungs, eyes, throat, skin, bones, brain and nerves (neurological problems), liver, kidneys, etc. The environmental pollution particularly phenols, polycyclic aromatic hydrocarbons (like pyrenes, fluoranthene, and anthracenes), beryllium, arsenic, mercury, nickel, chromium, and radiation, may cause genetic diseases, gene mutation, tumors, cancer etc. The Traffic Pollution has even been linked to childhood obesity amongst other bad effects. Air pollution puts a large number of people at risk for respiratory tract infections, pneumonia, Chronic Obstructive Pulmonary Disease (COPD), asthma, lung cancer, etc.

Household air pollution puts a large number of people worldwide at risk of ill health and early death.

1.2.4 Air Pollution Indication by Effects on Materials

Apart from the ill effects of air pollution on plants, animals and human beings as mentioned above, the pollution is detrimental to many materials including metallic parts, paints, dyes, document/books, rubber, leather, marble, bricks, etc. The heritage monuments, paintings, and various articles, original handwritten documents, electronic parts, electrical insulation, electronic/electrical contacts are very adversely affected by air pollution.

It may be noted that the ill effects of air pollution on plants including food crops, human and animal health, machinery, buildings and monuments, and various materials cause huge economic losses to the country besides various problems.

Note: Sampling and Measurements of gases in ambient air

The exact determination of quantity of each pollutant present in ambient air in a particular location can be done by sampling the air and analysis. Various methods of measurements have been standardized for specific pollutants by US, EPA.

1.2.5 General Atmospheric Effects of Air Pollution

The pollution affects the atmospheric turbidity, precipitation, fogs, and solar radiation.

1. The turbidity increase in atmosphere due to suspended particulate matter decreases the solar intensity at ground level. In extreme case of air pollution by particulate matter, the ground level solar intensity can drop by 50%.
2. The smog, ozone, nitrogen oxides, decrease the ultraviolet light reaching the earth surface, which may affect the production of vitamin D by skin of living beings. The nitrogen dioxide absorbs the wavelength of radiation in range 0.36–0.4 μm, while the ozone absorbs wavelength less than 0.32 μm. The gases and particulate matter also decrease ultraviolet radiation as well as the visible solar intensity.
3. The increased quantity/numbers of nuclei in atmosphere can cause dense persistent fogs due to many small droplets formed. Fog formation is also dependent on humidity, suspended particulate matter, temperature, wind velocity, etc. [The moisture is also released to atmosphere along with the exhaust gases by industries, transport vehicles and others.]
4. Addition of some particles (to atmosphere) that act as ice nuclei can cause ice particles to grow at the expense of supercooled water droplets, resulting in ice particles large enough to fall as precipitation. This is used in commercial scale cloud seeding by releasing silver iodide particles from air craft to induce rain. But, if too much particulate matter is added none of ice particles would grow sufficiently to cause precipitation. The effect of pollution on precipitation would depend upon concentration of pollution (among many other things); higher concentration of pollution would tend to decrease precipitation. However, the precipitation is dependent on a large number of other variables.

1.3 NATURAL SINKS FOR CERTAIN ATMOSPHERIC POLLUTANTS

- Carbon monoxide: Soil absorption, ultimate oxidation to carbon dioxide.
- Sulphur dioxide: Scavenging, chemical reaction with many things (and it gets consumed), soil and surface water absorption, dry deposition.
- Hydrogen sulphide: Chemical reaction and formation of sulphides, oxidation to sulphur dioxide.
- Nitrogen dioxide: Scavenging, photochemical reactions, conversion to nitrates.
- Nitric oxide: Oxidation to nitrogen dioxide and sink as above.
- Nitrous oxide: Photochemical reaction/dissociation in stratosphere, but very little absorption elsewhere.
- Ozone: Photochemical reactions, Oxidation of many things (and then gets consumed), absorption by soil, vegetation, surface waters.

- Nonreactive hydrocarbons—Biological action, ultimate conversion (by oxidation) to carbon dioxide.
- Reactive hydrocarbons: Photochemical reactions.

1.4 STRATEGIES FOR REDUCTION OF POLLUTION

1. Control of generation of pollutants in various processes
2. Control of emission of pollutants at the source
3. Conservation of natural resources,
4. Reduction of wastage of energy and materials
5. Recovery and recycling or utilization of the waste products from the processes
6. Reduction of leakages or inadvertent escape of polluting gases/vapours
7. Selection and use of efficient processes and machineries.
8. Raising efficiency of various processes by required modifications, better maintenance and operation practices. Achieving optimum 'Combustion efficiency' so as to reduce fuel consumption which would automatically reduce emission of pollutants.
9. Stress on utilization of solar and wind energy for generation of power.
10. Utilization of solar and wind energy resources for various requirements in industrial and domestic sector.
11. Using less polluting fuel as far as possible
12. Prevention, control of any untoward / accidental fires.

1.5 TOPICS COVERED IN SUBSEQUENT CHAPTERS OF THIS BOOK

Subsequent chapters in this book are devoted to various aspects of air pollution and its control and monitoring.

Chapter 2 deals in detail with sources of various pollutants and their effects, while Chapter 3 is devoted to meteorological effects on air pollution. The thermal power stations, industries and oil refineries being the main sources of air pollution, Chapters 4, 5, 6, and 7 are devoted to the control of air pollution from thermal power stations, while Chapter 8 is devoted to the control of air pollution in industries and Chapter 9 deals with air pollution control in oil refineries. The control of mercury emission which has gained big importance now-a-days, is dealt with in Chapter 10.

Because of extensive use of nuclear energy and threat of nuclear warfare/nuclear accidents, the safety and control of pollution by nuclides/ radioactivity has gained high importance; this topic is exclusively covered in the Chapter 11.

The air pollution due to municipal solid waste and biomedical wastes, which has become very important with increasing urbanization and population explosion in cities, is covered in Chapter 12. The control of emissions of air pollutants from mobile sources /surface transport, aviation/ shipping) have been amply covered in 'Chapter 13 along with standards applicable for control of emissions. As the air pollutants in the environment cause damage to electrical and electronic equipments, the protective measures for the same to be taken are dealt with in Chapter 14. The emission limits imposed by Indian regulations for various industries are mentioned in Chapter 15. Chapter 16 describes the methods of measurements of various air pollutants. The control of noise and the control of odours are dealt with in Chapters 17 and 18 respectively, while Chapter 19 is devoted to environmental impact assessment of projects without which no new project can be taken up in any sector. Chapter 20, deals with some miscellaneous topics like Stratospheric Ozone layer, Acid rain, Spontaneous combustion, Utilization of solar energy and wind energy as an anti-pollution measure and lichens.

REFERENCES

1. Anirban Roy, Sanjay Baxla, Tarun Gupta, Rajdip Bandyopadhyaya, Sachchida Nand Tripathi., "Particulates Emitted From Indoor Combustion. Sources: Measurement of Size Distribution and Chemical Analysis", Inhalation Toxicology, 21 (10), 837-848 (2009).
2. Avinash Kumar Agarwal, Tarun Gupta, Neelabh Dixit, and Pravesh Chandra Shukla, "Assessment of toxic potential of primary and secondary particulates/aerosols from biodiesel visa-vis mineral diesel fuelled engine", Inhalation Toxicology, DOI: 10.3109/08958378.2013.782515 (2013).
3. DK Upretti, et al., 'Species of lichen genus Usnea; most suitable organism for long term pollution monitoring', Vth International Conference on Plants and Environmental Pollution; ICPEP-5, Feb 2015 NBRI, Lucknow.
4. J. Jaidevi, Tarun Gupta, Sachchida Nand Tripathi and Kamal Ujinwal, "Assessment of Personal Exposure to indor and Outdoor Particulate Matter for Residents of an Academic Campus (IIT-Kanpur)." inhalation Toxicology, 21 (14), 1208-1222 (2009).
5. Sanjay Baxla, Anirban Roy, Tarun Gupta, Sachchida Nand Tripathi and Rajdip Bandyo padhyaya "Analysis of Diurnal and Seasonal Variation of Total Outdoor Aerosol Mass and Size Distribution in a Northerm Indian City and its Correlation to Black Carbon", Aerosol and Air Quality Research, 9, 458-469 (2009).
6. Santosh R. et al. 'Determination of concentration of heavy metals using MOSS as bioindicator in the high vehicular traffic zone of western express highway of Mumbai. .
7. Vartika Gupta et al.,'Air pollution monitoring in around Faizabad city utilizing Lichen distribution pattern', Vth International Conference on Plants and Environmental Pollution; ICPEP-5, Feb 2015 NBRI, Lucknow.
8. Wikipedia free encyclopaedia.

QUESTIONS

1. Air pollution may be caused by (answer correct statement)
   (a) Human activities,
   (b) Natural disastrous events,
   (c) Both (a) and (b)
   (d) None of the above.
2. (a) Define air pollution, and differentiate between natural and manmade air pollutions.
   (b) Describe the historical background and causes of air pollution.
   (c) How does air pollution affect the different zones of the atmosphere?
3. (a) Describe the harmful effects of important air pollutants on materials and services, as well as on aquatic life.
   (b) Enumerate the important local, regional and global impacts of air pollution.

Answer: **1.** (c)

CHAPTER 2

SOURCES OF AIR POLLUTION AND THEIR ILL EFFECTS

2.1 CARBON MONOXIDE (CO)

2.1.1 Sources of Carbon Monoxide (CO)

Along with carbon dioxide (CO_2), small amount of carbon monoxide (CO) is, generally, produced on burning or oxidation of carbon containing materials. Incomplete oxidation of carbon mostly during combustion of fuels like coals, lignites, liquid or gaseous hydrocarbons, wood and wood products, biomass, organic substances, etc. results in formation of carbon monoxide. It may also be formed / emitted during chemical or biochemical reactions. Carbon monoxide (CO) is also a constituent of gases emitted from natural sources like decay of organic matter, volcanic eruptions, etc.

The combustion of carbon takes place in two stages:

$$2C + O_2 = 2\,CO \qquad (1^{st}\ \text{step}) \text{ ---- Exothermic reaction}$$

$$2\,CO + O_2 = 2\,CO_2 \qquad (2^{nd}\ \text{step}) \text{ ---- Exothermic reaction}$$

The 1st reaction is ten times faster than the 2nd reaction and so CO is an intermediate product, but with insufficient oxygen or insufficient mixing of fuel may result in carbon monoxide remaining in the products of combustion of fuel. At high temperature, some carbon dioxide may also dissociate into carbon monoxide and oxygen. At 1745 °C about 1 % dissociation of carbon dioxide and at 1945 °C, about 5 % dissociation may take place.

$$2\,CO_2 = 2\,CO + O_2 \qquad \text{(However, this reaction is reversible)}$$

In reducing atmosphere also, carbon dioxide is reduced to carbon monoxide by reaction with carbon.

$$CO_2 + C = 2\,CO \qquad \text{Endothermic reaction}$$

Most of the combustion of fuels takes place in thermal power stations, industries, transport sector (including all types of transport (road / rail / marine / inland water transport / aviation) and domestic sector for cooking and heating, while the forest fires, any accidental fires or intentional fires, gas flaring, etc. also account for substantial combustion of fuel.

The carbon monoxide is a constituent of water gas, producer gas, coke oven gas / gaseous products in coking of coal, blast furnace gas, etc. and any leakage of these gases would add to atmospheric pollution by the gas. The composition of the above mentioned gases is approximately as given in **Table 2.1 A.**

Table 2.1 A: Percentage by Weight of Different Constituents in the Specific Gases

	% CH_4	% Cm Hn	% H_2	% CO_2	% CO	% N_2
Water gas	0.6	1.19	6.35	13.87	70.43	7.56
Producer gas	0.12	-	0.99	10.19	29.09	59.61
Coke oven gas	35.7	6.38	10.43	8.84	13.77	23.72
Blast furnace gas	-	-	0.19	28.59	22.22	49.0
CH_4 — methane; Cm Hn — hydrocarbons; H_2 — hydrogen; CO_2 — carbon dioxide; N_2 — nitrogen						

The biggest source of carbon monoxide is probably the exhaust gases from internal combustion engines driven machines and vehicles. Briefly it may be stated that the 'Carbon monoxide' is emitted from various sources including the following:

- Exhaust gases from vehicles, railway diesel or diesel- electric locomotives, prime-movers of ships, air crafts, agricultural machineries, construction machineries etc.; Flue gases from stacks of fossil fuel burning industries including cupola furnace, blast furnace, brick kilns, industrial diesel engines, etc.
- Flue gases from thermal power stations;
- Emissions from coke ovens; calcination process in aluminum plants; smoke in welding process; petroleum refineries; etc.
- Smoke of forest fires, coal or lignite mine fires, oil well or gas drilling site fires and accidental fires; smoke of fires in biomass in agriculture sector;
- Emissions from solid waste disposal operation, incinerators;
- Chemical processes and fermentation operations in industries
- Domestic cooking and heating
- Spontaneous combustion of coal / lignite in fuel stacks / bunkers / mines
- Carbon monoxide may also come from natural sources like volcanic gases, decay of organic matter

2.1.2 Ill effects of Carbon Monoxide (CO)

Effects on Humans and Animals

Carbon monoxide is a highly toxic gas and affects human beings and animals and may cause headache and breathlessness. It can adversely affect central nervous system causing lack of muscular coordination, and may even cause asphyxiation, stoppage of pulse and heart attacks. Carbon monoxide has a high affinity to combine with 'Hemoglobin' of blood and forms complex with it and thus reduces the ability of hemoglobin to carry oxygen to tissues of the body. The tendency of CO to attach with hemoglobin is about two hundred times greater than the tendency of oxygen to attach with hemoglobin. The increase of carbon monoxide in atmosphere particularly in confined spaces is, therefore, dangerous.

CO an Indicator of inefficiency of Fuel Burning Machines /equipments

Higher content of carbon monoxide in combustion gases in boilers or internal combustion engines is an indicator of inefficient combustion and consequent heat losses.

2.2 OXIDES OF NITROGEN

2.2.1 Various Oxides of Nitrogen

Nitrogen forms various oxides in which its valency varies from 1 to 6. These are mentioned below:

1. *Nitrous oxide (N_2O)*—It is a colourless and very toxic gas and is called 'laughing gas'. It is neutral, but it is a very potent greenhouse gas. Its molecule is very stable, having a very long life (120 years) in atmosphere.
2. *Nitric oxide (NO)*—It is a colourless gas having paramagnetic properties. The molecules of nitric oxide are stable, but, in contact with atmospheric oxygen, they get oxidized into nitrogen dioxide, but the reaction is reversible. Nitric oxide is very slightly soluble in cold water (at 15° C, 1 volume of gas can dissolve in 20 volumes of water).
3. *Di-nitrogen trioxide or Nitrogen sesquioxide (N_2O_3)*—It is viewed as anhydride of nitrous acid ($2HNO_2 = N_2O_3 + H_2O$). It exists only at very low temperatures as solid, but on rise in temperature it dissociates into '$NO + NO_2$.
4. *Nitrogen dioxide (NO_2)*—It is the most common oxide of nitrogen and is of brown colour having paramagnetic properties. It is an acidic gas and when treated with water, it forms mixture of Nitrous acid and Nitric acid ($2NO_2 + H_2O = HNO_3 + HNO_2$), but the reaction is slow and also reversible. Under the action of light (Photolysis), NO_2 dissociates into NO and atomic oxygen (O). The atomic oxygen reacts with oxygen molecule (O_2.) to form ozone (O_3) in troposphere.
5. *Di-nitrogen tetra-oxide (N_2O_4)*—It is a colourless solid at (-)10°C but on rise of temperature to 10°C, it melts to a colourless liquid which on further rise in temperature changes its colour to yellowish, then orange until at 22°C when it boils off to a brown vapour and starts dissociating into NO_2. The 'N_2O_4 and NO_2' system remains in equilibrium, but the dissociation is complete at 140°C.
6. *Nitrogen pentoxide (N_2O_5)*.— Nitrogen pentoxide is viewed as an anhydride of nitric acid ie nitric anhydride ($2\ HNO_3 = N_2O_5 + H_2O$). It is a colourless crystalline solid below 0°C but, on rise of temperature the crystals melt to a yellow liquid which rapidly dissociate at 40°C to give off oxygen (O_2) and nitrogen dioxide (NO_2).

2.2.2 Sources of Oxides of Nitrogen

The sum total of all other oxides of nitrogen except nitrous oxide 'N_2O' is designated by NOx (This includes $NO + NO_2 + N_2O_3 + N_2O_4 + N_2O_5$) and is reported as ppm of NO_2. Most of the NOx in the atmosphere is in the form of NO and NO_2. As the concentration of ($N_2O_3 + N_2O_4 + N_2O_5$) is not significant, therefore they are not of much concern. About half of the amount of oxide of nitrogen in the atmosphere is of natural consequences, while 3/4th of the balance is due to combustion of fossil fuels particularly at higher temperatures; the generation of NOx increases with the temperature during combustion.

It may be mentioned that there can be two sources of nitrogen in the formation of oxides of nitrogen during combustion of fuels – the atmospheric nitrogen and / the nitrogen contained in the fuel. The solid -, liquid- and gaseous fuels contain some amount of nitrogenous compounds as impurity. *This can be seen from the analysis of coal, lignite, fuel oil given in **Tables 2.1 B, and 2.1 C.*** The coal, lignite, and dry wood may contain nitrogen in the range of 0.1 to 3.0 % (average 1.7 %), 0.1 to 0.7 % (average 0.5 %), and about 0.1% respectively. The heavy fuel oils may contain about 0.6 % nitrogen, while natural gas may have varying nitrogen content depending upon the source. The derived fuel gases like blast furnace gas, coke oven gas, producer gas, water gas, etc. may have varying nitrogen content depending upon the quantity of air or oxidant used in their production. The important sources of oxides of nitrogen are mentioned below:

Nitrous oxide

The emission of nitrous oxide (N_2O) is prominent during low temperature oxidation of fuels / biomass. As an example, nitrous oxide is released prominently during spontaneous combustion of coal in stacks of coal which is a low temperature oxidation. It is released during the burning of biomass as the flame temperature in such a case is low because of presence of large amount of moisture in biomass. Nitrous oxide is also released from some man-made devices like certain types of *catalytic converters (platinum - palladium catalyst)* used in vehicles.

Very small amount of nitrous oxide may also be emitted during high temperature combustion along with other oxides of nitrogen designated as NOx whose emission is predominant at high temperatures.

In nature, nitrous oxide (N_2O) is released by microbes that act on nitrogen present in the manure or fertilizer in the soil* and nitrogenous matter in disposed refuse / waste. The nitrous oxide is also released due to biochemical reactions going on in natural environment like in oceans and aquifers.

[* Soils are a major source of nitrous oxide. Emission of nitrous oxide takes place during both the 'Denitrification' and 'Nitrification' of the soil. Moisture, temperature and nitrogen status of soil affect the emission of nitrous oxide. The increase in temperature promotes release of nitrous oxide. Fertilizer nitrogen say ammonium sulphate increases the emission of nitrous oxide.]

Table 2.1 B: Analysis of Typical Indian Coals and Ash

Area and Rank	**Jharia Mvb**	**Hvb (1)**	**Hvb (2)**	**Renusagar Subb**	**Singrauli Subc**	**Neyveli Lig A (Lignite)**
As Received basis:	---					
% Ash	38.9	31.6	28.0	28.6	31.5	4.5
%Moisture	1.1	6.9	10.0	14.9	7.9	53.1
Moist & Ash Free:						
% Volatile (VM)	25.3	37.2	41.0	45.1	47.4	57.1
% Carbon (C)	83.6	-	-	74.1	71.9	70.3
% Hydrogen (H)	4.5	-	-	4.8	5.0	5.2
% Oxygen (O)	9.9	-	-	18.6	20.3	23.1
% Nitrogen (N)	1.3	-	-	1.4	2.0	0.5
% Sulphur (S)	0.7	1.8	0.8	1.1	0.8	0.9
HHV, Btu/lb	14,635	13,500	13,630	12,690	12,230	11,820
HHV, MJ/kg	34.0	31.4	31.7	29.5	28.4	27.5
Presence of trace metals	yes	yes	yes	yes	yes	yes
Ash Fusibility						
I.D.Temperature. °F	2250	2600	2100	2700 +	2700+	2170
(°C)	(1232)	(1427)	(1149)	(1482)	(1482)	(1188)
S.Temperature. °F	2670	2700 +	2280	2700 +	2700 +	2200
(°C)	(1465)	(1482)	(1249)	(1482)	(1482)	(1205)
Fluid Temperature. °F	2700 +	2700 +		2700 +	2700 +	2310
(°C)	(1482)	(1482)		(1482)	(1482)	(1266)

Ash Composition :						
% SiO_2	65.9	60.8	66.5	64.3	66.5	41.3
% Al_2O_3	23.7	24.8	26.1	25.2	26.1	12.1
% Fe_2O_3	6.0	6.8	2.8	5.0	2.8	3.3
% CaO	1.1	2.1	0.9	0.8	0.9	12.7
% MgO	0.6	0.5	0.7	0.7	0.7	2.9
% Na_2O	0.1	0.1	0.1	0.1	0.1	0.9
% K_2O	1.4	1.0	0.8	0.9	0.8	0.2
% TiO_2	2.2	---	1.5	1.7	1.5	0.4
% P_2O_5	---	---	0.2	0.2	0.2	---
% SO_3	0.3	1.7	0.4	0.2	0.4	24.3
Trace Metals	Yes	Yes	Yes	Yes	Yes	Yes
Base / Acid Ratio	0.10	0.12	0.06	0.08	0.06	0.37
Hardgrove Grinding Index	63	60	50	56	50	---

Subb – Sub bituminous coal B; Subc – Sub bituminous coal C; Lig A – Lignite A
Mvb – Medium volatile bituminous coal; Hvb – High volatile bituminous coal
(Two alternative compositions of Hvb coals are given here i.e. 1 &2)
I.D. Temp.—Initial Deformation Temperature, S. Temp.— Softening Temperature
Note: The ranges of various quantities not given in above table are as follows
Hub (1) : C–79 to 85%, H_2–4 to 5%, O_2 9 to 12%, N_2–1 to 1.4%
Hub (2): C–80 to 85%, H_2–4 to 4.5%, O_2–9 to 11% , N_2–1.1 to 1.2%

Table 2.1 C: Typical Properties of Fuel Oils (Crude Oil, Diesel Oil, Residual Oil)

Se. No.	Property	Crude Oil	Diesel Oil	Residual Oil
1.	Specific gravity at 15°C	0.87	0.83	0.97
2.	(i) Viscosity 20°C cst.	32	6	-
	(ii) Viscosity 50°C cst.	11	3.4	100
	(iii) Viscosity 100°C cst.	3.8	-	16
3.	Pour Point °C	-10	-30	20
4.	Analysis			
(a)	Carbon. % wt.	86.0	85.6	87.0
(b)	Hydrogen % wt.	12.1	13.5	9.2
(c)	Sulphur %wt.	1.9	0.4	2.8
(d)	Nitrogen ppm	3,500	500	5400
(e)	Water+ Sediments by Vol.%	0.2	0.05	1.2
(f)	Ash ppm	100	20	500
(g)	Sodium +Potassium	5	0.5	50

(h)	Vanadium ppm	7	0.5	35
(i)	Lead ppm	0.2	0.1	4
(j)	Nickel ppm	2	0.1	3
(k)	Zinc ppm	2	0.1	5
5.	High Heat value kcal/kg	10,050	10,240	9,780

Source: *ABB Review Sept. 1995. Dr.Dilip K Mukerjee ABB Power Generation Ltd. Switzerland.*

NOx emission

Briefly it may be mentioned that NOx is emitted during combustion of fuels at high temperature such as from:

- Combustion turbines i.e. gas turbines in power stations / industries; combustion turbines of mobile sources like shipping, aircrafts, etc.
- Boiler furnaces of fossil fuel fired power stations,
- Industrial boiler furnaces/ other industrial furnaces, incinerators, coke ovens, etc.
- Internal combustion engine drives/ prime movers in industrial / agricultural / irrigation / domestic / power sectors.
- Internal combustion engines of mobile sources like automobile vehicles, tractors, earthmoving / material handling machines, ships / boats, railway diesel or diesel- electric locomotives, etc.
- Domestic fires; agricultural fires, use of gas, oil, or solid fuel anywhere; intentional gas flaring, etc.
- Emission from some industrial processes, like manufacture of nitric acid, glass manufacturing, etc.
- The oxides of nitrogen are formed in atmosphere under the influence of lightning discharges as shown below:

$$N_2 + O_2 = 2\,NO \text{ (Nitric oxide)};\ 4NO + 3O_2 = 2\,N_2O_5\,;$$

$$2\,NO + O_2 = 2\,NO_2;\ 2\,NO_2 = N_2O_4\,;\ N_2O_4 = 2\,NO_2$$

The oxides of nitrogen so formed may combine with moisture to form nitric or nitrous acid. About 250,000 tons/day of acid is said to be formed in this way.

- Accidental fires anywhere including fires in oil wells and oil / gas installations, war / bombardment, volcanic eruptions, or other natural disasters, forest fires, coal mine fires, etc.
- Soil also emits NO_2 because of bacterial action on substances / matter containing ammonical nitrogen.
- Oxides of nitrogen in atmosphere due to leakages from processes

It has been estimated that about 25 to 30 kg of oxides of nitrogen may be emitted on burning of one tonne of petroleum products while burning of coal may produce 5 to 10 kg nitrogen oxides per tonne of coal.

2.2.3 Ill Effects of Oxides of Nitrogen

As already mentioned Nitrous oxide is a greenhouse gas and is also very toxic.

Effects of NOx on Plants Animals and Human beings

These gases (NOx) can cause many ill effects on health of animals and humans. Even a small quantity of 15 ppm of NOx in air can cause irritation in eyes and nose, and more than this may cause respiratory disorders and pulmonary discomfort even on brief exposure. It may cause pulmonary congestion and oedema, emphysema, asthma, heart disease, diabetes, hypertension, dermatitis, 'Alzheimer's disease', 'Parkinson disease', etc. In combination with particulate matter, NOx becomes more harmful to human health and even to unborn children during pregnancy. People working on electric arc welding and those working in mining where explosives having nitrogen are used, or workers in manufacture of nitric acid, etc. are more often affected by occupational diseases caused by such pollution.

As Nitrogen dioxide and nitric oxide form acids (nitric acid and nitrous acid) on reaction with water, these gases are also a cause of acid rains with consequent adverse effects on agriculture / plants and they are corrosive to metallic surfaces and some other materials. The nitrogen dioxide may cause bleaching of plant leaves and suppress the growth. Some specific ill effects of 'Nitrogen dioxide' are mentioned in **Table 2.1 D.**

Formation of Smog and Tropospheric Ozone Pollutants

The nitrogen dioxide and nitric oxide emission and their presence in atmosphere play a part in the formation of ozone and or smog* in troposphere which have very harmful effects on humans, animals and agriculture / plants. *[* The nitrogen dioxide plays an important role in formation of secondary pollutants like nitrate aerosols, PAN, besides ozone. Also see* ***Para 2.14 of this chapter.****]*

Depletion of Stratospheric Ozone by NOx

NOx depletes stratospheric ozone permitting more ultraviolet radiation reaching earth with consequent ill effects. *[Refer to* ***Para 20.1. Chapter 20.****]*

Table 2.1 D: Some Effects of Nitrogen Dioxide

Area		Concen-tration ppm	Duration time	Doses in ppm minutes	Remarks
1. Health					
	a	0.12			Threshold for ill effects by short term exposures
	b	5	10 min.	500	Measurable increase in resistance to breathing
	c	15	2 hrs.	1800	Tissue changes in lungs, heart, liver and kidney of monkeys/ human being
	d	90	30 min.	2700	Pulmonary oedema found 18 hrs. after exposure accompanied by a 50 % reduction in normal vital capacity
	e	0.062 -0.109 Daily mean	Continuous	600-1100 per week	Increased incidence of serious respiratory disease in family groups
	f	0.063 -0.083 Daily mean	Continuous	630-830 per week	Increase in frequency of acute bronchitis among infants and school children
	g	5	540 hrs to 8760 hrs.	162,000 to 2628000	Toxicological changes in mice resembling swelling in lungs and breathing problem in humans.
2. Vegetation					
	a	0.25	Continuous 8 months	86500	Shedding of leaves & decrease in yield of naval oranges

	b	0.5	Continuous 35 days	25200	Shedding of leaves and whitening of green leaves of citrus trees
	c	1.0	1 day	1440	Visible leaf injury to plants sensitive to NOx
3. Photo-chemical oxidant production	0.04–0.16	In presence 0.3–1.4 ppm of non-methane hydrocarbons (VOC).			This level of Nitrogen dioxide can cause smog and ozone concentration of 0.1 ppm (max. daily average)

2.3 OXIDES OF SULPHUR (SOx)

2.3.1 Sources of Oxides of Sulphur

There are two oxides of sulphur namely sulphur dioxide (SO_2), and sulphur trioxide (SO_3). The oxides of sulphur 'SO_2 and SO_3.' are together designated as 'SOx'. Sulphur dioxide is generated along with a small percentage of Sulphur trioxide on oxidation or burning of sulphur compounds or sulphur containing fuels, or materials. Sulphur dioxide forms sulphurous acid when dissolved in water, while sulphur trioxide forms sulphuric acid with water.

The fossil fuels i.e. coal, lignite, and petroleum fuel oils contain some amount of sulphur compounds. Generally, the fuel oils available for burning in boiler contain more sulphur than solid fuels like coal or lignite and therefore, the problems of SOx emission are more acute with oil firing than with coal firing.

The emission sources of oxides of sulphur (SOx) include:

- Thermal power stations using solid, or liquid fuels
- Industrial boiler furnaces/ other industrial furnaces/ combustion turbines, etc. using fuels containing sulphur
- Internal combustion engine drives/ prime movers in industrial / agricultural / irrigation / domestic / power sectors. using fuels containing sulphur
- Internal combustion engines of mobile sources like automobile vehicles, tractors, earthmoving / material handling machines, ships / boats, railway diesel locomotives or diesel- electric locomotives, etc. using fuels containing sulphur
- Domestic fires; agricultural fires, burning of biomass, oil, or solid fuel anywhere;
- Incinerators for destruction of wastes
- Intentional gas flaring from refineries or oil wells, etc.
- Emission from some industrial processes like: copper smelting; zinc smelting; lead smelting; material recovery furnaces of Kraft paper mill; sulphuric acid manufacture; lead acid type battery operations; various units of petroleum refineries like distillation, catalytic cracking, sulphur recovery; etc.
- Iron and steel plant; ceramic industry; foundries (cupola furnaces); coke ovens; some chemical industries specially involving use of petroleum / sulphur compounds, etc.
- Volcanic eruptions; various accidental fires.

Some metallic ores have sulphur compounds in them. Metallurgical operations for recovery of metals from ores containing sulphur compounds emit oxides of sulphur mostly (SO_2). Sulphur compounds are present as impurities in 'Iron ore'. Almost all ores of copper, tin, antimony, arsenic, lead, etc. contain sulphur as a constituent or as impurity. Examples of sulphur compounds in ores are: Iron pyrites (FeS_2 / Fe S), copper pyrites (CuS), arsenical pyrites ($FeS_2.AsS_2$), cobalt and arsenic containing ore—cobaltite (CoAsS), lead ore— galena (PbS) or anglesite ($PbSO_4$).

Generally, non-coking coals available for power generation in India have low sulphur content (0.2 to 1.1 %). In India the coals of north eastern region are however higher in sulphur content (>2 %). The heavy fuel oil (HFO) has sulphur content of about 3.0 %, while the low sulphur heavy stock (LSHS) has a sulphur content of about 1.0 %. The light diesel oil (LDO) has sulphur content of about 1.8 %, while high speed diesel oil (HSD) has sulphur content of about 0.4 to 0.5 %. The surface transport vehicle engines use high speed diesel oil, while other diesel engines can be designed to burn other oils.

Sulphur dioxide is fully oxidized by oxygen to sulphur trioxide (SO_3)*in presence of catalytic agent and the reaction is exothermic. Ordinarily without catalytic agent there is little amount of conversion of SO_2 into SO_3. The sulphur dioxide emitted in some cases can be utilized for manufacture of sulphurous acid (H_2SO_3), sulphuric acid (H_2SO_4), / other chemicals.

*** Note:** *Inside a furnace, sulphur dioxide is converted to sulphur trioxide through reactions with molecular oxygen, atomic oxygen and catalytic action of the deposits or corrosion products on boiler tubes in presence of oxygen locally available there. The extent of conversion depends on fuel composition, the flue gas temperature, the amount of excess air in the furnace and characteristics of deposits on the boiler tubes. The conversion takes place in the 'radiant section of the furnace' and 'back pass of the boiler'. As already known that 'Vanadium based catalyst' is used to oxidize sulphur dioxide into sulphur trioxide for production of sulphuric acid; the 'Vanadium' content in the deposits and in boiler tube metal may be acting as catalyst in the formation of sulphur trioxide.*

2.3.2 Ill Effects of Oxides of Sulphur

Effects on Human and Animal Health

The oxides of sulphur are strongly acidic in character and therefore their presence in air has very bad effects on health of animals and human beings particularly on their respiratory organs. Sulphur dioxide may cause broncho- constriction, cough, tightness in chest and on continued exposures can cause lung cancer also. It has also been found that the air pollution by oxides of sulphur causes damages to plants / vegetation / agriculture. The problems of health and plant damages by SOx pollution of air attenuate in presence of particulate matter in atmosphere. The oxides of sulphur in the vehicle exhausts cause many health problems to urban population. The people working in thermal power stations and factories may be very adversely affected by the combination of particulate matter and oxides of sulphur. [*Also see Para 2.16 of this chapter for effects on respiratory system.*]

Effects on Equipments and Materials

The corrosive effects of oxides of sulphur on the plant and equipments of power stations and factories are very damaging and create many operational and maintenance problems. Electric and electronic parts, metallic parts and surfaces, paints, etc are particularly affected.

The oxides of sulphur in air can deteriorate various materials like rubber, leather, paper/ documents/ books, etc. The heritage monuments/ buildings, paintings, articles, etc. are damaged by them.

Damages to Vegetation by SOx

The oxides of sulphur being acidic in nature cause many ill effects on plant life. Emission of SOx also causes acid rains. Surveys were undertaken by various authorities to investigate the damages to crops, trees, and other plants in the vicinity of thermal power stations (the damages to vegetation were presumed on account of air pollution by oxides of sulphur and particulate matter).

(a) Vegetation survey around thermal power stations at Obra, Shaktinagar, Renusagar, in District Sonebhadra , Harduaganj in District Aligarh (all these power stations are located in Uttar Pradesh) showed the following detrimental effects. The names of vegetation under specific study were Tabernaemontana / Ervatamia (Tagar), coronaria weed, calotropis (Ak / Akada / Madar), Croton (chucka / jamalgota / joufra/ nanbhantur), Sparciflorus, Moringa (Sajana), etc.

- Increase in frequency of epidermal cells, stomata and trichomes of leaves of some plants.
- Decrease in size of epidermal cells and stomata of leaves
- In some plants necrotic lesion (death of living cells) and injury symptoms on upper epidermis of leaves.
- In some plants 'Cuticular striation' were not found prominent, 'Stellate hairs' lost their typical shape and size and became disorganized and 'Stomatal openings' which exchange gases were affected.

(b) Study around thermal stations in Singrauli region district Sonebhadra showed increase in 'Sulphate deposits' on plants, which were attributed to sulphur dioxide emissions from power stations

(c) Case study around Panki thermal power station, Kanpur, UP, indicated that the plants in the area have 'Poor growth', decreased 'Chlorophyll content', a significantly decreased 'Ascorbic acid and Protein content'. The damages were attributed to pollution by emission of sulphur dioxide, particulate matter, etc.

(d) Experimental studies have shown that sulphur dioxide causes damages to coconut plants – lowers foliar pH, reduces photosynthesis pigments, and causes fluctuations in metabolite content (protein, free amino acids, sugars, etc.) and causes fluctuations in activities of enzymes (protease, invertase, peroxidase, etc.).

(e) Experiments have shown that sulphur dioxide pollution can cause decrease in phytomass, total chlorophyll, ascorbic acid, catalase activity (enzymic activity) in paddy plants.

(f) In Brazil near a copper smelter, a strong accumulation of sulphur, arsenic, and copper was detected in mango leaves (of nearby mango trees), and surface soil. There was a reduction in microorganism and modification of enzymatic reactions at polluted sites. The air pollution affected the plants directly by uptake through leaves rather than through contamination of soil and subsequent uptake through roots. *[Regarding bio-indication of sulphur dioxide pollution, see Chapter 1.] [Ref#6]*

2.4 HYDROGEN SULPHIDE AND MERCAPTANS

2.4.1 Hydrogen Sulphide (H_2S)

Hydrogen sulphide (H_2S) is a colourless, poisonous, very foul smelling (like that of rotten eggs) acidic gas emitted both from anthropogenic and natural sources. It is slightly soluble in water (1 vol. of gas dissolves in 2.6 vol. of water at 20 °C. It burns in air forming sulphur dioxide. Metallic sulphides like Iron sulphide (FeS), Copper sulphide (CuS), Lead sulphide (PbS), Arsenic sulphide (As_2S_3), Antimony sulphide (Sb_2S_3), Tin sulphide (SnS), etc. are formed when H_2S is passed in aqueous solution of salts of metals in presence of acid, but the formation of Zinc sulphide (ZnS), Manganese sulphide (MnS), Nickel Sulphide (NiS), Cobalt sulphide (CoS), etc. require alkaline medium.

In nature, it is formed by anaerobic biological putrefaction / decay of organic matter containing sulphur compounds as constituents or as impurity. The decay process goes on in presence of moisture. It also occurs in volcanic gases, in waters of many mineral springs, in intestinal gases of animals / humans, and in emission gases from decaying organic matter in ponds, tanks, oceans, etc. Hydrogen sulphide is emitted along with sulphur dioxide in several metallurgical processes, especially when, ores containing sulphur as constituent or as impurity are processed for recovery of metals. The hydrogen sulphide is formed when materials containing sulphur are heated in presence of moisture in reducing atmosphere. Hydrogen sulphide is emitted when metallic sulphides are treated with acid in a chemical process.

As the hydrogen sulphide (H_2S) is formed in reducing atmosphere, it is emitted during pyrolysis of wood, biomass and municipal solid wastes, incineration of wastes, coal gasification, petroleum refining, and coking of coal (coke oven plants). Some industries like kraft pulp industry, rubber industry, tanneries, manufacture of sulphur dyes, viscose rayon plants, etc. may emit hydrogen sulphide along with other gases. H_2S is present in sewer gases, decay gases from municipal solid waste dumps, etc.

In uncontrolled situations, the hydrogen sulphide levels of 5000 ppb around industrial areas having industries like steel plants, pulp and paper mills, and chemical processing plants would not be uncommon and hence it is necessary to control the emission to safeguard against its bad effects.

2.4.2 Mercaptans

The mercaptans have 'thiol' ie 'SH' group present in them and have a general formula RSH. They have strong foul odour and cause air pollution by their odour. These may be Methyl mercaptan also called methanethiol (CH_3SH), Isobutyl mercaptan also called dimethyl propanethiol [$(CH_3)_2CHCH_2SH$)] or others. With the exception of *methyl mercaptans (CH_3SH) which is a gas,* other thiols are liquids. The mercaptans (thiols) have acidic characteristic and are soluble in water to some extent. It may be noted that even the liquid mercaptans would also give out very foul odour.

There are other sulphur compounds which smell like mercaptans such as 'Thioethers or Alkyl sulphides, with general formula 'RSR, (like CH_3 SCH_3). These are insoluble in water, but soluble in organic solvents.

The thiols / mercaptans when exposed to oxygen atmosphere in presence of moisture and acidic oxides of nitrogen (like nitrogen dioxide or nitrogen pentoxide) may get converted to 'Sulphonic acids (RSO_3H)'. They have strong acidic character.

$$2RSH + 3O_2 = 2RSO_3H$$

$$2CH_3SH + 3O_2 = 2CH_3SO_3H$$

Notes:

1. *The nitrogen dioxide / nitrogen pentoxide with moisture forms nitric acid which aids the reaction of mercaptans with oxygen.*
2. *R denotes alkyl group like CH_3, $(CH_3)_2CHCH_2$.*

The mercaptans are generally emitted, along with hydrogen sulphide from decay of organic matter, sewage, municipal solid waste, etc. The emission gases from sewage treatment plants, decaying organic matter in ponds, lagoons, certain industrial effluents, etc. contain mercaptans. The mercaptans may be emitted in petroleum refining process; tanneries, paper and pulp industry, petrochemical industry, incinerators; industrial fermentation processes, etc.

2.4.3 Ill Effects of Hydrogen Sulphide and Mercaptans

Effects on Materials and Equipments

Since the hydrogen sulphide gas (H_2S) is an acidic gas, it is very corrosive for silver, copper and many other nonferrous metals. It is very detrimental to electric and electronic contacts.

Effects on Plants Life

It can negatively affect the plant growth at higher levels (30 ppb and higher). At more than 300 ppb it results in visible injury to plants, but there are wide variations in susceptibility between species. At low concentration levels the atmospheric H_2S may be usefully utilized by plants for their growth.

Effects on Human Health

Both the hydrogen sulphide and mercaptans, at low levels in atmosphere may not affect health apart from creating foul smells. Hydrogen sulphide in higher concentration can cause headache, conjunctivitis, pain in eyes, enzyme poison, and nervous problems.

2.5 FLUORINE, HYDROFLUORIC ACID, AND OTHER GASEOUS COMPOUNDS OF FLUORINE

2.5.1 Sources of Fluorine and its Compounds

Fluorine containing materials are quite abundant in nature and often mixed up with many natural ores and minerals. It is member of halogen group (Fluorine, Chlorine, Bromine, and Iodine). The important minerals containing fluorine are *Fluorspar* (CaF_2), Cryolite ie Sodium fluoroalumrate ($Na_3Al\ F_6$), *Fluorapatite* ['CaF_2. 3 Ca_3 $(PO_4)_2$' or sometimes '3 Ca_3 $(PO_4)_2$ $(CaF,Cl)_2$']. The fluorapatite is a source of phosphorous and phosphoric acid. Several compounds containing fluorine are gases like Carbon fluoride / carbon-tetra-fluoride (CF_4), Silicon Fluoride (Si F_4), Sulphur hexafluoride (SF_6), various chlorofluorocarbons, etc. Some fluorine gaseous compounds like Sulphur hexa-fluoride, carbon-tetra-fluoride, Chlorofluorocarbons, etc are greenhouse gases. Fluorine is a highly reactive gas and acts as an oxidizing agent. Fluorine attacks glass vessels to form silicon fluoride gas and attacks carbon to form carbon fluoride gas. Fluorine attacks iron, tin and even platinum to form PtF, a chocolate colour powder. Some fluorides are volatile and the fluorine compounds including various fluorides are therefore potentially air pollutant. Fluorine combines with oxygen to form 'Oxygen- di- fluoride (OF_2)' which is a colourless gas, soluble in water having strongly oxidizing property.

With hydrogen, the fluorine forms 'Hydrogen fluoride (H_2F_2)' which is a colourless strongly fuming liquid (in moist air). Its boiling point is 19° C. At about 100° C, the hydrogen fluoride (vapour) becomes monomer with formula as HF. The hydrogen fluoride is dangerous acidic substance and can cause severe burns which are difficult to heal. In presence of moisture / water, the hydrogen fluoride attacks glass, silica, silicates, most of the metals. The vapours of hydrogen fluoride dissolve in water, but ionize only slightly and therefore the solution behaves as a weak acid. There are many industrial uses of hydrogen fluoride such as: (i) pickling of alloy steel, (ii) making 'Teflon (C_2F_4) and fluorine dyes, (iii) removal of silicon from graphite, (iv) as an antiseptic in dyeing and brewing industry, (v) as etching agent for glass, (vi) also used in some chemical industries.

Since hydrogen fluoride is used in some industries, the presence of fluorides in emissions from these industries cannot be ruled out. Fluorides (including hydrogen fluoride vapours) are emitted by various sources like ceramic industry, Ferro-enamel works, brick kilns, glass industry, aluminum refining/ smelting, steel manufacture, oil refineries, phosphatic fertilizer plants, incinerators for burning wastes, coal fired boilers (coals contain small amount of fluorides), processing of certain ores, minerals, and fuels containing fluoride, engines and boilers using oil fuel containing fluorides, etc.

2.5.2 Ill Effects of Fluorides

The pollution of air by hydrogen fluoride adversely affects the plants, humans and animals. It may interfere with photosynthesis in some plants. Once absorbed by plants, the fluoride effect on the plants lasts longer than that of sulphur dioxide as its natural removal is very slow. Accumulation of Fluoride in plants can cause necrosis (death of a part of living body) at leaf tip.

Hydrogen fluoride as air pollutant may cause diseases of respiratory system. The concentration of fluorides in forage of more than 50 ppm (as F) may make it unsuitable for animal feed. Fluoride accumulation in human and animal bodies would cause bone and dental diseases (sponging/ fluorosis) and it may alter metabolism and cause metabolic problems in soft tissues. (*Also see* ***Table 2.14 in Para 2.17 of this chapter.***)

The fluorides / fluorine is corrosive even to glass and ceramics as they react with silicon present therein. In plants they may cause necrosis at leaf tips on continued exposure.

Fluorocarbons and sulphur hexafluoride are also capable of depleting stratospheric ozone. Carbon tetra fluoride is a very strong greenhouse gas and does not get destroyed and keeps on accumulating in the atmosphere. *[Regarding bio-indication of fluoride pollution, see* ***Chapter 1.]***

2.6 CHLORINE, HYDROGEN CHLORIDE, AND OTHER GASEOUS COMPOUNDS OF CHLORINE

The chlorine (Cl_2) is a greenish yellow gas with pungent irritating odour causing many adverse effects on human and animal health. It is soluble in water, highly reactive and corrosive for metals. Chlorine is powerful oxidizing and bleaching agent. With water the chlorine reacts to form hydrochloric acid and hypochlorous acid (HClO). The nascent oxygen content of hypochlorous acid is mainly responsible for the bleaching action.

$$Cl_2 + H_2O = HCl + HClO$$

$$HClO = HCl + O \text{ (Nascent oxygen)}$$

Chlorine compounds are quite abundantly found in nature. These compounds are present as impurities in fuels, minerals, ores, etc. Chlorine is widely used as biocide ie for sterilizing to kill and control the population growth of microorganisms (bacteria / pathogens, fungi, protozoa), worms, algae, etc. in water. Chlorine is a constituent of pesticides whose air borne vapours and fumes are very poisonous. Chlorine is used in the manufacture of 'grease solvents' (by chlorination of paraffin), and in the manufacture of plastics, synthetic rubber, carbon tetra chloride (CCl_4), chloroform ($CHCl_3$), hypochlorides, bleaching powder [Calcium chlorohypochlorite – Ca (OCl) Cl], hydrochloric acid, some pesticides, and many other useful compounds. Chlorine is used in bleaching of cotton textiles, paper and rayon. It is also used in several chemical processes including the extraction of gold and platinum.

Chlorine also forms many gaseous compounds like hydrogen chloride gas, chlorofluorocarbons*, etc. The chlorofluorocarbons have industrial uses, but are green house gases and deplete ozone layer in stratosphere.

The hydrogen chloride gas is strongly acidic, soluble in water forming hydrochloric acid. About 450 volumes of HCl gas dissolves in one volume of water, but the solution cannot be concentrated much. The concentrated HCl is always fuming and has to be handled and used with great care. The Hydrogen chloride is used in textile and dyeing industry and leather industry. It is used as cleaning agent for various purposes such as cleaning of iron sheets before tinning and galvanizing. The HCl is used in the extraction of glue from animal tissues and bones. It is widely used as laboratory chemical and is used to form aquaregia when mixed with Nitric acid (HNO_3) in the ratio of '3 : 1 (i.e. 3 HCl + HNO_3)' to dissolve gold and platinum.

With such wide uses of 'Chlorine' and 'Hydrogen chloride' as indicated above, leakages of chlorine gas and hydrogen chloride are not uncommon. Such leakage and emissions have been taking place

during manufacturing, uses, transport, etc. causing serious air pollution and health problems.
Sources of air pollution by chlorine and hydrochloric acid gas include:

- Leakage during manufacturing and transport of chlorine and hydrochloric acid.
- Leakage during uses of chlorine/ hydrochloric acid.
- Emissions from fossil fuel fired power stations.
- Emission from metallurgical plants
- Emission from certain chemical industries such as chlor-alkali plant, caustic soda, etc.
- Paper and pulp mills
- Combustion of fuels containing chlorine compounds in industrial boilers
- Burning of chlorinated plastics like PVC
- Combustion of gasoline containing ethylene chloride.
- Waste fired and biomass fired boilers and incinerators
- Exhaust gases from engines
- Natural sources like volcanic gases, forest fires and biomass waste fires, accidental fires, decaying of biomass, etc.

Ill Effects of Chlorine

The chlorine in air directly affects eyes, nose, and throat. It is poisonous, and causes edema, and also emphysema (distension or stretching of lungs with breathing. difficulties). *Also see* ***Table 2.14 in Para 2.16 of this chapter.***

Chlorine is a biocide and chlorine pollution of air very adversely affects the plants. Chlorine pollution of air has been found to deform the epidermal cells, decreases chlorophyll content and sometimes also causes mutagenic effects. It also kills the useful microorganisms in the soil. The cases of chlorine exposure are generally found in cases of accidental leaks and in excessive use of gas used as biocide.

Sources and Effects of Chlorofluorocarbons

The chemical compounds of carbon with chlorine / fluorine like $CFCl_3$, CF_2Cl_2, CHF_2Cl, CF_4, etc. are entirely of anthropogenic origin and are very toxic and extremely harmful for health. The chlorofluorocarbons are generally used in pesticides, refrigeration, air- conditioning, etc.
Besides being strong greenhouse gases, these compounds destroy the protective ozone layer in stratosphere and are termed as very strong stratospheric ozone depleting compounds. Due to their ozone layer depletion potential, these are being replaced by less damaging compounds. ***(See Chapter 20, Para 20.1.4.)***

2.7 METHANE, VOLATILE ORGANIC COMPOUNDS (VOC), AND DIOXINS

2.7.1 Methane (CH_4)

Methane (CH_4) is the simplest 'alkane'. Methane is an air pollutant and is also a potent greenhouse gas. In high proportion, it is toxic and can catch fire easily and therefore, it is hazardous. It is a good gaseous fuel and is main constituent of natural gas, biogas, coal bed / coal mine gases, coal gas, etc. Methane is produced, or recovered or mined for fuel purposes. It is also used for the production of hydrogen, and various chemicals.

In nature, it evolves in marshy places (hence the name marsh gas) e.g. in rice fields, stagnant water, seas / oceans, and wherever some vegetable matter / weeds /organic matter / sewerage / night soil / animal waste / municipal solid waste dumps is decaying. It is formed by action of methanization bacteria on dead vegetable matter / organic matter. Paddy fields, Landfills and Wet lands are significant sources. Some methane is produced during cud chewing by animals. Methane is released from intestines of animals and human beings as a consequence of digestion of food.

Methane along with carbon dioxide is a component of gases from fermentation and methanization of organic matter. Methane is the gas lying trapped in virgin coal / lignite or as gas trapped in seams that are being mined. When methane gets released to atmosphere during coal mining, it threatens the life of miners besides leading to emission of greenhouse methane gas. In the abandoned mines, the methane may come out and also it is a component of ventilation exhaust of mine fans. Methane is present in the form of gas hydrates (CH_4 Clathrate ie $CH_4 \cdot n\,H_2O$) deep in the oceans particularly in Arctic-ocean or in oceans in south polar zones. Due to global warming, there is great danger of uncontrolled distributed release of Methane. There is also some uncontrolled escape of methane (Natural Gas or Biogas) during production or utilization:

2.7.2 Non-Methane Hydrocarbons / VOC / Dioxins

The non-methane hydrocarbons and volatile organic compounds (VOCs) including poly nuclear hydrocarbon (PNH)' / Polycyclic Aromatic Hydrocarbons (PAH), dioxins, furans are air pollutants and also play a role in the formation of ozone in troposphere.

PNH, and PAH including poly nuclear biphenyls (PNB) are emitted in the manufacture of carbon black, and in coal tar industry (including coal tar dyes), pyrolysis of coal, coke ovens, etc. Municipal solid waste fired boilers / incinerators may be the source of 'Dioxins and Dibenzofurans'. The main sources of non-methane hydrocarbons and VOCs are:

- Petrol and diesel vehicles, engine driven boats / ships, diesel or diesel- electric locomotives (in exhaust gases).
- Diesel generating sets or industrial internal combustion engines (in exhaust gases)
- Fossil fuel based power plants (partially burnt gaseous matter in exhaust flue gases under certain conditions).
- Oil refineries
- Coke ovens, coal tar industry, and carbon black industry
- Industries using solvents
- Certain chemical industry
- Wastes fired boilers and incinerators
- Vapours and fugitive emission from liquid fuel tanks, engines, leakages during use or production.

Dioxins:

Dioxins are a class of chemical contaminants that are formed during combustion processes such as 'Waste incineration', 'Forest fires, and backyard trash burning', as well as during some industrial processes such as 'Paper pulp bleaching' and 'Herbicide manufacturing'. The most toxic chemical in the class is 2,3,7,8-tetra-chloro-di-benzo-para-dioxin (TCDD). The highest environmental concentrations of dioxin are usually found in soil and sediment, with much lower levels found in air and water.

2.7.3 Ill Effects of Dioxins

Human beings are primarily exposed to dioxins by eating food contaminated by these chemicals. Dioxin accumulates in the fatty tissues, where they may persist for months or years. People who have been exposed to high levels of dioxin have been found to have developed 'Chloracne', a skin disease marked by severe acne-like pimples. Studies have also shown that chemical workers who are exposed to high levels of dioxins have an increased risk of cancer. Other studies of highly exposed populations show that dioxins can cause reproductive and developmental problems, birth defects, immune system disorder, disruption of prostrate development and an increased risk of heart disease and diabetes. The emissions of dioxins affect the local environment and may affect communities hundreds or thousands of miles away. Dioxin is one of the most toxic chemicals known to humankind. Dioxins have been linked to cancer. International Agency for Research on Cancer (IARC) - an arm of WHO, has acknowledged cancer causing potential of dioxins and classified it as human carcinogen. However, some more research may be needed to determine the long-term effects of low-level dioxin exposures on cancer risk, immune function, and reproduction and development. [Ref: 8]

2.8 AMMONIA

Emission of ammonia (NH_3) takes place in many chemical processes and industries, refineries, coke ovens, pyrolysis of organic matter, incineration of wastes, fertilizer plants, paper and pulp mills, agricultural process, leakage of gas during its manufacture or use, etc. It is a very useful chemical and is used in the production of many chemicals. It is an intermediary chemical in the production of fertilizer. In nature, ammonia is emitted by decay of vegetable matter, sewage, night soil, animal dung, animal urine, human urine and other organic matter.

Note: *On industrial scale it is generally produced by 'Haber's process' synthesis of N_2 and H_2 at about 500–550 °C in presence of catalysts. ($N_2 + 3H_2 = 2\ NH_3$). The catalyst may be 'finely divided iron mixed with materials like Alumina and potassium oxide', 'finely divided iron containing traces of molybdenum or calcium' or 'ferric oxide containing traces of silica and potassium oxide', etc.*

Ill Effects of Ammonia

Ammonia is low toxicity gas having very pungent smell. Its adverse effects on vegetation may appear in very heavy pollution on continued exposure by yellowing of leaves, but in low concentration there is no effect. it is sometimes used as gas for dispersal of unruly crowd by police. [*Also see* ***Table 2.14 in Para 2.16 of this chapter*** *for effects of ammonia on respiratory system.]*

2.9 OZONE (O_3) IN TROPOSPHERE

The ozone is mainly formed in troposphere due to photolytic reaction on the nitrogen dioxide (NO_2) present (due to emission sources) in the natural environment. The nitrogen dioxide and oxygen in the atmosphere form ozone and nitric oxide (NO) in presence of light (photolytic reaction). The volatile organic compounds (VOC) and non- methane hydrocarbons, if present in atmosphere due to air pollution, accelerate the oxidation of nitric oxide to nitrogen dioxide and thus create more 'Ozone' by further photolytic reaction of nitrogen dioxide. Partially burnt hydrocarbons and oxides of nitrogen create photo chemical smog of which ozone is one of the main components.

Note: *The hydrocarbon emissions are mostly due to incomplete combustion of liquid fuels in oil engines / petrol engines or incomplete combustion of 'liquid fuels' or 'solid fuels, biomass and wastes containing volatiles' in boilers/ furnaces/ domestic fires/ other fires.*

Ill Effects of Ozone in Troposphere

The Ozone in troposphere and smog adversely affects the respiratory system, and causes coughing, short breath, chest constriction, headache, eye/ nose/ throat/ irritation etc. Ozone may cause lung cancer and subsequent mortality. (*See also* ***Table 2.14 in Para 2.16 of this chapter.***). Besides being a greenhouse gas, 'Ozone of troposphere', is very damaging for health and agriculture.

Presence of ozone in atmosphere impairs the primary production. (The leaf area of plants decreases (smaller leaves), photosynthesis decreases, grain formation is delayed, and production and yield decreases). In mild concentration, ozone causes flecks, spots or streaks on upper surface of leaves, but in severe concentration it may cause bleaching, collapse of leaves, and necrosis. To cope up with elevated ozone concentration in atmosphere,

Some natural chemical defence mechanisms are triggered by the plants (such as production of 'VOCs' and phenolic compounds to enhance their level of anti-oxidative capacity). Plants natural defence also tries to decrease entry of ozone to leaves by decreasing size of leaves or by structural modification of leaves by thicker leaves.

The yield of wheat, rice, mung dal, spinach, Indian yellow mustard, potato, corn, soybeans and peanuts, is adversely affected by ozone stress. The ozone damages foliage of plants and trees in general and also reduces growth of certain trees. Increasing ozone levels are reported to cause delayed flowering, impaired flower production, impaired seed quality and impaired seed germination in several crop species.

Recent studies have indicated that level of ozone in atmosphere at the surface in most places in Indian region is above critical levels. The highest ozone concentration occurs in March (when it is the main wheat growing season). Studies are being conducted on ozone sensitivity of Indian crops by CSIR-NBRI, Lucknow. *[Regarding bio-indication of ozone pollution in atmosphere, see* ***Chapter 1 of this book.] [Ref: 6].***

2.10 PARTICULATE MATTER

2.10.1 Sources of Suspended Particulate Matter (SPM)

The Suspended particulate matter (SPM) is emitted from various sources including the following:

- Fossil fuel fired thermal power stations—stacks, coal / lignite handling plants, fuel pulverizing plants, ash handling plants, etc.
- Exhaust of diesel engines, vehicles, earth / material handling machines, agricultural machines, diesel locomotives, ships, aero planes, etc. Stacks of various industries like iron and steel industries, (coke ovens, various furnaces, foundries, blast furnace, sintering plants, boilers if any), aluminum plants, glass industry, lime kilns, brick kilns, fertilizer industry, calcium carbide kilns, carbon black industry, non-ferrous metal smelting, cement plants, refractory manufacturing plants, etc.
- Emitted during mining, handling, and crushing of coal, lignite, stones, ores, etc.—mainly material dust
- Emitted as smoke by accidental fires, burning of biomass / wastes, domestic fires, hot water boilers, incinerators.
- Emitted in nature by 'Volcanic eruptions, jungle fires, dust storms, etc.'

2.10.2 Respirable Suspended Particles (RSP) / Respirable Particulate Matter (RPM)

The 'Respirable suspended particles (RSP)' or 'Respirable particulate matter (RPM)' is the finer part of the suspended particulate matter [size less than 10 μm] which is respired along with air causing many ill effects on health of animals and human beings. The particles of size less than less than 10 μm do not generally agglomerate and settle down and therefore remain suspended and enter lungs with air during inhaling. Finer particles which enter lungs get lodged therein and do not come out during exhaling.

The RSP / RPM may contain fine particles of toxic material (solid or liquid), heavy metals, fumes, mist, dust, smoke from various industries (including chemical industries, metallurgical industries / operations), pesticide spray operations, pollens (the fertilizing powdery material formed in the anthers of flowers), etc. Lead or lead compounds are contained in the smoke of various fuels such as coal, oils, and other substances.

Aerosols which to the dispersed microscopic size particles of solids or liquids in gaseous media are part of RSP/RPM. The aerosols widely differ in their nature, properties, air pollution potential and their ill effects. Their size may be in the range of '0.01 μm–100 μm'. The aerosol may consist of dust, smoke, mist, fog, fumes, nitrates, and sulphate particles.

Asbestos mining, processing, manufacturing of asbestos products, using asbestos powdery products have high potential for air pollution by respirable particulate matter which has very grave effect on human health.

The smoke from industrial processes / furnaces / boilers, thermal power plants, vehicles, or the dust generated in mining and processing of materials / minerals / fuel etc. may contain certain suspended solid toxic micro-particles of arsenic, lead, mercury, barium, beryllium, boron, cadmium, chromium, vanadium, manganese, nickel, phosphorus, sulphur, selenium, zinc, iron, cadmium, etc. or their compounds / salts.

The presence of a particular pollutant in the emission depends on the materials and impurities present in the reactants, fuels, and/ or catalysts used in the process.

Some Examples

(a) The smoke from burning of fuel oils may contain compounds of vanadium, sulphur, aluminum, asphalt, silicon and lead.

(b) The diesel oils may contain compounds of lead and vanadium, sulphur, fluorine, chlorine, etc., besides sodium, potassium and calcium compounds which may be present in all liquid fuels.

(c) The smoke of solid fuel like coal and lignite would contain compounds of sulphur, iron, titanium, phosphorus, silicon, aluminum, manganese, chromium, nickel, mercury, calcium, sodium, potassium, chlorine, and fluorine.

(d) The smoke from sintering operations of copper ores may contain compounds of lead, arsenic, antimony, tin, sulphur and phosphorus.

(e) Brassware industry is a prominent source of air pollution as well as water pollution by toxic metals like copper, cadmium, zinc, and lead as revealed by air sampling in Moradabad, (UP), which is famous for it brassware industry. ***[Ref:1]***

2.10.3 Ill Effects of SPM, RSP / RPM, Air borne Micro particles of Toxic / Heavy Metals in Combination with other Air Pollutants

The suspended particulate matter causes the dust deposition on all things and is a great nuisance and spoils everything. The respirable dust in the air is a great health hazard especially for respiratory organs of humans and animals and may cause many diseases. In combination with sulphur dioxide and ozone, the particulate matter becomes very harmful. Dusty atmosphere causes asthma. The long exposure of pregnant women to such atmosphere may retard development of human fetus / embryo and cause birth of a baby with lesser weight.

Research by the World Health Organization has found that there is greatest concentration of particulate matter in countries with low economic status, high poverty and high birth rates in the world. The children are more susceptible to the dangers of air pollution which include pollution by ozone, particulate matter, SO_2, nitrogen dioxide, CO and lead. Children living in cities around the world with high exposure to air pollutants may develop asthma, pneumonia and other respiratory infections. Children born under such conditions may have low initial birth weight.

In a study Gauderman et al. followed 1,759 children (of age group of 10 to 18 years) in a dozen Southern California communities. According to it, the pollutants considered came primarily from car exhaust. The investigators found that 7.9 percent of the 18-year-olds in the highest pollution areas had 20% reduced lung capacities. Among those subjected to the least-polluted air, 1.6 percent had under-performing lungs. The investigators also found that the lung effects were similar to those that occur when children live in the home of a mother who smokes.

In the vicinity of a cement factory, the response of plants to the cement dust is reduction of plant height, and reduced yield of foliage, flowers, fruits, and seeds. The micro air borne particles of toxic or heavy metals may pollute drinking water or food and cause same diseases as mentioned above or even worst sometimes.

Heavy metals are phytotoxic (toxic to plants) and cause adverse effects on plants, besides deleterious effects on animal and humans. The ill-effects of some of the toxic and heavy metals are given in the **Table 2.2.** [*Also see **Table 2.14, Para 2.16 of this chapter** for ill effects on respiratory system].*

Table 2.2: Ill Effects of Toxic Metals/ Compounds as Air Pollutants

1.	**Zinc (Zn)**	Zinc is an essential element for humans, animal and plants. It is also an important cell component in several metallo-enzymes. Infants need 3 to 5 mg/day, adult males 15 mg/day, pregnant and lactating females 20-25 mg of Zn / day. However heavy dozes of Zn salts (165 mg) for 26 days causes vomiting, renal damage, cramps etc. Zinc in air would adversely affect respiratory organs
2.	**Copper (Cu)**	Excess of Cu in human body (more than 470 mg) is toxic, may cause hypertension sporadic fever, Uremia, Coma. Copper also produces pathological changes in brain tissue. Copper pollution of air would damage respiratory organs. However, Cu is an important cell component in several metallo-enzymes and lack of Cu causes Anemia, growth inhibition and blood circulation problem.
3.	**Barium (Ba)**	Excess of Ba (more than 100 mg) in human body may cause excessive salivation, colic, vomiting, diarrhea, tremors*, paralysis of muscles or nervous system, damage to heart and blood vessels.
4.	**Iron (Fe)**	It is one of the essential minerals for humans and animals, and is a component of blood cells and liver metallo-enzymes. But it is not absorbed for body growth from air. More than 10 mg per kg of body weight causes rapid respiration and pulse rates, congestion of blood vessels, hypertension and drowsiness. Excess Iron from air or water/food increases hazard of pathogenic organisms, as many of them require Fe for their growth.
5.	**Cadmium (Cd)**	Cd is very toxic, 50 mg may cause vomiting, diarrhea, abdominal pains, loss of consciousness. It takes 5 to 10 years for chronic Cd intoxication. During first phase, discoloration of teeth, loss of sense of smell, mouth dryness occurs. Afterwards it may cause decrease of red blood cells, impairment of bone marrow, lumber

		pains, disturbance in calcium metabolism, softening of bones, fractures, skeletal deformations, damage to kidneys, hypertension, tumor formation, heart disease, impaired reproductive function, genetic mutation, etc.
6.	**Mercury (Hg)****	Mercury is very toxic, Excess mercury in human body (more than 100 mg) may cause headache, abdominal pain, diarrhea, destruction of hemoglobin, tremors*, very bad effects on cerebral functions and central nervous system, paralysis, inactivates functional proteins, damage of renal tissues, hyper coagulability of blood, minamata disease, and even death. May cause impairment of vision and muscles and even coma. It disturbs reproductive and endocrine system. Also causes insomnia, memory loss, gum inflammation, loosening of teeth, loss of appetite, etc.
7.	**Lead (Pb)**	Lead as an air pollutant is quite damaging as it accumulates in the body. It can be absorbed through respiratory track. More than 400 mg of lead in human body can cause mental retardation, reduced attention, reduced IQ, learning difficulties, hearing loss, behavior effects, vomiting, loss of appetite, convulsions, uncoordinated body movements, brain damage, helplessly amazed state, and coma. It is retained in liver, kidney, brain, muscle, soft tissues, bones, leads to high rate of miscarriages. affects skin, and respiratory system, damages kidneys, liver and brain cells. Disturbs endocrine system, causes anemia, and long term exposure may cause even death.
8.	**Arsenic (As)**	Arsenic when inhaled or absorbed through skin causes many problems. More than 25 mg of Arsenic causes vomiting, diarrhea, nausea, irritation of nose and throat, abdominal pain, skin eruptions (including inflammations, scaling of skin, pigment changes), and even death. It binds globulin of blood hemoglobin in erythrocytes. May causes cancer of skin, lungs and liver, chromosomal aberration and damage, gangrene, loss of hearing, injury to nerve tissue, liver and kidney damage. Minor symptoms of As poisoning – weight loss, hair loss, nausea, depression, fatigue, white lines across toe nails and finger nails.
9.	**Vanadium (V)**	Very toxic, may cause paralysis.
10.	**Metal Fluorides**	Excess Fluoride intake in body results in progressive crippling scourge (sponging)/ Fluorosis of bones, teeth. May cause metabolic alternations in soft tissues and their functional mechanism.
11.	**Selenium (Se)**	Signs of Se poisoning (more than 4 mg) are fever, nervousness, vomiting, falling of blood pressure, causes damage to liver, kidney and spleen, loss of nails and hair, and causes blindness to animals. Cats are most susceptible. It affects enzyme systems and interferes with sulphur metabolism. It can cause growth inhibition, skin discolorations, bad teeth, psychological problem, gastro intestinal problems, but trace amount of Se is protective against poisoning by Hg, Cd, Ag.
12.	**Chromium (Cr)**	Chromium pollution in air is irritant, corrosive, toxic, and carcinogenic and may cause dermatitis. Any chromium compound is toxic, but Hexavalent Cr greater than 70 mg is very toxic. It causes cancer, anuria, nephritis, gastro-intestinal ulceration, perforation in partition of nose. It penetrates cell membrane and badly affects central nervous system. Causes respiratory trouble, lung tumors when inhaled. May cause complications during pregnancy. It has adverse effects on aquatic life. Trace amount of CrIII is essential for normal glucose, protein and fat metabolism & hence it is a essential trace element in diet.

13.	**Manganese (Mn)**	Mn is essential for mammals, but in concentration greater than 100 ppm Mn is toxic and causes growth retardation, fever, sexual impotence, muscles fatigue, and eye blindness.
14.	**Cobalt (Co)**	High dose (27 mg or above) can cause paralysis, diarrhea, low blood pressure, lung irritation, and bone defects.
15.	**Nickle (Ni)**	More 30 mg may cause changes in muscle, brain, lungs, liver, kidney and can also cause cancer, tremor*, paralysis and even death.
16.	**Boron (B)**	Essential for plant growth in traces. Harmful to crops and affects metabolic activities of plants in higher concentration. Affects central nervous system.
17.	**Chlorine (Cl)**	Destroys plant and aquatic life and is a biocide.
18.	**Sulphide**	Gives bad odour, and is toxic to many aquatic organisms and animals.
19.	**Phenols**	Toxic and imparts objectionable odour. Also subdues plant growth generally. Some phenols (Nitrophenyl, etc) are carcinogens.
20.	**Pesticides/ Insecticides**	Highly poisonous for humans and animals. Also they lower seed germination, play a role in development of Parkinson's disease, and destruction of nerve cells in certain regions of brain resulting in loss of dopamine which is used by nerve cells to communicate with brain. Some of these are physical poisons, some are protoplasmic poisons causing liver damage, some are respiratory poisons and some are nerve poisons.
21.	**Aluminum (Al)**	Toxic especially for brain, sometimes may lead to 'Alzheimer's' disease in humans.
22.	**Beryllium (Be)**	For toxic effects of beryllium see **Para 2.10.5** of this chapter
23.	**Silica dust**	Causes Silicosis (see **Para 2.10. 6** of this chapter)
24	**Asbestos**	Causes Asbestosis (see **Para 2.10.7** of this chapter)

2.10.4 Ill Effects of Fly Ash Air Pollution from Coal Fired Power Stations

The toxic metals, which adhere to fly ash emitted from the stacks of thermal power stations, have the same effects as given **Table 2.2**. The adherence of toxic metals to fly ash particles is shown in **Table 2.3**. The percentage of respirable matter increases with the decrease in size of fly ash particles as seen from **Table 2.4.** The deposition of particles in pulmonary area of a person breathing such polluted air also increases with the decrease in size of particles as seen from **Table 2.5.**

Table 2.3: Toxic Metals in Adherence to Fly-ash Particle Depending on Particle Size

Trace elements in fly ash		Particle size-wise concentration of trace metals in ppb				
		25 μm	12.5 μm	10 μm	3.5 μm	1.5 μm
Vanadium	V	200	200	200	200	200
Lead	Pb	300	200	300	300	500
Nickel	Ni	300	200	200	300	300
Manganese	Mn	200	240	290	390	500
Iron	Fe	40,000	59,000	43,500	35,500	32,300
Copper	Cu	150	150	200	200	200
Chromium	Cr	130	130	130	300	300

Cadmium	Cd	55	5	5	5	100
Beryllium	Be	2	1	2	2	2
Boron	B	300	500	500	500	500
Aluminum	Al	67,000	54,300	57,300	63,600	59,300

Table 2.4: Percentage of Respirable Fly-ash Particles from Stacks of Coal Fired Power Stations

Particle size µm	2	2.5	3.5	5	10
% respirable	100	75	50	25	0

Table 2.5: Percentage Deposition in Pulmonary Area of a Person Breathing the Air

Particle size µm	0.2	0.6	1	2	3	4	5	6	10
% deposition	33.4	22	24	33	25	15	6	3	0

Note: *Any type of fine particulate matter other than fly ash can also create this type of pulmonary deposition.*

2.10.5 Toxic Effects of Beryllium

The Beryllium is a very toxic and harmful metal and urban population gets exposed to it.

Beryllium can enter body by breathing air, through food, or drinking water containing beryllium. When one breathes air containing beryllium particles, the beryllium may get deposited in the lungs. It slowly dissolves in the lungs and moves into the blood stream. Some of the beryllium deposited in the lungs can move to the mouth and get swallowed; the rest remains in the lungs for a long time. The small amount of beryllium moves from the lungs, stomach, and intestines into the blood stream. Some beryllium carried by the blood to the kidneys leaves the kidneys with urine. Some beryllium can also be carried by the blood to the liver and bones where it may remain for long periods. On inhaling beryllium, it may take months to years before the body gets rid of beryllium by itself.

Beryllium can damage the lungs. It acts particularly on alveoli of lung. With more than 1mg / cubic meter concentration, it may give symptoms like that of pneumonia, and cause reddening and swelling and edema of lungs. In chronic cases it may lead to pulmonary fibrosis, progressive dyspnea, interstitial granulomatosis, etc. The symptoms may subside on stopping the exposure to beryllium. If not treated the disease may be fatal. Some people may be allergic to 'Beryllium' containing air. The white cells of blood may gather around beryllium molecule and form chronic inflammatory reaction, but this would take place on very long exposure. This condition is very bad because of weight loss, weakness, difficulty in breathing, and heart enlargement or heart disease. Long term exposure may cause cancer. Beryllium may get transferred to child through mother's milk when mother is exposed to beryllium.

Beryllium may enter the atmosphere during ore extraction, manufacture of alloys, and ceramics, and automobile smoke. Also ***see Para 11.1.2, Chapter 11.***

2.10.6 Adverse Effects of Inhaling Silica Dust (Silicosis)

The workers employed on stone crushers, mining/ crushing of various ores, sand quarries or people working with substances like glass, concrete, etc. are at increased risk of inhaling small particles of silica dust. The tiny silica particles travel through the smallest airways into the 'Alveolar sacs' which may cause a serious disease called *'Silcosis'*.

During normal respiration, air travels through the nose, down the trachea, and into smaller and smaller airways called bronchi. The bronchi divide into bronchioles and finally into tiny grape-like clusters of thin, fragile sacs called 'Alveoli'. The alveoli are where oxygen is exchanged for carbon dioxide in the blood. Inside the alveoli, the silica particles are engulfed by immune cells called macrophages, which are designed to destroy foreign organisms in the body with special enzymes. The macrophages do not destroy the silica, but the silica proves lethal to these immune cells, which release enzymes as they die. When the enzymes are released, they act as an irritant in the lungs and initiate an inflammatory process. Thus, the immune system is badly affected. The silicosis causes asthma, bronchitis, pulmonary fibrosis, various lung diseases, tuberculosis, and eventual deaths.

2.10.7 Adverse Effects of Inhaling Asbestos Dust (Asbestosis)

Asbestos particles cause *'Asbestosis'* which is a chronic inflammatory medical condition affecting the tissue of the lungs. It occurs after long-term, heavy exposure to asbestos from asbestos-containing materials in structures, manufacture of asbestos containing materials, asbestos mining. The exposure to these particles causes severe dyspnoea (ie shortness of breath), pulmonary fibrosis, pleural calcification, pleural mesothelioma and an increased risk of several different types of lung cancers.

2.10.8 Other Occupational Diseases due to Inhalation of Various Particulates

Apart from silicosis and asbestosis, some other occupational diseases due to exposure and inhaling various types of particulates are mentioned below:

(a) **Anthracosis/ Pneumoconiosis:** Disease of lungs due to breathing of air laden with coal dust which is likely to happen to people working in coal/ lignite mines and thermal power plants / industries based on such fuels. The disease affects lung parenchyma, lymph nodes, etc. leads to pulmonary fibrosis.

(b) **Byssinosis:** Disease of lungs due to exposure and inhalation of air laden with cotton fibre which is likely to happen to persons working in Textile mill / Ginning mills.

(c) **Bagassosis:** Disease on account of exposure /inhaling of dust from sugar cane & bagasse which is likely to be found in workers engaged in crushing of sugar cane or labour engaged in loading/ unloading/ handling of sugar cane/ bagasse.

(d) **Siderosis:** Lung disease on account of exposure/ inhalation of Iron dust or other metal dust or fragments which is likely to be found in workers engaged in factories / workshops for metal cutting, milling, drilling, molding, casting, smelting etc.

(e) **Tobaccosis:** Disease due to exposure / inhaling tobacco dust which is likely to be found in workers engaged in tobacco factories or for harvesting or for handling tobacco and its products.

(f) **Farmers Lung:** Disease of lungs caused by breathing of hay or grain dust which is likely to be found amongst farmers.

[For more professional respiratory system diseases due to industrial toxicants see ***Table 2.14, Para 2.16*** *of this chapter]*

2.11 BIOLOGICAL POLLUTANTS AND THEIR ILL EFFECTS

Biological pollutants such as aero-allergens, bacteria, fungus, virus, etc. are present in air, specially, near the municipal solid waste dumps, night soil, sewage, drainage, hospital wastes, animal wastes, fish waste, dairy waste, and decaying organic wastes of any type. The indoor air of crowded places such as trains, buses, slums, cinema halls, crowded markets, underground places, etc. may, sometimes, be risk areas for biological pollution because of concentration of such pollutants. Bacterial molds wherever present may create various diseases. Pollen from plants may also create various health problems.

The biological pollution in air may lead to various diseases in humans, animals and also in plants including agricultural crops, seeds, fruits, etc. The biological pollution of inside air is dealt with in **Para 2.13 (vii)** of this chapter. The lack of air circulation allows biological airborne pollutants to accumulate and concentrate in indoor environment.

2.12 PESTICIDES, INSECTICIDES AND HERBICIDES

2.12.1 Pollution by Pesticides

The manufacturing and use of pesticides, insecticides, and herbicides can pollute air by way of their aerosols. The main use of pesticide and herbicide sprays is in agriculture or horticulture to save the crops from pests and weeds respectively. These pollutants also pollute water and enter into the food cycle of animals and human beings. The insecticide sprays are used to eliminate mosquitoes and flies to mitigate diseases spread by them. All this causes deleterious effects on health. The biggest sources of this type of pollution are the agricultural and horticultural use of these chemicals.

Some of these products which were considered extremely polluting because of their long time persistence and deleterious effects on health have been banned, but they may still be available in Indian market. It may be mentioned that as far as possible bio- pesticides should be used instead of chemical pesticides. A lot of research is going on in developing bio-pesticides. As examples *of bio-pesticide are Bacillus, Pseudomonas and Trichoderma species.*

The pesticides mentioned in **Table 2.6**, called 'Dirty' Dozens' though banned in Western countries, were being exported from USA between 1991—94 and imported by developing countries. The **Table 2.7 (a)** shows the pesticides banned in June 1993 by Government of India

BHC (Benzene Hexa Chloride) was banned in April 1997 in India. Although DDT was banned in 1992 for agricultural use, but it is still persisting (2005) in food chain and soil and there-by appears in some ground waters. According to the Indian Council of Agriculture Research (ICAR), the pesticides mentioned in **Table 2.7 (b)** were banned with effect from 25th March 2002 due to being non-biodegradable.

Table 2.6: List of Dirty Dozens

(1) D.D.T Dichloro-Diphenyl Trichloro ethane	(7) Chlordane
(2) DBCP (Dibromochloro propane)	(8) Aldrin
(3) Benzene Hexa Chloride (BHC)	(9) Paraquat
(4) Hexa Chloro-cyclo Hexane (HCH)	(10) Ethyl Parathion
(5) Penta Chlorophenol (PCP) 2 – 4 – 5 – T	(11) Camphechlor
(6) Ethylene Dibromide (EDB)	(12) Chlordimeform

Table 2.7 (a): Pesticides Banned in India in June 1993

1. DBCP	8. Nitrofen
2. PCP	9. Tetradifon
3. Penta chloro Nitro benzene (P.C.N.B.)	10. Paraquat
4. Chlordane	11. Aldrin
5. Heptachlor	12. Endrin
6. Toxafen	13. DDT (banned earlier in 1992)
7. Ethyl Parathion	

Table 2.7 (b): Pesticides Banned w.e.f. 25-3-2002 as per ICAR

1. Benzene Hexachloride (BHC)	12. Chlordane
2. Dibromochloro propane (DBCP)	13. Endrin
3. Pentachloro Nitrobenzene (PCNB)	14. Ethyl Parathion
4. Penta Chloro phenol (PCP)	15. Heptachlor
5. Phenyl Mercury Acetate (PMA)	16. Menazone
6. Sodium Methane Arsonate (MSMA)	17. Nicotine Sulphate
7. Paraquat Dimethyl Sulphate	18. Tetradifon
8. Copper Aceto Arsenite	19. Toxafen, Methomyl 12.5 % L
9. Ethyl Mercury Chloride (DDT Already banned since 1992 for Agricultural use)	20. Methomyl 24 % formulation
10. Aldrin	21. Phosphamidon 85 % SL
11. Calcium Cyanide	22. Nitrofen (already in list of banned pesticide of 1993)

Environmental Persistence of Pesticides:

Some of the pesticides persist in the environment for a long time because of their persistence in the soil which acts as reservoirs as given in the following **Table 2.8.** Organo phosphorous based pesticides / insecticides are nowadays used more often than organo chlorine based compounds because of lower half-life of organo phosphorous based compounds.

Table 2.8: Persistence of Pesticides in Soil

Pesticide	Duration	Pesticide	Duration
DDT	> 10 years	BHC	> 11 years
Chlordane	> 12 years	Heptachlor	> 9 years
Aldrin	> 9 years	Toxafen	> 6 years
Dieldrine	> 9 years	HCH	> 10 years
Monuron	> 3 years	Diuron	> 19 months
Atrazine	> 18 months	Simazine	> 17 months
Arsenic based Pesticides	> 30 years	Triazine Herbicides	> 2 years

Examples of half life of insecticides:

Dieldrine, BHC, DDT > 9 years; Organophosphorous based insecticides — 0.02 to 0.2 years; Carbamate/carbaryl insecticides — 0.02 years.

Examples of half-life of Herbicides:

Triazine Herbicides — 1 to 2 years; Urea Herbicides — 0.3 to 0.8 years; Benzoic Acid Herbicides — 0.2 – 1 year; 2,4, D and 2,4,5 – T Herbicides — 0.1 – 0.4 years.

2.12.2 List of Pesticides Banned in India up dated to 2011

An updated list of pesticides banned in India (up to 2011) is given in **Table 2.9. (i).**

Table 2.9 (i): List of Pesticides / Pesticide Formulations Banned in India

A	Pesticides Banned for manufacture, import and use (27 Nos)	
	1	Aldrin
	2	Benzene Hexachloride
	3	Calcium Cyanide
	4	Chlordane
	5	Copper Acetoarsenite
	6	CI-bromochloropropane
	7	Endrin
	8	Ethyl Mercury Chloride
	9	Ethyl Parathion
	10	Heptachlor
	11	Menazone
	12	Nitrofen
	13	Paraquat Dimethyl Sulphate
	14	Pentachloro Nitrobenzene
	15	Pentachlorophenol
	16	Phenyl Mercury Acetate
	17	Sodium Methane Arsonate
	18	Tetradifon
	19	Toxafen
	20	Aldicarb
	21	Chlorobenzilate
	22	Dieldrine
	23	Maleic Hydrazide
	24	Ethylene Dibromide
	25	TCA (Trichloroacetic acid)
	26	Metoxuron
	27	Chlorofenvinphos Lindane (Banned vide Gazette Notification No S.O. 637(E) Dated 25/03/2011)
B	Pesticide / Pesticide formulations banned for use but their manufacture is allowed for export (2 Nos.)	
	1	Nicotine Sulfate
	2	Captafol 80% Powder

C		Pesticide formulations banned for import, manufacture and use (4 Nos)
	1	Methomyl 24% L
	2	Methomyl 12.5% L
	3	Phosphamidon 85% SL
	4	Carbofuron 50% SP
D		**Pesticides Withdrawn (7 Nos)**
	1	Dalapon
	2	Ferbam
	3	Formothion
	4	Nickel Chloride
	5	Para-dichlorobenzene (PDCB)
	6	Simazine
	7	Warfarin

Pesticides listed in **Table 2.9 (ii)** have been refused registration in India

Table 2.9 (ii): List of Pesticides Refused Registration

1.	Calcium Arsonate
2.	EPM
3.	Azinphos Methyl
4.	Lead Arsonate
5.	Mevinphos (Phosdrin)
6.	2, 4, 5-T
7.	Carbophenothion
8.	Vamidothion
9.	Mephosfolan
10.	Azinphos Ethyl
11.	Binapacryl
12.	Dicrotophos
13.	Thiodemeton / Disulfoton
14.	Fentin Acetate
15.	Fentin Hydroxide
16.	Chinomethionate (Morestan)
17.	Ammonium Sulphamate
18.	Leptophos (Phosvel)

The pesticides for only *'Restricted use'* are listed in the **Table 2.9 (iii)**

Table 2.9 (iii): Pesticides for only 'Restricted Use'

1.	Aluminum Phosphide
2.	DDT
3.	Lindane
4.	Methyl Bromide
5.	Methyl Parathion
6.	Sodium Cyanide
7.	Methoxy Ethyl Mercuric Chloride (MEMC)
8.	Monocrotophos
9.	Endosulfan
10.	Fenitrothion
11.	Diazinon
12.	Fenthion
13.	Dazomet

2.12.3 Problems Caused by Pesticide Pollution

This is already stated that the pesticides, insecticides, herbicides cause a lot of health problems to human beings and animals by entering food chain and also by direct inhalation of these chemicals along with air. These chemicals are extremely toxic and may contribute to a large number of diseases including damage to liver and kidney, reproductive problems, birth defects, Parkinson disease. etc. These may cause cancer also. The more the persistence of these chemicals, more is the harm caused by them. Some of these chemicals can cause even gene mutations in vegetation and humans. The herbicides like pesticides also cause deleterious effect on 'cyanobacteria' which is primary producer of our ecosystem.

According to poisoning effects, there are mainly four categories of pesticides.

(a) Physical Poisons
(b) Nerve poisons
(c) Proto plasmic poisons
(d) Respiratory poisons

Physical Poisons: These kill the living organism by physical action. An example of this is Endrin which penetrates percutaneously through the epidermis of the skin producing lethal effects.

Nerve Poisons: These pesticides initiate extreme nervous excitation causing release of excessive neuroactive substance disrupting nerve activity. Example is DDT which can cause tumor, cancer, leukemia, sterility, and hypertension even in small quantities say 5-25 ppm. Other similar pesticides are Malathion, Diazinon, Systox, M.I.C., etc.

***Protoplasmic Poisons*:** Insecticides and chemicals like Lead arsenate, Endrin, Sodium Arsenite, Ziram, etc. cause precipitation of protein in the body resulting in liver damage and ultimately death.

Respiratory Poisons: These types of chemicals inactivate the respiratory enzymes like oxidases, peroxidases and reductases. For example; fumigants such as hydrocyanic acid gas, methyl bromide, ethylene dichloride and ethylene dibromide cause acute suffocation and block the respiratory tract.

2.12.4 Bhopal Gas Disaster (World's Worst Pesticide Factory Accident)

The Bhopal disaster or Bhopal gas tragedy was a disaster that resulted from an accident at a Union Carbide subsidiary pesticide plant in the city of Bhopal, India on December 3rd , 1984, when the plant released 42 tonne of toxic Methyl Iso Cyanate (MIC) gas, exposing more than 500,000 people to toxic gases and immediately killing at least 3,800 people and causing significant morbidity and premature death for many thousands more. A more probable figure is that 8,000 died within two weeks and it is estimated that an additional more Than 8,000 have since died from gas-related diseases. Bhopal disaster is frequently cited as the world's worst industrial disaster.

An International Medical Commission on Bhopal was established in 1993 to respond to the long term health effects of the disaster.

Probable Cause of Accident

The plant in Bhopal where the disaster happened started to produce Carbaryl in 1977. Carbaryl is mainly used as an insecticide. At first, the production was at 2.500 tonne a year. This was no problem, as the plant had been designed for an output of 5.000 tonne, that is to say, double that amount. At the beginning of the 1980s, insecticide, and Carbaryl did not sell very well. For this reason, the owners of the plant started to cut costs. This included employing fewer people, doing maintenance less frequently and using parts that were made of lower-grade steel. Closing the plant was being considered as well. When the disaster happened, there was no production at the plant because there was a surplus on the market.

The disaster happened because water could get into a tank containing Methyl isocyanate, an intermediate chemical . This caused a chemical reaction which resulted in the buildup of much Carbon dioxide, amongst other things. The resulting reaction increased the temperature inside the tank to reach over 200 °C (392 °F). The pressure was more than the tank was built to withstand. The tank had valves to control the pressure. These were triggered in an emergency, which reduced the pressure. As a result, a large amount of toxic gases was released into the environment. The pipes were rusty. The iron in the rust made the reaction faster.

All the contents of the tank were released over a period of about two hours. The water had entered the tank because of a sequence of events. The tank had been maintained badly. When cleaning work was done, water could enter the tank. A mixture of poisonous gases flooded the city of Bhopal, causing great panic as people woke up with a burning sensation in their lungs. Thousands died immediately from the effects of the gas. Many were trampled in the panic that followed.

Factors leading to this huge gas leak include:

- The use of hazardous chemicals (MIC) instead of less dangerous ones
- Storing these chemicals in large tanks instead of over 200 steel drums.
- Poor maintenance after the plant ceased production in the early 1980s
- Failure of several safety systems (due to poor maintenance and regulations).
- Safety systems shut down to save money - including the MIC tank refrigeration system which alone would have prevented the disaster.

Aftermath of Bhopal Disaster

The disaster indicated a need for enforceable international standards for environmental safety, preventative strategies to avoid similar accidents and industrial disaster preparedness. Since the disaster, India has experienced rapid industrialization. While some positive changes in government policy and behavior of a few industries have taken place, but major threats to the environment from rapid and poorly regulated industrial growth still remain.

One of the reasons why many people are still suffering is that the terrain where the Bhopal plant stands is still contaminated with mercury and other carcinogenic substances. Dow Chemical who owns Union Carbide refuses to decontaminate the soil. Greenpeace had estimated that decontamination would only cost around 30 million USD.

2.13 INDOOR AIR QUALITY

A lack of ventilation indoors concentrates air pollution where people often spend the majority of their time. The lack of air circulation allows airborne pollutants to accumulate more than they would otherwise occur in nature. The indoor pollutants can cause many diseases including disorder of respiratory organs, impaired lung function, asthma, allergies, eye irritation, skin irritation, etc. The indoor pollution may also lead to even cardiac ailments. The indoor pollutants may include variety of gases and vapours some of which are mentioned below.

(i) Radon (Rn) gas, a carcinogen, is exuded from the 'Earth' in certain locations and trapped inside houses.

(ii) Certain building materials, carpeting, grocery bags, paper towels, particle boards and plywood emit formaldehyde (HCHO) gas. This chemical is also released when oil burning appliances are used or during smoking of cigarette.

(iii) Paint and solvents give off volatile organic compounds (VOCs) as they dry. Lead paint can degenerate into dust and may get inhaled. Clothing emits tetra-chloro-ethylene, or other dry cleaning fluids, for days after dry cleaning. Benzene vapours may be given out by solvents, inks, latex paints, cigarette smoke, etc. Benzene is a known carcinogen. Trichloroethylene, a central nervous system depressant, is found in paints, adhesives, lacquers, carpet -shampoos, spot removers, etc. and its vapours may contaminate indoor air. Indoor air pollution by VOC also gets introduced with the use of air fresheners, incense, and other scented items.

(iv) Wood / coal/ oil fires in stoves and fireplaces can add significant amounts of smoke particulates, NOx, carbon monoxide etc. in indoor air. Carbon monoxide (CO) poisoning and deaths are often caused by faulty vents and chimneys. Burning of charcoal in closed space indoors while sleeping, is very dangerous.

(v) Indoor air pollution, health problems and even death may be caused by using pesticides and other chemical sprays indoors without proper ventilation.

(vi) Biological sources of air pollution are also found indoors, some examples are: Pets produce dander; people produce dust from minute skin flakes and decomposed hair. Dust mites in bedding, carpeting and furniture produce enzymes and μm-size fecal droppings. Inhabitants emit methane. Mold forms in walls generate mycotoxins and spores. Air conditioning systems can incubate bacterial molds and spores. Houseplants, soil and surrounding gardens can produce pollen, dust, and mold. Lack of gas exhaust from domestic sewers and bathroom/ toilet manholes may lead to indoor air pollution by sewer gas, hydrogen sulphide, bacteria, etc.

2.14 SECONDARY AIR POLLUTANTS AND SMOG: SOURCES AND ILL EFFECTS

The secondary pollutants created in atmosphere as mentioned earlier are 'Acid fumes and vapours', 'Aldehydes', 'Peroxy acetyl nitrate (PAN)', and 'Ozone' etc. These together with other pollutants comprise smog. Their source is the reactions / inter-reactions amongst various pollutants and atmospheric gases.

The smog may be of two kinds: *Photochemical smog or Industrial smog.* Both forms of smog cause extensive health problems depending on their concentration, and exposure time. These mostly attack the respiratory organs, eyes, and skin. As already mentioned earlier, 'Ozone' itself which is a part of smog causes eye irritation, impaired lung function, and damage to trees and crops.

Photochemical smog

It may appear in regions where biomass of trees / grasses is burnt periodically. Smog results from the irradiation by sunlight of oxides of nitrogen and non-methane hydrocarbons caused primarily by unburnt gasoline/ fuel oil emitted by automobiles, oil engines, and other combustion sources. The products of 'Photochemical reactions' include: *free radicals, ozone, aldehydes, ketones, PAN, organic acids, and other oxidants or other organic particles.*

Industrial Smog

The industrial smog is created by burning coal and heavy oil containing sulphur impurities (in power plants, industrial plants, etc.). The smog consists mostly of a mixture of sulphur dioxide, suspended droplets of sulphurous acid / sulphuric acid (formed from a part content of the oxides of sulphur), and fog with a variety of suspended solid particles. This smog is common during the winter in big cities. In London, 4,000 people died as a result of this form of smog in 1952.

The sulphur dioxide in presence of atmospheric oxygen and solar light is slowly photolyzed to sulphur trioxide which in presence of fog/ water vapour/ water gets converted to sulphuric acid aerosol. The reaction is enhanced in presence of non-methane hydrocarbons and oxides of nitrogen.

2.15 OTHER MISCELLANEOUS AIR-POLLUTANTS LIKE PAH, PCB, PYRENES, NITRO-PAH, MERCURY, ALLERGENS, ETC. AND THEIR EFFECTS

2.15.1 Examples of Toxic and Hazardous Pollutants

A large number of **toxic** or hazardous air-pollutants are known or suspected to cause cancer or other serious health effects, such as reproductive defects or birth defects, neurological problems, etc. These include poiycyclic aromatic hydrocarbons (PAH), polychlorinated biphenyls (PCB), dioxin, most of the benzo- pyrenes, dibenzo pyrenes, Indeno pyrenes, nitro-poly aromatic hydrocarbons (Nitro-PAH), benzene, benzo-Fluoranthene, per-chloro-ethylene, methylene-chloride, toluene, and metals such as cadmium, mercury, chromium, and lead compounds, asbestos etc.

Polycyclic Aromatic Hydrocarbons (PAH)

The major PAH compounds listed in **Table 2.10** are present in the most of the areas near the point sources of pollution and mobile sources of pollution.

Table 2.10: Molecular Formula and Molecular Weight of Important PAHs

Name	Molecular Formula	Molecular Weight
Naphthalene	$C_{10}H_8$	128
Phenanthrene	$C_{14}H_{10}$	178
Anthracene	$C_{14}H_{10}$	178
Fluoranthene	$C_{16}H_{10}$	202

Pyrene	$C_{16}H_{10}$	202
Chrysene	$C_{18}H_{12}$	228
Benzo(a) anthracene	$C_{18}H_{12}$	228
Benzo(b) fluoranthene	$C_{20}H_{12}$	252
Benzo(k) fluoranthene	$C_{20}H_{12}$	252
Benzo(e) pyrene	$C_{20}H_{12}$	252
Benzo(a) pyrene	$C_{20}H_{12}$	252
Perylene	$C_{20}H_{12}$	252
Benzo(ghi) perylene	$C_{22}H_{12}$	276
Dibenzo(ah) anthracenes	$C_{22}H_{14}$	278
Indeno(cd) pyrene	$C_{22}H_{12}$	276
Coronene	$C_{24}H_{12}$	300

***Note:** Lichens are good biomonitors for PAH (see **Para 20.5, Chapter 20**).*

2.15.2 Sources of PAH, PCB, Pyrenes, Nitro-PAH, Mercury, etc.

Most of the toxic air- pollutants originate from man-made sources, including mobile sources (e.g., cars, trucks, buses) and stationary sources (e.g., refineries, petrochemical industries , power plants, steel plants), as well as indoor sources (e.g., some building materials and cleaning solvents). Some of these air pollutants are also released from natural sources such as volcanic eruptions and forest fires.

Such pollutants may be specific to a particular industry; in larger sense, they may not be very significant, but locally (such as inside a particular industry, big cities urban atmosphere with traffic concentration, etc.) these may be quite important and damaging. These may include the following:

- Carbon sulphide—from 'Viscose manufacturing industry,
- 2-Butene more aptly 'But-2-ene' ($CH_3CH{=}CHCH_3$), But-1-ene ($CH_3CH_3CH{=}CH_2$), Isobutene [$(CH_3)_2C{=}CH_2$]—from cracked petroleum, gasoline, petrochemical operations,
- Ethyl benzene [C_6H_5-CH_2-CH_3], Butane [C_4H_{10}], Hexane [C_6H_{14}] and other oxygenated hydrocarbons—from solvents operations
- Methyl ethyl ketones, 2- Propanol—from surface coating operations,
- Phenols—from petroleum operations,
- Tetra-chloro ethylene / Per-chloro-ethylene—from 'Dry cleaning process'
- Trichloroethylene—from 'De- greasing operations'
- Mercury pollution—by Mercury cells used in caustic soda and other industries
- Mercury pollution—by coal fired thermal power stations
- Benzene—is found in 'Gasoline'. Benzene compounds are present in smoke/ exhaust gases of petrol vehicles and also diesel vehicles, :
- Phenanthrene , Fluoranthene and pyrene;—from coal combustion
- Anthracene, Phenanthrene and benzo(a)pyrene—from coke production
- Pyrene, Phenanthrene and Fluoranthene—from incineration
- Benzo(a)pyrene and Fluoranthene—from wood combustion
- Fluoranthene pyrene and chrysene—from burning of oil by industries

- Benzo(ghi)perylene, Indeno (123-cd)pyrene and coronene—from exhaust gases of petrol vehicles/ engines.
- Fluoranthene and pyrene with higher ratios of benzo, (b)Fluoranthene and benzo(k) Fluoranthene, plus thiophene compounds—from exhaust gases of diesel vehicles /engines.
- Methylene chloride—from solvents and paint stripping operations.
- Pyrene with high ratio of benzo(b)Fluoranthene & benzo (k) fluoranthene—in the Diesel vehicle Exhaust
- More than 70 PAH compounds—in cigarette smoke.*

Nitro PAHs

Both petrol and diesel fueled vehicles produce PAHs and nitro - PAHs. Diesel exhaust contains higher concentration of carcinogenic nitro-PAHs and low toxicity 2, 3 and 4 ring PAHs. Petrol engine exhaust gases tend to produce higher concentrations of the 5 and 6 rings PAHs [Benzo(a)pyrene, Benzo (g, h, I) perylene, Indeno (1,2,3-cd) pyrene coronene and Anthracene] which are more carcinogenic than the 2 and 3 ring PAHs . Nitro-PAHs are either formed from combustion process or via nitro substitution of the parent PAH. In later case, these are initiated by chemical reactions in the atmosphere with OH radicals in the presence of NOx and N_2O_5 and sunlight.

* *The average total BaP content in the main stream smoke of 1 cigarette was18 ng in 1978-1979. Modern low tar cigarettes deliver 10 ng BaP. The concentration of BaP in a room extremely polluted with cigarette smoke was 22 ng/BaP (WHO).*

2.15.3 List of Important 'Urban Toxic and Hazardous Air Pollutants'

The Environmental protection agency (EPA) of the United States of America has listed 33 toxic pollutants in urban air as important as mentioned in **Table 2.11.**

Table 2.11: EPA's (US) List of Important 33 Urban Air Toxics/Pollutants

Se. No.	Air Pollutant	Se. No.	Air Pollutants
1	Acetaldehyde	18	Ethylene Oxide
2	Acrolein	19	Formaldehyde
3	Acrylonitrile	20	Hexachlorobenzene
4	Arsenic Compounds	21	Hydrazine
5	Benzene	22	Lead Compounds
6	Beryllium Compounds	23	Manganese Compounds
7	1, 3-Butadiene	24	Mercury Compounds
8	Cadmium Compounds	25	Methylene Chloride
9	Carbon Tetrachloride	26	Nickel Compounds
10	Chloroform	27	Polychlorinated Biphenyls (PCBs)
11	Chromium Compounds	28	Polycyclic Organic Matter (POM)
12	Coke Oven Emissions	29	Quinoline
13	Dioxin	30	1, 1, 2, 2-Tetrachloroethane
14	Ethylene Dibromide	31	Perchloroethylene
15	Propylene Dichloride	32	Trichloroethylene
16	1, 3-Dichloropropene	33	Vinyl Chloride
17	Ethylene Dichloride		

Exposure to Humans—(PAHs)

The urban atmosphere is a very complex and dynamic system containing a large variety of interacting chemical species in both the gas and particulate phases. PAHs compounds can reach to the human body by four different modes of exposure:

(i) Direct inhalation of the polluted air
(ii) Direct inhalation of tobacco
(iii) Ingestion of contaminated and processed food and water
(iv) Dermal contact

Air Inhalation – The intake of BaP by inhalation of polluted ambient air depends on the occurrence of PAHs in air. For example, an exposure to relatively high concentration of 50 ng BaP/m^3 and a deposition rate of 40 % from 20 m^3 air inhaled per day, the daily intake would be 400 ng of BaP.

2.15.4 Effects of Diesel Engine Exhaust on Health

The popularity of the diesel engine in heavy duty applications in trucking, railways, marine transport, DG sets and construction industry is due to both its fuel efficiency (therefore lower carbon dioxide emission) and long service life relative to the gasoline engine. Compared with gasoline engine, diesel emissions are lower in carbon monoxide (CO), and hydrocarbons (HC), but higher in oxides of nitrogen (NOx) and particulate matter (PM). Diesel exhaust is a complex mixture of both particulate and gaseous phase.

Diesel exhaust has particulate matter with mass median diameter of 0.05 to 1.00 µm, a size rendering them easily respirable and capable of depositing in the airways and alveoli. The particles consist of a carbonaceous core with a large surface area to which various hydrocarbons are absorbed, including carcinogenic PAHs and Nitro-PAHs that have drawn forth big concern with respect to human health. The gaseous phase contains various products of combustion and hydrocarbons including some of the PAHs present in the particle phase. Once emitted, components of diesel exhaust undergo atmospheric transformation in ways that may be relevant to human health. For example, nitro-PAHs, created by the reaction of directly emitted PAHs with hydroxyl radicals and oxides of nitrogen (NOx) in the atmosphere can be more potent mutagen and carcinogen and more bio available than their precursors.

A study undertaken by Ahlvik and Branberg, (1999) showed that the cancer potency of diesel vehicles is more than two times than that of petrol vehicles in India. If only the most harmful of the exhaust emissions, that is 'particulate emission', is considered, the carcinogenic effect of one new diesel car is equivalent to 24 petrol cars and 84 new CNG cars on the road.

The Honorable Supreme Court of India has restricted the use of commercial diesel driven vehicles in Delhi due to its harmful effects. **Table 2.12** shows the potential carcinogenicity and bioactivity of PAH.

EPA has given an upper bound lifetime cancer risk estimate of 62 per 100000 exposed people per µg benzene soluble coke-oven emission per m^3 of ambient air. They are public health concerns because they are spread over a large urban population.

Table 2.12: Potential Carcinogenicity and Bioactivity of PAH

Se. No.	PAH	Carcinogenicity potential	Bioactivity
1.	Anthracene	0	0
2.	Fluoranthene	0	Carcinogen with BaP CC
3.	Pyrene	0	
4.	Benzo (a) anthracene	+	Tumor initiator TI
5.	Chrysene	+	Tumor initiator TI
6.	Benzo (b) Fluoranthene	++	Complete carcinogen, Also tumor initiator C,T I
7.	Benzo(k) Fluoranthene	0	0
8.	Benzo (a) pyrene	+++	Complete carcinogen, Also tumor initiator C,T I
9.	Dibenzo (ah) anthracene	+++	Complete carcinogen, Also tumor initiator C,T I
10.	Benzo (ghi) pyrene	0	Carcinogen with BaP CC
11.	Indeno (1,2,3 –cd) pyrene	+	Tumor initiator TI
*Activity for carcinogenicity-----from + **to** +++; No activity 0*			
*Causing cancer--**CC**; Tumour Initiator--**TI**, Cancer and tumour initiator----**C,T I***			

2.15.5 Sources of Mercury Pollution and its Ill-effects

Deadly mercury pollution and contamination can be caused by emissions of mercury into air, water and land. It may be noted that mercury compounds on heating give out elemental mercury in air which is not easy to remove unless these are oxidized into solid oxide or other compounds. Apart from air pollution by elemental mercury, the dust (mainly RPM) of the solid compounds of mercury can cause serious 'mercury air pollution'. The source of mercury in air and also elsewhere are particularly:

- Coal fired boilers of thermal power stations and industries. (Mercury is released in gaseous form when coal is burned and it falls to earth and enters food chain.)
- Some industries like Caustic-Chlorine industry (manufacturing process utilizing mercury cells as in manufacturing of caustic soda).
- Artisanal Small-scale Gold Mining (ASGM) operations where the mercury is added to gold ore and sediments to bind gold (forming amalgam of gold and mercury. Gold is separated by heating when mercury vaporizes.)
- Metallurgy, mining, manufacture, storage, handling, and transport of products having mercury in their composition / content.
- Used Compact Fluorescent lamps (CFL), tube lights, mercury vapour lamps, etc.

Mercury in air when comes down to earth, pollutes water and land wherefrom, it enters the food chain. The mercury in environment concentrates biologically (bio-concentration). All the humans and animals accumulate mercury in their bodies while breathing, drinking, and eating etc.

It can cause numbness of limbs and lips, slurred speech, constricted vision, brain damage, death of children in womb, birth defects, weakening of immune system, serious health damage to animals and humans, etc.

An industrial disaster occurred more than 60 years ago named infamous *'Mina Mata tragedy'**. Many severe problems were traced to a factory that dumped mercury contaminated waste into the bay at Mina Mata, Japan. The mercury in environment concentrated biologically (bio-concentration) as it worked its way through the food chain back to people. Human beings of the area accumulated mercury in their bodies while breathing, drinking, and eating. That tragedy continues to haunt the caustic – chlorine industry sector even today. ***[For control of mercury see Chapter 10]***

***Note: Mina Mata Tragedy**

The Chisso Corporation, one of the main employers of Mina Mata, was making petro-chemicals, plastics and other products. From 1932 to 1968, Chisso Corporation dumped an estimated 27 tonnes of mercury compounds into Mina Mata Bay. The destruction of large scale fishing areas following the dumping simply saw the exchange of money to buy people off. The logic of the company was to pay people in exchange for polluting.

It was not till mid 1950s that people began to notice a strange phenomenon in animals and humans. People began to experience numbness in their limbs and lips. Their speech slurred and their vision constricted. Some people had serious brain damage. Birds started to drop dead from the skies.

The innocent fishing folk in a seaside town of Japan suffered the most. It killed children in wombs and affected innumerable people of which statistics could not put an estimate to the suffering as it spanned three decades.

The valiant effort of a doctor from Chisso Corporation itself, Dr Hosokawa, brought the reasons for the disease to light, but he faced resistance to his theory that linked the disease with the dumping of mercury compounds from the company into the sea. The company 'Chisso Corporation' initially bought the silence of people, but soon the incident came into national and international limelight. Though the victims testified at the United Nations Environmental Conference in Sweden, the UN did not intervene. Till more than a decade ago, the Japanese courts were still resolving suitable compensation for the victims.

It was Mina Mata which ultimately forced the Japanese government to ban mercury in all processes and products. It also heralded a new technology for the caustic-chlorine industry - the membrane cell technology which gave it a new life.

2.15.6. Allergies due to Air pollution

Allergens

Some solids, liquids or gases in environment may cause allergies. Such materials are termed as 'Allergens'. Inhalation, contact, or ingestion of allergens can cause many health problems including asthma. Allergens may affect the immune system. It may cause inflammation of airways of respiratory system. Pollution is important cause of allergies. The allergens may include the following:

Pollens, yeast / fungal / bacterial moulds, fine particles of animal hairs/ wool/ fur, cotton or artificial textile fibre, leather dust, tobacco dust, wood dust, particulate matter/ dust, smoke of vehicles, smoke from any type of fire, smog, dust or fine particles of heavy and toxic metals and their compounds, coal or ash dust, sulphur dioxide or any acidic gas, certain dyes, aerosols, etc. Particle sizes of some common Allergens are mentioned in the **Table 2.13**

Table 2.13: Some Common Allergens and their Particle Size

Se. No.	Allergens	Diameter in μm
1	Fungal spores	1–200
2	Viruses	< 0.004

3	Dust mites	>10
4	Bacteria	0.3–50
5	Rat urine	0.8
6	Pollens	1–200
7	Ragweed	20–30
8	Guinea pig pelt	0.8–4.9
9	Tobacco smoke	0.01–1
10	Myco-bacterial tuberculosis	Length 0.3–0.7

Asthma

Asthma is a chronic and debilitating disease causing swollen/ inflamed airways (of respiratory system) prone to sudden and violent constrictions and increase in lung resistance even 20 fold. Asthmatic attacks are characterized by shortness of breath and wheezing and can be life threatening. This disease occurs both to professionals and non-professionals and therefore may not be called a professional disease as such. It may most often be caused or propagated by *allergens* present in the working atmosphere at place of work, place of living, place of visit / travel or others.

2.16 GENERAL EFFECTS OF AIR POLLUTION ON RESPIRATORY SYSTEM

There are very grave effects of air pollution on animal and human respiratory system. Many gases, vapours of industrial chemicals, particulate matter, etc. adversely affect the bronchial system. They may cause constriction or irritation of airways of respiratory system, giving a feeling that one can not breathe deeply enough to satisfy respiratory demand (Dyspnoea). Exposure may or may not cause permanent damage to the respiratory system. High concentration of certain pollutants like Chlorine, Sulphur dioxide, Ozone, Formaldehyde, etc. may result in deaths.

The **Table 2.14** indicates the adverse effects of certain toxicants on respiratory system. A schematic view of human respiratory system is shown in **Fig. 2.1.**

Table 2.14: Some Industrial Toxicants and Respiratory System Diseases

Toxicant	Big Sources	Disease	Chronic affect
Ammonia	Production of Ammonia, Fertilizers, Chemicals, Explosives	Affects upper airways Bronchitis	Bronchitis, Laryngitis, Lung Cancer
Chlorine	Production of Pulp & Paper, Plastics, Chlorinated chemicals	Affects upper airways, Cough, dyspnea, trachea-bronchitis, Bronchi-pneumonia	
Arsenic/ & compounds	Production of pigments, glass, alloys, pesticides, etc.	Affects upper airways, Bronchitis,	Bronchitis, Laryngitis, Lung Cancer
Aluminum, Alumina	Alumina crushing, smelting & other operations, Production of abrasives	Affects upper airways, Alveolar interstitial, Cough, Shortness of breath	Interstitial fibrosis

Hydrogen fluorides	Manufacture of some chemicals, solvents, plastics, photographic films	Affects upper airways, Respiratory irritation, pulmonary edema, metabolic problems in soft tissues.	
Iron Oxides	Foundry, steel plants, iron ore mining, welding, etc.	Coughing, Lung diseases: (i) affecting pulmonary vessels & alveolar walls (Silver finisher's lung disease); (ii) affecting upper lobes, bronchi, alveoli (Haematite miner's lung); (iii) affecting bronchi (Arc Welder's lung); (iv) lungs affected by breathing iron oxide (Siderosis).	(i) Subpleural & perivascular aggregations of macrophages; (ii) Diffused fibrosis-like pneumoconiosis; (iii) Bronchitis
Ozone	Welding, bleaching, deodorizing	Affecting bronchi, & alveoli terminally; pulmonary edema	Emphysema (distention of lung with difficulty in breathing)
Talc [$Mg_3\ Si_4\ O_{10}\ (OH)_2$]	Rubber industry, cosmetics	Affecting lung parenchyma, & lymph nodes (Tracheitis)	Pulmonary fibrosis
Chromium, chromates	Reduction of chromite ore, production of chromium compounds, & paints pigments,	Affects upper airways; affects nose (Nasal pharynx); nasal irritation; bronchitis	Lung tumors and cancers
Coke oven emissions (HC*, SOx, NOx, PM**, Heavy metals	Coke production	Affects upper airways, Bronchitis	Tracheo-bronchial Cancers
NiCO (Nickel carbonyl); $Ni_2\ S_3$ (Nickel sub sulphide) NiO; Ni	Nickel ore mining, processing, & smelting; Electroplating; etc.	NiCO affects parenchyma, $Ni_2\ S_3$ affects nasal mucosa; NiO affects bronchi Pulmonary edema	Squamous cell carcinoma of nasel cavity and lung

Note: *Ill-effects of some chemicals on respiratory system have already been covered in the various above paragraphs of this chapter.*

****HC- hydrocarbons, **PM- particulate matter***

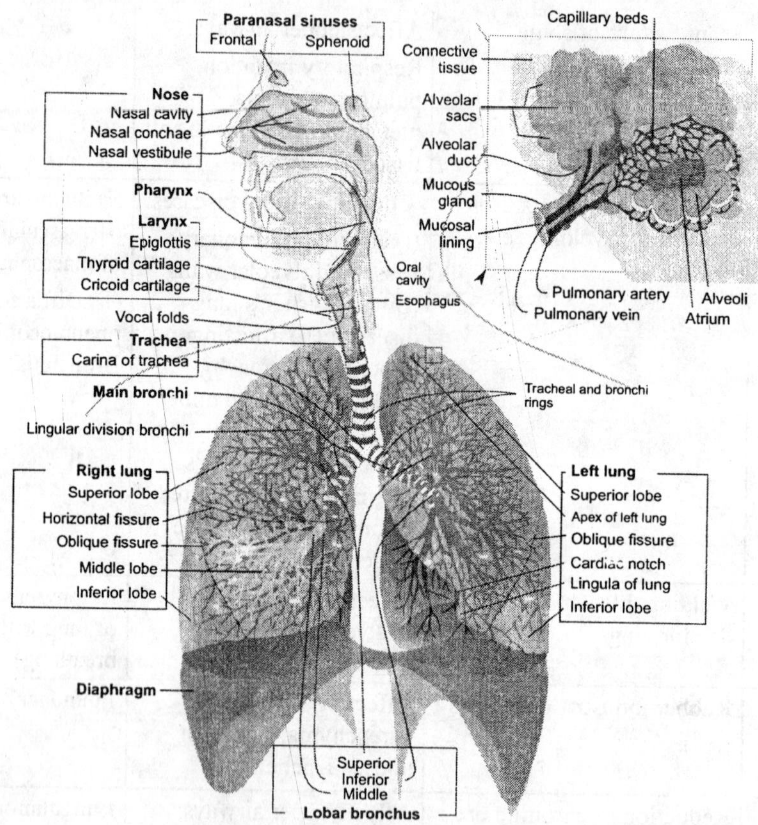

Fig. 2.1: Schematic View of Human Respiratory System

REFERENCES

1. Aameeyata and Sharma M., polycyclic Aromatic Hydrocarbons, Elemental and Organic Carbon Emissions from Tire-wear. Science of Total Enviroment, Vol. 408(20), pp. 4563-4568, 2010.
2. Aameeyata, Kaul D.S. and Sharma M., Traffic Generated Non-Exhaust Particulate Emmission from Concrete Pavement: A Mass and Particle Size Study for Two-Wheelers and Small Cars. Atmospheric Environment, 43, pp.5691-5697, 2009.
3. Adems, Braley & Senior, Constance, 'Curbing the blue plume: SO_3 formation & mitigation 'Power' May 2006.
4. Ahlvik, Peter and Branberg, Ake, 'Cancer Potency of Diesel vehicles as compared to those of Petrol Vehicles in India', Swedish Consultancy, Ecotraffic 1999.
5. Anirban Roy, Sanjay Baxla, Tarun Gupta, Rajdip Bandyopadhyaya, Sachchida Nand Tripathi, "Particulates Emitted From indoor Combustion Sources: Measurement of Size Distribution and Chemical Analysis", Inhalation Toxicology, 21 (10), 837-848 (2009).
6. Bjorseth, A., in Polynuclear Aromatic Hydrocarbons (P. W. Jones and P. Leber, Eds.), Ann Arbor Science Publishers, Ann Arbor, Michigan, pp. 371-381 (1979).
7. Case study around Panki thermal power station, Kanpur ,UP, [A M Khan 'Phlox drummondi Hook &Verbena officinalis as indicator of thermal power emission- a Case Study', International conference on plants & environment ICPEP-I , Nov. 1996, NBRI, Lucknow].

8. Environment protection rules 1985 (India)
9. Garg A and Sharma M., Contribution of Traffic Generated Non-exhaust PAHs, Elemental Carbon, and Organic Carbon Emission to Air and Urban Runoff Pollution. journal of Environmental Engineering, ASCE, American Society of Civil Engineers, Vol.136(12), 2010.
10. Gaudermann, James et al., university of southern California, The New England journal of medicine
11. Hintemann Therese et. al., 'Bio-monitoring of air pollution effects near a copper smelter in Brazil using mango tree and . .soil microbiological parametres' (SIII/L-4) , 2nd Int. conf. on plant and environmental pollution, Feb. 2002, Lucknow.
12. http://www.niehs.nih.gov/research/supported/sep/2008/dioxin.cfm.
13. Kant, R., 'Remedial Strategy-Drinking Water Pollution', Chemical Engineering World, Vol. 40, no. 1, Jan. 2005
14. Oksanen, Elina; Pandey, Vivek ; Sari Kontunen-soppela; Sarita Keski-Saari , 'Ozone sensitivity of Indian plant Species', Environews, April, 2011.
15. P.V. Sundareshwar, R. Murtugudde, G. Srinivasan, S. Singh, K.J. Ramesh, D. Agarwal, D. Baldocchi, C. Baru, K. K. Baruah, V.K. Dadhwal, C.B.S. Dutt, J. Fuentes, P. Gupta, M. Howard, C. S. Jha, S. Lal, W. H. Michener, A. P. Mitra, J. T. Morris, R. Myneni, M. Naja, R. Nemani, S. Raha, R. Ramesh, S. Santhana Vanan, M. Sharma A. Subramaniam, R. Sukumar, R. Twilley, S. Verma. 2007. Environmental monitoring Network for India. Science 2007: Vol. 316 no. 5822, 204 – 205.
16. Rule of Environmental protection agency (EPA) of United States of America
17. Raza, S. H. & Gouri K., 'Impact of certain pollutants and growth regulating substances on paddy', Int. conf. on plant and environment Nov. 1996, NBRI, Lucknow].
18. Shaji, P. K. & Suja, S., 'Experimental studies on impact of sulphur dioxide on coconut', Int. conf. on plants & environmental pollution, Nov. 1996, NBRI, Lucknow].
19. Sharma, B R and Bhatnager, P R (ICAR), 'Pollution of ground water through agricultural practices and agro industries in India', 2nd conference on water quality managemnt CBIP, Feb, 2003, New Delhi.
20. Sharma M., McBean E A and Glowing A., Bioconcentration of Dioxins and Furans in Vegetation. Journal of Water, Air, and Soil and pollution 179 (1-4): 117-124.Feb 2007.
21. Sharma, M., Maheshwari Mayank and morisawa, S., Dietary and Inhalation Intake of Lead and Estimation of Blood Lead Levels in Adults and Children in Kanpur, India. Risk Analysis: An International Journal Vol. 25 No. 6;1573:1588, 2005.
22. Sharma M., Narendra Kumar V, Katiyar S.K., Sharma R, Shukla B.P. and Sengupta B., Cohort-Based Acute Health Effect Study of PM10 and PM2.5 Pollution in the City of kanpur, India, Archives of Enviromental Health: An International Journal Vol. 59, No. 7; 348:358, 2004.
23. Sharma M. and Maloo Shaily, Assessment and Characterization of Ambient Air PM10 and PM2.5 in the City of Kanpur, India. Atmospheric Environment 39 (2005) 6015-6026, 2005.
24. Sen, indra, S.N. Tripathi, et al., Emerging airborne contaminants in India: Platinum group elements from catalytic converters in motor vehicles, Applied Geochmistry, 75, 100-106, 2016.
25. Survey of Indian plants in relation to atmospheric pollution – a research report by NBRI, International conference on plants & environment, ICPEP-2, Feb2002, NBRI, Lucknow.
26. Tripathy, Anamica, 'LAIV/P Toxic heavy metal concentration in the brass city of India', 3rd Inter. conf. on plants & environmental pollution Nov./ Dec. 2005.

27. Vreeland, H., J.J. Schauer, S.N. Tripathi et al., Chemical characterization and toxicity of particulate matter emissions from roadside trash combustion in urban India, Atmospheric Enviroment, 147, 22-30, 2016.
28. Waffa M, Ame aqnd Mohamad A. Soliman, 'Impact of chlorine gas pollutant on cultivated crops', 2nd Int. conf. on plant and environmental pollution, Feb. 2002, Lucknow].

QUESTIONS

1. Which of the following statements is incorrect ?
   (a) Air pollution causes local, regional, as well as global adverse impacts on the environment
   (b) Air pollution has been in existence at some places even in the ancient times
   (c) Air pollution does not cause adaptations and environmental imbalances in biological species
   (d) Non of the above -
2. The factor which is primarily responsible for causing air pollution in modern days is:
   (a) Industries & automobiles (b) Dust storms
   (c) Forest fires (d) None of them.
3. About 90% of world's air pollution is caused by:
   (a) Primary air pollutants (b) Secondary air pollutants
   (c) Volcanism (d) None of them.
4. The primary air pollutant which is formed due to incomplete combustion of organic matter is:
   (a) Methane (b) Sulphur dioxide
   (c) Ozone (d) Carbon monoxide.
5. Which among the following is not a secondary air pollutant?
   (a) Sulphuric acid (b) Formaldehyde
   (c) Peroxy acetyl nitrate (d) None of the above.
6. Which among the following is not a primary air pollutant ?
   (a) Oxides of nitrogen (b) Volatile organic compounds like hydrocarbons
   (c) Suspended particulate matter (d) PAN.
7. The chemical compounds which are responsible for production of photochemical smog are:
   (a) Hydrocarbons (b) Oxides of nitrogen
   (c) Both (a) and (b) (d) None of them.
8. The poisonous gas responsible for causing the catastrophic Bhopal gas tragedy in 1984 in India was:
   (a) Sulphur dioxide (b) Methyl isocyanate
   (c) Carbon monoxide and methane both (d) Laughing gas.
9. Which of the following is not a major constituent of a photochemical smog ?
   (a) PAN (b) CO (c) HC.
10. More than 4000 people died in December 1952 in 'London smog', which was caused by:
    (a) Fog in the city (b) Evolution of smoke containing sulphur dioxide in the city,
    (c) Arrival of anticyclone in the city (d) All of the above.
11. The secondary pollutant among the following is:
    (a) Sulphur dioxide (b) Methane
    (c) Carbon dioxide (d) Peroxy acetyl nitrate.

12. The most hazardous gaseous air pollutant for human health is:
    (a) Carbon dioxide (b) Oxygen
    (c) Nitrogen (d) Sulphur dioxide.
13. The primary air pollutant, which is formed due to incomplete combustion of organic matter, is:
    (a) Methane (b) Sulphur dioxide
    (c) Ozone (d) CO
14. The Respiratory Suspended Particulate Matter (RSPM) concentration includes all particles in an air mass of size up to:
    (a) 10 μm (b) 25 μm
    (c) 50 μm (d) none of them.
15. The RSPM is also designated as:
    (a) PM 2.5 (b) PM 10
    (c) PM 25 (d) none to them.
16. Monitoring of particles of smaller than 10 μm (PM 10) or even smaller than 2.5 μm (PM 2.5) is a better indicator of air pollution, because:
    (a) Such smaller particles tend to be deposited in nasopharyngeal region, causing serious health problems,
    (b) Such smaller particles are mostly deposited in trachea bronchial and pulmonary regions causing bronchitis and other lung disease.
    (c) None of them.
17. The air pollutant which causes severe damage to plants, even at much lower concentrations than what may be harmful to human health is:
    (a) Fluorine (b) Ozone
    (c) PAN (d) None of them.
18. Sulphur dioxide exposures to quantities greater than the permissible in the ambient air may cause:
    (a) Breathing and respiratory illnesses (b) Hay fever
    (c) Typhoid (d) Malaria.
19. Higher exposures of humans to nitrogen oxides may cause:
    (a) Bronchitis, (b) Pneumonia (c) Allergies and viral attacks, (d) All of the above,
20. Carbon monoxide is hazardous to health, because:
    (a) It causes loss of sense of smell
    (b) it is carcinogenic in nature
    (c) It reduces oxygen carrying capacity of blood
    (d) It may cause conjunctivitis.
21. Which of the following is not an anthropogenic cause of air pollution, (a) Burning of fossil fuels, (b) burning of firewood, (c) agricultural activities, (d) Burning of forests due to lightening.
22. (i) List out the air pollutants and state which are more rampant in (a) Industrial areas, (b) Urban areas, (ii) Enumerate the sources of emission of CO. What is the disadvantage of CO emission with regard to efficiency of fuel combustion?
23. Enumerate the sources of emission of nitrous oxide, nitric oxide and nitrogen dioxide. Enumerate the ill effects on plants, crops, and human health due to the presence of various oxides of nitrogen in atmosphere? How do the oxides of nitrogen affect the generation of other pollutants in atmosphere?

24. Enumerate the sources of emission of oxides of sulphur. What are the ill effects of air pollution by oxides of sulphur on human and animal health and on the plants?
25. Enumerate the sources of emission of hydrogen sulphide and mercaptans. Where are the pollutants more likely to be found in air?
26. Write short notes on air pollution by: (a) Fluorides (b) Chlorine (c) Dioxins.
27. What are the causes of tropospheric and stratospheric ozone pollution? What are the ill effects of these types of pollution on plants, humans and animals.
28. Enumerate the sources of emission of suspended particulate matter/ respirable suspended particles/ respirable particulate matter. What is the likely content of this particulate matter from different souces?
29. Write short notes on ill effects of air borne micro particles of (a) Heavy and toxic metals, (b) Fly ash of fuel and silica dust?
30. Write short notes on: (a) Occupational diseases due to air pollution; (b) General effects of air pollution on respiratory system; (c) Allergens.
31. How do the pesticides, insecticides, and herbicides cause air pollution and what do you understand by physical poisoning, nerve poisoning, proto plasmic poisoning, and respiratory poisoning caused by various pesticides?
32. Enumerate the pollutants which can be found in indoor air in homes.
33. Write short notes on: (a) Photochemical smog; (b) Industrial smog.
34. Write short notes on the sources and ill effects of air pollution by: (a) Polycyclic Aromatic Hydrocarbons; (b) Diesel engine exhaust; (c) Mercury.
35. What are the causes and effects of acid rains?
36. Define and explain indoor air pollution. What adverse effects are caused by indoor air pollution?
37. Differentiate between the primary and secondary air pollutants, and enumerate the various such pollutants of both the categories, explaining each in detail.
38. How does the diameter of airborne pollen grains compare with the diameter of a human hair?
39. What are the source and nature of the condensable organic vapors in unpolluted air?
40. (a) Explain how CO interacts with the circulatory system, (b) Why are individuals with heart disease at greater risk when exposed to elevated CO levels?
41. How is the particle deposition and removal from the lung influenced by the size of the particles?

Answers:

1. (d)	**2.** (a)	**3.** (a)	**4.** (d)	**5.** (d)	**6.** (d)	**7.** (c)
8. (b)	**9.** (b)	**10.** (d)	**11.** (d)	**12.** (d)	**13.** (d)	**14.** (a)
15. (b)	**16.** (b)	**17.** (a)	**18.** (a)	**19.** (d)	**20.** (c)	**21.** (d)

CHAPTER 3

EFFECTS OF METEOROLOGICAL CONDITIONS ON AIR POLLUTION

3.1 METEOROLOGICAL PARAMETERS AND DISPERSION OF POLLUTANTS

Meteorological conditions such as wind direction, wind speed, temperature inversions, atmospheric stability and wind turbulence, wind pattern, mixing height, precipitation / rain / snow /fog, humidity, and solar radiation (related to photochemical reactions) influence the pollution at a particular location. The meteorological conditions vary with geographical location of a place, season, topography, time of the day, etc.

The pollutants are emitted in the form of a puff (as by automobiles / diesel engines) or in a plume from a chimney (as in the case of industries / fossil fuel fired power plant). The pollutants are emitted and then according to meteorological conditions prevailing get carried away horizontally and vertically and diluted. The pollutants then descend after some distance and get dispersed.

3.1.1 Atmospheric Dispersion

Atmospheric dispersion occurs over a wide range of horizontal length scales. Traditionally, it has been broadly classified as macro scale, meso scale and micro scale processes. The horizontal length scales associated with the macro scale processes are above 1000 km, micro scale phenomena have horizontal length scales of less than 5 km while in between the two lay the meso scale events. The pollutants, while getting dispersed rarely move above the troposphere, which is the lowest level of the Earth's atmosphere. The troposphere is characterized by a nearly constant rate of decrease in temperature (as we go up), known as the adiabatic lapse rate and it extends up to an altitude of 9 to 16 km from the ground. The lowest part of the troposphere is known as the Planetary Boundary Layer (PBL); this region is directly influenced by the Earth's surface in the form of frictional drag (from topology and obstacles like buildings etc.), solar heating (radiation) and evapotranspiration. The PBL may extend up to 100 m to 3 km from the ground. Almost all the pollutants (except from aircraft and rockets) are released in this layer, so this is very important for short range pollutant transport. Dispersion in the micro scale range, almost always, takes place wholly in the PBL.

3.1.2 Turbulence

Turbulence in atmosphere is one of the most important phenomenon for dissipation of pollutants. 'Turbulence' is a phenomenon associated with fluid flows of all types. Turbulent flows are characterized by motions as in vortex and high rates of dissipation. The type of motion causes increased dispersion and strong rates of mixing. High dissipation rates contribute to substantial increase in pressure drop across flows in internal channels and also result in high drag across obstacles in flows.

The turbulence in the atmosphere is the reason behind the dispersion of pollutants. If there were no turbulence, the pollutants would have just followed the streamlines of mean wind velocities, so the point sources of pollutants would have given rise to long lines of pollutant transport with minimal mixing (due to molecular viscosity) in the lateral directions.

3.1.3 Characterization of Atmospheric Turbulence

The amount of turbulence in the ambient atmosphere has a major effect on the dispersion of air pollution plumes because turbulence increases the entrainment and mixing of unpolluted air into the plume and thereby acts to reduce the concentration of pollutants in the plume (i.e., enhances the plume dispersion). It is, therefore, important to categorize the amount of atmospheric turbulence/ stability present at any given time.

The Pasquill Atmospheric Stability Classes

The oldest and, for a great many years, the most commonly used method of categorizing the amount of atmospheric turbulence present was the method developed by Pasquill in 1961. He categorized the atmospheric turbulence into six stability classes named A, B, C, D, E and F with class A being the most unstable or most turbulent class, and class F the most stable or least turbulent class.

The Pasquill stability classes are as follows:

Very unstable A; Unstable B; Slightly unstable C

Neutral D; Slightly stable E Stable F

For air dispersion modeling exercises, the conditions of dual stability classes like A – B, B – C and C – D can be considered as B, C and D respectively.

Table 3.1: Meteorological Conditions that Define the Pasquill Stability Classes

Surface wind speed in m/s	Day time incoming Solar Radiation			Night time Cloud Cover	
	Strong	Moderate	Slight	More than 50%	Less than 50%
Less than 2 m/s	A	A – B	B	E	F
2–3 m/s	A–B	B	C	E	F
3–5 m/s	B	B–C	C	D	E
5–6 m/s	C	C–D	D	D	D
More than 6 m/s	C	D	D	D	D

Note: D class applies to Overcast Skies at any wind speed in Day or Night.

It may be noted that some more advanced air pollution dispersion models no longer use Pasquill classes as given above. A theory called 'Monin-Obukhov similarity theory' is often applied to the models using characteristics like surface roughness or boundary layer height, etc., and wind speed to categorize the atmospheric turbulence.

Atmospheric Stability

Stability of atmosphere is related to the vertical profile of density in the atmosphere. Density stratification / variation of density is due to 'variation of temperature and specific humidity in space'. Stability plays a direct role in the amount of turbulence present in the atmosphere and therefore significantly affects atmospheric dispersion

If cooler (and hence heavier) parcels of air reside above hotter (and hence lighter) parcels, then there is always a tendency of the heavier parcels to move down and the lighter parcels to move up. This kind of stratification is unstable and is associated with high turbulence.

If heavier particles are already there below the lighter particles, it results in 'Stable stratification' and is therefore associated with low turbulence. Uniform density of air results in neutral stratification.

3.1.4 Types of Plume according to Density of Emitting Gases

There can be three types of plumes (of air pollution) according to density of gases / matter being emitted:

(i) *Buoyant plumes* – Plumes which are lighter than air because they are at a higher temperature and lower density than the ambient air which surrounds them, or because they are at about the same temperature as the ambient air, but have a lower molecular weight and hence lower density than the ambient air. [For example, the emissions from the flue gas stacks of industrial furnaces are buoyant because they are considerably warmer and less dense than the ambient air. As another example, an emission plume of methane gas at ambient air temperatures is buoyant because methane has a lower molecular weight than the ambient air.]

(ii) *Dense gas plumes* – Plumes which are heavier than air because they have a higher density than the surrounding ambient air. [A plume may have a higher density than air because it has a higher molecular weight than air (for example, a plume of carbon dioxide). A plume may also have a higher density than air if the plume is at a much lower temperature than the air. For example, a plume of evaporated gaseous methane from an accidental release of Liquefied Natural Gas (LNG) may be as cold as–161 °C.]

(iii) *Passive or neutral plumes* – Plumes which are neither lighter nor heavier than air.

3.1.5 Ways of Designating Emission Sources

Emission sources may be characterized as follows:

(i) Sources may be designated as either stationary or mobile. Flue gas stacks are examples of stationary sources and buses/ trucks/cars are examples of mobile sources.

(ii) Sources may be designated as either urban or rural because urban areas constitute a so-called heat island and the heat rising from an urban area causes the atmosphere above an urban area to be more turbulent than the atmosphere above a rural area.

(iii) Sources may be designated in accordance with their elevation relative to the ground as either surface or ground-level, near surface or elevated sources.

(iv) Sources may be designated according to their time duration.

- Short-term sources (for example, many accidental emission releases are short term puffs)
- Intermittent Puffs
- Continuous: A long-term source (for example, most flue gas stack emissions are continuous)

Moreover the pollution sources may be (a) Point source, (b) Line source, (c) Area source.

3.1.6 Wind Rose

Wind rose shows the prevalent direction of wind at a location over a considerable period during a year in different seasons. The general form of wind rose has sixteen lines showing direction such as North (N), North of North-East (NNE), North East (NE),

East of North- East (ENE), East (E), East of South- East (ESE), South East (SE), South of South-East (SSE), South (S), South of South-West (SSW), South West (SW), West of South-West (WSW, West (W), West of North-West (WNW), North West (NW) , North of North-West (NNW). The length of line proportionally shows the 'Frequency (period)' of wind from that direction during a year in percentage. As an example, the period or frequency during the year may be typically shown in percentage of a period of year proportional to the length of line in centimeters / millimeters such as:

North 2%, NNE 2%, NE 3%, ENE 4%, E 4%, ESE 2%, SE 2%, SSE 2%, S 2%, SSW 5%, SW 20%, WSW 5%, W 2%, WNW 3%, NW 5%, NNW 4%, Calm wind 33%

The speed of wind is also indicated by appropriate shades in a few ranges such as:

(Calm---as circle); (1–3-------- as a line); (4–7----------as blank rectangle); (8–15-----as single hatched rectangle); (16 or more-------as double hatched rectangle). The length of the line and these rectangles are proportional to the frequency /period of such a wind speed. The 'Calm state' of wind is denoted at the centre by a circle. The wind speed below a certain speed say 1 km per hour may be taken as calm condition.

Sometimes pollution roses are constructed in which the 'degree of pollution' instead of 'wind speed' is attached to the 'Direction of wind'. Separate rose may be constructed for each type of pollution such as 'Sulphur dioxide wind rose', 'Smoke wind rose', 'Hydrocarbon/ VOC wind rose'. Similarly,'Precipitation/ rain fall wind rose' can also be made.

3.2 PLUME BEHAVIOUR OF A POLLUTING SOURCE AND DISPERSAL OF POLLUTANTS

3.2.1 General

Gases that are emitted from stacks are often pushed out by fans. As the turbulent exhaust gases exit the stack, they mix with ambient air. This mixing of ambient air into the plume is called '*Entrainment*'. As the plume entrains air into it, the plume diameter grows as it travels downwind. These gases have momentum as they enter the atmosphere. Often these gases are heated and are warmer than the outdoor air. In these cases, the emitted gases are less dense than the outside air and are therefore buoyant. A combination of the 'gases *momentum and buoyancy*', causes the gases to rise. This is referred to as '*Plume rise*' and allows air pollutants emitted in this gas stream to be lifted higher in the atmosphere. Since the plume is higher in the atmosphere and at a farther distance from the ground, the plume will disperse more before it reaches the ground level.

The final height of the plume, referred to as the '*Effective stack height (H)*', is the sum of the physical *stack height (hs)* and the plume rise (Δh). Plume rise is actually calculated as the distance to an imaginary centerline of the plume rather than to the upper or lower edge of the plume.

Plume rise depends on the stack's physical characteristics and on the effluent's (stack gas) characteristics. The difference in temperature between the stack gas (Ts) and ambient air (Ta) determines the plume density which affects plume rise. Also, the velocity of the stack gases which is a function of the stack diameter and the volumetric flow rate of the exhaust gases determines the plume's momentum.

The behaviour of plume of gases emitted from a stack of an industry / fossil fuel fired power plant depends on the condition of stability of local atmosphere, wind velocity, turbulence, inversion, cloudiness, temperature inversions, etc. Dispersal of pollutants is dependant on the plume behaviour. The topography of the land and local condition of precipitation are also important factors in dispersal of pollutants.

The pollutants from high stack (300 m and above) may get injected into the atmosphere on a higher plane on account of velocity of gases / plume rise and effect of buoyancy and the pollutants get dispersed in larger air volumes and larger area.

The lower height of stacks may create typical plume behaviour. The vertical temperature gradient (rate of fall of temperature) of the gases in the plume is an important factor in the spread of plume.

3.2.2 Lapse Rate

Apart from wind speed, the transport of pollutants emanating from a stack are mainly affected by a stability factor which largely depends on atmospheric temperature gradient.

The 'Adiabatic Lapse Rate' is the rate of temperature variation (cooling or heating) of a parcel of dry air with no heat exchanges when it moves up (in a hydrostatically stable environment) expanding slowly to a lower pressure. When air expands, it cools at the 'adiabatic lapse rate' if no heat exchange takes place. If surrounding air is cooler, the parcel of air will continue to rise and shows the unstable conditions and existence of strong lapse rate. The 'adiabatic lapse rate' i.e. 'La' is about '1°C per 100 m'.

The 'Temperature lapse rate', L is defined as:

L = 'Change in temperature in °C / 'Change in height in 100s of meters'.

The Environmental Lapse Rate (ELR) is defined as '(La – L)'. The strong lapse is defined as '–2°C' per 1000 m or more negative. This condition is favourable for removable of air pollution at a given location. Stronger the environmental lapse rate, the more unstable is the condition of atmosphere and more is the mixing of air. When the environmental lapse rate is less negative or is towards positive, the stability of atmosphere increases and turbulence / mixing decreases. The change of environmental lapse rate bringing the 'stable atmospheric condition' from 'unstable atmospheric condition' is called 'Inversion'. The 'temperature of inversion' is a meteorological condition, when the temperature actually 'increases with the height' instead of 'decreasing'. In such a case, the exhaust gases and pollutants, would not rise much from the point of release and dispersal of pollutants would be limited to a small area near the point of release with consequent concentration of localised pollution. If such temperature inversion persists for longer duration, it may become very harmful for the locality.

Note:

In order to understand the nature of vertical mixing in the atmosphere, one may consider an air parcel of infinitesimal dimensions which (as mentioned above) is assumed to be:

- Thermally insulated from its environment, so that its temperature changes adiabatically as it rises or sinks
- It is always at the pressure of the environmental air at the same level, which is assumed to be in hydrostatic equilibrium.
- It is moving very slowly such that the macroscopic kinetic energy of the air parcel is a negligible fraction of its total energy.

This idealized model is helpful in understanding some of the physical processes which influence the distribution of vertical motions and vertical mixing in the atmosphere.

We will now derive an expression for the rate of change of temperature with height of a parcel of dry air as it moves about in the Earth's atmosphere. Since the air parcel is thermally insulated, $dq = 0$, one can write for a unit mass of air in the parcel from the Ist law of Thermodynamics. *If,* v = *specific volume;* dv = *change in volume;* dT = *change in temperature;* R = *Gas constant* = $C_p - C_v$; . C_p = *specific heat of air at constant pressure;* C_v = *Specific heat of air at constant volume;* g = *acceleration due to gravity*

$$dq = C_v\, dT + p\, dv = 0$$

$$\text{or} \quad C_v\, dT + d(pv) - v\, dp = 0$$
$$C_v\, dT + d(R\,T) - v\, dp = 0$$
$$(C_v + R)\, dT - v\, dp = 0$$

p = Pressure at an altitude of z ; $(p + dp)$ = Pressure at an altitude (z+ dz)

For a column of unit cross section, $(p + dp) + \rho g\, dz = p$

$$-(1/\rho)\, dp = g\, dz$$
$$\text{Or} \quad -v\, dp = g\, dz$$

Therefore, Eq. (1) becomes

$$C_p\, dT + g\, dz = 0$$

Dividing through by dz, we obtain $-\dfrac{dt}{dz} = \dfrac{g}{C_p} = \Gamma d$ where Γd is called the dry adiabatic lapse rate.

Since an air parcel expands as it rises in the atmosphere, its temperature will decrease with height so that Γd is a positive quantity.

Substituting $g = 9.81$ m/s^2 and $C_p = 1004$ J/kg K gives $\Gamma d = \dfrac{g}{C_p} = 0.0098$ K/m = 9.8 K/km ≈ 10 K/ km which is the dry adiabatic lapse rate.

For adiabatic conditions, dry adiabatic lapse rate $\Gamma = 9.8$°C/ km or approximately 1°C per 100 m.

3.2.3 Temperature Inversion and Dispersal of Pollutants

Normally, the air near the Earth's surface is warmer than the air above it because the atmosphere is heated from below as solar radiation warms the Earth's surface, which in turn warms the layer of the atmosphere directly above it. Thus, the atmospheric temperature normally decreases with increasing altitude. However, under certain meteorological conditions, atmospheric layers may form in which the temperature increases with increasing altitude. Such layers are called *'inversion layers'. When such a* layer forms at the Earth's surface, it is called a surface inversion. When an 'inversion layer' forms at some distance above the earth, it is called an *'inversion aloft or a capping inversion'*.

The air within an 'inversion aloft' is very stable with very little vertical motion. Any rising parcel of air within the inversion soon expands, thereby adiabatically cooling to a lower temperature than the surrounding air and the parcel stops rising.

Any sinking parcel soon compresses adiabatically to a higher temperature than the surrounding air and the parcel stops sinking. Thus, any air pollution plume that enters an inversion aloft will undergo very little vertical mixing unless it has sufficient momentum to completely pass through the inversion aloft. That is one reason why an inversion aloft is sometimes called a 'capping inversion'.

Sometimes when a high pressure condition is created in upper layers of atmosphere, the upper cool air descends ('subsidence condition'). The descending air warms up due to contraction, but the contraction and warming of air is greater in the upper layers than in lower layers near the ground. Thus a 'temperature inversion' would be created. If such a condition persists over a longer time, there would be a concentration of air pollutants and smog over a localised area which may become dangerous. During temperature inversion in atmosphere, air pollutants tend to accumulate above the inversion layer.

When there is a collision of air masses of warm air and cool air in nature, the cool air due to its higher density would get pressed under the warmer mass of air and thus temperature inversion condition would be created at the boundaries of such air masses. This is generally called 'Frontal inversion.

The *'temperature inversion'* is commonly occurring during night or early morning. The ground is heated during the day by solar heat, but at night it loses its heat to the atmosphere. The air at the ground level may be cooler than at higher level thus creating '*temperature inversion*'. The cold air has higher density and has tendency of collecting at bottom specially, in valleys. The low wind speed is favourable for such temperature inversions and accumulation of pollutants in localised area. This inversion condition would generally disappear during the day when there is sun shine and the earth starts warming up and there is air movement due to thermal convection in the lower atmosphere.

3.2.4 Mixing Height

When an *'inversion aloft'* is formed, the atmospheric layer between the Earth's surface and the bottom of the *'inversion aloft'* is known as the *'mixing layer'* and the distance between the Earth's surface and the bottom of inversion aloft is known as the *'mixing height'*. Any air pollution plume dispersing beneath an '*inversion aloft*' will be limited in vertical mixing. Even if the pollution plume penetrates the inversion, it will not undergo any further significant vertical mixing.

Rarely, a pollution plume may pass completely through an *'inversion layer aloft'*. This occurs only when the pollution plume's source stack is very tall and the *'inversion lid'* is fairly low. Maximum Mixing Depth (MMD) is the height to which the unstable air can mix. The *'maximum mixing depth'* varies during the day as well as varies from season to season. The variation of mixing depth also depends on the topographical features of land. The vertical dispersion of pollutants is limited by these features and therefore in estimating the amount of vertical diffusion of pollutants in the atmosphere, it is necessary to estimate mixing depth. In absence of radio sound observation, the 'maximum mixing depth' may be estimated as being the height of the *'top of the low altitude cloud layer'*.

3.2.5 Influence of Ground Topography on Dispersal of Pollutants

The local atmospheric conditions are affected by topography of land, especially valleys, hills, sea coast, etc. In valleys, generally the wind would tend to flow axially. In the evening, when the air near the ground is cooled, *'slope winds'* occur. Therefore, if the air pollutants are released on the slope would flow to the valley, they will stay in the valley if inversion is present.

The roughness of hilly track plays a part in dispersal of air pollutants because the rougher the track, the more is air turbulence. If there is a smooth transition of heights in the hilly track, there would be little influence on air turbulence.

The wind movements on the shore are a usual feature. The absorption of heat during sunshine by land is faster than that by water mass of sea and also the rate of cooling of land is faster than that of sea. Hence the sea breeze flows from sea to land on sunny days, while wind would flow from land to sea during night and cloudy days.

3.2.6 Influence of Precipitation on Pollution

It is obvious that any kind of precipitation / rain / snow would wash out the pollutants from air (scrubbing) to a large extent especially particulate matter and soluble gaseous pollutants. It may also be mentioned that the pollution level can also influence the quality and characteristics of precipitation. This type of deposition is wet deposition of pollution components from plume.

Hot Spots of Radioactivity

A typical influence of precipitation on pollution happens when there is wet deposition of radio- nuclides in a pollution plume by a burst of rain often forming so called '*hot spots of radioactivity*' on the underlying surface.

3.2.7 Effect of Tall Building on Pollutant Dispersal

When an air pollution plume flows over nearby buildings or other structures, turbulent eddies are formed in the downwind side of the building. If a stack source is located within about five times the height of a nearby building or structure, the turbulent eddies would force down the plume to the ground much sooner than it would if a building or structure were not present. The effect can greatly increase the resulting near-by ground-level pollutant concentrations downstream of the building or structure. The pollutants in the plume are subject to depletion because of depositing particulates on the ground downstream of the building. There would be a comparative decrease of ground level concentration further downstream than it would have been the case if the building did not exist.

3.2.8 Dry Deposition of Pollutants from Plume

It is the removal of gaseous or particulate material from the pollution plume by contact with the ground surface or vegetation (or even water surfaces) through transfer processes such as absorption and gravitational sedimentation. This may be calculated by means of a deposition velocity, which is related to the resistance of the underlying surface to the transfer.

3.2.9 Types of Plume Conditions

Various plume conditions described below are shown in **Fig. 3.1:**

(i) ***Conning***—The horizontal cone shaped plume may occur when there is a moderate to strong wind and there is largely mechanical turbulence rather than thermal, under lapse rate between dry adiabatic and isothermal. The pollutants would disperse not so rapidly and the pollutants would descend at large distance in the direction of the wind.

(ii) ***Fanning***— The plume of exhaust gases rises (according to their temperature) on release from stack slowly and then drifts horizontally up to a large distance from the source in the wind direction with very little vertical spread and would take narrow vertical shape. It may occur at night or in the early morning when there is light wind, little turbulence, inverted and isothermal lapse rate, and stable atmospheric conditions. The dispersion of pollutants is slow and the concentration of pollution is high at higher levels from the ground at great distance from the source. The increase in turbulence may result in pollution hitting the ground.

(iii) ***Looping*** — The plume of irregular loops shape may occur when there is light wind and intense thermal turbulence, the temperature profile is unstable, and lapse rate is strong during the day along with intense thermal heating. The pollutants disperse rapidly with high concentration of pollutants within short distance from the source.

(iv) ***Lofting*** — The plume may have loops or may be conical, but would have a well defined bottom line and there is rapid upward diffusion to the top. There is very little chance of emission reaching the earth surface and the ground level pollution in this type of plume condition is smaller. Typically, it may occur during sunset on fair days, when there is change from lapse to inversion condition. This may occur under following conditions occurring at a time:

The lapse rate is inverted below the top of the stack, while there is an adiabatic lapse rate at the top of the stack (there is a strong lapse rate above the surface inversion).

In lower layers of air there is light wind, very little turbulence and stability, but in upper layers of air there is considerable turbulence, moderate wind and instability.

(v) ***Fumigation*** — The plume is conical with diffused bottom inclined towards earth. There is high probability of high concentration of pollutants coming to the ground after plume has stagnated aloft. This may typically occur when lower level layers of air are unstable (or neutral) with thermal turbulence and light winds, while the upper layers are stable with very little turbulence and moderate winds. There is adiabatic lapse rate (or super adiabatic) below the top of the stack, while there is inverted lapse rate above it. The ground level pollution increases in fumigation type of plume condition.

(vi) *Trapping* — The plume is caught between two inversions—one inversion is somewhere above the top of stack, while the other is somewhere below near the middle of the stack. Here the diffusion of pollutants can take place within a limited height and ground level pollution increases in this condition of plume.

3.3 DISPERSION MODELS

3.3.1 General

There can be three types of pollution sources namely Point Source, Line Source, and an Area Source. When gases are released from a stack, they undergo vertical rise called 'Plume rise', a horizontal movement due to wind called 'Advection of the gases' and dispersion due to turbulence. The dimensions of a plume of gases released from a stack depend upon many factors like: wind direction and wind speed, stability, turbulence, momentum and buoyancy, terrain effect, building-wake effect, etc. The dimension of the plume (which determines the concentration at any point in space) is determined by plume-spread parameters Sigma-y and Sigma –z, which in turn depend on the turbulence state or stability of atmosphere.

For a dispersion model to be effective, it is necessary that reliable meteorological data are available and that there is a good understanding of dispersion characteristics of an area or more precisely of the particular site chosen for the project.

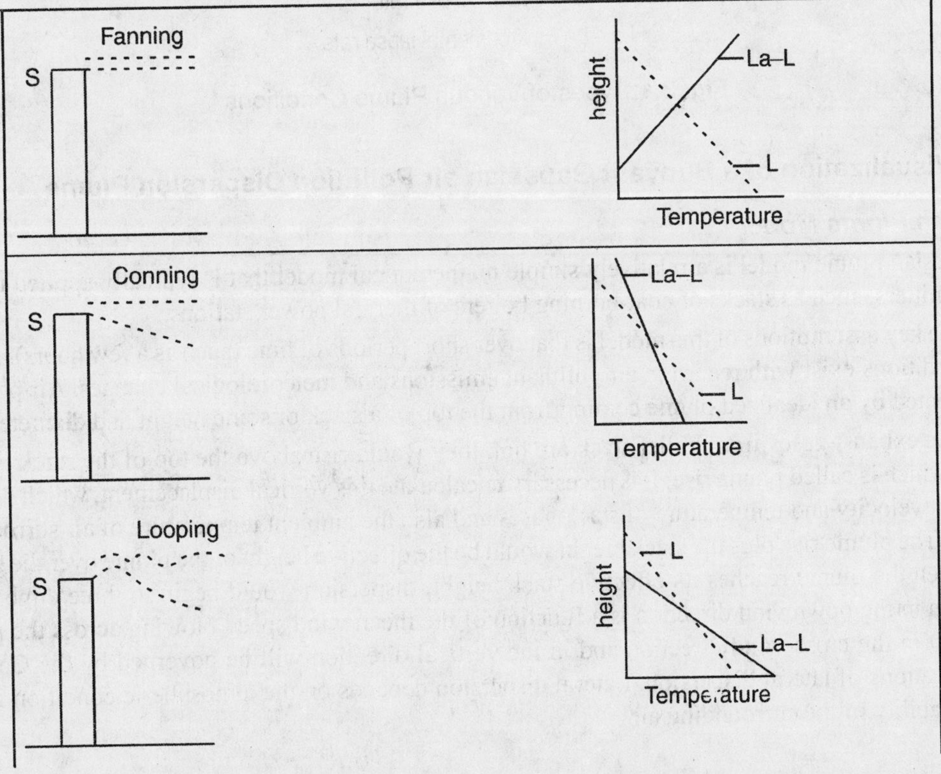

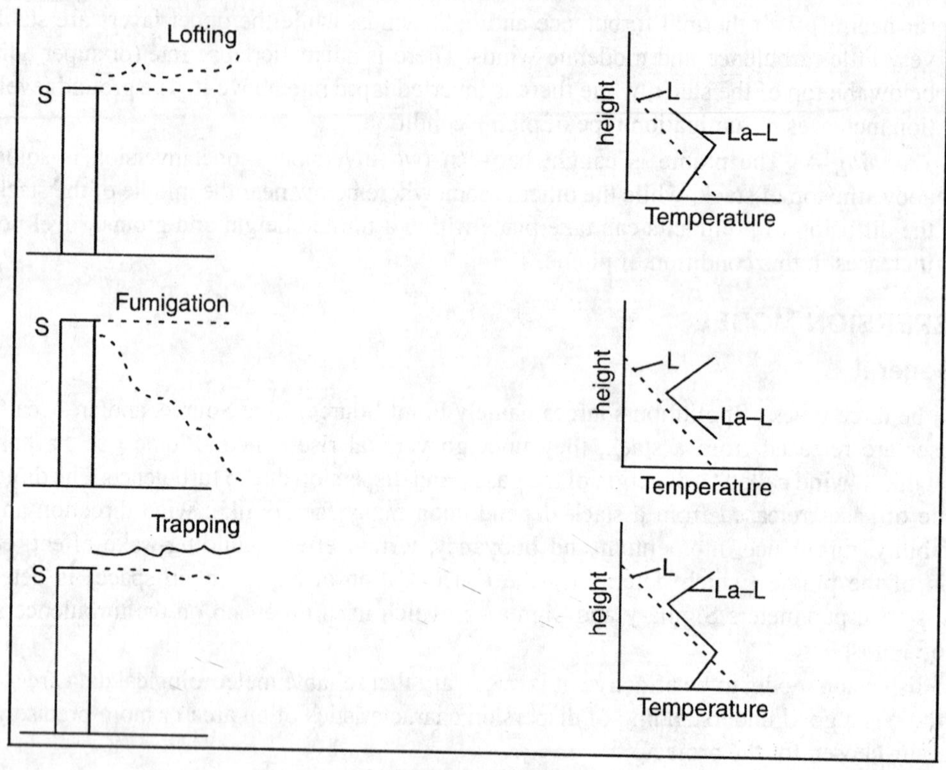

Fig. 3.1: Types of Various Plume Conditions

3.3.2 Visualization of a Buoyant Gaussian air Pollution Dispersion Plume

Gaussian Plume Model

The Gaussian plume model is a relatively simple mathematical model that is typically applied to point source emitters, such as stacks of coal-burning boilers of thermal power stations.

One of the key assumptions of this model is that over short periods of time (such as a few hours), steady-state- conditions exist with regard to air pollutant emissions and meteorological changes. Air pollution is represented by an idealized plume coming from the top of a stack of some height and diameter.

As the exhaust gases from boiler stack are hot, they would rise above the top of the stack to some distance which is called plume rise. It is necessary to calculate this vertical displacement, which depends on the exit velocity and temperature of stack gases and also the ambient temperature of air surrounding the stack. The plume rise plus the stack hieght would be the effective height of the plume over the ground level. When the plume reaches its effective stack height, dispersion would begin in three dimensions. Dispersion in the downwind direction is a function of the mean wind speed blowing across the plume. Dispersion in the cross-wind direction and in the vertical direction will be governed by the Gaussian plume equations of lateral dispersion. Lateral dispersion depends on the atmospheric condition i.e. the relative stability of the surrounding air.

The model assumes that dispersion in these two dimensions will take the form of a normal Gaussian curve, with the maximum concentration in the center of the plume. The "standard" algorithm used in plume studies is the Gaussian plume model, developed in 1932 by O.G. Sutton. The algorithm is as follows (Ref: **elte.prompt.hu/sites/default/files/tananyagok/.../ch10s03.html):**

$$c(x,y,z)=\frac{Q}{2\pi\sigma_y\sigma_2 u}\exp\left(\frac{-y^2}{2\sigma_y{}^2}\right)\left(\exp\left(\frac{-(z-h)^2}{2\sigma_z^2}\right)+\exp\left(\frac{-(z+h)^2}{2\sigma_z^2}\right)\right)$$

where:

- $C(x, y, z)$ is the concentration of the emission (in $\mu g/m^3$) at any point x meters downwind of the source, y meters laterally from the centerline of the plume, and z meters above the ground level.
- Q is the quantity or mass of the emission (g/s).
- u is the wind speed (m/s)
- H is the height of the source above ground level (m)
- σ_y and σ_z are the standard deviations of a statistically normal plume in the lateral and vertical dimensins, respectively

The above formula has been shown to be fairly predictive in a number of studies by investigators for a variety of conditions. This model considers dispersion in all the three directions (x, y, and z) [in x direction longitudinally along a centerline of maximum concentration running downwind from the source; in y direction laterally on either side of the centerline, as the pollution spreads out sideways and in z direction vertically above and below a horizontal axis drawn through the source].

The stability categories were developed in the late 1970s, and are based on wind speed, insolation, and extent of cloud cover. The values of standard deviations are to be calculated for any such stability condition

[Initially, Gaussian plume model was used for pollutants such as carbon monoxide and other non-reactive species. The model has serious limitations when trying to account for pollutants that undergo chemical transformation in the atmosphere. Coupled with its dependence on steady state meteorological conditions and its short-term nature, this model has many limitations for use as a long-term airshed pollutant evaluator.]

3.3.2.1 *Brigg's Formula for Plume Rise*

Many scientists have studied plume rise over the years. The most common plume rise formula is the one developed by Gary A. Briggs. Plume rise formulas are to be used on plumes with temperatures *greater* than the ambient air temperature. *Briggs' plume rise formula* is as follows:

$$\mathbf{\Delta h}\, x = 1.6\,[F^{1/3} . x^{2/3}]/u$$

Where: Δh = plume rise (above stack); h = Physical height of stack

H = $h + \Delta h$; u = average wind speed; F = *Buoyancy Flux;*

x = downwind distance from the stack/source

g = acceleration due to gravity (9.8 m/s^2)

V = volumetric flow rate of stack gas

T_s = temperature of stack gas (K), T_a = temperature of ambient air (K)

F (Buoyancy flux) = $(g/\pi) . V. [(T_s - T_a)/T_a]$

[The plume rise formulas determine the imaginary centerline of the plume. The centerline is where the greatest concentrations of pollutants occur.]

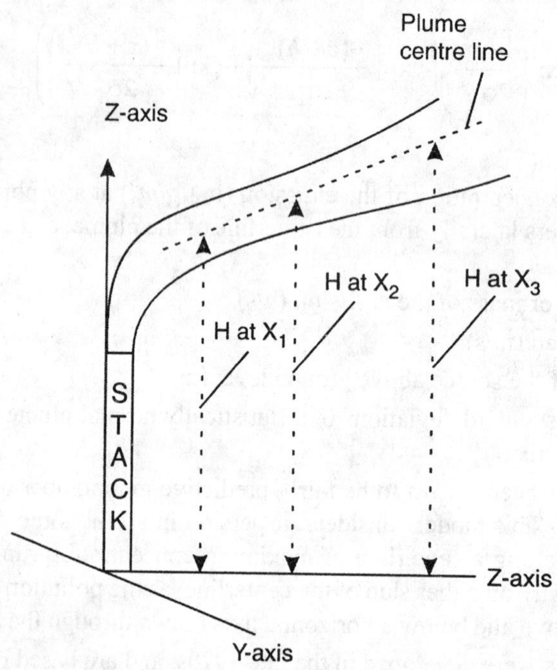

Stack height –h; Height of plume—dh

Effective or actual height of plume—H = stack height + plume rise (h + Δh) [m]

Fig. 3.2: General Representation of Dispersion Shape from Stack

3.3.2.2 Estimate of Emission Dispersion

As mentioned in the previous section, plume rise formulas are used to determine the imaginary centerline of the plume. Although the maximum concentration of the plume exists at this centerline, plume rise formulas provide no information about how pollutant concentrations vary from this centerline. Dispersion estimates are made in order to determine pollutant concentrations at a point of interest.

Dispersion estimates are determined by using distribution equations and/or air quality models. These dispersion estimates are typically valid for the layer of the atmosphere closest to the ground where frequent changes occur in the temperature and distribution of the winds. These two variables have an enormous effect on how plumes are dispersed. Therefore, the distribution equations and air quality models mentioned above must include these parameters. The basis of most of the models for air dispersion of pollutants is the complete Gaussian equation given below.

Complete Equation for Gaussian Dispersion Modeling of Continuous, Buoyant Air Pollution Plumes

$$C = \frac{Q}{u} \frac{f}{\sigma_y \sqrt{2\pi}} \frac{g_1 + g_2 + g_3}{\sigma_z \sqrt{2p\pi}}$$

Where

f = crosswind dispersion parameter

$= \exp[-y^2/(2\sigma_y^2)]$

g = vertical dispersion parameter = $g_1+g_2+g_3$

g_1 = vertical dispersion with no reflections

$= \exp[-(z-H)^2/(2\sigma_z^2)]$

g_2 = vertical dispersion for reflection from the ground

$= \exp[-(z+H)^2/(2\sigma_z^2)]$

g_3 = vertical dispersion for reflection from an inversion aloft

$$= \sum_{m=1}^{\infty} \{\exp[-(z-H-2mL)^2/(2\sigma_z^2)] + \exp[-(z+H-2mL)^2/(2\sigma_z^2)] + \exp[-(z-H+2mL)^2/(2\sigma_z^2)] + \exp[-(z-H+2mL)^2/(2\sigma_z^2)]\}$$

C = concentration of emissions, in g/m³, at any receptor located:

x meters downwind from the emission source point

y meters crosswind from the emission plume centerline

z meters above ground level

Q = source pollutant emission rate, in g/s

u = horizontal wind velocity along the plume centerline, m/s

H = height of emission plume centerline above ground level, in m

σ_z = vertical standard deviation of the emission distribution, in m

σ_y = horizontal standard deviation of the emission distribution, in m

L = height from ground level to bottom of the inversion aloft, in m

exp = exponential function, where ***e*** represents a value = 2.71828

; *Stack height—**h**; Plume rise—**Δ h***

*Effective or actual height of plume,—**H** = **(h+ Δh)** [meters]*

The above equation not only includes upward reflection of the pollution plume from the ground, it also includes downward reflection from the bottom of any temperature inversion lid present in the atmosphere. The sum of the four exponential terms in g_3 converges to a final value quite rapidly. For most cases, the summation of the series with m = 1, m = 2 and m = 3 will provide an adequate solution.

It should be noted that σz and σy are functions of the atmospheric stability class (i.e., a measure of the turbulence in the ambient atmosphere) and the downwind distance to the receptor. Two most important variables affecting the degree of pollutant emission dispersion are the height of the emission source point and the degree of atmospheric turbulence. More the turbulence, better is the degree of dispersion. The resulting calculations for air pollutant concentrations are often expressed as an air pollutant concentration contour map in order to show the spatial variation in contaminant levels over a wide area under study. In this way, the contour lines can overlay sensitive receptor locations and reveal the spatial relationship of air pollutants to areas of interest.

In general, Briggs's equations for bent-over, hot buoyant plumes are based on observations and data for hot flue gas plumes from boilers of thermal power stations where the flue gas exit velocities from stack may be in the range of 6 to 30 m/s (20 to 100 ft/s) with exit temperatures ranging from 120 to 260°C (250 to 500 °F).

Logic Diagram for Brigg's equations

A logic diagram for using the Briggs equations to obtain the plume rise trajectory of bent-over buoyant plumes is shown **Fig. 3.3:**, where E and F refer to Pasquill's atmospheric stability as disceussed in para 3.1.3.

The following symbols are used in the logic diagram:

Δh = Plume rise, m
F = Buoyancy factor, in m^4/s^3
x = Downwind distance from plume source, m
x_f = Downward distance from plume source to point of maximum plume rise, m
u = Wind speed at actual stack height, m/s
s = Stability parameter, s^{-2}

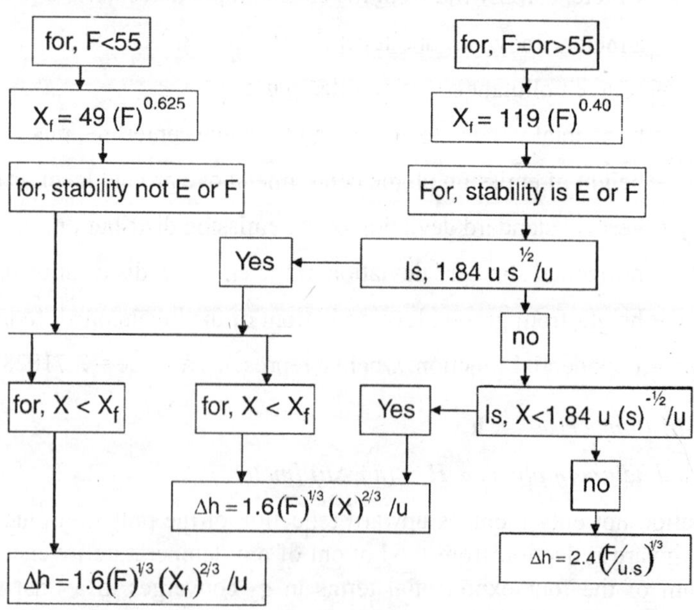

Fig. 3.3: Logic Diagram for Brigg's Equation for Application to Plume Rise of Buoyant Plume

3.3.2.3 Influence of Speed and Wind Direction on Gaussian Formulation

Wind direction is a very important parameter in determining the ground level concentration of emitted pollutants. Stern et al. (1984) noted that in the Gaussian formulation, with wind directly blowing towards the receptor, an error in wind direction of approximately five degrees (which may be a typical accuracy for wind direction measurements) may lead to:

(i) A drop in predicted value of ground level concentration of 10 % under unstable conditions when the plumes are generally broad,

(ii) A drop of approximately 50 % under neutral conditions and 90 % under stable conditions (when plumes are narrow).

In case, it is seen that 'a change in wind direction with the height of substantially more than five degrees is very commonly taking place in the atmosphere' and that 'plume from a buoyant point source is rising through several hundred meters', the performance of Gaussian models would be poor. In such cases, the prediction made by Gaussian model using wind data collected from a single height when compared with the measurements and time periods would be unreliable. However, for the purpose of Environmental Impact Assessment (EIA), it is necessary to know the maximum concentration that would occur at any point over a large area during an extended period in a year.

Wind speed affects the dilution of the plume, plume rise, and the atmospheric turbulence. Wind speed generally increases with height up to the top of mixed layer and this increase is usually assumed to vary approximately according to the following equation:

$$u(z) = u(z_m)\,(z / z_m)^p.$$

Where, $u(z)$ = Wind speed at a height z, m,

$u(z_m)$ = Wind speed at measurement height, m

p = Exponent between 0.1 and 0.5 depending on atmosphere stability

3.3.2.4 Influence of Stability / Turbulence

Turbulence causes the plume to spread and dilution of the pollutants. Turbulence may be caused by convection and also by blowing of wind over rough ground resulting in eddies in the wind. Eddies of large size are very important in causing dilution of the plume. Eddies with scales much larger than the plume size may transport the plume even without diluting it, while the eddies much smaller than the plume may simply cause dilution at the edges of the plume. In India, the convection is quite strong than in countries located in higher latitudes and also the convective cells much larger than the plume size would often be developing during the day. These convective cells may transport undiluted plume material from higher altitudes down to ground and cause high short-term concentrations at ground level. In countries located in higher latitudes, mechanical turbulence is more important factor in creating eddies which are normally of smaller sizes resulting in the material from the plume being diluted by the time it reaches the ground level.

The manner in which plumes disperse depends on the strength of turbulence and the size of eddies in the turbulence stability classes. Pasquill's scheme of categorization of meteorological conditions defining atmospheric stability category can also be determined from measurements of standard deviation of horizontal wind direction (σ_a). These data can be measured electronically or by reading the width of the direction trace from continuous records of wind direction, provided the damping ratio of the wind vane is approximately 0.6. The US EPA, provides a classification scheme that relates stability category to σ_a, wind speed and time of the day. It is important to note that the same σ_a can relate to quite different categories depending on whether the observation relates to day or night conditions. The reason for this is that at night, stable flows can be established because of the wind meanders (sinuosity or winding course), giving rise to large values of σ_a, even though the vertical dispersion might be very limited. The present stability classifying schemes based on σ_a observations take this also into account.

Alternative to above Approach for Estimating Plume Spread Parameters

An alternative to the above approach for estimating plume spread parameters is to use directly measurements of 'standard deviation' of horizontal and /or vertical wind direction.

It was suggested (Draxler, 1976) that the horizontal and vertical plume spread parameters can be related to the standard deviation' of horizontal wind direction (σ_a), and standard deviation of vertical wind direction (σ_c) by the equations of the form:

$$\sigma_y = x\, \sigma_a f_y.$$

$$\sigma_z = x\, \sigma_c f_z.$$

Where f_y and f_z are empirically determined functions, which include the effects of release height. Further Irwin(1983) suggested that f_y and f_z should take the form as follows:

$$f = [1 + 0.9(t / t_0)^{0.5}]^{-1}$$

Where, t is the travel time = x / u and t_0 is 1000 for f_y (horizontal direction), t_0 is 500 for f_z (vertical direction) for unstable and day-time neutral conditions; t_0 is 50 for f_z (vertical direction) for stable and night-time neutral conditions.

Alternative Plume Rise Calculations

The plume rise is an important variable in dispersion calculations. As the emissions from the stack are with a finite upward momentum and are also generally warmer than the surrounding atmosphere, the plume of exhaust gases rises up due to effects of buoyancy and momentum.

Following procedure is widely used for calculating the plume rise.

First step is to calculate the momentum flux (F_m) and buoyancy flux (F_b) of the plume by the equations:

$$F_m = (vd/2)^2\, T / T_s.$$

$$F_b = gv\,(d/2)^2\,(1 - T / T_s)$$

where, d = Internal diameter of stack at the tip, m

T = Ambient absolute temperature, K

T_s = Absolute temperature of emission, K

v = gas exit velocity, m/s.

g = acceleration due to gravity 9.8 m/s^2.

For sources with small buoyancy flux, i.e. $F_b < 55$, the plume is assumed to rise out to a limiting downward distance from the stack. The limiting distance x_m is given by

$$x_m = 49\, F_b^{5/8}.$$

For large buoyant sources i.e. $F_b > 55$, the plume is assumed to rise to a limiting distance x_m given by

$$x_m = 119\, F_b^{2/5}.$$

For neutral and unstable conditions (Pasquil stability A, B. C, and D), plume rise dh is taken to be:

$$dh = [C_3 x + C_4 x^2]^{1/3} \quad x \le x_m$$

Where, $C_3 = 27\ F_m / [\ u\ (1 + 3\ u/v)]^2$;

$C_4 = 1.5\ F_b / (B_1^{\ 2}\ u^3)$

The parameter B_1 is entrainment coefficient which is usually taken to be 0.6. For downwind distances $x \geq x_m$, the plume rise is assumed to remain constant. For stable conditions (Pasquill stability E and F) plume rise dh is taken to be:

$$dh = [C_5 \sin (S^{1/2}\ x/u) + C_6 \{1 - \cos (S^{1/2}\ x/u\ )\}]^{1/3} \text{ for } x \leq 3.14\ u\ /\ S^{1/2}.$$

Here $C_5 = 27\ F_m / [u\ S^{1/2}\ (1 + 3u/v)^2]$ and

$C_6 = 3\ F_b / u.S.B_2^{\ 2}$.

B_2 is the entrainment coefficient for stable conditions and **S** is the stability parameter given by $S = g\ \Gamma\ /T$, where Γ is the potential temperature gradient (K/m).

For $x \geq 3.14\ u\ /\ S^{1/2}$, the plume rise is taken to be a constant

3.3.3 Plume Rise Formula (Moses & Carson)

An Empirical formula suggested by Moses and Carson for predicting Plume rise is as follows:

$$dh = C_1\ (v\ d\ /\ u) + C_2\ (Q_h^{\ 1/2}\ /\ u)$$

Where: dh = Plume rise; d = stack diameter at exit (both in m).

u = wind speed, m/s

v = exit velocity of gases from stack, m/s

Q_h = rate of emission of heat, kcal/s;

C_1 and C_2 are regression coefficients for plume rise which depend on condition of atmospheric stability.

The first term of the equation is linked to vertical momentum of gases, while the second term relates to buoyancy of the plume.

The plume rise as seen from above is directly proportional to the sum of mass ejection rate and square root of heat content of gases, and inversely proportional to wind speed.

3.3.4 Plume Rise Formula (Bureau of Indian Standards):

(i) For large release of hot effluents from a stack with heat emission of the order of 1000 kcal/s or more, the plume rise (dh) can be estimated by the following formula:

$$dh = [0.84\ (12.4 + 0.09h)Q_h^{\ 1/4}]\ /\ u$$

Where: dh = plume rise and h = stack height (both in m)

u = wind speed in m/s; Q_h = heat release in cal/s

(ii) For sources of lower heat release, where the momentum is very predominant, the plume rise can be estimated by the formula:

$$dh = 3\ W_0\ d\ /u$$

Where: dh = Plume rise and d = Stack diameter at exit (both in m); u = Wind velocity in m/s; W_0 = Efflux velocity

3.3.5 Maximum Ground Level Concentration

The height of release of pollutants is the height of stack plus the plume rise. The plume height is a variable depending on many factors as discussed earlier. The height of stack which is fixed for a particular factory or power station therefore plays a major role in restricting ground level concentration of pollutants. For a known stack height, there is a critical velocity when the ground level concentration attains its maximum value

The maximum concentration (C_m) may be estimated from the formula:

$$C_m = (M.K / h^2)\,[1 / V.A_T]^{1/3}.$$

Where: C_m = the maximum concentration of pollutants on ground level;

h = stack height,

V = volume of pollutant gases discharged per unit time,

A_T = difference of temperature between stack gases and surrounding atmosphere

M = mass of the pollutant gases discharged per unit time,

K = constant which accounts for factors such as, horizontal and vertical mixing of pollutants in the air, settling velocity of polluting substances, and discharge characteristics of gases from stack,

3.3.6 Other Factors Affecting Dispersion of Pollutants

Particle Settling out of Plume-- In some cases, there is emission of small particles which can settle out of plume under gravity. For Particle size of the order of 10 μm and smaller, the settling out of plume may not take place **(Also refer Section 3.2.6 and 8, of this Chapter).**

Terrain Effects-- When the ground is not flat, the surface will either rise towards an elevated plume or fall away from it. With the rising land the receptor would gradually get closure to the plume centre-line and could expect higher ground –level concentrations. In practice, the plume centre-line would also rise in response to rise in the land surface and so, the decrease in distance between the plume centre-line and the receptor would not be the same as the increase in height of the terrain over the height at the base of stack **(Also refer Section *3.2.5, of this Chapter*).**

Building wake Effects-- The effect of nearby high rise building would depend on the ratio of height of discharge of pollutants to the height of building, exit velocity of plume, buoyancy of plume, shape of the building, distance of building from stack, etc. Turbulence increases as the wind blows over the building / obstacle. This enhanced turbulence affects the plume dispersion. Models may include the effects of building-wake turbulence by adjusting the plume spread parameters to take into account the extra dispersion that occurs as the turbulence is increased due to blowing of wind over the building. The form of adjustment is determined experimentally from wind tunnel studies. However, if the stack is very high compared to the building (height), say 2.5 times the building height as in thermal power stations, the 'building–wake effect' need no consideration. ***(Also refer Para. 3.2.7, of this Chapter).***

3.3.7 AERMOD Dispersion Model

'AERMOD' dispersion model was developed by the American Metrological Society (AMS), US Environmental Protection Agency (US, EPA) and Regulatory Model Improvement Committee (RMIC) short named as AERMIC—a collaborative working group of scientists from the AMS and the EPA. The AERMIC was initially formed in 1991. The AERMOD was developed in the following steps:

- Initial model formulation
- Developmental evaluation

- Internal review by peer and beta testing
- Revised model formulation
- Performance evaluation and sensitivity testing
- External peer review
- Submission to the EPA for consideration as a regulatory model.

On 21st April 2000, the EPA proposed that AERMOD be adopted as the EPA's preferred regulatory model for both simple and complex terrain and on 9th Nov. 2005, AERMOD was adopted by the EPA and promulgated as their preferred regulatory model effective as of December 9, 2005. The entire developmental and adoption process took 14 years (from 1991 to 2005). AERMOD becomes the regulatory air dispersion model of EPA, Victoria, Australia from 1st Jan 2014.

The **AERMOD** dispersion modeling system is an integrated system including the following modules.

- A steady-state dispersion model for short-range (up to 50 km) dispersion of emitted air pollutants from stationary industrial sources. In the stable boundary layer, the distribution of pollutant's concentration is assumed to be Gaussian in both the vertical and horizontal direction. In the convective boundary layer, the horizontal distribution of pollutant's concentration is assumed to be Gaussian, but its vertical distribution is described with a bi-Gaussian probability density function. Such a behavior of the concentration distributions in the convective boundary layer was demonstrated by (Willis, and Deardorff, 1981) and (Briggs, 1993).
- Also, in convective boundary layer AERMOD takes into consideration 'Plume lofting' (whereby a portion of plume mass, released from a buoyant source, rises to and remains near the top of the boundary layer before getting mixed into the convective boundary layer). AERMOD also would track any plume mass which penetrates into elevated stable layer and then may allow it to re-enter the boundary layer if and when appropriate.
- A metrological data pre-processor (AERMET)* accepts surface metrological data, upper air surroundings and also optionally the data from on-site instrument towers. It then calculates atmospheric parameters required by dispersion model — turbulence characteristics, mixing heights, friction velocity, 'Moin-Obukov length, and surface heat flux.
- A terrain pre-processor (AERMAP) whose main purpose is to provide a physical relationship between terrain features and the behavior of air pollution plumes. It generates location and height data for each receptor location. It also provides information that allows the dispersion model to simulate the effects of air flowing over hills or splitting the flow around hills.
- AERMOD also includes PRIME (Plume Rise Model Enhancements) which is an algorithm for modeling the effects of downwash created by the pollution plume flowing over nearby buildings.

Briefly the primary features and capabilities of AERMOD are:

- Source types: Multiple point, area and volume sources
- Source releases: Surface, near surface and elevated sources
- Source locations: Urban or rural locations. Urban effects are scaled by population.
- Plume types: Continuous, buoyant plumes
- Plume deposition: Dry or wet deposition of particulates and/or gases.
- Plume dispersion treatment: Gaussian model treatment in horizontal and in vertical for stable atmospheres. Non-Gaussian treatment in vertical for unstable atmospheres
- Terrain types: Simple or complex terrain

- Building effects: Handled by PRIME downwash algorithms
- Meteorology data height levels: Accepts meteorology data from multiple heights
- Meteorological data profiles: Vertical profiles of wind, turbulence and temperature are created

***Notes on AERMET**

(Source:http://wwwweblakes.com/products/aermod Lakes Environmental Guide)

Sensible Heat Flux in the Convective Boundary Layer

The fluxes of heat and momentum drive the growth and structure of the planetary boundary layer (PBL). In order to properly characterize the PBL, a good estimate of the surface sensible heat flux, H is needed which depends on the net radiation, Rn. When the net radiation is positive for a given hour, AERMET defines the PBL as convective. The net radiation is either measured or estimated as a function of cloud cover, solar angle, albedo, and surface temperature. In the convective boundary layer (CBL), a simple energy balance approach (as in Oke, 1978) is used to derive the following expression for H:

$$H = [0.9R_n] / [1 + (1/B_0)] \quad (1)$$

Where: B_0 = *Bowen Ratio*

H = *Sensible Heat Flux*

R_n = *Net Radiation*

The underlying assumption in eq.(1) is that the soil heat flux is 10% of the net radiation. If the measured values for R_n are not available, R_n can be estimated from the insolation and the thermal n n radiation balance at the ground following the method of Holtslag and Van ulden (1983) where the insolation for clear skies, R, is calculated from Collier and Lockwood (1975).

Solar radiation, R, corrected for cloud cover is taken from (Kasten and Czeplak, 1980):

$$R_n = \frac{[1 - r\{\phi\}]\,R + c_1 T_{ref}^6 - \sigma_{SB} T_{ref}^4 + c_2 n}{1 + c_3} \quad (2)$$

Where: C_1 = 5.31×10^{-13} $\mathrm{Wm^{-2}\,K^{-6}}$

C_2 = 60 $\mathrm{W/m^2}$

C_3 = 0.12

σ_{SB} = Stefan Boltzman constant (5.67×10^{-8} $\mathrm{W/m^2\,K^4}$)

T_{ref} = Ambient air temperature

R_n = Net Radiation

$r\{\phi\}$ = Albedo $= r^1 + (1 - r^1) \exp [a\Phi + b]$

Where:

$$a = -0.1 \quad (3)$$

$$b = -0.5(1 - r^1)^2$$

$$r^1 = r\{\Phi = 90°\}.$$

Solar radiation, R, corrected for cloud cover is taken from (Kasten and Czeplak, 1980):

$$R = R_0(1 - 0.75\, n^{3.4}), \tag{4}$$

Where,

R = Solar radiation

n = Cloud Cover {0.0 – 1.0}.

$$R_0 = 990 \sin \phi - 30, \tag{5}$$

where,

$$\Phi = \frac{\Phi\{t_p\} + \Phi\{t\}}{2}$$

t_p = Previous hour

and t = present hour

R_0 = Clear Sky Insolation (Wm^{-2})

Φ = Solar Elevation angle.

Friction Velocity, and Monin Obukhov Length(L):

In the CBL, AERMET computes the surface friction velocity, u*, and the Monin-Obukhov length (MOL), L, using the value of H estimated from eq. (1). Since the friction velocity and MOL depend on each other, an iterative method, similar to that used in CTDMPLUS (Perry, 1992) is used. AERMOD initializes u* assuming neutral conditions, calculates L, then proceeds with subsequent estimates of u* and L until convergence is reached (i.e., there is less than a 1% change between the successive iterations). The expression for u* (e.g., Panofsky and * Dutton, 1984) is.

$$u_0 = \frac{ku_{ref}}{\ell n(z_{ref}/z_0) - \Psi_m\{z_{ref}IL\} + \Psi_m\{z_0IL\}}, \tag{6}$$

$$\Psi_m\left\{\frac{z_{ref}}{L}\right\} = 2\ell n\left(\frac{1+\mu}{2}\right) + \ell n\left(\frac{1+\mu^2}{2}\right) - 2\tan^{-1}\mu + \pi/2$$

$$\mu = \left(1 - 16\frac{z_{ref}}{L}\right)^{1/4}$$

$$\Psi_m\left\{\frac{z_0}{L}\right\} = 2\ln\left(\frac{1+\mu_0}{2}\right) + \ln\left(\frac{1+\mu_0^2}{2}\right) - 2\tan^{-1}\mu_0 + \pi/2$$

$$\mu_0 = \left(1 - 16\frac{z_0}{L}\right)^{1/4}$$

The initial step in the iteration solves equation (6) for u assuming m = 0 (neutral limit) and *m setting u = u. Having an initial estimate of u, L is calculated from the following definition for ref *L (eg. see Wyngaard, 1988):

$$L = -\frac{\rho c_p T_{ref} u_*^3}{kgH} \tag{7}$$

Where,

g = *acceleration due to gravity*

c_p = *specific heat of air at constant pressure*

ρ = *density of air*

k = *0.4; von Karman's constant.*

u is recalculated using eqs. (6) and (7) and with L from eq. (8). This procedure is continued until * values of u and L change by less than 1%.*

The reference heights for wind speed and temperature to be used in the determination of friction velocity and MOL are optimally chosen to be representative of the surface layer in which the similarity theory has been formulated and tested with experimental data. Typically, a 10-m height for winds and a temperature within the range of 2 to 10 m is chosen. However, for excessively rough sites (such as urban areas with z_0 *in excess of 1 m), AERMET has a safeguard to accept wind speed reference data in the vertical height range between 7* z_0 *and 100 m. Below 7* z_0 *(roughly, the height of obstacles or vegetation), measurements are unlikely to be representative of the general area. A similar restriction for temperature measurements is imposed, except that temperature measurements as low as zero, are permitted. Above 100 m, the wind and temperature measurements are likely to be above the surface layer, especially during stable conditions. Therefore, AERMET imposes an upper limit of 100 m for reference wind speed and temperature measurements for the purpose of computing the similarity theory friction velocity and MOL each hour. Of course, other US EPA guidance for acceptable meteorological site should be consulted in addition to keeping the AERMET restrictions in mind*

3.3.8 Other Dispersion Models

There are dozens of models for air dispersion of pollutants. Some of the air dispersion modeling systems are mentioned below:

1. **Box Model**

   The 'Box model' of air pollution dispersion is a very simple model which assumes that in a geographical area there is a given volume of atmospheric air in the form of a box (air- shed) and the air pollutants inside the box are homogeneously distributed. On this assumption the average pollutant concentration at any point within the air-shed is estimated. This model has a very limited ability to accurately predict dispersion of air pollutants over an air shed because of the assumption of homogeneous pollutant distribution.

2. **Lagrangian Model**

   In the 'Lagrangian dispersion model, it is assumed that the motion of pollution plume parcels (particles) in the atmosphere is like a random walk process. The model mathematically follows pollution plume parcels (also called particles) as they move in atmosphere. This model uses a moving frame of reference as the parcels move from their initial location. It is assumed that an observer of a Lagrangian model moves along with the plume. The dispersion calculation are then made by computing the statistics of the trajectories of a large number of the pollution plume parcels.

3. **Eulerian Model**

   The Eulerian dispersion model is somewhat similar to the Lagrangian model as it also tracks the movement of a large number of pollution plume parcels as they move from their initial location. The most important difference between these models is the use of 'fixed three-dimensional Cartesian grid' as a frame of reference by 'Eulerian model', while the Lagranian model uses a 'moving' frame of reference. It is said that an observer of an 'Eulerian model' watches the plume go by.

4. **Dense Gas Models**

These models are for simulating the dispersion of 'dense gas pollution plumes'. The 'dense gas' signifies that the gases of the plume are heavier than air. Three commonly used dense gas models are: (a) DEGADIS model, (b) SLAB model, and (c) HEGADAS model

The 'DEGADIS model' was developed by 'Dr. Jerry Havens and Dr. Tom Spicer at the University of Arkansas under commission by the US Coast Guard and US EPA'. The 'SLAB model' was developed by the 'Lawrence Livermore National Laboratory funded by the US Department of Energy, the US Air Force and the American Petroleum Institute', while the 'HEGADAS model' was developed by 'Shell Oil's research division'.

5. **CALPUFF Model**

The CALPUFF model was originally developed by the Sigma Research Corporation (SRC) in the late 1980s under contract with 'California Air Resource Board (CARB)'. The Sigma Research Corporation subsequently became part of Earth Tech, Inc. After the US EPA designated CALPUFF as a preferred model in their *Guideline on Air Quality Models*, Earth Tech served as the designated distributor of the model. In April 2006, ownership of the model switched from Earth Tech to the TRC Environmental Corporation, who are currently (August 2013) responsible for maintaining and distributing the model. It is a non-steady-state puff dispersion model that simulates the effects of time- and space-varying meteorological conditions on pollution transport, transformation, and removal. CALPUFF can be applied for long-range transport and for complex terrain.

CALPUFF is an advanced, integrated Gaussian puff modeling system for the simulation of atmospheric pollution dispersion distributed by the Atmospheric Studies Group at TRC Solutions. The model has been adopted by the US Environmental Protection Agency (EPA) in its '*Guideline on Air Quality Models* as a preferred model for assessing long range transport of pollutants and their impacts on Federal Class I areas and on a case-by-case basis for certain near-field applications involving complex meteorological conditions'.

The CALPUFF model is designed to simulate the dispersion of buoyant, puff or continuous point and area pollution sources as well as the dispersion of buoyant, continuous line sources. The model also includes algorithms for handling the effect of downwash by nearby buildings in the path of the pollution plumes.

6. **BLP Model**

It is a Gaussian plume dispersion model designed to handle unique modeling problems associated with industrial sources where plume rise and downwash effects from stationary line sources are important.

7. **CALINE3 Model**

It is a steady-state Gaussian dispersion model designed to determine pollution concentrations at receptor locations downwind of highways located in relatively uncomplicated terrain.

8. **CAL3QHC and CAL3QHCR**

CAL3QHC is a CALINE3 based model with queuing calculations and a traffic model to calculate delays and queues that occur at signalized intersections. CAL3QHCR is a more refined version based on CAL3QHC that requires local meteorological data.

9. CTDMPLUS –

It is a Complex Terrain Dispersion Model (CTDM) plus algorithms for unstable situations (i.e., highly turbulent atmospheric conditions). It is a refined point source Gaussian air quality model for use in all stability conditions (i.e., all conditions of atmospheric turbulence) for complex terrain.

10. OCD—Offshore and Coastal Dispersion Model

The OCD is a Gaussian model developed to determine the impact of offshore emissions from point, area or line sources on the air quality of coastal regions. It incorporates over-water plume transport and dispersion as well as changes that occur as the plume crosses the shoreline.

Note:

1. *There are other models like (a) ADMS and NAME developed in UK, (b) AUSPLUME and LADM developed in Australia, (c) DISPERSION 2 developed in Sweden, (d) ISC3 developed in US, (e) MERCURE developed in France, (f) OML developed in Denmark.*
2. *The European Topic Centre on Air and Climate Change, which is a part of the European Environment Agency (EEA), maintains an online Model Documentation System (MDS) that includes descriptions and other information for almost all of the dispersion models developed by the countries of Europe. The MDS currently contains about 140 models developed in Europe (excluding the United Kingdom).*

REFERENCES

1. Beychok, Milton R. (2005). Fundamentals of Stack Gas Dispersion, 4th Edition. Author-published. ISBN 0-9644588-0-2.
2. Bosanquet, C.H. and Pearson, J.L., “The spread of smoke and gases from chimneys”. Trans. Faraday Soc., 32:1249, 1936.
3. Briggs, G.A., “A plume rise model compared with observations”. JAPCA, 15:433-438, 1965.
4. Briggs, G.A., “CONCAWE meeting: discussion of the comparative consequences of different plume rise formulas”. Atmos. Envir., 2:228-232, 1968.
5. Briggs, G.A., “Plume rise”, USAEC Critical Review Series, 1969.
6. Briggs, G.A., “Some recent analyses of plume rise observation”, Proc. Second Internat'l. Clean Air Congress, Academic Press, New York, 1971.
7. Briggs, G.A., “Discussion: chimney plumes in neutral and stable surroundings”. Atmos. Envir., 6:507-510, 1972.
8. Briggs, G.A., 1984: Plume rise and buoyancy effects. In Atmospheric Science and Power Production.
9. Briggs, G.A., Journal of Applied Meteorology, 07/1993; 32(8); 1388-1425.
10. Carson, J.E., and H. Moses, 1969: The validity of several plume rise formulae, J. Air Pollu. Control Assoc., 19, 11, 862-866.
11. Draxler, R.R. 1976. Determination of atmospheric diffusion parameters. Atmospheric Environment., 10, 99-105.
12. Hanna, S.R., D.G. Strimaitis, and J.C.Chang 1991: Hazard response modelling uncertainty (A quantitative method), Vol. I: User's guide for software for evaluating hazardous gas dispersion models Sigma Research Corporation, Westford, USA.

13. http://en.citizendium.org/wiki/File; BriggsLogic.png
14. http://pandora meng.auth.gr/mds; and http://en.citizendium,org/wiki/special
15. Holtslag, A. A. M. and A. P. van Ulden, 1983: A simple scheme for daytimeestimates for the surface fluxes from routine weather data. J.Climate Appl.Meteor., 22, 517-529.
16. Irwin, John S. 1983: Estimating Plume Dispersion-A Comparison of Several Sigma Schemes. J. Climate Appl. Meteor., 22, 92–114.
17. Kumari Manju, and O.P. Sharma, 1988: A model for ambient air quality analysis using routine meteorological observations. Atmos. Environ., 22, 651-661.
18. Moses, Harry & James E. Carson (1968) Stack Design Parameters Influencing Plume Rise, Journal of the Air Pollution Control Association, 18:7, 454-457.
19. Nigal Holms. PhD ' Environmental Services, Australia.
20. Oke, T. R., 1978: Boundary Layer Climates. John Wiley and Sons, New York, New York, 372pp.
21. Panofsky, H. A. and J. A. Dutton, 1984: Atmospheric Turbulence: Models and Methods for Engineering Applications. John Wiley and Sons, New York, 417pp.
22. Ragland, K.W., 1973: Multiple box model for dispersion of air pollutants from area sources. Atmos. Environ.,7, 1017-1031.
23. Randerson (Ed.) DOE/1L 27601, Technical Information Center, Oak
24. Rao, K.S., 1985: User's Guide for PEM-2. NOAA Tech. Memo. ERL ARL-137.
25. San.hufsa.ac.krl-gwlee/session9/gaussian.html developed by the shodor education foundation inc.
26. Slade, D.H. (editor), "Meteorology and atomic energy 1968", Air Resources Laboratory, U.S. Dept. of Commerce, 1968.
27. Stern, A.C., R.W. Boubel, D.B. Turner, and D.L. Fox (1984), Fundamentals of Air Pollution, (2nd ed.), Academic Press Inc., New York, 300 pp.
28. Stull, Roland B., Meteorology for Scientists and Engineers, Brooks/Cole, 2012.
29. Sutton, O.G., "The problem of diffusion in the lower atmosphere", QJRMS, 73:257, 1947 and "The theoretical distribution of airborne pollution from factory chimneys", QJRMS, 73:426, 1947
30. Teri's Model of air pollution dispersal.
31. The Model Documentation System (MDS).
32. Turner, D.B. (1994). Workbook of Atmospheric Dispersion Estimates, 2nd Edition. CRC Press. ISBN 1-56670-023-X. www.crcpress.com.
33. van Ulden, A. P. and A. A. M. Holtslag, 1983: The stability of the atmospheric surface layer during nightime. Preprints, Sixth Symposium on Turbulence and Diffusion, American Meteorological Society, Boston, 257-260.
34. Willis, G.E., J.W. Deardorff, A laboratory study of dispersion from a source in the middle of the convectively mixed layer, Atmospheric Environment (1967), Volume 15, Issue 2, 1981, Pages 109-117.
35. Wyngaard, J. C., 1988: Structure of the PBL. Lectures on Air Pollution Modeling. Venkatram, A. and Wyngaard, J. C., American Meteorological Society, 9-57 pp.

QUESTIONS

1. Write short notes on:
   (a) Atmospheric turbulence and its classification.
   (b) Influence of density of emitting gases on types of plume.
   (c) Methods of designating sources of emission.
   (d) Wind Rose.

(e) Mixing Height

2. What is meant by environmental lapse rate and adiabatic lapse rate? What is negative laps rate?
3. Explain how the temperature inversion takes place.
4. What are various plume conditions for a buoyant plume?
5. Describe Gaussian air dispersion model
6. Write short notes on:
   (a) Influence of wind speed and wind direction on dispersion of pollutants
   (b) Influence of Turbulence on dispersion of pollutants
   (c) AERMOD model
   (d) CALPUFF Model
7. How do you determine the maximum ground level concentration of a pollutant?
8. During temperature inversion in atmosphere, air pollutants tend to:
   (a) accumulate above inversion layer
   (b) accumulate below inversion layer
   (c) disperse laterally
   (d) disperse vertically.
9. What is meant by the effective height of a chimney and how is it computed? What is its use?

Answer: 8. (a)

CHAPTER 4

CONTROL OF EMISSION OF SUSPENDED PARTICULATE MATTER IN COAL FIRED THERMAL POWER STATIONS

4.1 PARTICULATE EMISSION AND CONTROL OPTIONS

The coal fired power generation has a considerable potential impact on air quality as most of the power generation world over (including India) is coal based. The sustained development of power would largely depend upon the proper utilization of coal with efficiency in power generation keeping in view the air quality, water quality, land use, ecology, health and safety. The coal and its utilization requires the effective environmental management as most power is to be generated by coal fired power stations for a long time till other sources of energy are adequately harnessed. All possible strategies have, therefore, to be adopted for prevention of pollution by thermal stations. The scope of this chapter is related to the abatement of emission of *Suspended Particulate Matter* (SPM) from the chimneys / stacks of coal burning stations.

Like the adverse effects of other types of pollution, there are adverse effects of SPM emission also. As already described in *chapter 2*, the emission of SPM and particularly the *Respirable Particulate Matter* (RPM) / *Respirable Suspended Matter* (RSM) / fly ash* causes many problems. Apart from silica, there are other metallic oxides in fly ash, some of which are very toxic. The amount of toxic metals absorbed on the surface of 'fly ash particles' generally increase with a decrease in size of particles. Toxicity, thus, increases with fineness of the dust. ***[See, Para. 2.10, Chapter 2]***

Note: Coal Ash from Boilers

There are two portions of total ash of coal in a boiler, namely 'Bottom ash' and 'Fly ash'. The bottom ash is the portion of ash which gets collected in the hopper at the bottom of boiler. The bottom ash consists of coarse ash including clinkers of ash. The fly ash is the fine ash which is carried by the flue gas. The fly ash is the SPM which we are concerned with at the moment. Typically in a pulverized coal fired boiler, the bottom ash may be 20 % and fly ash may be 80 %.

The environmental regulations are likely to become more stringent as the industrial development proceeds. It is, therefore, necessary to design new plants so that the emissions may remain below the prescribed upper limits, while the existing plants would be required to be retrofitted with systems to bring pollution below the safe limits.

The present norms in India for emission and ambient air quality regarding SPM are mentioned in ***Table 4.5, and Para 4.7, of this Chapter***.

The options available for control of SPM emission are:

- Ordinary dust trap (These are used in many locations in industries and power stations, but are not used for flue gas cleaning. A typical dust trap is shown in **Fig. 4.1.**

- Gravity settling chambers and baffled chambers [these are not used at thermal power stations for flue gas cleaning. ***[For details refer Chapter 8]***
- Mechanical Precipitators (MP)
- Electrostatic Precipitators (ESP)
- Bag house
- Wet scrubbers [used in power station at certain locations only, like coal /lignite handling plant, pulverized fuel preparation plant and sometimes as a constituent of flue gas desulphurization.]

The severity of the problem of SPM emissions differs with different types of boilers due to factors such as:

- Proportion of fly ash (to be handled by dust collector) to the total ash.
- Particle size of fly ash.
- SO_X content of flue gases at the inlet of dust collector as a result of any in-furnace flue gas desulphurization.
- Design flue gas inlet temperature to dust collector and volume flow at that temperature.

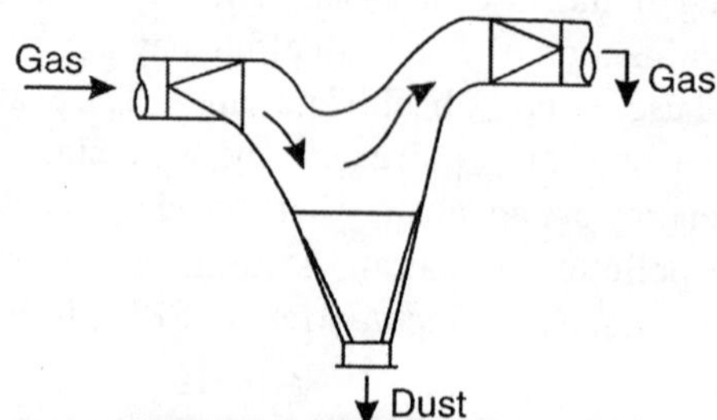

Fig. 4.1: Inertial Dust Trap (with change in direction of gas velocity)

Indian coal is low in sulphur content and very high in ash content. The gross calorific value (GCV) is low and so a much larger amount of coal is burnt and ash produced, when compared to thermal plants of same capacity in other countries where coal has high GCV and low ash.

Fly ash from Indian coals has high electrical resistivity. Typically, it may be between 2.4×10^{12} and 1.5×10^{13} ohm cm which is very high in comparison to the fly ash from imported coals. Electrical resistivity higher than 1.5×10^{13} up to 1×10^{14} ohms cm has been encountered in fly ash from some sub-bituminous coals. This property of fly ash causes major problems in the design of electrostatic precipitators. Typical coal and ash characteristics and fly ash resistivity is given in the ***Annexure 4 [Tables (i)/4, (ii)/4, (iii)/4]***.

Many developments have taken place in the technology for combating the problems encountered in control of stack emission of SPM. The problem of high resistivity fly ash of coals can be effectively dealt with and 'Collection Efficiencies*' of more than 99 % can be achieved in dust collecting equipments to restrict the SPM emission to 50 to 100 mg/m3. The 'Collection efficiency' of a dust collection equipment is the ratio of mass of dust collected to total mass of dust in the flue gas per unit volume of flue gas at the inlet to the precipitator. The Collection efficiency (η) of a dust collector in percent can be determined in accordance with the formula:

$$\eta = [(Z_i \,.\, V_i - Z_o \,.\, V_o) / (Z_i \,.\, V_i)] \times 100$$

Where Z_i is the dust loading of flue gases at the inlet of dust precipitator, Z_o is the dust loading of flue gases at the outlet of dust precipitator in grams per cubic meter of flue gases, V_i is the volume of flue gases at the inlet of the dust precipitator, V_o is the volume of flue gases at the outlet of dust precipitator in cubic meters. For achieving minimum emission, generally electrostatic precipitators (ESP) or bag filters (bag house) are adopted. In order to effectively deal with high resistivity fly ash, modern ESP designs have incorporated measures like:

- Increased number of fields in series;
- High specific collection area m^2 / Nm^3 / sec of gas flow;
- Lower flue gas velocity through ESP and longer residence time;
- Better integrity of electrode system against misalignment of electrodes and wire snapping;
- Integrated control of rapping and its optimization; intermittent / semipulse / multi-pulse energisation technique for electrodes;
- Injection of conditioning agents to reduce fly ash resistivity;
- Reducing temperature of flue gas at inlet to ESP by "Gas to Gas Heat exchangers" or other means.

Bag filters (bag house) are now being increasingly adopted for power stations as the final emission from bag house does not increase due to change in operating parameters. New fabrics have been developed which have adequate strength to withstand adverse service conditions. Pulse jet cleaning is found to be a more efficient method of filter bag evacuation. Bag filters suffer however, from certain problems like blinding, high pressure drop, bag erosion / abrasion, damage, sensitivity to volume flow, filter cake resistance and above all the high cost of bag replacements.

The bag house are attractive for fluidized bed combustion (FBC) boilers, or boilers with 'Spray dry process' of 'Flue gas desulphurization (FGD) process' or sometimes for 'Retrofitting with existing ESP'.

4.2 MECHANICAL PRECIPITATOR (MP)

Cyclone Separator

Cyclone separators are of two designs, i.e., 'Straight through cyclones' and 'Reverse flow cyclones'. Both types of mechanical separators use centrifugal force to throw solids from gas stream for separation of solids.

In straight through design, the flue gas enters the top and leaves from the bottom, discharging solids at the lower periphery. In reverse flow type, the flue gas enters at the upper periphery, and leaves overhead, solids being discharged from the bottom.

The 'Reverse flow cyclones' offer higher 'Collection efficiency' and have wider use than straight through type which, however, has a 'Higher gas handling capacity' and incurs a 'Low pressure drop'. The straight through type is therefore appropriate for large gas flow rates and low pressure drop requirements, but because of lower efficiency these are seldom used.

The principle of cyclone separation has many applications in power stations and in industries. Cyclones are integral to 'Circulating Fluidized Bed Combustion (CFBC) boilers', 'Coal gasification',

'Wet type flue gas desulphurization' technologies, etc.

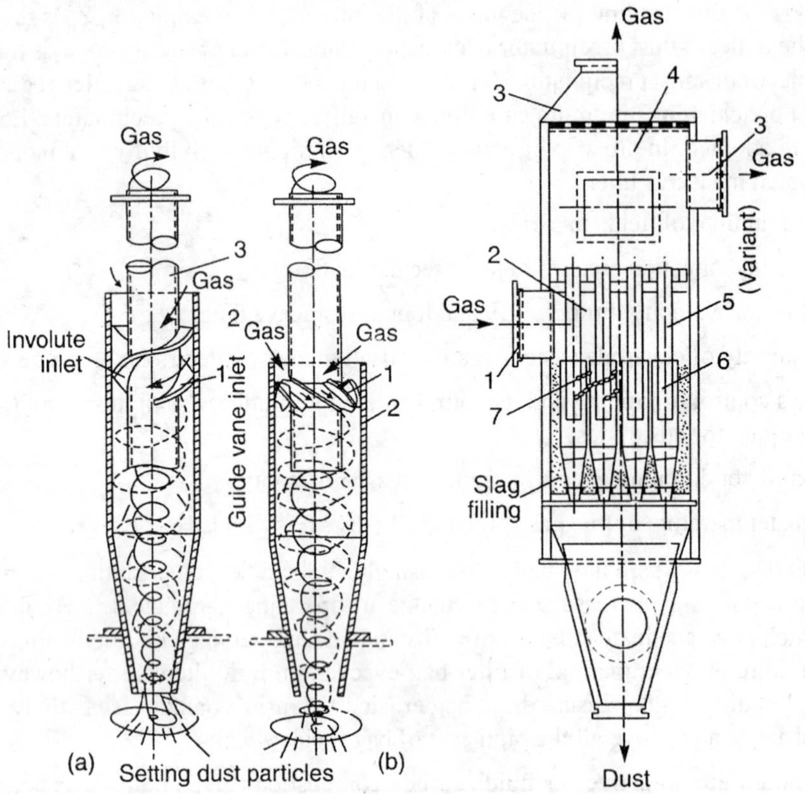

(a) Involute Inlet type Cyclone Units of 'Multiclone';
(b) Guide Vane Inlet type Cyclone Unit of 'Multiclone'
Right hand side: 'General view of Typical Multiclone'

1—Dirty gas inlet; 2—Casing; 3—Clean gas outlet; 4—Clean gas chamber; 5—Clean gas pipes; 6—Individual cyclone units; 7—Guides for gas.

[Dirty gases enter the common divider 2 through inlet 1. Gases pass through guides 7 (to acquire rotary motion into individual cyclone units 6. The cleaned gases ascend via outlet pipes 5 into cleaned gas chamber 4] (Drawing is not to the scale).

Source: *Miscellaneous Literature of old Russian Power stations in India*

Fig. 4.2: Diagram Showing Principle of Typical Multiclone Type Mechanical Dust Separator

Multiclones or Battery Cyclones

In older power stations, 'Multiclone (Multi-cyclone) separators/ battery cyclones separators' are installed. When large quantities of flue gases are to be handled for dust separation, a large group of small diameter cyclones (of diameter say 200 or 250 mm) would be more effective than single large diameter cyclone for effective dust separation.

Each cyclone is provided with gas guide of helical type or vane type. The multiclone has a common inlet chamber and a common outlet plenum and a common dust collecting system. The principle of multiclone dust separator is illustrated in **Fig. 4.2.**

They are effective in collecting coarse particles, their collection efficiencies being 95 % for particles above 50 μm, but only 80% for particles below 20 μm. Their efficiency increases with the density of particulate matter.

Battery cyclones type of dust separators are generally unable to contain the SPM emission from power station within the present day prescribed limits and are not installed now as main dust collectors for the power stations.

Battery cyclone can be used as primary dust separators before secondary dust removal equipment (ie bag filters or wet type scrubbing- sometimes used in flue gas desulphurization) to reduce duty requirement on the secondary dust removal equipment. The 'Battery cyclones' are, however, not used before 'Electrostatic Precipitators (ESPs)', as the ESP without a primary separator is found to be more efficient than an ESP with a primary separator before it.

4.3 ELECTROSTATIC PRECIPITATOR (ESP)

4.3.1 ESP System

Flue gas laden with fly ash after having passed through air heater is led to the Electrostatic Precipitator (ESP) where an electrostatic charge is applied to dust particles with a corona discharge. The electrostatic field is created by high voltage transformer rectifier. In the ESP, there are suspended positive electrodes (called collection plates) and negative electrodes alternately, each set is called a field. There are 6 or 7 fields in the series from which the flue / exhaust gas passes through and after passing through the last field of the series, the gas should be clean enough to conform to the regulations.

The electrical potential between discharge electrodes (negative) and collecting plates (earthed) is high (4000 Volts per cm). (The operating voltage may be between 15 kV to 100 kV depending upon the configuration of the precipitator.) When the high voltage is applied at a value which causes a molecular breakdown of gas at the discharge electrode, gas is ionized and is referred to as corona discharge. The corona discharge occurs when the electric field potential across the space between the discharge electrode and collecting surface reaches a value where conductivity or the flow of free electrons (current flow) has been achieved. The current flow would continue to increase until there is a catasphoric electric discharge consisting of sparks or arcs. The corona discharge is primary effect which imparts the driving force or electric charge into the particulate matter. Maximum precipitator performance is achieved when the highest effective charging current and voltage are applied to precipitator. Field charging and diffusion charging are the two main mechanisms which affect the charging of particulate inside the precipitator. Field charging, also known as 'Impact charging', will dominate particles larger than 1.0 μm *by* impaction of free negative ions when they bombard and collide with the suspended particulate. The diffusion or ion diffusion charging will dominate particles below 0.3 μm. Ion diffusion depends on the thermal energy or velocity of ions and not on electric field. The negative ions will collide with suspended particulate due to the thermal motion of ions. Collision of particle and the ion will impart the charge on the particle. Both the field and diffusion charging affect the particles between 0.3 and 1.0 μm.

The discharge electrodes (which are in the shape of wires of non-corrosive materials like stainless steel) are negatively charged, while the collecting electrodes (which are in the shape of plates) are earthed. The discharge electrodes are hung midway between the parallel collecting electrodes; the gas flow is parallel to fields / collecting electrodes. The negatively charged particles of dust are then attracted to collecting plate electrodes where they lose their negative charge to earth. The positively charged ions of gases neutralize their positive charge at the discharge electrodes itself. The collecting plate electrodes are periodically rapped to cause collected ash to drop in the ash hoppers.

The electric field produces a force of the charged particles proportional to the magnitude of the field and the charge.

Force due to electric field F = q.E

Where q is the charge on the particle and E is the Strength of electric field.

ESP design would consider: (i) Inlet dust level, (ii) Temperature of flue gas, (iii) SPM emission limit, (iv) Volume flow, (v) Fly ash particle sizes, (vi) Fly ash resistivity, (vii) Physical and chemical properties of flue gas and fly ash, (viii) Number of series fields and parallel passes, (ix) Gas velocity and treatment time, (x) Voltage and mode of its application and control, (xi) Requirement of specific collection area*, (xii) Rapping mechanism and its control, etc.

The ESP may have several series fields. It may be noted that the maximum collection of ash is in the first field (about 70 %). In the next field it is about 60 to 65 % of the remaining and like this in the following fields also.

For higher resistivity dust, higher specific collecting area and higher number of fields are required. Slow velocity of gases through the ESP is better for dust removal. The 'Residence time' of the gases in the ESP depends on the length of ESP and the velocity of gases through it; the longer the 'Residence time' the better is the dust removal. Higher velocities of gases also create problem of 'Re-entrainment of dust' in the gas stream. Uniform distribution of gases across the ESP is better for dust removal performance. It is better to confine the difference in gas velocity at different points in ESP to about 12%. Gas distribution screen is put in the first stage on the inlet side of gas. The screen also acts as a baffle and some dust separation takes place at it which falls in the hopper below.

The intensity of electrostatic field is governed by the voltage applied to discharge electrodes. The charge on the dust particles and their velocity of motion (after charging) are largely affected by the field intensity (particularly for larger particles). The supply and maintenance of highest possible voltage on discharge electrodes is one of the essential conditions for the maximum dust collection by the ESP.

It may also be noted that a rise in gas temperature decreases the limit of 'Voltage' which can be maintained without flash over on discharge electrodes ie breakdown voltage. However, the moisture content in flue gas has beneficial effect on limit of breakdown voltage ie the higher the moisture content, the higher is the limit of break down voltage.

It may be mentioned that thick deposit of ash (cake) on collecting electrode is very detrimental for the operation of ESP. Even the deposits on the 'Discharge electrodes' is detrimental to ESP performance. Therefore, the provision of good rapping mechanisms for both the collecting electrodes and discharge electrodes and their proper performance is very important.

Note: **The specific collecting area is the area in square meters of collecting electrodes per normal cubic meters per second of flue gas cleaned in the ESP ($m^2/Nm^2/s$) ie. (total area of collecting electrodes in m^2 Volume flowof gas through ESP in Nm^3/s)*

4.3.2 Effect of Fly Ash Resistivity on ESP Performance

Dust precipitation rate of ESP decreases as the resistivity of fly ash increases. The fly ash resistivity above 1×10^8 ohm cm (at 150°C) and below 2×10^{10} ohm cm (at 150°C) is best suited for ESP performance. As the resistivity increases beyond 5×10^{10} ohm.cm, it becomes more difficult for ESP performance.

The trend of variation in the rate of precipitation in ESP and that in the emission of SPM at ESP outlet with the variation of resistivity is shown graphically in **Annexure Fig. (i)/4 and Annexure Fig. (ii)/4.** The actual variation in a particular case would depend on 'Design of boiler and ESP', 'Composition of coal' and other 'Operating Parameters'.

When the fly ash resistivity is high, the charge transfer to the collecting electrode is slow and the dust cake acquires a certain potential thus affecting the collection. The voltage across a dust layer (U) is expressed as

$$U = b \rho j V$$

Where b = thickness of dust cake; ρ = Resistivity in Ohm- cm;

j = current density ampere/ cm^2; V = Volume flow of gas per second.

The Field intensity, E in the dust cake is given by

$$E = U / b = \rho j V \text{ per cm}$$

4.3.3 Factors Affecting 'Fly Ash Resistivity'

The 'Fly ash Resistivity' decreases with the increase in concentration of 'Sulphur trioxide SO_3' and content of 'Moisture', Sodium (Na)', 'Potassium (K)' and 'Carbon' in the 'fly ash'. The 'Fly ash Resistivity' increases with the increase of content of 'Calcium', 'Magnesium', 'Alumina' and 'Silica' in 'fly ash'.

'Fly ash Resistivity' varies with the 'Temperature', the peak value occurring between 150 °C and 170 °C for most Indian coals (depending on chemical composition) as shown in **Annexure 4**. This annexure also shows analysis of typical Indian coal and ash.

The 'Fly ash Resistivity' is approximately 'Directly' proportional to the 'Absolute Temperature' in the temperature range of 120 - 150 °C and 'Inversely' proportional to the 'Absolute temperature' in the range of 230 to 290 °C.

The resistivity versus temperature curve beyond this range shall depend on the 'Sodium content' of fly ash. Low sodium coal, say less than 0.4% Na in fly ash irrespective of 'Sulphur content' shall present problem of 'High fly ash Resistivity' in the temperature range of 340 °C to 425 °C in which the '*Hot side Precipitators'** are designed to operate. The fly ash of Indian coal has a significantly lower Sodium content and hot side precipitators are, therefore not considered suitable.

***Notes: (i) *Hot side precipitator*–** *Normally the ESP is located after the 'Air Pre Heater' where temperature of gases is about 150 °C. The ESP located in between economizer and air preheater is called 'Hot Side Precipitator' as the temperature of gases in the precipitator is higher. Hot side precipitators can be preferred due to lower resistivity of fly ash at higher temperatures provided the sodium content in ash is higher than 0.4%. However, in the hot side precipitators, more expensive materials suitable for higher temperature would have to be used.*

(ii) Wet ESP for use in industries— *See Para 8.2.3 and Para 8.4.5 of Chapter 8*

4.3.4 Important Operating Factors Affecting ESP Performance

4.3.4.1 Maintenance Steps

The performance of the ESP would depend on the proper maintenance and adjustment of controls. It is therefore necessary to take the following steps regularly for which a schedule should be made and stuck to.

- Mechanical inspection which would involve the inspection of: 'Collecting electrode and discharge electrodes, rapping mechanisms of both the electrodes, high voltage support insulators, gas distribution screens, casings, ductings, hoppers, inlet and outlet transitions, dust removal system, corrosion, cleanliness, misalignments, bows and bends, wire snapping, air leakage, etc.'.

[Misalignment of Electrodes — The integrity of discharge and collecting electrodes is an important point to be checked during maintenance. Misalignment and snapping of discharge electrodes brings down the ESP efficiency badly.]

- Electrical inspection which would cover the inspection of: 'Transformer-rectifiers and its control system, rapper control system, rapper motors, coils, vibrators, interlock system, hopper heaters, insulator heaters, insulator compartment and penthouse ventilation system, data management systems, changes in current voltage curves for each electrical bus section, etc. *(Initially the curves are drawn under air load and clean gas load conditions; the changes in these curves would be useful to identify many problems diagnostically such as dust build up/ full hopper/ high resistivity.)*
- Process review which involves review of actual operating conditions, fuel characteristics and fuel sizing range, gas loading, gas temperature, ash resistivity, particle size range, etc. and comparison with the design conditions of the plant. The review would enable the engineers to take steps for modification, if necessary.

4.3.4.2 Operating Steps

Apart from the fly ash resistivity, the following operating factors affect the performance of ESP.

(a) *Temperature of Flue Gas*

The 'Flue gas volume flow' through ESP is 'directly proportional' to the 'Absolute temperature' and therefore the 'Velocity' through ESP shall be lower and 'Treatment time' higher when flue gas temperature in ESP is lower. It is therefore necessary that optimum gas temperature is maintained and is not allowed to rise by taking suitable steps.

The lower temperature shall increase the collection efficiency because of increased 'Treatment time' of flue gas through the ESP and reduced resistivity of fly ash assuming that the temperature is less than 150°C. In view of this, in Japan, the 'Heat exchangers' are being recommended to be installed between air preheater and ESP to lower the flue gas temperature by 30°C to 40°C before ESP inlet.

As already indicated in earlier paragraph, the 'Limit of break down voltage of an ESP' decreases with the 'rise in gas temperature' and increases with the 'decrease in gas temperature'.

(b) *Ash Accumulation*

Ash accumulation in hoppers or on electrodes and in-effective dust removal may lead to even short-circuit of electrodes at high voltage.

A cake of ash on collecting electrodes consists of only partly of dust; the remaining part may be pores and channels / cracks filled with gas. Because of a difference in dielectric permeability of dust and gas, the lines of force concentrate inside the channels/ cracks. When the voltage is high, an electric break down may take place across the dust cake ionizing the gas inside the pores / channels / cracks. The outlet of each crack/ channel acts as a point source on the collecting electrode and is accompanied by the release of positive ions which move counter current to dust particles, partly neutralizing the negative charge of the dust particles and thus adversely affecting the dust collection. This phenomenon is called '*Back corona*'. When the back corona occurs, it becomes necessary to reduce the voltage to prevent breakdowns and the dust collection is seriously affected.

The dust deposition on discharge electrodes affects the operation of ESP adversely. If the dust has a high resistivity (low conductance), the layer of ash on discharge electrode acts as an insulating layer and may cause corona to cease. In case the deposit on discharge electrode has low resistivity (high conductivity), it would be equivalent to increase in diameter of discharge electrode which requires higher corona starting voltage that may not be possible to provide.

Condensation of moisture blocks the ash flow from fly ash hoppers of boilers during evacuation. 'Fluidizing air' is used to aid free flow of fly ash from the hoppers. The 'Valley angle' of fly ash hoppers should not be less than 60° for the same reason. For avoiding condensation of 'Moisture' and consequent caking of fly ash, 'Heaters' in fly ash hopper are incorporated.

It is also important that ash is cleared from electrodes, but without causing significant

'Re-entrainment of dust' in the flue gas stream. The ash should drop out in an 'Agglomerated' condition. The rapping process should, therefore, be optimized and 'Integrated controller' should be provided.

(c) ***Inefficient Combustion / Overloading of Boiler:***

Optimised combustion with optimum quantity of excess-air, also helps to improve the ESP performance. Overloading of boiler on account of outage of HP heater or otherwise also creates problems in ESP. Change of coal characteristics may cause problems for boiler and also for ESP. *[Optimization of ESP performance during operation would be possible after all the above steps have been taken].*

4.3.5 Effect of Electrode Energisation Technique on ESP Performance

For high resistivity fly ash, the collection efficiency of ESP can be improved by:

- Semipulse or intermittent energisation (of ms duration).
- Multi pulse energisation (Pulse is of μs duration)

Since a high voltage can be applied for an extremely short time in the 'pulse energisation system', the corona discharge created from the discharge electrode would be uniform without being affected by the contamination of discharge electrode. The negative ions released (from discharge electrodes side) intermittently are very dense and are uniformly dispensed because of repulsion force and consequently the current density on the collecting electrode is uniform.

In the intermittent / semi pulse energisation technique , a pulsating corona with pulse width of some ms is formed as a result of intermittent charging of electrodes, while in the multi pulse energisation technique, a steep pulse voltage is superimposed on the normal voltage creating pulse of '2 to 100 μs duration'. The above techniques have been experimented with coals having different resistivity by MHI (Japan) and their results show that:

The higher the resistivity of fly ash, better is the improvement in the performance of ESP by the above energisation techniques. For fly ash with resistivity of 5×10^{13} ohm cm, the improvement is two times. For fly ash with resistivity of 2×10^{13} ohm cm, the improvement is only 30 to 40 % and for lower resistivity of 1.0×10^{12} ohm cm, there is no significant improvement.

Relative performance is highest when pulse energisation is applied to the first section. The relative benefit is in declining order for pulse energisation application in subsequent sections. It may, however, be pointed out that significant power saving is achieved by application of pulse energisation technique even for lower resistivity coal. Saving may vary from 65% to 83%.

4.3.6 Specific Collecting Area and Treatment Time / Residence Time

The collection efficiency of ESP improves with the increase in 'Total area of collecting plate-electrodes' provided per Nm^3 of gas flow per sec (called, **'Specific Collecting Area'). Treatment time available for the gas in ESP depends on the flue gas velocity and length of path through ESP comprising number of series fields. The efficiency of collection improves with the increase in number series fields/ treatment time. In coal fired power stations, the ESP usually has 6 series fields.

For boilers burning Indian coal (which mostly has high resistivity), the ESP should have specific collecting area of the order of 280 m^2/Nm^3 of gas flow per sec, with 'Treatment time' / 'Residence time' of more than 40 s, gas velocity of less than 0.75 m / s and 6 or 7 series fields.

The efficiency or 'Collection efficiency' of ESP may be expressed by:

$$\eta = 1 - e^{-wf}$$

Where η is the Collection efficiency of ESP,

w is the velocity of charged particle as it drifts towards the collecting electrode (m/s)

L = length of active part of electric field of ESP (m)

w_g = velocity of gas stream in ESP (m/s)**

d = distance between collecting electrodes and discharge electrodes

b = height of collecting electrode; a = breadth of gas pass as shown

$a \times b$ = cross section of ESP at right angles to gas flow

V = volume of gas

n = number of channels of the active cross section formed by collecting electrodes A, B, C, D

$$w_g = V/a.b = V/n.2.d.b$$

$$wL/d\,w_g = w\,L\,n\,2\,d\,b/V\,d = w\,L\,n\,2\,b/V = wf$$

$$f = L\,n\,2\,b/V = \text{(area of collecting electrodes in m}^2\text{)} / \text{(gas load on ESP in m}^3\text{/s)}$$

f is the specific collecting area of collecting electrodes per m^3/s of flue gas cleaned in the ESP (m^2/ $m^3$of gas /s)

f = (total area of collecting electrodes in m^2) / (gas volume flow in ESP m^3 /s)

Therefore, largher the specific collecting area, higher is the efficiency.

An arrangement of collecting electrodes and discharge electrodes is shown in **Fig. 4.3 (a) and Fig. 4.3 (b)** where in the various quantities in the above equations are shown *[**For the sake of comparison, it is better to express gas flow in normal m^3/s]*

4.3.7 Conditioning Agents for Enhancing ESP Performance

Injection of certain 'Conditioning agents' in flue gas stream can lower the 'Fly ash resistivity' thereby considerably improving the 'Collection efficiency' of ESP. In this process heated sulphur trioxide (SO_3) well above dew point is injected and dispersed into gas stream. The SO_3 combines with moisture and condenses on fly ash particles to reduce ash resistivity.

It is very important that only an optimum dose of SO_3 injection is given so that there is no increase in SO_3 emission, but the risk of higher SO_3 emission would always remain if this method is used. This method is generally, therefore, not used.

Indirect conditioning takes place when coal is blended with high sulphur coal or when there is mixed firing of coal and fuel oil (which normally has higher sulphur content). By this method also, there would be an increased SO_x emission.

Injection of water also has a beneficial effect as it combines with SO_x already present in the flue gases and lowers resistivity. Water injection also gives the benefit of bringing down the gas temperature, but it is also very important that the injection has to be properly controlled as excessive moisture may hinder the free flow of fly ash from hoppers during evacuation.

Injection of ammonia (NH_3) in gas stream reacts with SO_2 existing in the stream to produce bi-sulphates and sulphates which increase the agglomeration of fly ash particles reducing re-entrainment during rapping. The NH_3 injection does not decrease fly ash resistivity but slightly increases ESP efficiency because of reduction of re-entrainment by it. Interestingly, it may be mentioned that injection of ammonia may also reduce oxides of nitrogen (NOx) by Selective Non-Catalytic Reduction (SNCR).

Since the conditioning agents have a recurring cost, an economic evaluation is needed to find out their acceptability for future projects

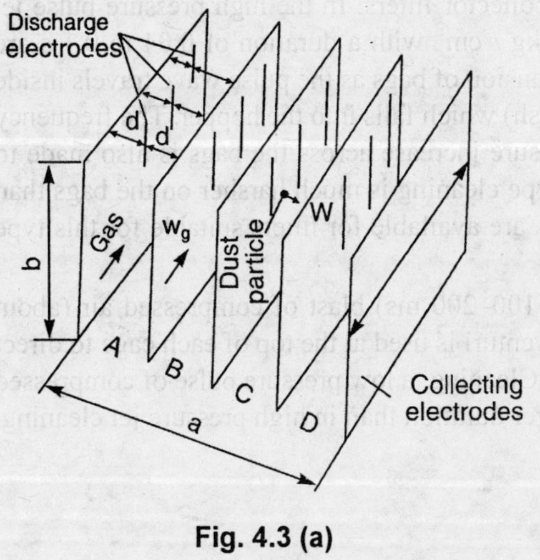

Fig. 4.3 (a)

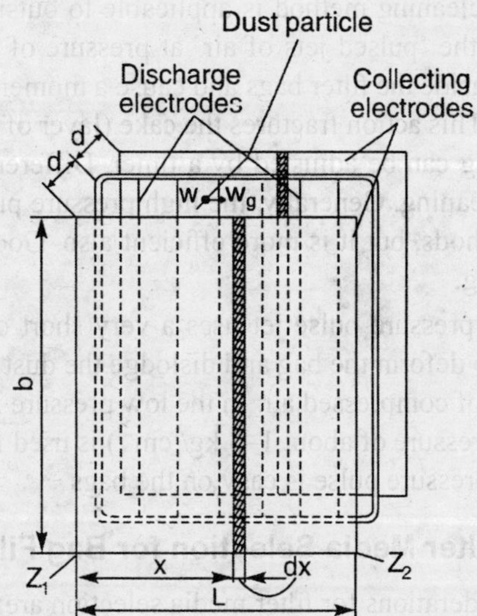

Fig. 4.3 (b)

Height of collecting electrodes = b;
Velocity of charged particle = w
Length of active part of electrical field of ESP = L;
Distance between collecting and discharge electrodes = d
Dust loading at distance x from the inlet (in kg/m^3) = Zx

Figs. 4.3 (a) and (b): Electrostatic Precipitator—Arrangement of Collecting & Discharge Electrodes

Source: Miscellaneous Literature of old Russian Power stations in India

4.4 BAG FILTERS

4.4.1 Dust Filtration System

Bag filter or bag house comprises of an array of fabric filter tubes/bags suspended inside a housing.

According to design, the dirty gases have either to pass from inside of bag to outside or from outside to inside thereby depositing dust on incident face of bag. A general view of a typical sleeve type bag house / bag filter is shown in **Fig. 4.4.**

There can be 'Multi compartment' type bag house which offers more flexibility in operation and maintenance. The fly ash can be periodically removed by one of the following methods:

(1) Rapping; (2) Shaking by mechanical or high frequency acoustic vibration.

(3) Reverse air flow / deflating; (4) Pulse jet of compressed air.

The gases at the inlet strikes baffle plates and larger dust particles fall due to gravity (in the hopper below) on change of momentum before the gases travel through the bags. In case of design where gases travel from outside to inside, the gases after striking the baffle plate encounter the suspended fabric tubular bags and travel through it (from its outside to inside) and emerge into a space from where the clean gases go out.

Pulse Jet Cleaning

Pulse jet cleaning may be of two types, namely 'High pressure pulse jet' or 'Low pressure pulse jet'. Pulse jet cleaning method is applicable to outside dust collector filters. In the high pressure pulse jet cleaning, the 'pulsed jets of air' at pressure of 5.6 to 7 kg / cm^2 with a duration of 0.04 to 0.1 s are injected inside the filter bags and cause a momentary expansion of bags as the pulse wave travels inside the bags. This action fractures the cake (layer of deposit ash) which falls into the hopper. The frequency of cleaning can be adjusted by a timer. Differential pressure increase across the bags is also made to initiate cleaning. Generally, the high pressure pulse jet type cleaning is much harsher on the bags than other methods, but it is more efficient also. Good fabrics are available for filters suitable for this type of cleaning.

High pressure pulse jet uses a very short duration (100–200 ms) blast of compressed air (about 5–7 kg) to deform the bag and dislodge the dust cake. A venturi is used at the top of each cage to direct the pulse of compressed air. In the low pressure Pulse Jet Cleaning, a low pressure pulse of compressed air (at a pressure of about 1–3 kg/ cm^2) is used for a longer duration than in high pressure jet cleaning. The low pressure pulse is easy on the bags

4.4.2 Filter Media Selection for Bag Filters

The considerations for filter media selection are:

- Temperature of gases
- Presence of oil mist and agglomerating materials
- Particle size and abrasiveness
- Chemical properties
- Method of cleaning

Table 4.1: Properties of Various Filter Media *[E - Excellent, G - Good, F - Fair, P - Poor]*

S.No.	Material	Temp. limit	Chemical resistance		Strength		Filteration property
		°C	Acid/Alkali	Moisture	Flexure	Cleaning	
1.	Poly propylene felt	88	E / E	P	E	G	E
2.	Polyester felt	135	G / F	P	E	E	E

Contd...

S.No.	Material	Temp. limit	Chemical resistance		Strength		Filteration property
		°C	Acid/Alkali	Moisture	Flexure	Cleaning	
3.	Acrylic felt	127	G / F	E	G	F	E
4.	Nomex felt	204	F / G	E	G	G	E
5.	P 84 felt	238	E / P	E	E	E	E
6.	Teflon felt	246	E / E	E	F	P	E
7.	Fiber glass woven	260	G / P	E	F	P	P
8.	Ryton felt*	190	E/F	E	E	E	E
9.	Dacron (E > I > DU Point)	135	G/F	P	E	E	E
10.	Cotton	180–200	P/G	P	G	P	F
11.	Ceramic candles	1650–1850	G/G	E	----	-----	E

Note: *As can be seen from the table, filter materials namely Nomex felt, P84 felt, Ryton felt, at Serial Nos. 4, 5 and 8 are very suitable for usual boiler conditions?*

4.4.3 Advantages of Bag Filters and Their Operation Problems

The rise in volume of flue gas over the design gas volume or the rise in particulate loading over the design particulate loading increases the final emission after the dust collectors other than bag house. In case of bag house, even though the pressure drop through filter may change, but the final emission is not affected and it is therefore a constant emission control device. The performance of bag house is not affected by ash resistivity which is an important factor in ESPs.

When a suitable sorbent is injected into the gas stream ahead of filters, the removal of some pollutants like SO_2, Mercury, etc., can be done. However, the commercial viability of this method of removal of pollutants needs to be established for each case.

Operating Problems

The following operating problems are encountered with bag filters:

- Bag blinding may be caused by oil mist, agglomerating materials, moisture, operation below acid dew point, non-effective cleaning, etc.
- The bag house is very sensitive to the increase in acid dew point*. The effect of blinding is the development of high pressure differentials beyond the capability of 'Induced Draft (ID) Fans' and subsequent loss of generation. ***[For acid dew point see note in Para 6.1, Chapter 6 & Para 20.2, Chapter 20]***
- Bag erosion/ abrasion may be caused by high velocity streams of fly ash laden gases especially at the entrance. Damage of fabric of even one bag or any hole caused in the bag would pass a large quantity of untreated gas to stack causing increase in emission.
- One of the most important concerns for bag filters is the cost of replacement of bags which could be about 15 % of total cost of bag house.
- The range of efficiencies, advantages, and disadvantages of various dust collection equipments are mentioned in **Table 4.2** and their other characteristics are mentioned in **Table 4.3. The relative cost of dust collection equipments is shown in Table 4.4.**

Table 4.2: Equipment Selection

Device	Efficiency	Advantages	Disadvantages
Battery cyclones	85 – 90 %	Low capital cost Low operating cost Simple operation	Low efficiency High pressure drop High erosion of cyclone
ESP	More than 99 %	(i) High efficiency for recovery of very fine particles. (ii) Low operating costs. (iii) Low pressure drops (iv) Can handle large volumes of gas well. (v) Has very small treatment time. (vi) Can collect corrosive or abrasive dust. (vii) Easy cleaning (viii) Low power requirement (ix) High overall efficiency	High initial cost and high space requirement for high resistivity of fly ash.
Bag house	More than 99.9%	Can collect fine electrically resistive particles Simple operation High efficiency	(i) High maintenance for fabric bags. (ii) Problems in handling abrasive, corrosive or dust which blind the fabric (iii) Limits imposed by high humidity and high temperature

Table 4.3: Characteristics of Various Dust Collecting Equipments

*[The material used in construction must be suitable for the working temperature range]

Collector type	Applicable Particle size range µm	Pressure drop in water column	Degree of cleaning to be expected	*Max. working temp.°C	Dew point sensitivity	Effect of particle density
Battery cyclones	>10	25 – 75 mm	(i) 80% for particles below 20 µm (ii) 95% for 1 particles above 50 µm	500°C	Critical	Efficiency increase with density
Electrostatic precipitators	> 0.001	6 – 12.5 mm	80% to > 99% on all particles	500°C	Critical	Little effect on efficiency
Bag house	> 0.5 – 1.0	25 – 250 mm	95% to 99% on particles below 5 µm	200 – 250°C	Very critical	No effect on efficiency

Table 4.4: Relative Cost of Dust Collection Equipments

Type	Typical overall efficiency	For 1020000 Nm^3/h gas treatment		
		Capital cost	Operating cost	Pressure drop

Battery cyclone	85 – 90%	2	2	5x
Electrostatic precipitator	above 99%	10	1	x
Bag house	above 99%	7	9	15x
[Number '1' denotes lowest and number '10' denotes highest cost]				

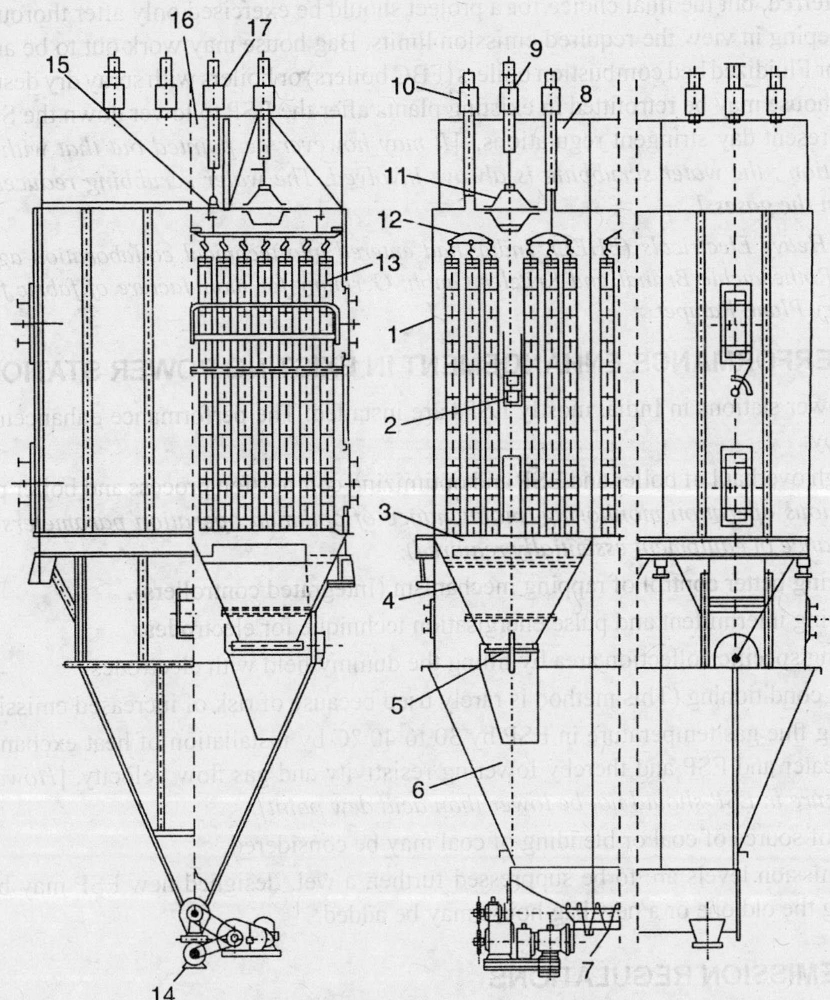

Fig. 4.4: Typical Sleeve Type Bag Filter—General view
Cross Section (a); Longitudinal Section (b)

[Filter casing 1 is divided in several compartments which terminate into hoppers. Gas enters the sleeves via diffuser plate 3. Each compartment has many sleeves 13 which are clamped to to the plate inlet pipes. Platform 2 is between rows of sleeves. Sleeves are suspended from a frame 12 which is carried by rod 10 of shaking pneumatic cylinder. Blowing air is supplied through valve 15 controlled by a pneumatic cylinder 17. Clean gas flows via valve 11 actuated by a pneumatic cylinder 9 to clean gas header 8 The loosened dust from fibre bags passes compartments through shut off valve 5 into dust hopper which is carried by screw conveyer 7 via rotary discharge valve 14]

Source: Miscellaneous Literature of old Russian Power stations in India

New Power Projects— Choice of Type of Dust Collection Equipments

For new stations, the choice is between 'a liberally designed ESP along with better control and energisation techniques 'and 'a bag house'. With experience available within India, ESPs are being generally preferred, but the final choice for a project should be exercised only after thorough economic evaluation keeping in view the required emission limits. Bag house may work out to be an economical proposition for Fluidized bed combustion boilers (FBC boilers) or boilers with spray dry desulphurization process. Bag house may be retrofitted in existing plants after the ESP to lower down the SPM emission to meet the present day stringent regulations. .[*It may however be pointed out that with 'Wet type of Desulphurization', the water scrubbing is always involved. The water scrubbing reduces most of the dust load from the gases.]*

***Note:** Bharat Heavy Electricals (BHEL), India, had entered into technical collaboration agreement with Apparatebau Rothemuehle Brandt and Kritzler Gmbh, Germany for manufacture of fabric filters at their Boiler Auxiliary Plant, Ranipet.*

4.5 ESP PERFORMANCE ENHANCEMENT IN EXISTING POWER STATIONS

In existing power stations in India, mostly ESPs are installed. The performance enhancement of ESPs can be made by:

- Thorough overhaul of boiler and ESP and optimizing combustion process and boiler performance. *(Continuous operation monitoring, maintenance of optimum operation parameters and prompt maintenance of equipment essentially required).*
- Introducing better control of rapping mechanism (Integrated controllers).
- Introducing intermittent and pulse energisation technique for electrodes.
- Increasing specific collection area by filling the dummy field with electrodes.
- Flue gas conditioning (This method is rarely used because of risk of increased emission of SO_3.
- Lowering flue gas temperature in ESP by 30 to 40 °C by installation of heat exchanger between air preheater and ESP and thereby lowering resistivity and gas flow velocity. [*However, the gas temperature in ESP should not be lower than acid dew point].*
- Change of source of coal or blending of coal may be considered.
- If the emission levels are to be suppressed further, a wel. designed new ESP may be retrofitted replacing the old one or a new bag house may be added.

4.6 DUST EMISSION REGULATIONS

Ambient Air Quality Standards in India – SPM limits

Maximum limits for SPM in ambient air as 24 hrs average is 100 μg/Nm^3, 200 μg/Nm^3, 500 μg/Nm^3 respectively for (i) sensitive areas (ii) residential, rural and other areas and (iii) industrial areas. [**Also see *Table 15.11, & Table 15. 12, Chapter 15.]***

SPM Emission limit in India: Accoding to notification Deted 7/12/2015 Ministry of environment forest and Clinlate change. New emission limits for Thermal Power Plants (TPP) have been issued which are given in **Table. 4.5**

Table 4.5: SPM Limits for Thermal Power Plants (TPP) in India (Revised)

	SPM Limits
1. TPP installed before 2003	100 mg/Nm^3
2. TPP installed from Jan. 2003 to Dec. 2016	50 mg/Nm^3
3. TPP installed or to be installed after 1st Jan. 2017	30 mg/Nm^3

World Bank Guidelines Regarding Particulate Matter

- Limits for emission of suspended particulate matter – 500 mg/Nm^3
- Limit for mean 24 hrs. peak for suspended particulate matter in ambient air – 500 µg/Nm^3

However, it is desirable that for all 'New plants or units', the particulate matter emissions (all sizes) should not exceed 50 mg/Nm^3.

Specific attention should be paid to particulates smaller than 10 µm size.(PM10) in the air shed, since these are inhaled into the lungs and are associated with the most serious effects on human health. Wherever possible, ambient levels of fine particulates (less than 2.5 µm in diameter) should also be measured. Recent epidemiologic evidence suggests that much of the health damage caused by exposure to particulates is associated with these fine particles, which penetrate most deeply into the lungs. Emissions of PM10 and fine particulates include ash, soot, and carbon compounds (which often are the results of incomplete combustion), acid condensates, sulphates, and nitrates, as well as lead, cadmium, and other metals. Fine particulates, including sulphates, nitrates, and carbon compounds, are also formed by chemical processes in the atmosphere, but they tend to disperse over the whole air-shed. [In this connection also refer *to **Para. 2.10.2 to Para 2.10.4 of Chapter 2.***]

The New Draft Guidelines of world Bank / IFC ProPosed in May/ June 2017 are as Followrs:-

(i) SPM Limit for Plant Size <600 MWth in non-degraded airshed and degraded airshed—50 and 30 mg/ Nm^3 respectively.

(ii) SPM Limit for Plant Size >600 MWth in non-degraded airshed and degraded airshed—40 and 25 mg/ nm^3 respectively.

REFERENCES

1. Hiroyuki katayama et. al ' ESP technology with new energization system ' MHI technical review, Japan, vol. 32 no. 1 Feb 1995.
2. James Hanly et.al., 'Equipment Selection for solid gas separation', June 1993.
3. K Sampeth, et. al, 'Capture or destroy toxic air pollutants' Chemical engineering June 1993.
4. Kazua Uoya et al., 'Technology on flue gas treatment of fine harmful compounds from municipal incinerators' MHI technical review vol. 29 no. 2 Jan 92.
5. Op Rao, 'Eco friendly practices for coal utilisation', 2nd Int. R&D conf. 97 Vadodra, Central board of irrigation and power(CPIB), New Delhi.
6. Pablo Ruiz, Tarun Gupta, Choong-min Kang, Joy Lawrence, Stephen Ferguson, Milk Wolfson, Annette C. Rohr and Petros Koutrakis, "Development of an Exposure System for the Toxicological Evaluation of Particles Derived from Coal-Fired Power Plants." Inhalation Toxicology, 19(8), 607-619 (2007).
7. Pablo Ruiz, Joy Lawrence, Milk Wolfson, Stephen Ferguson, Tarun Gupta, Choong-min Kang and Petros Koutrakis, "Development and Evaluation of a Photochemical Chamber to examine the

Toxicity of Coal Fired Power Plant Emissions", Inhalation Toxicology, 19 (8), 597-606 (2007).

8. R. Kant, 'Abatement of emission of particulate matter from stacks of coal based power stations', Water & Energy International July – Sept 2000, CBIP, NEW DELHI.
9. Sharma Mukesh and Vyas praveen, (2001) Identification and Quantification of Enviromental Issues of Aging Coal-based Power Plant - A Case Study. Dec. 2001 Vol. 127 No.3 ASCE Journal of Energy Engineering.

QUESTIONS

1. Enumerate the options available for control of emission of particulate matter from boilers.
2. Write short notes on the following types of dust precipitators pointing out their limitations in controlling emission of suspended particulate matter: (a) Multiclones; (b) Electrostatic precipitators.
3. Write short notes on: (a) Fly ash resistivity and its effects on ESP performance; (b) Operating factors affecting the performance of ESP; (c) Conditioning agents for enhancing the ESP performance; (d) Hot side precipitators.
4. Write short note on the importance of 'Specific collecting area' of ESP and 'Residence time/ treatment time' of flue gases in ESP.
5. Describe the importance of semi pulse and multipulse energizing system of ESP.
6. Describe briefly the pulse jet cleaning system. What are the main advantages of using bag filters (bag house) in comparison with other dust removal systems and what may be the probable difficulties?
7. Write short note on Fabric Filters Bag Filters pointing out their operating problems.

ANNEXURE 4

(Source: Author's own Paper—Ref:6)

Annexure Table (i)/ 4: Typical Characteristics of Indian Coal and Ash

Coal Analysis	
GCV	3000 – 4200 kcal / kg
Total moisture	8 – 15%
Ash	38 – 45%
Volatile matter	19 – 27%
Sulphur	0.3 – 0.6%

Annexure Table (ii)/ 4

Ash analysis							
Constituent	%	Constituent	%	Constituent	%	Constituent	%
SO_4	60 – 68	Al_2O_3	18 – 28	Fe_2O_3	4 – 8	TiO_2	1.4 – 1.7
CaO	0.4 – 2.0	MgO	0.4 – 0.9	P_2O_5	0.2 – 0.3	Na_2O	0.1 – 0.2
K_20	0.9 – 1.0	SO_3	0.1 – 0.4				

Annexure Table (iii) /4: Fly Ash Resistivity at Various Temperatures

	Fly ash sample—Anpara A Thermal Power Station, UP, India		**Fly ash sample—Panki Thermal Power Station, Kanpur, UP, India**	
	Tested by BHEL Ranipet		**Tested by BHEL Ranipet**	
	Dew point 40°C		**Dew point 52°C**	
	Temp. °C	**Fly ash resistivity ohm cm**	**Temp. °C**	**Fly ash resistivity ohm cm**
Increasing temperature	120°C	1.95×10^{11}	120°C	$8.05\ 10^{12}$
	150°C	3.90×10^{11}	150°C	1.335×10^{13}
	180°C	1.075×10^{12}	180°C	1.03×10^{13}
	210°C	1.0×10^{12}	210°C	5.95×10^{12}
Decreasing temperature	180°C	2.15×10^{12}	180°C	1.135×10^{13}
	150°C	2.40×10^{11}	150°C	1.55×10^{13}
	120°C	8.4×10^{11}	120°C	9.65×10^{12}

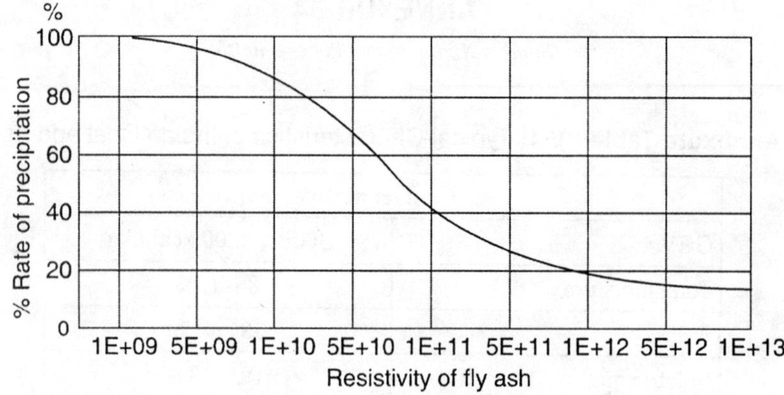

Annexure Fig. (i)/4: Variation of Rate of Precipitation with Fly Ash Resistivity

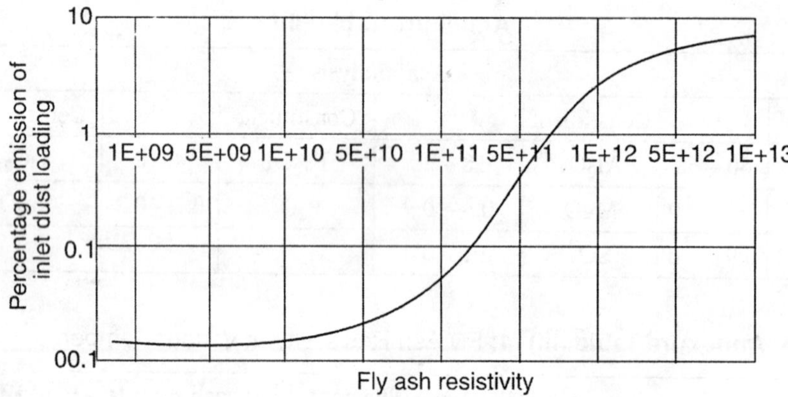

Annexure Fig. (ii)/4: ESP Performance Variation with Fly Ash Resistivity (Source: Author' Own Paper Ref # 6)

CHAPTER 5

CONTROL OF OXIDES OF NITROGEN (NO_X) IN COMBUSTION OF FOSSIL FUELS IN POWER GENERATION AND INDUSTRIES

5.1 NO_x EMISSION AND CONTROL OPTIONS

Oxides of Nitrogen (NO_x) are the principal air pollutants besides Oxides of Sulphur (SO_x) and Suspended Particulate Matter (SPM) emitted by thermal power stations and industries. NO_x are generated during the combustion process of various fuels (coal, oii, gas, lignite/ brown coal, or emulsion, etc.). Exhaust gases from boilers, gas turbines, and Internal Combustion Engines (ICE) all have oxides of nitrogen (NO_x). The contribution of thermal stations and industries in air pollution is quite high, although air pollutants are emitted by transport vehicles and are also generated due to other causes including natural causes. It may be mentioned that about 37 % of oxides of nitrogen present in atmosphere are due to combustion of fossil fuels in fuel burning utilities and partly in transport vehicles.

The ill effects of air pollution by NOx have already been enumerated in ***Para. 2.2.3 Chapter 2. The*** role of NOx in formation of ozone in troposphere and depletion of ozone layer in stratosphere is dealt with in ***Para 2.9 of Chapter 2 and Para.20.1 of Chapter 20.***

Because of rising pollution level in the atmosphere and its evident ill effects, it is essential that NO_x control measures are introduced both in the existing thermal stations and industries as well as newly planned projects. To curb the emission of NO_x, the design of modern boilers may incorporate some of the features like: 'Low NO_x burners'; 'Over fire air'; 'Air staging'; 'Flue gas recirculation'; 'Re-burn technology'; 'Selective Non-Catalytic Reduction (SNCR) of NO_x; etc. to reduce NO_x in the furnace itself. The 'Selective Catalytic Reduction (SCR) system' for NO_x is installed after the exit from furnace. The 'Wet scrubbing' used after the exit of gases from boiler, sometimes in 'Flue Gas De-sulphurisation (FGD)', besides removing Oxides of Sulphur (SO_x) and particulate matter from flue gas also dissolves away other soluble pollutant gases including small part of NO_x. The quantity of NO_x emission also depends on the type of boiler besides other things. The ascending order of emission of Oxides of Nitrogen (NO_x) in different types of boilers under similar conditions would be as follows:

(i) Pressurized Fluidized Bed Combustion (PFBC) boilers;

(ii) Fluidized Bed Combustion (FBC) boilers;

(iii) Travelling grate type boilers,

(iv) Spreader stoker type boilers;

(v) Tangential corner fired PF (Pulverized fuel) boilers;

(vi) Wall Fired PF boilers;

(vii) Cyclone fired wet bottom boilers.

5.2 FORMATION OF OXIDES OF NITROGEN DURING FUEL COMBUSTION

The oxides of nitrogen (NO_x) formed by combustion of fossil fuels in boiler, furnaces, or in other equipments / machines consist of three parts:

(i) ***Thermal NO_x*:** Produced by fixation of the diatomic nitrogen and oxygen of air during combustion process at elevated temperatures.

(ii) ***Fuel NO_x*:** Formed from mono atomic nitrogen released from nitrogen compounds present in fuel during combustion process.

(iii) ***Prompt NO_x*:** Produced by reactions of hydrocarbon radicals ($C_x H_y$) with nitrogen (N_2) of air during combustion to produce hydrogen cyanide (HCN) which in turn reacts with hydrogen (H_2), hydroxyl ion (OH), Oxygen (O_2) to form NO_x.

Both coal and oil have inherent heavy compounds of nitrogen, which account for the generation of significant amount of oxides of nitrogen. Natural gas, however, has little or no mono atomic nitrogen. Fuel NO_x is, therefore, not releasesd from natural gas based stations / facilities / equipments. In case of coal, it would depend on volatile content of coal and its nitrogen content.

Formation of 'thermal NO_x' is largely dependent on temperature during combustion. In a boiler, it can be correlated to heat density existing within the burner zone (from furnace floor to 3 m above top row of burner). The combustion flame radiates heat to all heating surfaces within the zone, i.e., enclosing walls and division walls, horizontal planes at the bottom and top.

In coal fired boilers, the contribution of fuel nitrogen in formation of NO_x is more than in oil fired units as the coal contains about 1.2–1.4 % nitrogen against oil which contains about 0.38-0.44 % nitrogen on ash free moisture free basis. However, only about 15% of fuel nitrogen gets converted to NO_x in coal fired boilers. Nitrogen bound in the volatile portion of the coal is much more readily oxidized to NO_x than the nitrogen bound in the non-volatile portion. About 30% of total fuel nitrogen comes from volatile content of coal. With coal firing, NO_x is produced by both thermal generation as well as by fuel nitrogen. Prompt NO_x due to hydrocarbon reaction with nitrogen of air may be only up to 5% but in case of pulverized fuel firing without any secondary fuel, it may be negligible.

With coal firing, the NO_x formed in the flame envelope is a linear function of 'Oxygen' even in the sub-stoichiometric region. Reduced admission of air at the burner which decreases partial pressure of oxygen in the furnace, therefore, reduces the NO_x generation.

The generation of NOx is dependent upon various factors like fuel nitrogen, burner heat release density, air preheat temperature, equivalent flame temperature, type of fuel, etc. NO_x formed during combustion is mainly NO with a small part (10% or less) of Nitrogen Dioxide (NO_2). [See graphs in **Annexure 5** which are based on test results of *Foster Wheeler Corporation* (1973) *for oil and natural gas firing*]

5.3 MEASURES FOR REDUCTION OF NOx INTERNAL TO BOILERS

There are some technologies which reduce the NO_x in the furnace itself which is designated as '*in-furnace reduction of NO_x*'. There are other technologies which reduce the NO_x after the exit from the furnace. It is called '*post combustion reduction of NO_x*'.

5.3.1 Low NOx Burners, Overfire Air, Air-staging

Low NO_x burners, Overfire air

Special design of fuel burners called Low NO_x Burners (LNB) (schematically shown in **Fig. 5.1**) in the boiler can reduce the generation of NOx during combustion by about 30-50 %. The design differs from manufacturer to manufacturer.

It is based on a fuel rich burning in low excess air atmosphere and lower flame temperatures. The combustion of remnant fuel or CO is completed by 'Over Fire Air (OFA)'. Additional air ports over the 'Over Fire Air inlets' are also provided in some boilers. In order to reduce percentage carbon in fly ash, the pulverization has to be finer. The 'Coal mills' have to be specially designed for finer pulverization and may be fitted with rotary separators. The pulverized coal must at least pass 100% through 100 mesh (149 micron).

By this technology alone, MHI (Japan) has achieved the following results in their 1000 MW Haramachi unit No. 1 [2970 t/h, 260 kg/cm^2 , 570/595°C, fired boiler with coal of GCV 6750 kcal/kg, 14% ash, 35 % VM: NO_x - 150 ppm; Unburnt carbon in fly ash 3%; Boiler efficiency based on LCV 91.81%].

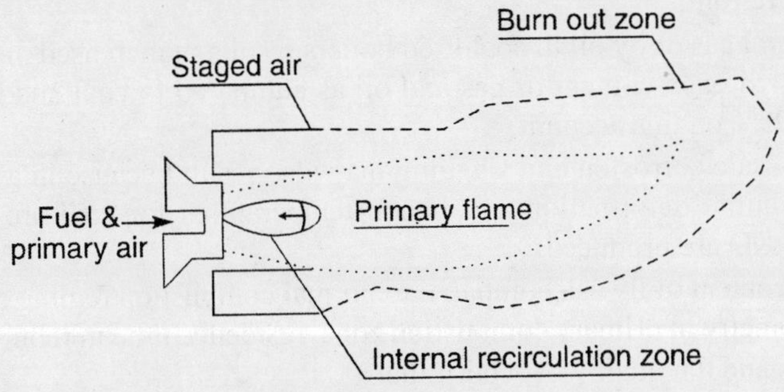

Fig. 5.1: Schematic of Low No_x Burner

Overfire Air, Air-staging

Furnace air- staging provides two-stage combustion. The technology can be applied to both wall-fired boilers and tangentially fired boilers. The main combustion zone is operated preferably in a reducing environment with a stoichiometry typically of 0.9 to 1.10. The over fire air provides air for completing the combustion above the primary combustion zone. In case of new boiler, there is no problem in the design and required residence time for completing the combustion can be provided. But for retrofit the residence time for 'burn out' gets reduced and percent carbon in ash inevitably would increase. The increase in temperature in upper regions (secondary combustion region) creates difficulty in the retrofits although the same can be taken care of in a new boiler in initial design. Schematic diagram of an advanced air staging system is shown in **Fig. 5.2.**

5.3.2 Flue Gas Recirculation

Flue Gas Recirculation (FGR) can be used in case of gas and oil fired boilers for lowering the furnace temperature to suppress the NO_x generation. In case of coal firing, flue gas recirculation is not favoured mainly because it causes severe erosion in the flue duct, dampers, recirculation fan, etc. involved in FGR system.

5.3.3 Re-burn Technology

Re-burn technology is now being employed to suppress NO_X. It is schematically shown in **Fig. 5.3.** Reburning may be with the help of gas, oil or even pulverised coal. This technology has so far not been considered for high ash coal (Indian coal). This technology requires a complete redesign of boiler. In this design, there are three zones of combustion:

(i) 80% of fuel is supplied in the main burning zone. Stoichiometry is kept at 1 to 1.1 (very low excess air).

(ii) 20% of fuel gas, oil or coal is supplied in the reburn zone over the main burning zone. Here the stoichiometry is 0.85–0.95 which provides a reducing atmosphere. Here the hydrocarbons generated by breakdown of reburn fuel react with NO_x from primary zone to form nitrogenous compounds like Hydrogen Cyanide (HCN), etc. The nitrogenous compounds decay to form HNCO, NCO, *and 'NH groups of intermediary compounds.* A 'substantial decrease in emission of NO_x' results in the reburn process.

(iii) Over fire air is supplied over the reburn zone for burn out, i.e., for completing the combustion. The reburn technology requires complete relocation of pressure parts and is, therefore not feasible for a retrofit.

Re-burning by natural gas or by oil in coal fired boilers results in increased moisture content of flue gas because of higher hydrogen content of gas and oil as compared to coal and there is some drop in efficiency say 0.7 to 0.95% on this account.

Increase in flue gas-side corrosion and slagging may also result because in a reducing atmosphere in reburn process, the initial deformation and fusion temperatures for coal are lower and corrosive products like H_2S, FeS, CO are produced.

Increase in carbon content of fly ash compared to normal combustion regime depends upon volatile content of coal, stoichiometry in primary combustion zone, residence time in reburn and burn out zones, pulverised coal fineness and the quality of reburn fuel.

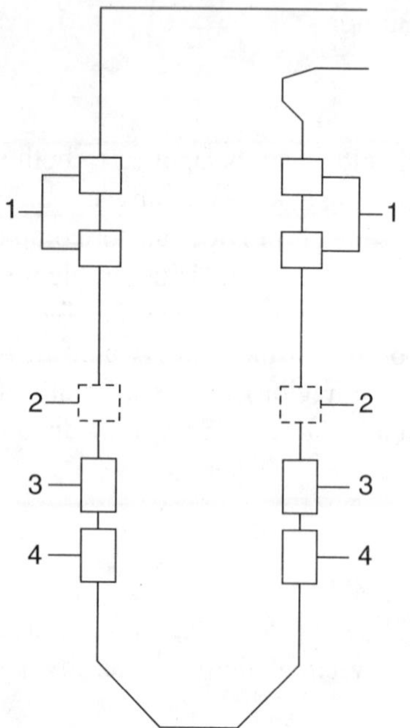

1. Multilevel Separated Over Fire Air
2. Closed Coupled Over Fire Air.
3. & 4. Coal Nozzel Tips With Combustion Air

Fig. 5.2: Typical Schematic of an Advanced Air Staging System

Source: Kant (2001, 2002)

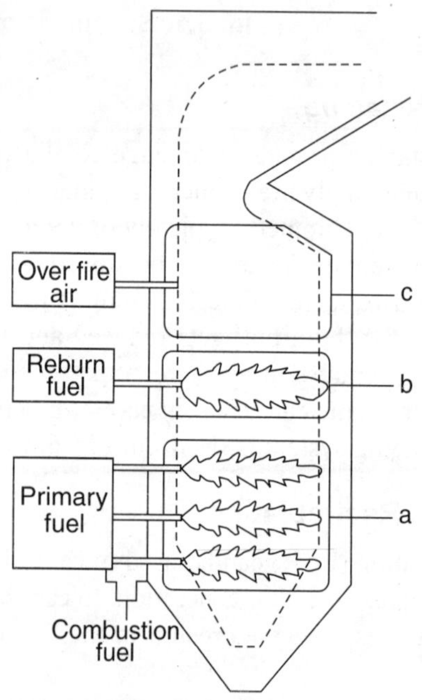

a. Primary Consumption Zone Pleduced Firing Rate Low Excess Air and lower NO_x
b. Reburning Zone Slighlty Fuel Nuck Box reducid to N_2
c. Burn Out Zone (Normal Excess Air)

Fig. 5.3: Typical Schematic of Reburn Process

5.3.4 Selective Non-catalytic Reduction of NO_x (SNCR)

The selective non catalytic reduction of NO_x in the boiler can be achieved by injection of ammonia or chemicals like urea. In absence of catalyst, the reaction takes place in a narrow high temperature range and therefore, it is essential to select appropriate injection points. For ammonia, the optimum temperature range is 950°–1050°C and for urea it is 1000°–1150°C. The NO_x reduction efficiency of the process is low and only about 35–60 % reduction can be achieved by SNCR.

(a) Desirable reaction for optimum temperatures mentioned above is:

$$4NO + 4NH_3 + O_2 = 4N_2 + 6H_2O; 6NO_2 + 8\ NH_3 = 7\ N_2 + 12\ H_2O$$

(b) Undesirable reactions whose pace increases as the temperature increases.

$$4NH_3 + 3O_2 = 2N_2 + 6H_2O \text{ (Combustion of } NH_3)$$

$$4NH_3 + 5O_2 = 4NO + 6H_2O \text{ (Formation of NO from } NH_3)$$

5.3.5 NO_x Reduction by Fluidized Bed Combustion (FBC) Technology

Coal fired boilers based on Fluidized Bed Combustion (FBC) technology generate much less NO_x as compared to pulverized fuel boilers. The 'Atmospheric Fluidised Bed Combustion (AFBC)' and 'Circulating Fluidized Bed Combustion (CFBC)' boilers may have about 225 - 400 mg/Nm^3 NO_x emission. Pressurized FBC combined cycle plants can achieve NO_x emission values less than 140 mg/Nm^3 as demonstrated by Vartan Plant Stockholm (***Ref:11***). These emissions can, however, be lowered by post-combustion NO_x control. ***[For description of FBC Technology see Chapter 6]***

5.4 POST COMBUSTION NOx REDUCTION TECHNOLOGIES:

5.4.1 Selective Catalytic Reduction (SCR) of NO_x

In the selective catalytic reduction, the NOx in flue gases, which is mainly Nitric Oxide (NO) plus Nitrogen Dioxide (NO_2), react with injected ammonia in presence of catalyst to form nitrogen (N_2) and Water Vapour (H_2O).

$4\ NO + 4\ NH_3 + O_2$ (*with catalyst*) $= 4\ N_2 + 6\ H_2O$

$NO + NO_2 + 2\ NH_3$ (*with catalyst*) $= 2\ N_2 + 3\ H_2O$

$6NO_2 + 8\ NH_3$ (*with catalyst*) $= 7\ N_2 + 12\ H_2O$

Reaction stoichiometry with ammonia, injected upstream, depends on the relative amount of each oxide and whether or not oxygen (O_2) is present. Typical aqueous ammonia concentrations range is 19–29 %.

The catalyst* used is titanium oxide (TiO_2), coated on thin plate type elements which are stacked in the reactor for flue gas treatment. Spare space is provided in the reactor for additional catalyst units. A reduction of 70 – 90% of NO_x can be achieved by SCR. The SCR may add about 6% to the cost of electricity. The catalyst, however, deteriorates over a time due to the following:

- Deposit containing high concentration of calcium, which plugs the pores of catalyst surface (in case of coal firing).

- Attack on catalyst by alkaline metals present in fuel oil when oil is fired.
- Heat sintering of titanium oxide (catalyst) in case of gas firing. The sintering causes increase in particle size affecting the catalyst performance.

* **Note:** *The other catalysts besides titanium oxide (T_iO_2) used are: Vanadium pentoxide (V_2O_5), Tungsten oxide (WO_3), Molybdenum oxide (MoO_3), etc.*

The catalysts generally work between 200°C – 400°C temperature range and their efficiency considerably decreases beyond 400°C. The SCR reactors placed after the hot side ESP before air preheater are designed for low dust loading. The SCR reactors designed for high dust loading can be placed after the air heater before ESP. *[The ESP placed between boiler and air heater is called hot side ESP]*

Norton Chemical Process Products, Akron, Ohio has reported that their Zeolite based catalyst assists in NO_x reduction. Well-defined crystalline structure of alumino - silicate zeolite provides uniform pore space and a large internal surface area that promotes the selective reduction of NO_x with ammonia. The proprietary NC-300 catalyst can handle NO_x concentration from 42 ppmv to 30,000 ppmv at inlet temperatures ranging from 220 - 520°C. The material is said to be resistant to abrasion, thermal cycling, and acidic corrosion and poisoning by common flue gas components.

MHI(Japan) developed a low temperature (range 100°C – 150°C) de-NO_x system using active carbon fiber using a supporting metal oxide for de-NO_x activity and regeneration. Below 100°C, de-NO_x activity of catalyst even without supporting metal oxide was found to be 100%. At 100 - 150°C, de-NO_x activity of catalyst with supporting metal oxide was found to be 80%.

As known, ammonia's transportation, storage, and handling is hazardous. Therefore several companies have introduced systems to convert urea to ammonia at the power plant site to eliminate the transport of ammonia. A urea based system, for example, is installed at 2600 MW Govin Power Station of American Electric Power Co. at Columbus (Ohio) to meet the target of 90% elimination of NO_x.

5.4.2 Wet Scrubbing of Flue Gases

The wet scrubbing is not used particularly for NO_x control. The flue gas after ESP if passed through a wet scrubber shall have the advantage of dissolving away a small percentage of the NO_2 content besides other gases like SO_2, SO_3, NH_3, etc. The wet scrubbing is used in the wet flue gas desulphurization system. *[For more details about scrubbing,* ***see Para 8.2.5 of Chapter 8 and Para 7.4.5 of Chapter 7]****.*

5.5 EVALUATION OF NO_X REDUCTION TECHNOLOGIES

Comparison of NO_x emission from different types of boilers (without special NO_x reduction measures) is given in **Table 5.1**

Table 5.1: Comparison of NO_x Emissions by Boilers of Different Types

	Type of boiler	NOx emission
(a)	Travelling Grate boiler Spreader stoker boiler	350–450 mg/m^3 at 7% O_2; 86–172 mg/MJ (0.2–0.4lb/10^6 btu) 450–700 mg / m^3 at 7% O_2
(b)	Pulverised fuel dry bottom boilers	

	Wall fired boiler	1000–1300 mg / m^3 at 6% 0_2
	Tangentially fired Boilers in France	660–1100 mg / m^3 at 6% O_2
	in U.K	650–1200 mg/m^3 at 6% O_2; 215–344 mg /MJ (0.5-0.8 lb/10^6 btu)
	in U.S.A	490–975 mg/m^3 at 6 % O_2
(c)	Pulverised fuel wet bottom type cyclone fired boiler	820–1850 mg-/ m^3 at 6 % O_2 [Medium value 1540 mg/m^3 at 6 % O_2]
(d)	FBC boilers	300 mg / m3 at 6 % O_2 (105 mg / MJ at 6 % O_2)
(e)	PFBC boilers	140 mg / m^3 at 6 % O2 (70 mg / MJ at 6 % O_2)

Actual results of measurements at a power station in Uttar Pradesh for NO_x and SO_x emission from 200 MW tangentially coal fired boilers (grade of coal E) not having any low NO_x 'in-built technology' are given in **Table 5.2,** while the comparative typical cost of various NO_x reduction measures is mentioned in **Table 5.3.**

Table 5.2: Measured Values on 200 MW Units without any NO_x Reduction System

200 MW Unit No.	Unit 9		Unit 10		Unit 12		Unit 13	
Oxygen % V / V	5.0	4.8	5.2	4.6	5.6	5.6	5.4	5.6
NOx as NO_2 mg/Nm^3	875	884	888	879	782	784	836	828
SOx as SO_2 mg/Nm^3	994	1110	1059	975	994	926	906	926

Table 5.3: Comparative Cost of NO_x Reduction Measures

Serial No.	Technology	Differential capital cost in percentage	NOx reduction
1.*	**New boiler**		
(a)	Low NO_x burner corner firing system	3	25–40 %
(b)	Low NO_x corner firing + over fire air	5	30–60 %
(c)	Reburn + Low NO_x burners	10	40–75 %
(d)	S.N.C.R. + Low NO_x burners	9	50–75 %
(e)	S.C.R.	27	80–90 %
(f)	Combined SO_x and NO_x removal	100	80–90 %
2.**	**Retrofitting in boiler**		
(a)	Low NO_x burners. (LNB).	33	
(b)	Over fire air (OFA)	40	
(c)	Other measures	34	
(d)	LNB + OFA. + others	100	Up to 50 %

Notes: ** Assuming that the cost of combined 'NOx and SOx' removal measures in case of New Boiler is 100, the comparative cost of various measures is indicated as a percentage of this against item 1 in column 3*

above.

Assuming that the cost of ' Low NOx burner + Over Fire Air + Other measures' in case of retrofitting is 100, the comparative cost of various measures as a percentage of this in retrofitting is shown against item no 2 in column 3 above.

The increase in carbon content of ash by the use of low NO_x technologies may be about 1.5 – 2 times the normal value of carbon content of ash without low NO_x technologies. It may, however, be noted that the increase in carbon content of fly ash by the use of in-furnace low NO_x technologies mentioned above would depend upon many factors including:

- Volatile content and fineness of coal,
- Stoichiometry in the primary combustion zone,
- Residence time in reburn and burn out zones,
- Quality of reburn fuel, etc.
- In a specific case, it is also affected by factors like furnace turbulence, coal reaction processes such as devolatilization, char oxidation, gas / particle interchange, particle heat convection, radiation, absorption, emission and scattering, gas-particle energy exchange, ash deposition on surfaces, air distribution and furnace geometry.

Introduction of 'Low NO_x burners' and 'Over Fire Air' induce external corrosion on water walls. Deposits rich in FeS decompose to corrosive compounds containing sulphur leading to severe sulphidation of tube alloy. The problem can be overcome to large extent by burner system optimisation and burner settings, along with protection of selected areas by corrosion-resistant coatings, claddings or weld overlays.

Lowering of emission of pollutants has an impact in many areas of power plant. Increase of carbon in ash and external corrosion as a result of low NO_x technology adoption is a challenge. Increase in carbon in ash has a detrimental effect on ESPs performance. Coal milling systems need improvements in order to produce finer pulverised coal for lowering carbon in ash. Accuracy in measurements, uniformity in fuel quality, and optimization of controls has become all the more important.

It may be pointed out that the use of higher steam parameters which increase the thermal efficiency of the station would ultimately result in reduction in emission of NOx and oxides of sulphur (SOx) from the power station, other conditions remaining the same.

In the SCR (selective catalytic reduction) and SNCR (Selective Non Catalytic Reduction) techniques some injected reagent say ammonia (NH_3) escapes even under optimal conditions. This is called ammonia slip (unreacted NH_3) which would mainly deposit on fly ash in the form of solid ammonium compounds. These compounds and fly ash can cause air heater fouling* or even problem of plugging at narrow locations in boiler. Ammonia contamination of fly ash may pose odour and personnel problems during fly ash handling/transport/industrial applications.

[* *If SCR is placed after hot side ESP before a tubular air heater, the tubes of tubular air heater would be affected. In case SCR is located before ESP after 'Air heater', there is no such problem. In case of SNCR, both the boiler tubes and air heater tubes can be affected.]*

5.6 NO_X REDUCTION BY IGCC

In *'Integrated Coal Gasification Combined Cycle (ICGCC)' Plant*, gas from gasifier (which gasifies coal / solid fuel) is cleaned before it enters the coal gas turbine where the only pollutant created during combustion of gas is NO_X. The formation of NO_x is suppressed by injection of steam into the gas turbine combustor where flame temperature is already low due to dilution by CO_2 and N_2. The NO_x emission from ICGCC may be below 50 ppm. [*For ICGCC, see Para 6.4, Chapter 6]*

5.7 NO_X REDUCTION IN COMBINED CYCLE (CC) POWER PLANTS

5.7.1 Selective Catalytic Reduction (SCR) in CC Power Plants

The exhaust gases from the Combined Cycle (CC) power station are passed through a reactor where it is subjected to Selective Catalytic Reduction (SCR) of NO_x by ammonia injection. The reactor is stacked with catalyst coated plates. It is similar to that used for coal based power station.

In case of combined cycle plants, the catalysts working in the range of 200–400°C are used. The SCR reactor for ammonia injection with such a stack of catalyst is positioned in the middle of 'Waste Heat Recovery Steam Generator (WHRSG)' so that the catalysts perform well and reduce / eliminate NO_x.

High Temperature Catalyst

There are high temperature catalysts which have good performance in the range of 500–450°C. The performance of high temperature catalysts deteriorate considerably below 450°C. Such catalyst may be able to withstand temperatures higher than 500°C. These high temperature catalysts can be employed for ammonia injection and in the SCR reactor located at the exhaust end of gas turbine after the 'Silencer' if the exhaust temperatures are within the range. The Exhaust gases first pass through the reactor and then go further.

MHI (Japan) has developed a high temperature catalyst suitable for a temperature of 550°C. This catalyst is very useful for SCR. installation just before the WHRB. There is no evidence of undesirable reaction of conversion of injected NH_3 to NO and NO_2 while using this high temperature catalyst up to 550°C. The high temperature catalyst can be used for open cycle gas turbine plants also.

5.7.2 NO_x Reductions with Various Other Technologies in Gas Turbines

Water or Steam Injection

Injection of Demineralized Water (DM water) or steam into combustion area of combustion turbine (gas turbine) to lower the combustion temperature lowers the level of NO_x emission. The normal NO_x emission from combustion turbine (gas turbine) in open cycle is about 360 ppmv which can be suppressed to a level of about 50 ppmv by DM water / steam injection. The DM water consumption can be estimated to be of the order of 0.6 tonne / hour / MW.

Proprietary Systems for Gas Turbine for Lowering NO_x Emission

(i) ***Dry Low NO_x (DLN) System***

The Dry Low NO_x (DLN) design gas turbine combustor even with distillate fuel may suppress NOx to a level of 90 – 150 ppmv (without water / steam injection). The 'Selected Catalytic Reduction (SCR)', if provided, may reduce the NO_x by 60%. Provision of SCR is relevant only when NO_x suppression to a level 'below 20 ppmv' is required. However, water / steam injection in addition to DLN system can cut the NO_x to a level of 65 ppmv or lower. DLN (dry low NO_x) system developed by GE for their heavy duty gas turbines (combustion turbines) covers hardware features and operational methods to achieve low NO_x, proper burn out of CO, and stability at part load in models like MS6001B MS 7000 B/E, MS 7001 EA, MS 9001 E and later/models FA, EC, G and H.

Without diluent ie water or steam, the above models of combustion turbines achieve the following NO_x and CO levels [O_2 = 15%, ambient 30°F–100°F (0–38°C] for different fuels:

- Natural gas fuel— NO_x level about 25 ppmv or even less, and CO level 15–25 ppmv

- Distillate fuel— NO_x level 100–150 ppmv, and CO level 20–30 ppmv. *[With diluent (water or steam), the distillate fuel fired gas turbine with DLN combustor can achieve a level of NO_x as low as 42-65 ppmv and CO level 20-30 ppmv].*

(ii) ***Dry Low Emission (DLE) Combustor for Small Gas Turbines***

Various manufacturers have their own design of low NO_x combustor. M/s Allied Signal Engines (Arizona, USA) and M/s Optimal Radial Turbine B.V.OPRA (Netherlands) together Developed Dry Low Emission (DLE) combustor for small aeroderivative gas turbines which give very low NO_x (10 ppmv for natural gas and 25 ppmv for liquid fuel). The turbines have high efficiency and a 'Low CO + hydrocarbons' levels (say about 50 ppmv). They operate on the basis of lean premix - prevaporization technique. Excess air is premixed before entering combustion chamber resulting in lower and uniform combustion temperature.

(iii) ***Dry Low Emission (DLE) Combustor with Embedded Catalyst***

In the DLE combustor, 'Lean fuel-air mixture' is burnt in a 'Catalytic embedded combustor'. To initiate the combustion, a portion of fuel is burnt in the presence of embedded catalyst, while the balance is burnt downstream. This process holds the maximum temperature below 2730°F (1499°C) and limits the residence time, thereby reducing the NO_x emission with low dynamic pressure pulsations reportedly up to 5 ppm. Catalytic Combustion Systems Inc., (Mountain View, CA, USA), installed such systems on various makes of gas turbines. They installed the first system at Gianera power station of Silicon Valley Power (Santa Clara, California). This system is called Xonon System and was also tested on GE 9E turbine. **(Ref:18)**

(iv) ***Low NO_x Combustor' with 'Steam Cooling for Walls'***

MHI, Japan uses 'steam cooling for walls' of 'low NO_x combustor' in addition to conventional premixing of fuel and air in the '270 MW gas turbine' to limit the gas temperature from combustor to 1450°C. This limits the NO_x generation during combustion. This technology was used in many units eg. 'Higashi Niigata thermal power station of Tohoku Electric Co. Inc.' and Unit No. 4 (805 MW) using M701 G gas turbine.

(v) ***LEC System***

A Low Emission Combustion (LEC) system was developed by Power system manufacturing LLC, Boca Raton, (FL, US). This system achieved 5 ppm NO_x and 2 ppm CO corrected to 15% oxygen, in a base load GE frame 7EA gas turbine at 50°F ( ) ambient **(Ref:18)**.

5.8 ADVANCED MONITORING AND CONTROL SYSTEMS FOR NO_x

A number of advanced NO_x control systems based on the 'On-line optimization' of plant by means of 'Expert Systems' have been devised and are commercially available to minimize NO_x and achieving a specified target percentage carbon in ash. Some systems worth mentioning are: (i) Generic NO_x Control intelligent system (GNOCIS), (ii) NO_x Adviser; (iii) NO_x Emission advisor and automation system (NO_x EA), (iv) NO_x Smart.

The application of any of the above monitoring systems would depend on the real requirement to justify the expenditure involved. Continuous monitoring of emission of NO_x may be required for future installations in India. However, a regular measurements and monitoring of ground level NO_x in air is a must even at the existing stations. *[For continuous NO_x monitoring systems see **Para 16.10 of chapter 16**]*

REFERENCES

1. Dhar, T.K., "Environmental assessment of technology options for control of stack emission in thermal power plant", 3rd International Conference of Power Development in Afro-Asian

Countries, Kathmandu, Nepal 1996.

2. Environmental News, ISEB, Vol. 4, No. 2, April 1998 and April 1996.
3. Foster Wheeler Corporation, "Nitrogen oxides emission, an evaluation of test data for design", Heat Engineering March-April 1973.
4. Gupta, N.D., "Eco-friendly practices of thermal projects", 3rd International Conference, Afro-Asian Countries, Kathmandu, 1996.
5. Hessel maur, G. et al., 'In furnace NO_x reduction strategies', International Conference Future Strategies and Tech. for Development of Thermal Power,- December 1999, New Delhi, Council of Power Utilities.
6. Kaneko, Shozo et al., "Features and operating results of Tohoku Elec. Power Co. Inc., Haramachi 1000 MW boiler unit No. 1", MHI Tech. Review, Vol. 35, No. 3, October 1998.
7. Kant, R. NO_x Control Thermal Power Houses', Water and Energy International Journal, Oct-Dec 2001.
8. Kant, R., 'Control of Emission of Oxides of Nitrogen (NO_x) from Thermal Power Stations' International Conference on Plant and environmental pollution, ISEB, NBRI, Lucknow, 4/2/2002 to 9/2/2002.
9. Kuroda, Hiroshi et al., "Recent technological development in SO_K and NO_x removal system for flux gas from thermal plants", Hitech Review, Vol. 36, No. 6, December 1987.
10. Lida, Kozo et al., "NO_x removal technology development of high temperature SCR system for gas turbine" MHI Tech. Review, Vol. 30,.No. 1, February 1993.
11. Lofe, J.J. and Par Almhem, "Environmentally sound high efficiency PFBC plants rated to 750 MW", ABB Review 4/94.
12. Mahi, P. of EW bank, Preece Ltd., UK, "Current combustion technologies for design" Heat Engineering, March-April 1973, Their environmental performance - A review.
13. MHI Tech. Review, Vol. 31, No. 2, June 1994.
14. MHI Tech. Review, Vol. 37, No. 3, October 2000.
15. Special report - Distributed Generation, Power USA November-December 1999.
16. Rao, Nagaraja, V.N., CPRI, "Environmental audit thermal power plant - Two case studies'. International Association on Electrical Generation, T&D, Afro-Asian Journal October-December 1993.
17. Rao, O.P., CSIR, "Eco-friendly, practices for coal utilization", 2nd International R&D Conf. 1997, Vadodara, CBIP (India).
18. Special Report, Power, May – June 2002. p 3. & Power Nov – Dec. 1999, p.40.
19. Stull, Roland B., 'Meteorology for Scientists and Engineers', 3rd Edition, 2011, Brooks/Cole
20. Takaishi, Tatsuo et al., "Development of new concept pre-chamber diesel engine for low NO_x emission". MHI Tech. Review, Vol. 32, No. 2 January 1995.
21. Toyada et al., "State-of-art SCR tech and update R&D trend", MHI Tech. Review, Vol. 28, No. 1, 1991.

QUESTIONS

1. Write short notes on the formation of oxides of nitrogen during combustion of fossil fuels in boilers and combustion turbines (gas turbines). Clearly differentiate between Thermal NO_x, Fuel NO_x, and Prompt NO_x.
2. Describe the following measures which can be taken to reduce the production of NO_x in the Pulverized fuel boilers internally and point out the comparative advantages and disadvantages of use of these technologies
   (a) Use of 'Low NO_x burners';
   (b) Over Fire air and air staging;
   (c) Reburn Technology
   (d) Flue gas recirculation
3. How does the NO_x emission affected by use of different types of boilers. How do you rate the various

NO_x reduction technologies?

4. What is 'Selective Non-Catalytic Reduction (SNCR) of NO_x' in boilers?.
5. Describe the 'Selective Catalytic Reduction (SCR) technology for control of NO_x in fossil fuel fired boilers and waste heat recovery boilers.
6. What measures can be taken for control of NO_x emission for internal combustion turbines/ gas turbines. What are the main principles of NO_x reduction involved in them?

ANNEXURE 5

(Source: Ref. # 7 and 8)

RELATION OF NO_X GENERATION WITH BURNER HEAT RELEASE FOR NATURAL GAS AND OIL FIRING

Conditions: Air preheat temp 343°C, excess air

(i) 8% for gas firing

(ii) 25% for oil firing

Curve 1. Gas firing with no flue gas injection

Curve 2. Gas firing with 15% flue gas injection and 2nd stage fining

Curve 3. Oil firing with no flue gas infection

Curve 4. Oil firing with 15% flus gas injection and 2nd stage firing

Base: (a) Burner heat Release of 400 MBTU/hr/ft^2 as 100%

(b) Zero flue gas injection and no second stage firing as 100%

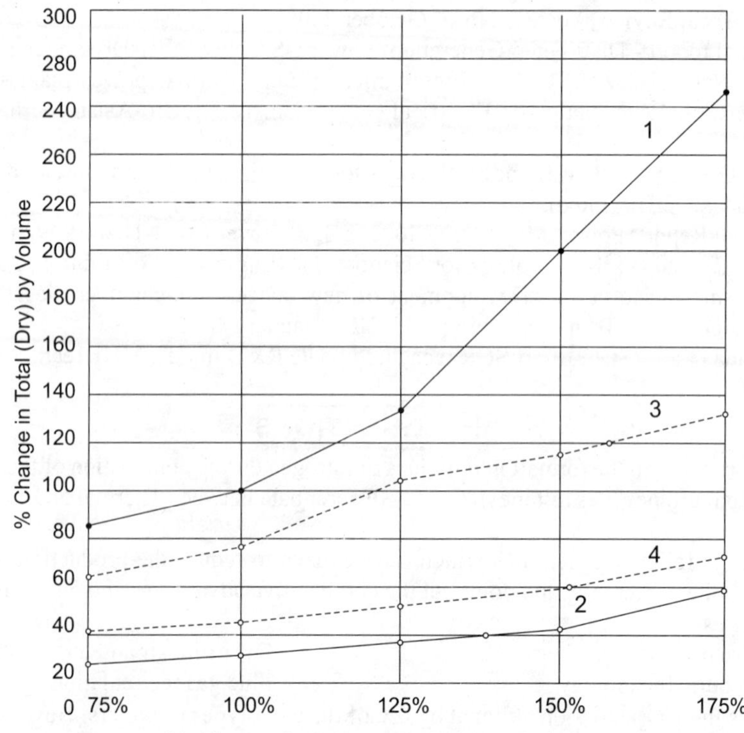

EFECT OF AIR PREHEAT TEMPERATURE ON TOTAL NO_x GENERATION FOR NATURAL GAS FIRING AND OIL FIRING

(Base: Air preheat temp 600°F i.e. 315.5°C)

% Reduction of NO_x

40%
30%
20%
10%
0%

600 550 500 450 400 350 300

Air Preheat Temprature °F

CORRELATION BETWEEN EQUIVALENT TEMPERATURE (i.e. PSEUDO FLAME TEMPERATURE) AND THERMAL NO_x GENERATION AND GAS FIRING

(Base equvalent temp 3860°F i.e. 2120.6°C)

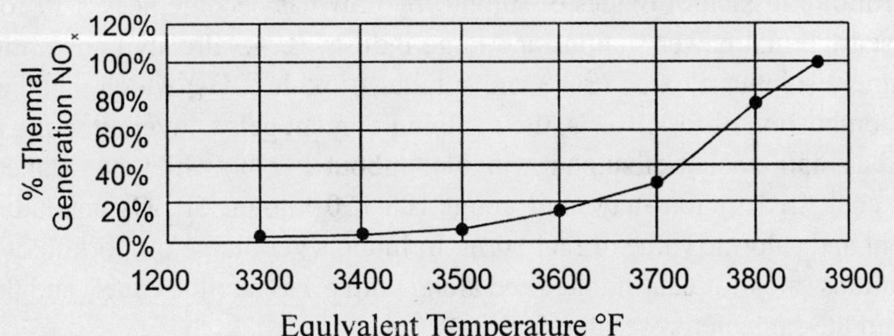

Equlvalent Temperature °F

CORRELATION BETWEEN PERCENTAGE N_2 IN FUEL AND FUEL NO_x FOR OIL AND COAL FIRING

(Base 0.8% N_2 in fuel)

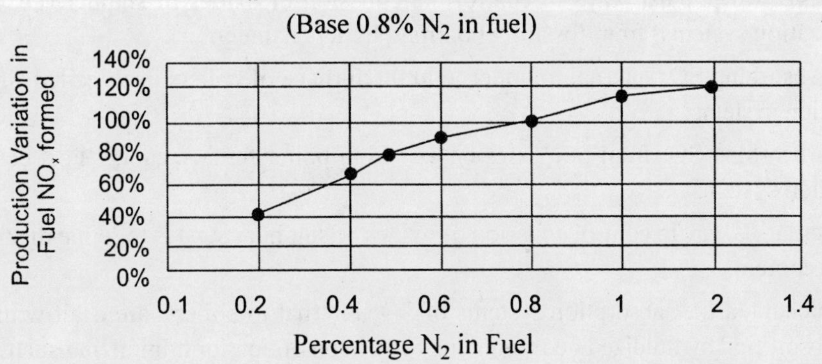

Percentage N_2 in Fuel

CHAPTER 6

CONTROL OF EMISSION OF OXIDES OF SULPHUR (SO_x) BY ABSORPTION SYSTEMS INTERNAL TO BOILERS

6.1 SO_x EMISSION AND OPTIONS FOR IN-FURNACE CONTROL

The pollution of air by emissions of oxides of sulphur from coal/lignite/oil fired boilers in power stations and industries has already assumed menacing scale and is causing much damage to health, plantations, agriculture, properties, metallic parts, etc.

The control of emission of oxides of sulphur by now has become necessary for all solid or liquid fossil fuel based thermal power stations and large industries. As already known, there are two oxides of sulphur namely sulphur dioxide (SO_2) and sulphur trioxide (SO_3) which together are referred to as SO_x. During combustion of fossil fuels, their sulphur content gets converted to mostly sulphur dioxide (90–95%) and a small amount of sulphur trioxide* (about 1–2 %) while the balance is retained in ash. A 2000 MW coal fired station may emit about 100-150 kiloton of SO_2 annually depending on the sulphur content and calorific value of fuel burnt. In India, a very large proportion of power is generated by thermal stations (mostly coal / lignite fired along with oil as auxiliary fuel) and the future scenario is likely to remain the same in foreseeable future.

Although coals in India have low sulphur content, but as their calorific value is low, large quantity of fuel is required to be burnt to produce power and therefore SO_x generated per MWh with Indian coals is also quite significant.

There are two popular ways to control the emission of oxides of sulphur (i.e. De-SO_x system / desulphurization systems) from fossil fuel burning boilers namely:

(i) Absorbing SO_x internal to boiler ie in the furnace of boilers itself called 'In-furnace absorption systems'

(ii) Absorbing SO_x from flue gases on exit from boiler furnace called 'Post combustion absorption systems'

An alternative way to control emission of oxides of sulphur (SO_x) is to refine the fuel itself to lower its sulphur content.

In this chapter, the 'absorption systems of SO_x internal to boilers' are dealt with. In-furnace SO_x and NO_x absorption by fluidized bed technology is a good option for minimizing such air pollution. The control of SO_x emission by in-furnace absorption systems for steam generators is preferable as it takes care of NO_x reduction also and is much cheaper than external flue gas desulphurization (FGD) ie 'Post combustion absorption systems'.

In the 'Fluidized Bed Combustion (FBC)' technology based boilers burning solid fuels, the SO_x is absorbed in the furnace itself with added substantial reduction in NO_x emission. FBC technology is suitable for combustion of various types of solid fuels including 'low grade high ash' coal and any percentage of sulphur content of the fuel.

Atmospheric 'Fluidized Bed Combustion (FBC)' boilers work on the 'Rankine cycle', while the 'Pressurised Fluidized Bed Combustion (PFBC)' boilers work on the 'Combined cycle'. The 'Fluidized Bed Combustion (FBC) Boilers' are of the following types:

- Bubbling Fluidized Bed Combustion Boiler - Atmospheric [BFBC- atmospheric]
- Circulating Fluidized Bed Combustion Boiler- Atmospheric [CFBC- atmospheric]
- Pressurised Fluidized Bed Combustion (PFBC) boiler may be of 'Bubbling type' or 'Circulating type'

***Note:** *The sulphur trioxide* (**SO_3**) *level is reduced during collection of dust by electrostatic precipitator (ESP). The extent of reduction primarily depends on the flue gas temperature and composition of fly ash. The* **SO_3** *reduction may range from 10 to 50 %.*

The SO_3 level also gets reduced in 'Air preheaters' depending upon flue gas temperature and the type of air preheater (i.e. whether the air preheater is tubular or regenerative type). In regenerative type of air preheater, faster the temperature reduction, lower is the air preheater outlet temperature and the greater is the **SO_3** *reduction. The SO_3 reduction rate may be from 10–50 %.*

The reduction, however, is tempered with the risk of formation of 'Sulphuric acid (H_2SO_4) at the lower-temperature- end of air preheater due to condensing of H_2SO_4 because of 'Acid dew point' being reached.*

Acid dew point

The acid dew point is the temperature at which the sulphuric acid condenses. The acid dew point normally depends upon the 'SO_3 content' of flue gases. The higher the 'SO_3 content' in flue gases, higher is the 'Acid dew point' temperature and the condensation of H_2SO_4 takes place at higher temperature causing corrosion in equipments / parts more upstream.

The atmospheric Bubbling Fluidized Bed Combustion (BFBC) and atmospheric Circulating Fluidized Bed Combustion (CFBC) boilers are being increasingly installed even for high ash low grade coal successfully.

Pressurised Fluidized Bed Combustion (PFBC) boilers which work on the combined cycle are also being set-up. CFBC boilers, which have a proven better adaptability for 'low grade Indian coals', are ideal for utilization of washery rejects and middlings at washeries and high ash coal at the pit heads without causing much air pollution problems. Coal gasification based hybrid generation system, like 'Topping cycles' have SOx absorption in coal gasifier and also in constituent CFBC boiler.

NOx emission in FBC technology based boilers is less than one-fourth than that in pulverized fuel boilers because of lower furnace temperatures.

6.2 IN-FURNACE SO_X ABSORPTION BY FLUIDIZED BED COMBUSTION

6.2.1 Principle of In-furnace SOx Absorption

A boiler based on FBC technology essentially has a hot fluidized material bed over which coal/lignite (size 0–6 mm) is fed continuously along with calculated quantity of lime stone or dolomite *(1–3 mm size) in proportion to the sulphur content of the fuel which gets rapidly distributed throughout the bed and is burnt. The initial bed material consists of above size of lime stone or dolomite, but after the start of operation, the bed material has calcium oxide ie lime *[along with some magnesium oxide if dolomite is used]*, ash, silica, sulphites, sulphates and fuel mixed up with lime stone. The combustion of fuel takes place partly in suspension and partly in the bed. The oxides of sulphur generated during combustion of solid fuels and secondary oil react with lime (formed from lime stone) in the bed material and get retained in ash mostly as calcium sulphite (and small quantity of calcium sulphate) which get disposed off mixed in bottom ash and fly ash. No pulverizing of fuel is required in FBC boilers and only good crushing is needed. Crushing is required for lime stone also. A molar ratio Ca:S of about 2 is maintained.

Following reactions with regard to absorption of SO_x take place in the furnace:

$Ca\,CO_3 = CaO + CO_2$; $SO_2 + CaO = Ca\,SO_3$;

$2Ca\,SO_3 + O_2 = 2Ca\,SO_4$ (small quantity); $SO_3 + CaO = Ca\,SO_4$ (small quantity)

Note: *Dolomite is '$MgCO_3$. $CaCO_3$' and when it is heated/ calcinated, calcium oxide (CaO) and magnesium oxide (MgO) are formed. The magnesium oxide (MgO) also acts like calcium oxide (CaO) and absorbs SOx.* ***[$MgO + SO_2 = MgSO_3$]***

6.2.2 Sulphur Capture Efficiency

The 'Sulphur capture efficiency' in FBC boilers depends on following factors:

- Temperature of fluidized bed
- Contact and mixing of gases with sorbent
- Calcium/sulphur molar ratio
- Fuel distribution and fluidization condition of bed
- Fluidizing velocity and gas residence time
- Combustion conditions
- Design of FBC boiler particularly combustor, cyclone, etc.

The 'Sulphur capture' for bituminous coal is optimum between 820° – 870°C temperature, but it falls off abruptly beyond 900°C for boilers with fluidized bed combustion at atmospheric pressure. In FBC boilers, the combustion temperature is, therefore, maintained in the above range.

Dolomite can also be used as sorbent instead of lime, but lime stone is cheaper. In pressurised FBC boilers (PFBC), the sulphur capture efficiency of dolomite is higher than that of lime stone, but in atmospheric FBC, dolomite has less efficiency as sorbent than lime stone.

It may be mentioned that optimum temperature for carbon burn out and sulphur capture for anthracite fuel is about 900°C while that for other fuels like bituminous coal, lignite, petroleum coke, wood waste and 'spent sulphite liquor' is about 850°C.

Lime stone consumption of a 250 MW boiler based on FBC technology using typical high ash low sulphur Indian coal shall be about 53,250 tonnes per year, assuming gross station heat rate of about 2350 kcal/kWh, GCV of coal (as burnt) 3860 kcal/kg, sulphur content 0.56 % and Ca:S molar ratio of 2.0. The calculation for the *Annual requirement of Limestone for SOx Absorption in 250 MW Unit is* shown in **Table 6.1**.

Table 6.1: Calculation of Annual Lime Stone Requirement

S. No	Items	Values
1.	Specific coal consumption	0.6064 kg per kWh
2.	Specific secondary oil consumption (assumed)	1 ml per kWh

S. No	Items	Values
3.	Coal consumption per year for a 250 MW unit at 85 % p.l.f.	1128810 tonne
4.	Secondary oil consumption per year	1730 tonne
5.	Sulphur in coal @ 0.56 % and sulphur in oil @ 4% burnt per year	6390 tonne
6.	SOx produced per year (95 % of S to SO_2 and 2 % of S as SO_3)	12460 tonne
7.	Lime stone (with 75% purity) required per year for Ca : S molar ratio of 2.	53250 tonne

6.3 VARIOUS TYPES OF BOILERS BASED ON FBC TECHNOLOGY

6.3.1 Atmospheric Bubbling Fluidized Bed Combustion Boilers (BFBC)

In this type of boiler, on account of fluidizing air through the bed, very turbulent and rapid mixing of particles occurs. The bed appears diffused and not well defined. Bubbles of gases rise through the bed like that from a boiling liquid and hence the name 'bubbling FBC'. Lower density solids float on the denser solids of the bed and some may be carried away by gas stream. Bubbling bed relies solely upon gravity for return of particles ejected from the bed surface. The boiler works on near about atmospheric pressure. Furnace of BFBC boiler therefore requires a larger plan area and overall height as compared to Circulating Fluidized Bed Combustion (CFBC) boiler to prevent carbon loss due to carry over.

In this boiler about 70 to 80 percent oxides of sulphur generated during combustion are absorbed by the lime present at a 'Ca:S' molar ratio of 2 to 2.3. Smaller installations up to 130 MW unit size with atmospheric FBC type boilers are commonly constructed.

Difficulties with atmospheric FBC boilers particularly with bigger sizes were experienced with regard to:

- Maintaining a uniform distribution of fuel over the bed,
- Dampening of fluidization called 'Slumping' at some locations on the bed causing local hot spots with poor combustion,
- Corrosion of tubes mounted on the fluidized bed.

An example of overcoming the above problems was set by 'EPDC Ltd. (Japan)' by construction and commissioning of a 350 MW bubbling bed FBC boiler 'which was a retrofit of an oil fired unit at Takehara power station', Japan in 1995. This boiler had twin furnace — one main bed cell and other carbon burn up cell which got return of carbon grit from cyclone provided at the gas outlet from the main bed cell furnace. Very low 'emission level' of SO_x and NO_x was achieved in this boiler.

6.3.2 Atmospheric Circulating Fluidized Bed Combustion Boilers

The Circulating Fluidized Bed Combustion (CFBC or simply CFB) boiler was developed to overcome the difficulties and limitations with simple bubbling type FBC boiler. The bed in CFBC boilers is fully fluidized and permits the escape of some solid particles from combustion chamber with flue gases. The coarser material consisting of unburnt fuel, lime stone, etc., carried by gases are, however, collected in the cyclones which are provided between combustion chamber and second pass. The material collected in the cyclones is re-injected into the furnace to check unburnt carbon loss and utilize the lime better. The furnace is of water wall construction with refractory lined lower portion. The superheaters, reheater and economizer are installed in the second pass. No tubes are located in the actual bed.

A simplified flow diagram of CFBC system is shown in **Fig. 6.1.** Primary air is injected at the bottom of furnace and secondary air is introduced at a higher level thus leading to staged combustion and lower furnace temperature of about 850° – 900°C, resulting in lower NO_x level. Because of recirculation of coarse solids (including coal grit and lime/ lime stone) from the cyclones, there is better utilization of fuel and sorbent material, and therefore better efficiency of boiler and also of SO_x removal (DSO_X). About 90% SO_x is absorbed at a Ca : S molar ratio of 2 to 2.3. This type of boiler also works at near about atmospheric pressure.

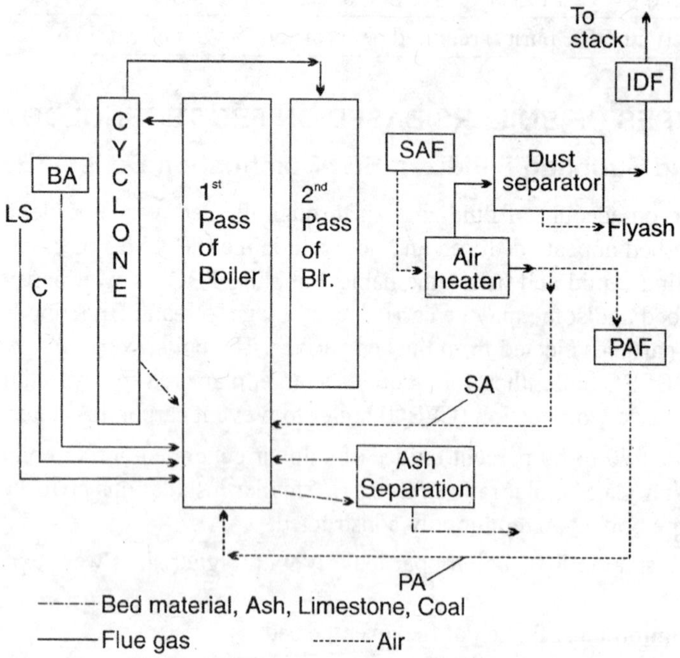

PA–primary air; PAF–primary air fan; SA–secondary air; SAF–secondary air fan IDF – Induced draft fan; C– coal; LS– limestone; BA– bed ash

Fig. 6.1: Circulating Fluidized Bed Combustion Boiler Flow Diagram

CFBC technology has matured and many plants have been constructed worldwide. These boilers can be designed to burn any type of solid fuel, (lignite, anthracite, coal washery middlings and washery rejects, low grade high ash coal, petroleum coke, etc.) with any percentage content of sulphur without causing concern about SO_x pollution. First CFBC boiler of 250 MWe size was commissioned in Gardane, France in 1996. It was a coal fired boiler based on 'LURGI' design and was supplied by GEC Alsthom for EDF (France). It may be noted that BHEL, India, had entered into technical collaboration with M/s Lurgi Babcock Energy Technik Germany for CFBC boilers.

The largest CFBC unit in Operation is the 460 MWE bituminous coal fired boiler at Lagisza Power Plant, Poland, which uses Foster wheeler's once- Through Super Critical (SC) boiler design (2009). Plants of bigger size are being designed and developed.

Earlier design of CFBC boiler for high ash coal suffered mainly from erosion of pressure parts, blocking and erosion of cyclones. These problems were minimized by taking into consideration the result of field trials and properties of particular coal in detail during design.

Due to the advantages of CFBC boilers, several old medium capacity Pulverized Fuel (P.F.) boilers are being converted CFBC boilers. This serves double purposes of dealing with deteriorating fuel quality and drastically reducing SO_x and NO_x emission. Conversions of pulverized fuel boiler to CFBC do not offer insurmountable difficulties. The combustion efficiency and height of 'pulverized fuel boiler' and 'CFBC boiler' are almost similar for the same energy input. Forced draft fan is replaced by 'High pressure blowers', and 'Coal pulverizers / coal mills' are replaced by less energy intensive 'Coal crushers', during conversion to CFBC type boiler.

6.3.3 Pressurized Fluidized Bed Combustion (PFBC) Boilers

Pressurized Fluidized Bed Combustion (PFBC) boilers may be either bubbling type or circulating type and are used in a combined cycle as shown in the simplified flow diagram **Fig. 6.2 and Fig. 6.3.** The flue gases from boiler at a pressure of 10 – 16 bars (or even more) and temperature of about 830°C pass through the dust removal system and are then led to expand through a gas turbine which drives an electric generator and a two stage air compressor to provide high pressure air to the boiler. The exhaust from the gas turbine goes to a heat exchanger/economizer where it heats up feed water supplied to PFBC boiler that generates steam to drive steam turbine generator set.

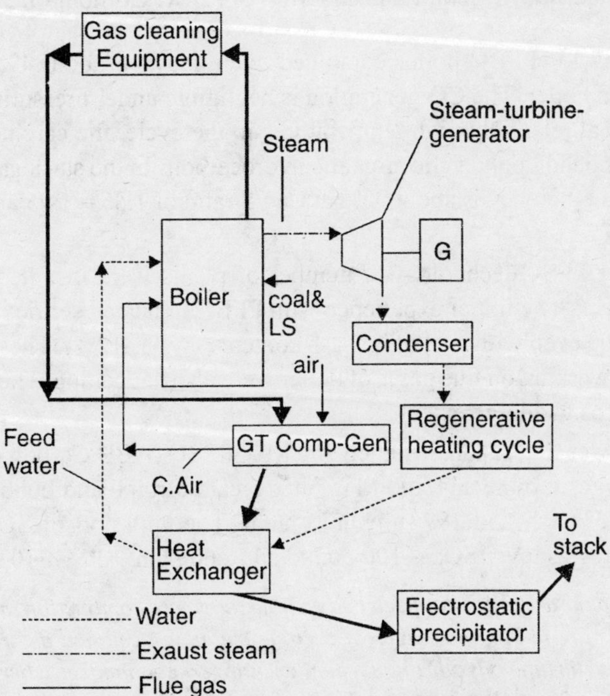

***Legend:** GT Comp-Gen–Gas turbine driven compressor and generator C.Air–Compressed Air; BF–Booster fan, G–Generator, Ls–lime stone*

Fig. 6.2: Simplified Diagram for Pressurised Fluidized Bed Combined Cycle (With PFBC Bubbling Type Boiler)

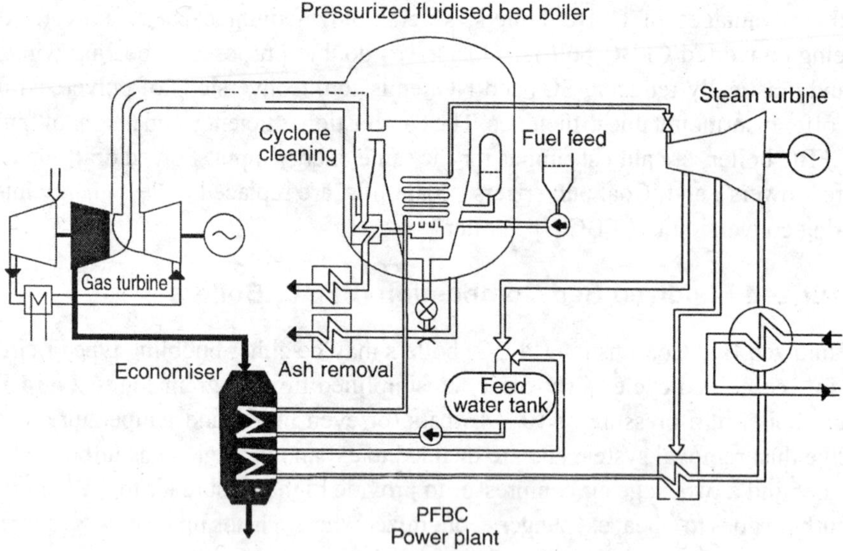

Fig. 6.3: Alternative Schematic Diagram of PFBC Combined Cycle Plant

The cycle efficiency for PFBC boiler combined cycle may be about 4% more than the simple Rankine cycle for CFBC boiler. The CO generation is negligible under pressurised combustion. Due to the long residence time at relatively high temperatures in the cycle, the carcinogenic compounds like polycyclic organic compound / polycyclic aromatic hydrocarbons in the stack gases are below detection levels. The SO_x removal efficiency is about 90% at Ca : S ratio of 1.85 – 1.95 and NO_x emission is also highly subdued.

For development of PFBC technology, a number of plants were installed mostly with bubbling type boiler to get feedback. The initial experience with PFBC indicated serious problems of water wall erosion, cyclone plugging even with coal of low ash content (say 11–12%). The occurrences of cracking of blades of gas turbine working on flue gases of boiler were also happening. These problems were taken care of in the later designs to a large extent.

Kyushu Electric Power Co., Japan, installed 350 MW PFBC (ABB Carbon design) combined cycle unit with 70 MW rated gas turbine and 290 MW rated steam turbine and bubbling bed boiler burning high grade coal with 15% ash and 1% sulphur content. The emission of SO_x, NO_x and suspended particulate matter (SPM) was low. [SOx—100 ppm; NO_x—60 ppm; SPM—10 mg/Nm^3].

[Coal and sorbent for sulphur are fed to fluidized bed combustor where combustion takes place at about 850°C from where the flue gases are led to two stages of cyclone cleaning designed as protection for gas turbine. The exhaust gases from the turbine pass through a heat exchanger/ economizer where gas temperature drops. The gases then are led to ESP and then to stack. The feed water is heated in the heat exchanger and then led to boiler. The steam output from boiler may be at about 180 bar, 540/540 °C which is sent to steam turbine-generator unit.]

A 250 MW class PFBC unit was developed by Hitachi Ltd with bubbling type once through boiler. The circulating type PFBC boilers have also been developed. There are many difficulties in developing a satisfactory design of PFBC unit for high ash coal as available in India, although efforts are being made.

6.4 COAL GASIFICATION BASED IN-FURNACE SO_X REMOVAL SYSTEMS

6.4.1 General Principles of Coal Gasification

Pressurized gasifiers are of interest for combined cycle plants as these operate at 20–30 bars pressure. The gasification process involves the following chemical reactions:

(a) Partial oxidation of carbon also generating some heat: $C + O = CO$ (exothermic)

(b) Carbon and steam reaction: $C + H_2O = CO + H_2$ (endothermic)

(c) Shift reactions in some percentage: $CO + H_2O = CO_2 + H_2$ (exothermic)

(d) Some hydrogen may get combusted with some oxygen present
$2 H_2 + O_2 = 2H_2O$ (exothermic)

(e) A small amount of coal may get fully oxidized to form CO_2 generating heat:
$C + O_2 = CO_2$ (exothermic)

(f) Reaction with sulphur in coal: $2S + H_2 + CO = H_2S + COS$ (Carbonyl Sulphide)

The oxygen blown gasifiers and air blown gasifiers have different compositions of 'Coal gas' as shown below:

(i) Oxygen blown gasifier: 60% CO, 30% H_2, 3% CO_2, 2% N_2, 0.5% CH_4, etc.

(ii) Air blown gasifier: 24% CO, 14.5% H_2, 5.5% H_2O, 1.4% CH_4, 48.5% N_2, 0.6% Ar, etc.

Normally coal gasifier uses external SO_x removal system as employed in Integrated Coal Gasification Combined Cycle (IGCC)*. Specific hybrid combined cycles like 'Topping cycle' have been developed with gasification of a part of solid fuels and burning the balance fuel in a Circulating Fluidized Bed Combustion (CFBC) boiler. The sulphur is removed by lime stone both in gasifier as well as in CFBC boiler. A major portion of hydrogen sulphide (H_2S) and carbonyl sulphide (COS) generated in the gasifier is internally absorbed by lime forming calcium sulphide (CaS) in the gasifier itself. The balance sulphur compounds are removed in CFBC boiler of 'Topping cycle'.

Note: *Integrated Gasification Combined Cycles (IGCC)

In this system, the coal is gasified in a pressurised gasifier and gas is led to gas cleaning unit. Sulphur is converted to hydrogen sulphide (H_2S) in the gasifier. Some of this H_2S is absorbed by lime in the gasifier while the balance is removed by special solvents which can be regenerated and the entrained acid gases are led to sulphur recovery unit. The cleaned gas is led to gas turbine driving air compressor and AC generator. Steam is generated in the 'gasifier cooler' which is used for cooling of 'gas from gasifier', and steam is also generated in the 'Waste heat recovery boiler' which utilizes the heat of 'gas turbine exhaust gases'. Both of these components of steam are led to steam turbine generator unit. A typical scheme is shown in **Fig. 6.4.**

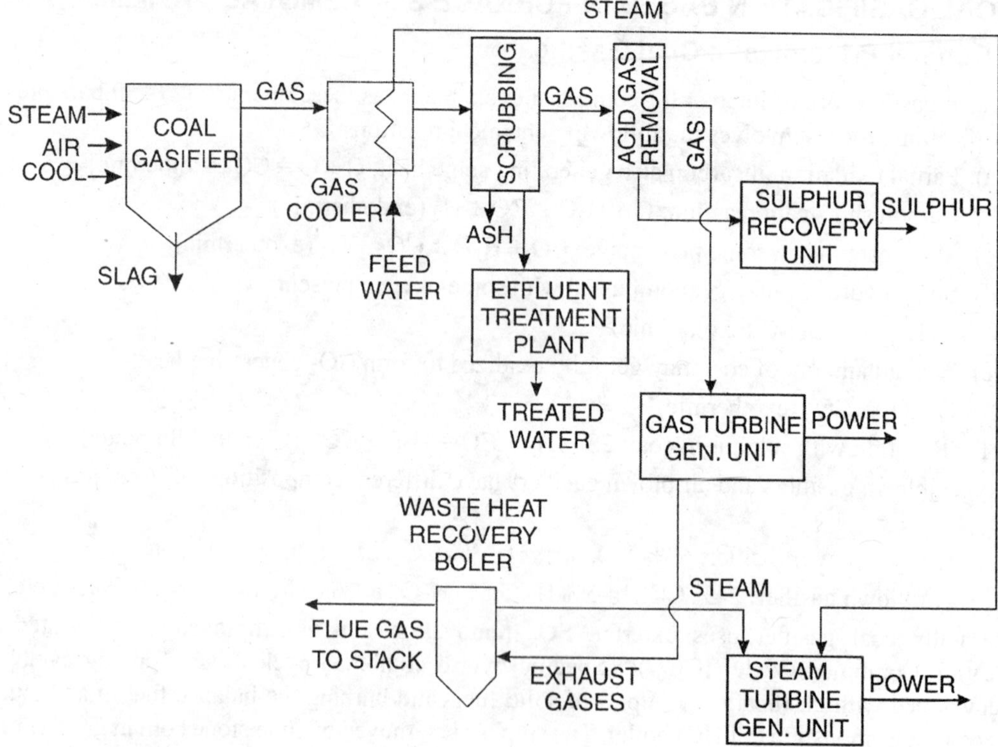

Fig. 6.4: Integrated Gasification Combined Cycle Scheme (***Ref:2***)

6.4.2 Topping Cycle

It is an advanced combustion cycle combining 'Coal gasification', 'Circulating FBC boiler', 'Gas turbine generator set driven by clean coal gas', and 'Steam turbine generator set driven by steam from CFBC', heat exchangers in the coal gasification circuits and waste heat recovery boiler of gas turbine exhaust. Typically 70% of coal is converted to coal gas at a pressure of about 15–20 kg/ and a temperature of about 1000°C, and is led to a gas cleaning unit where SO_x is absorbed by injection of lime and particulates are removed by cyclones and ceramic candle filters. The clean gas is led to gas turbine driving air compressor and AC generator. The remaining 30% coal along with char from gasifier is used in Circulating Fluidized Bed Combustion (CFBC) boiler to raise the steam.

Lime stone/lime is fed to the gasifier for sulphur removal as in CFBC boiler for removal of oxides of sulphur. The SO_x removal (DSO_x) efficiency of topping cycles is more than 90%. Various topping cycles have been developed such as British coal topping cycle, ABB Carbon process, Stien Muller Vew process, Foster Wheeler process, and Rheinbraun high temp Winkler process. A concept of topping cycle is shown in **Fig. 6.5.**

The exhaust gases from the gas turbine can also be led to a Waste Heat Recovery Boiler (WHRB) and the steam from WHRB can be used in the cycle. Alternatively, the exhaust from gas turbine can be led to the CFBC boiler already in the cycle, but in that case CFBC boiler design has to take this into account.

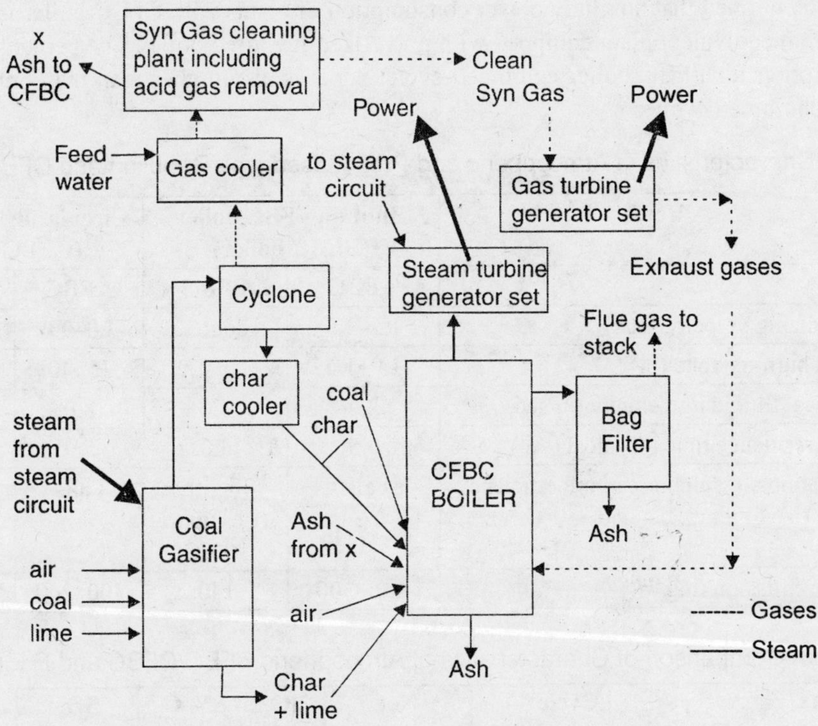

Fig. 6.5: Topping Cycle Scheme

6.5 LIME STONE FURNACE INJECTION FOR PULVERIZED FUEL BOILERS

Injection of powdered lime stone in the upper part of furnace of a Pulverized Fuel (P.F.) boiler absorbs the oxides of sulphur to some extent. This type of in-furnace D-SOx system has inherent limitation for P.F. boilers as sulphur capture efficiency is optimum at temperatures 800°-950°C while in bituminous coal firing, the temperature goes beyond 1200 °C which may cause sintering of sorbent thereby reducing its activity seriously. Also, it may be mentioned that reverse reaction, i.e., decomposition of calcium sulphite and calcium sulphate to release SO_2 and SO_3 takes place at this temperature. The SO_x removal efficiency of furnace injection of lime stone in P.F. boilers is hardly 30 % at Ca:S ratio of 2 and sorbent fineness of < 100. With some coals this may result in slagging.

The chemical reactions in the furnace injection of lime stone would be:

$CaCO_3 = CaO + CO_2$.; $CaO + SO_2 + 1/2\ O_2 = CaSO_3$

Furnace lime stone injection followed by an external spray dryer desulphurization can remove about 80 to 90 % SO_x.

6.6 EVALUATION OF VARIOUS 'IN-FURNACE DESULPHURIZATION SYSTEMS'

A comparison of characteristics of BFBC and CFBC boilers of both atmospheric as well as pressurised type is shown in **Table 6.2**. A comparison of atmospheric BFBC and atmospheric CFBC boilers with pulverised fuel boiler is given in **Table 6.3**. Environmental performance of various in-furnace desulphurization systems (D-SO_x) systems is given in **Table 6.4**. The percentage of power generated by gas turbines in the PFBC boiler combined cycle system is approximately 20%.

It may be mentioned that auxiliary power consumption of plants with CFBC boiler is about 1.5 % higher than for an equivalent plant equipped with pulverized fuel fired boiler. The percentage auxiliary power consumption for PFBC boiler combined cycle plants is also higher than the conventional gas turbine based combined cycles.

Table 6.2: Characteristics of Atmospheric and Pressurised type Bubbling and CFBC Boilers

S. No.	Items	Bubbling FBC boiler (BFBC boiler)		Circulating FBC boiler (CFBC boiler)	
		AFBC	PFBC	AFBC	PFBC
1	Fluidizing velocity (m/s)	1.3–3.5	About 1.0	4.0–9.0	About 5
2	Fuel burning zone (m)	1.0–1.4	3.5–4.0	15–40	15–40
3	Typical fluidizing velocity (m/sec)	2.0	1.0	8.0	5.0
4	Bed residence time - Typical (sec)	0.5	4.0	5.0	8
5	Variations of sulphur capture efficiency with rise of temp. beyond 850°C	Falls	Slightly rises up to 900°C	Falls	Slightly rises up to 900°C
6	NOx emission (mg/m^3)	140–300	140	100–200	100

Table 6.3: Comparison of Characteristics of Atmospheric BFBC/CFBC and P F boilers

S. No.	Systems	Cycle	Net efficiency %	NO_x at 6% O_2 limited to mg per m^3	SO_x removal %	Ca: S Mol ratio
1	FBC (Bubbling)	Rankine	35–37	140–300	70–80	2.0–2.3
2	CFBC (Circulating)	Rankine	34–36	100–200	90	2.0–2.3
3	PFBC (Bubbling)	Combined	39–41	140	90	1.85–1.95
4	PFBC (Circulating)	Combined	38–40	100	90	1.85–2.0
5	Topping cycle anticipated	Combined	41 approx.	100	>90	2.0–2.3

Table 6.4: Environmental Performance

SI. No.	Items	BFBC boiler	CFBC boiler	P.F. boiler*
1	Coal particle size (mm)	0–6	0–6	0.074 (200 mesh)
2	Mean particle size (mm)	0.03–3.0	0.05–0.5	0.02–0.08
3	Excess air (%)	20–25	10–20	15–30
4	Furnace temperature (°C)	830–870	830–870	1100–1300
5	Turn down (ratio)	3:1	5:1	2.5:1
6	Combustion efficiency (%)	90–96	95–99	98–99
7	SOx emission problem	Eliminated	Eliminated	No control
8	NOx emission mg/m^3	140–300	100–200	500–1200
9	Slagging problems	Almost nil	nil	Yes

**PF boiler*—Pulverised Fuel Boiler

It may be stated that 'In-furnace SO_x absorption system' is a promising option as it has added advantage of NO_x reduction. Popularization of fluidized combustion technology in India would depend upon actual field trials with low grade, high ash, Indian coals on various types of FBC boilers, their feedback and consequent design changes particularly to eliminate tube erosion, cyclone plugging, bed slumping, excessive loss of unburnt carbon, etc. It would also depend upon developments in the direction of increase of overall boiler availability and efficiency. However, the experience on smaller boilers indicate that Circulating Fluidized Bed Combustion (CFBC) boilers have a better adaptability with Indian coals for larger installations than other types of FBC boilers. These CFBC types of boilers may be ideal for utilizing washery rejects and middling at washeries and high ash coal at the pithead power stations without causing emission concerns.

From the point of view of control of pollution, FBC technology based boilers are to be preferred irrespective of coal quality for installation in industrial areas, sensitive areas and urban centres near large cities. CFBC boilers in the range 110–250 MW and BFBC boilers in the capacity up to 100 MW should be considered. The range may expand in future.

As regards PFBC technology and 'Topping cycle technology' although promising, but it needs development of satisfactory and economical design specifically for Indian high ash coals. A comparison of the installed costs of power plants with in-furnace D-SO_x system is given in **Table 6.5.**

Table 6.5: Comparative Installed Costs of Power Plants with in-Furnace D-SO_x System

Pulverized fuel boiler with D-NO_x and D-SO_x in Rankine cycle	**FBC Rankine cycle**	**CFBC Rankine cycle**	**PFBC (Bubbling) Combined cycle**	**PFBC (Circulating) Combined cycle**	**Topping cycle**
100 % (Base)	88%	90%	110%	114%	140%

REFERENCES

1. I.P. Research Digest, CBIP India, September-December 1995.
2. Kant, Rajni, 'In furnace SOx Absorption Systems for steam generators', Water and Power International Journal, Vol.60, No. 4, Oct-Dec. 2003.
3. Kant, Rajni, 'control of SOx Emission and clean Technologies for coal based thermal power plants', Journal of international association on electric Gen., T & D, Afro- Asian region, 'ARRO' Oct- Dec. 1993.
4. Kant, Rajni, 'NO_x Control - Thermal power houses'. Water & Energy International, Vol. 58, No. 4, Oct.-Dec. 2001.
5. Lofe, J.J. et al., 'Environmentally sound high efficiency PFBC plants rated to 750 MW', ABB Review, 4/1994.
6. Misra, Robindra et al.,'Indal Matures with CBC boilers', Int. Conf. Council of Power Utilities. New Delhi, December 1999.
7. Nanda Kumar, K. et al. - (BHEL), 'Fossil fuel based thermal power generation for coming decade - A perspective', Int. Conf. Council of Power Utilities, New Delhi, December 1999.
8. Nettel, M.D., EW Bank Preece Ltd. U.K., 'Types of advanced power generation systems'.
9. Schilling, H.D., 'Structural changes in power plant tech. of FR. Germany'.
10. Sharma, M., and Gosh, U., 'Assesment of Sulphur dioxide Emission Impact of 284 and 135 MW Power Stations', International Conference on Enviromental Impact Analysis for Developing Countries, New Delhi, India, Nov. 28 – Dec. 2, 1988.

11. Special Number of Indian Journal of Power & River Valley Development, Nov 1990. p.376.
12. Swanekamp, Robert, 'Turow furnishes model for Central European development', Elec. Power Inter., Vol. 9, Nos. 3 fall 1999.
13. Taneja, Akash, 'Clean env. tech. in power sector IGCC - A case study'. Water & Energy International, Vol., 58, No. 2, June 2001.
14. Vegetational surveys and studies carried out by National Botanical Research Institute, Lucknow around thermal power station (**References given, in Para. 2.3.2 Chapter 2**).

QUESTIONS

1. Describe the principles of Fluidized Bed Combustion (FBC) and how does it reduce the emission of NO_x and SO_x.
2. Write short notes on (a) Bubbling type FBC boiler (atmospheric) (b) Circulating type FBC boiler (atmospheric).
3. Write short note on 'Pressurized fluidized combustion boiler (PFBC boiler)'.
4. Enumerate the factors on which the sulphur capture efficiency depends in a fluidized bed combustion boiler.
5. Write a short note on coal gasification based power generation commenting upon the NO_x and SO_x emission.
6. What are the advantages and disadvantages of various in-furnace SO_x removal technologies

CHAPTER 7

POST COMBUSTION CONTROL OF OXIDES OF SULPHUR (SO_X) FOR FLUE GASES OF BOILERS

7.1 INTRODUCTION TO POST COMBUSTION SO_x REMOVAL

Air pollution by oxides of sulphur (SO_x) from thermal stations can be controlled either by in-furnace absorption as in Fluidized Bed Combustion (FBC) based boiler as dealt with in **Chapter 6** or by post combustion 'Flue Gas Desulphurization (FGD)'. It may be mentioned that high capacity boilers based on fluidized bed combustion technology are still under development, large power stations would still be equipped with pulverized fuel boilers. Post combustion removal of SO_x or Flue Gas Desulphurization (FGD) for the stations with pulverized fuel boilers or oil fired boilers is an obvious choice. For the stations to be located near the coal mines or the 'Coal Washeries', circulating fluidized bed boilers with capacity in the range of 250 MW or more would be preferred to efficiently use high ash coal and washery coal rejects with very low emission of oxides of sulphur (SO_x) and oxides of nitrogen (NO_x).

A number of Flue Gas Desulphurization (FGD) processes have emerged with newer techniques of improving SO_X removal efficiencies. Some processes have merits of not generating liquid effluents such as spray dryer and circulating fluidized bed lime reactor. Some processes convert SO_x into marketable by-products like gypsum, sulphuric acid or elemental sulphur. A few FGD processes regenerate the sorbent as in Wellmenlord process- Magox process. Because of high ash content of Indian coals, an efficient dust collector before the absorption tower or reactor is necessary.

This Chapter deals with post combustion control of SO_x / 'Flue Gas Desulphurization (FGD)' i.e. 'Post Combustion De-SO_x Systems'. Post combustion FGD technologies discussed in this chapter are equally applicable to exclusively oil fired or 'gas + oil' fired boilers with slight modifications in the scheme because of absence of ash.

Suitable FGD technology is available for stations using any type of fuel irrespective of its calorific value, ash content, and sulphur content (including Indian coals/lignite). Because of bad effects of air pollution, it is expected that all future power stations would be provided with systems for control of SO_x emission, and even the existing big stations may be required to install/ retrofit suitable FGD systems.

7.2 SO_X GENERATION WITH LOW GRADE INDIAN COALS AS COMPARED TO COALS OF WESTERN COUNTRIES

In India, only low grade coals are available for power generation. These have low calorific value (3000-4200 kcal/kg) and high ash content (35–45%). With the exception of north-eastern coals, Indian coals, in general, have quite low sulphur content (0.4–0.6%). Because of the use of low grade coals in India for power generation, the boiler efficiencies are lower and auxiliary consumption is higher than in many developed countries where high grade coal is used.

Most developed countries use coals of higher calorific value (6500–7900 kcal/kg) and lower ash content (4–15%) for power generation. These coals have sulphur content varying between 0.5–2.5% depending upon the source. (Most of these coals have sulphur content in the range of 0.7–2%). Because of low ash content and high calorific value the boiler efficiencies are higher and auxiliary consumption is lower. Based on the above facts and assuming that no desulphurization technique is used, SO_x generated / emitted has been compared with that of Indian coals and foreign high grade coals. The results (***based upon Ref. 6***) are shown in **Table 7.1.**

The comparison in the table is based on tentative net overall efficiencies. In India with Indian coals as indicated in **Table 7.1,** it is 32 %. whereas in developed countries with their coals as indicated it is 36 %

Table 7.1: Comparison of SOx Generated/MWh Using Indian & Foreign Coals

Source Country and state or area	Coal characteristics M %	A %	FC+V %	S %	GCV kcal/ kg	Coal Consumed kg/MWh	SO_x generated per MWh
Examples of foreign coals (Base: overall net efficiency 36%)							
United Kingdom							
Durham	1.5	3.8	94.7	0.87	7410	322.4	5.19
Cardiff	1.5	2	96.5	0.8	8000	298.6	4.42
South Africa Transvaal	2.2	13.3	84.5	0.7	6730	355	4.6
Australia New South Wales Western Field	2.6	14.6	82.8	0.7	6455	370.1	4.793
USA - Pennsylvania	3.0	9.7	87.3	0.7	7472	319.7	4.14
USA - New Mexico	2.9	8.9	88.2	0.8	7411	322.4	4.77
USA – Arkansas Maryland Oklahoma Virginia	2.6 to 3.4	7.2 to 8.7	88 to 89.7	1.0	7611 to 7795	313.9 to 306.5	5.67 to 5.807
USA – Pennsylvania	2.6	9.1	88.3	1.3	7561	316	7.6
USA – Pennsylvania	3.1	7.2	89.7	1.8	7950	300.5	10
USA – Texas	4	12.2	83.8	2.0	6794	351.6	13
Examples of Indian coals (Base: overall net efficiency 32%)							
Kaniha Block Mahanadi coal fields	5.5	42	52.5	0.41	3875	693.5	5.26
Talcher area SE coal fields Orissa	5	38.62	56.38	0.47	4193	641	5.573
Northern coal fields Singraluli area	6	43	51	0.41	3635	739.3	5.608
North Karanpura	6.1	38.9	55	0.5	4007	670.7	6.204
Neyveli Lignite	50	7.0	43	0.66	2664	1009	12.32

Notes*: M = % Moisture, A = % Ash, FC = % fixed carbon, V = % Volatile matter.*
(90% S converted to SO_2 and 2% S converted to SO_3 and balance retained in ash).

It is clear from the comparison that the foreign high grade coals having sulphur content of about 1% emit equivalent amount of SO_x as compared to Indian coals of grade E,F and G with sulphur content 0.4%–0.5%. However, lignite from Neyveli (Tamil Nadu) has very large moistures content, low calorific value, low ash content and 0.66% sulphur content and it would emit SO_x equivalent to foreign high grade coals having sulphur content of around 2%. Overall net efficiency is lower in case of lignite because of high moisture in fuel.

Developed countries are generally using desulphurization techniques even with their low sulphur coals (up to 1% sulphur content). With increasing coal based generation in India, it has become necessary to use techniques to curb SO_x emission as otherwise environmental consequences may be too grave because emission levels are already very high.

7.3 Various Flue Gas Desulphurization (FGD) / De-SO_x Systems

There are two types of flue gas desulphurization processes namely:

- Non-regenerative processes
- Regenerative processes

In the non-regenerative processes, the oxides of sulphur are absorbed in the sorbent and, the sorbent is not recovered. The oxides of sulphur are permanently bound in a chemical compound which may be disposed off as waste or sold as a by-product such as gypsum. The non-regenerative processes include:

- Spray dry process
- Hybrid process of In-furnace injection and spray dryer.
- Circulating fluidized bed lime reactor
- Lime stone gypsum process or sometimes lime gypsum process
- Sea water scrubbing
- Lime stone sludge process
- Dual alkali process
- Double Contact Flow Scrubber (DCFS)
- Selective catalytic Reduction reactor (SCR) to absorb both NO_x and SO_x

In the regenerative systems the absorbent material (sorbent) used for removal of SO_x is recovered for reuse / recycling. The SO_2 or SO_3 released during the recovery of sorbent can be used to produce more stable and useful saleable products (like elemental sulphur, sulphuric acid, etc.). The regenerative processes include:

- Wellman Lord process
- Magox process
- Activated charcoal process

There are some more processes for removal of oxides of sulphur which are used in industries such as wet scrubbing, chemical scrubbing (Absorption) by ammonium sulphite or zinc oxide). These have been described in **Para 8.2.5 and Para 8.3 3 of chapter 8.**

7.4 NON REGENERATIVE SYSTEMS OF DESULPHURIZATION

7.4.1 Spray Dryer Process of FGD

In this system of Flue Gas Desulphurization (FGD), the flue gas after primary dust collector is led to a spray dryer absorber tower in which hydrated lime [$Ca(OH)_2$] slurry is sprayed. The sulphur dioxide (SO_2) in the gas reacts with lime to form mainly calcium sulphite ($CaSO_3$) while sulphur trioxide (SO_3) forms calcium sulphate ($CaSO_4$).

$$Ca(OH)_2 + SO_2 = CaSO_3 + H_2O; Ca(OH)_2 + SO_3$$
$$= CaSO_4 + H_2O$$
$$2\,Ca(OH)_2 + 2SO_2 + O_2 = 2CaSO_4 + 2H_2O$$

$CaSO_3$ and small amount of $CaSO_4$ so produced are carried away by flue gases and get mixed up in ash precipitated in the final dust-collector after spray dryer. Solids precipitated in spray dryer tower, contain unreacted lime and are, therefore, dissolved in water, screened and recycled. Simplified flow diagram of spray dryer process is shown in **Fig. 7.1**.

Example of Development of Absorbent for Spray Dryer and Atomizer:

(a) ***Absorbent*** **:** It has been reported by MHI Japan that they along with Hokkaido Elec. Power Co. have developed an absorbent for spray dryer process named as 'Lively Intensified Lime Ash Compound (LILAC)', which is a mixture of lime, flyash, spent absorbent $CaSO_4$ (in a certain proportion not disclosed) cured in hot water at about 95 °C. Instead of hydrated lime, the solution of LILAC is sprayed in spray dryer by which it is possible to achieve SO_2 removal up to 90% with a Ca : S ratio of 2.0 and approach temperature ΔT of 15°C ***(defined as 'Spray dryer outlet gas temperature - Saturation temperature').***

(b) ***Rotary atomizer:*** This type of atomizer for spraying of absorbents/ sorbent if fitted on to the spray dryer greatly improves the efficiency of absorption (Takahashi et al., 1993).

SOx Removal Efficiency of Spray Dryer Process:

The efficiency of SO_x removal in the process depends upon the following factors:

- Ca: S molecular ratio. (The efficiency of SO_x removal increases with the Ca:S molar ratio.)
- Uniformity and adequacy of mixing of lime particles with gas.
- Completeness of evaporation of the sorber solution/slurry and fineness of the particles of sorber.
- Decrease in approach temperature ΔT ('Spray dryer outlet gas temperature - Saturation temperature').

It is also experienced that the use of bag filters as dust collector after absorber tower instead of ESP account for an increase of almost 15–20 % in SO_x removal efficiency probably because of absorption of SO_x from gases by unreacted lime in ash when the gases pass through the bag filter. Bag filters are also more effective as dust collectors than ESP and also for SO_x removal. Approximate variation of SO_2 removal efficiency with Ca:S molar ratio and approach temperature ΔT is shown in **Table 7.2.**

Table 7.2: SO_2 Removal Efficiency with Ca:S and ΔT

Ca : S molar ratio	**SO_2 removal efficiency**		
	ΔT 6 to 7°C	ΔT 10 to 12°C	ΔT 19 to 20°C
1.25	71 – 74 %	53 – 56 %	
1.5	75 – 78 %	58 – 61 %	
1.75	77 – 80 %	61 – 64 %	
2.0	79 – 82 %	65 – 68 %	48 – 53 %
2.25	82 – 85 %	68 – 71 %	55 – 59 %
2.50	83 – 86 %	70 – 73 %	58 – 62 %
2.75	85 – 88 %	74 – 77 %	60 – 64 %
3.0	87 – 90 %	77 – 80 %	62 – 66 %

Process Suitability to Indian Coals

Indian coals have high ash content up to 45% and it is advisable to have a primary stage of dust collection with a collection efficiency of about 85–90% before the spray dryer absorption tower. An additional fan (booster fan) would be required in the flue gas circuit to make up for the additional pressure drop. The spray dry process is simple and has less operating problems.

7.4.2 Hybrid Process of In-furnace Injection and Spray Dryer.

In - furnace injection of lime stone has already been described in before ***(Chapter 6, para.6.5)***. A hybrid process is created by combining the 'In-furnace injection of powdered lime stone' and 'spray dryer'. MHI (Japan) tested such a process and achieved an efficiency of SOx removal of about 80–90 % easily. In this process the lime stone powder is injected into upper part of the furnace (equivalent to a molar ratio of Ca: S as 2 : 3). A part of SO_x is absorbed in the furnace by this lime stone powder. The chemical reactions during primary stage of desulphurization by the in-furnace injection of lime stone would be:

$$CaCO_3 = CaO + CO_2.$$

$$CaO + SO_2 + 1/2\ O_2 = CaSO_4 \text{; and } CaO + SO_2 = CaSO_3$$

With some type of coals and furnace conditions, there may be slagging in the furnace in such a process. The secondary stage of desulphurization by spray dry injection of lime slurry would be according to the following chemical reactions:

$$Ca(OH)_2 + SO_2 = CaSO_3 + H_2O;\ Ca(OH)_2 + SO_3 = CaSO_4 + H_2O$$

$$2\ Ca(OH)_2 + 2SO_2 + O_2 = 2CaSO_4 + 2H_2O$$

The consumption of lime stone in this hybrid system is very high and the process is not generally favoured.

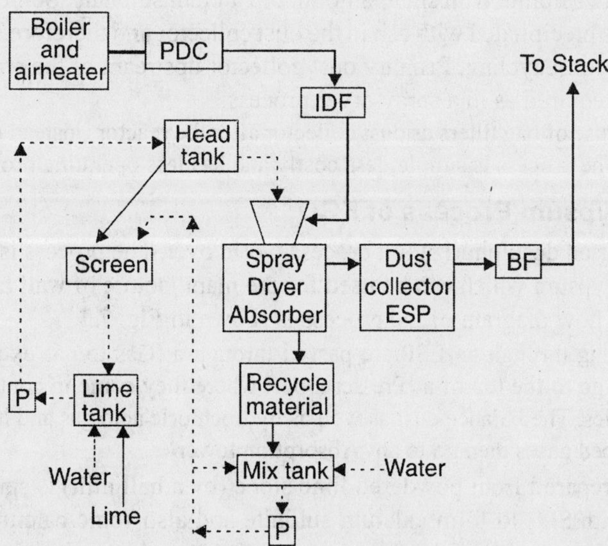

Legend: *P–Pump, IDF–Induced Draft Fan, BF–Booster Fan, PDC–Primary Dust Collector, ESP–Electronic Precipitator*

Fig. 7.1: Spray Dry Process

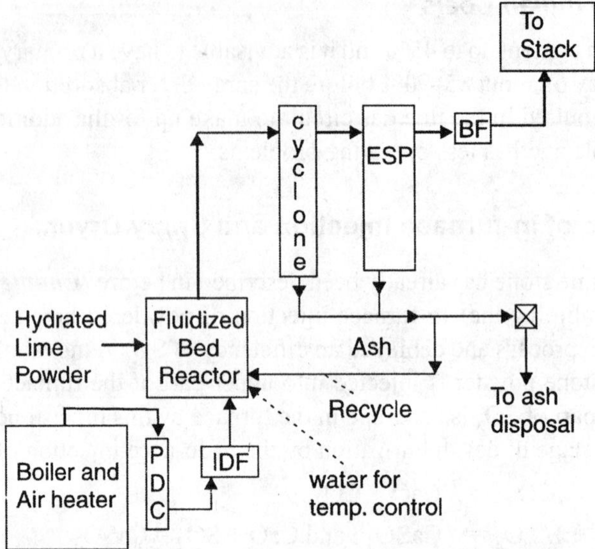

Legend: IDF–Induced Draft Fan, PDC–Primary Dust Collector, BF–Booster Fan, ESP–Electrostatic Precipitator, P–Pump

Fig. 7.2: Circulating Fluidized Bed Reactor

7.4.3 Circulating Fluidized Bed Lime Reactor Process (CFB Lime Reactor)

A simplified flow diagram of this system is shown in **Fig. 7.2.** Here, the flue gas is used to fluidize the hydrated lime (reagent) in a tall reactor, provided with a humidifier stage at the bottom to control the temperature in the reactor. The flue gas from the reactor is led to a dust collector. The reaction product of lime and SO_2 is calcium sulphite with small amount of calcium sulphate. Some unreacted lime passes through the reactor and is precipitated with ash in the dust collector and therefore, a part of dust collected is injected in the reactor for recycling. Primary dust collector upstream of the reactor and a booster fan in the flue gas circuit is required as in a spray dryer process.

As in case of spray dryer, use of bag filters as dust collector after the reactor (instead of an ESP) increases the SO_x removal efficiency. The process is simple, less costly and has less operating problems.

7.4.4 Lime Stone Gypsum Process of FGD

This is the most widely used desulphurization process world over. The process is efficient and produces marketable by-product gypsum which can be used for the manufacture of wall board, cement, building blocks, etc. A simplified flow diagram of the process is shown in **Fig. 7.3.**

The flue gases after passing through an ESP are passed through a 'Gas to gas exchanger' where they are cooled. These gases then go to the top of a 'Pre-scrubber' where they come in contact with water which is sprayed through the nozzles. The balance dust as well as hydrochloric acid gas and hydrofluoric acid gas are removed here. The scrubbed gases then go to an 'Absorption tower'

Lime stone slurry prepared from powdered lime stone (by a ball mill) is sprayed in the absorption tower where it reacts with SO_2 to form calcium sulphite and also some calcium bi-sulphite which is oxidized to calcium sulphate by blowing compressed air in the oxidation section of tower sump. MHI (Japan) has developed an Arm Rotating Air Sparger (ARS) which finely disperses air by utilizing energy of agitation in oxidation tank. Following chemical reactions take place in the process:

$$CaCO_3 + 2\ SO_2 + H_2O \ = \ Ca\,(H\,SO_3)_2 + CO_2$$

$$CaCO_3 + SO_2 = Ca\,SO_3 + CO_2$$
$$Ca(HSO_3)_2 + CaCO_3 = 2\,CaSO_3 + H_2O + CO_2$$
$$2\,CaSO_3 + O_2 + 4\,H_2O = 2\,[CaSO_4.2\,H_2O^*]$$
$$2\,SO_2 + 2H_2O + O_2 = 2H_2SO_4$$
$$CaCO_3 + H_2SO_4 = CaSO_4 + H_2O + CO_2 \; [^*Gypsum\ is\ CaSO_4.2\,H_2O]$$

Integrated oxidation section with absorber tower, as shown in the flow diagram, has advantages of maintaining high rate of oxidation over a wide range of pH values which in case of separate oxidation tank falls off abruptly on increase of pH beyond 5. The oxidation in integrated tower starts before crystallization of $Ca(HSO_3)_2$ and partial pressure of SO_2 is reduced due to oxidation of H_2SO_3 to H_2SO_4 which is neutralized by incoming $CaCO_3$

The process is capable of attaining SO_x removal efficiency of more than 95% or even up to 99%. A 'Gas to gas heat exchanger' is used for cooling the incoming hot gases by outgoing cool gases from absorption tower. The liquid effluents can be concentrated by evaporation using the waste heat of the flue gases and then can be used to make solid blocks by mixing some lime, cement and fly ash. It is also possible to completely evaporate waste water by spraying it in the gas stream and collecting solids in the dust collector along with fly ash. Various additives including magnesium oxide have been found to increase sulphur removal efficiency.

Automatic Calcium Carbonate Slurry Analyzer

An automatic calcium carbonate slurry (limestone slurry) analyzer (developed by MHI) for optimum control of desulphurization is used. The calcium carbonate ($CaCO_3$) sensor consists of a slurry sampler for sampling of absorbent slurry continuously at a fixed flow rate, a reactor for generating CO_2 from the sample by addition of sulphuric acid, a gas analyzer for measuring CO_2 being so generated, and a controller unit for $CaCO_3$ concentration calculation in the absorbent slurry.

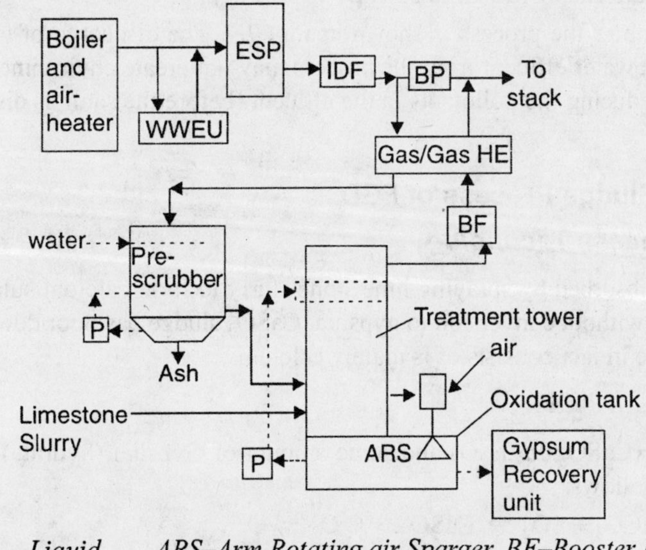

----- *Liquid* *ARS–Arm Rotating air Sparger, BF–Booster Fan,*
___ Flue gas *IDF–Induced Draft Fan,*
ESP–Electrostatic Precipitator
P–Pump, WWEU–Waste Water Evaporator Unit
Gas/Gas HE–Gas/Gas Heat Exchange
BP–By Pass

Fig. 7.3: Lime Stone Gypsum Process

By continuously controlling and measuring the concentration of $CaCO_3$ by its sensor, the absorption capacity of the absorbent and the purity of gypsum are kept constant. The lime stone concentration is maintained at a fixed value by controlling the raw limestone supply with continuous detection of absorbent slurry lime stone concentration with this type of sensor. The designed desulphurization efficiency is maintained by continuous control of the recirculated absorbent liquid flow rate with variable vane pitch pump. At low loads, there is power saving by such a control.

As a measure against delay in detection by SO_2 analyzer, a model prediction control method having a dead time compensation function is employed. The system can be integrated with DCS (distributed control system). It is claimed that it improves: (i) reliability of DSO_x control, (ii) follow up ability of boiler load, (iii) retention of purity of gypsum (by- product) and (iv) energy conservation **(Ref: 8)**.

7.4.5 Sea Water Scrubbing

Sea water scrubbing of flue gases is possible for plants located on sea shore for removal of SO_2 as sea water contains salts like bicarbonates and carbonates of calcium, magnesium and other metals which can react with SO_2 to form sulphites. CO_2 is released in the reaction which is stripped off to bring down the acidity and chemical oxygen demand (COD) by aeration.

$$MgCO_3 + SO_2 = MgSO_3. + CO_2$$

$$Mg(HCO_3)_2 + SO_2 = MgSO_3. + H_2O + 2CO_2$$

The cooling water from condenser outlet can be used for scrubbing. Additional quantity of sea water may be required according to the percentage of SO_X removal desired and maintenance of optimum temperature and pH value which should be more than 7. The metal sulphites, formed after absorption of SO_2, get converted to sulphates when air is blown.

A simplified scheme of the process is shown in **Fig. 7.4.** The discharge of toxic compounds and heavy metals in the sea water effluent from the process may not create environmental problems, but if it is so, treatment for reducing the pollutants in the effluent (before the same is discharged in sea) may be considered.

7.4.6 Lime Stone Sludge Process of FGD

7.4.6.1 Normal Process Technology

In this process SO_2 is absorbed by spraying lime stone slurry to form calcium sulphite ($CaSO_3$) which is disposed off as such without conversion to gypsum. $CaSO_3$ sludge has poor dewatering property and forms slush. The sludge in fact consists of is mainly calcium

To Stack

Sulphite hemi-hydrate ($CaSO_3.0.5\ H_2O$) and some amount of Gypsum (hydrate) [$CaSO_4.2H_2O$]. The reactions would be as follows:

$$CaCO_3 + SO_2 = CaSO_3. + C\ O_2$$

$$CaSO_3 + 0.5H_2O = CaSO_3.0.5\ H_2O$$

$$CaCO_3 + SO_3 = CaSO_4.$$

$$CaSO_4 + 2H_2O = [CaSO_4 \, . \, 2H_2O]\ (Gypsum)$$

The process also neutralizes the acids like hydrochloric acid (HCl) and hydrofluoric acid (HF) to form Calcium chloride ($CaCl_2$) and Calcium fluoride (CaF_2) respectively.

Since the sludge/ slush obtained in the process cannot be easily dewatered, it is essential to fix the sludge with some cheap waste material like fly ash. Because of this difficulty and problem of chemical scaling, plugging and erosion, the process is not used nowadays, although its efficiency of removal of oxides of sulphur and other soluble acid gases is quite high, say 95 %.

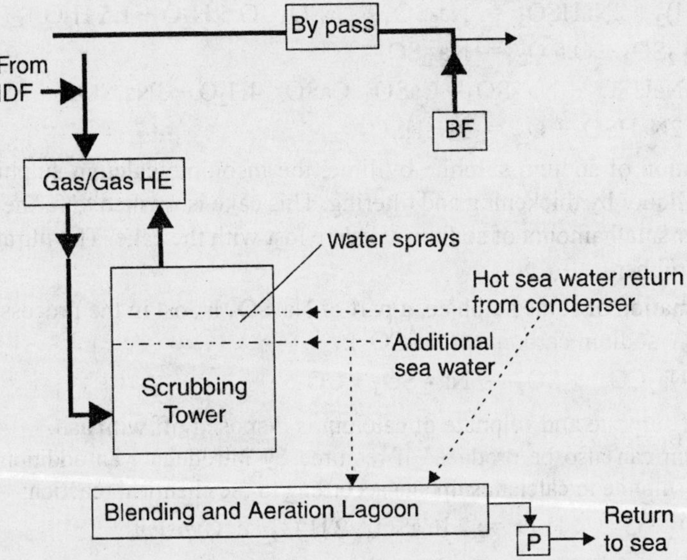

Legend: p—Pump; BF—Booster Fan; IDF—Induced Draft Fan; Gas/ Gas HE—Gas/ Gsd Heat Exchanger

Fig. 7.4: Sea Water Scrubbing Process

7.4.6.2 Modified Technology—Chiyoda Thorough-Bred 121 (CT-121) Process

Normally, the lime stone slurry is sprayed into flue gas for desulphurization, but Chiyoda Corp. Yokahama Japan have developed a FGD process based on dispersal of dust free flue gas into lime stone slurry through sparger tubes using a Jet Bubbling Reactor (JBR). This process is supposed to be more efficient than the conventional lime stone slurry spray tower and is called **CT-121** process. This process uses 10–15 % less auxiliary power than conventional spray towers and *reduces some 'Mercury'* also from flue gases.

The technology has been licensed in US to Black and Veatch Corp., Kansas City. The system is installed in many power stations. This FGD system has high efficiency of removal of SO_x, SPM, and also of '*Mercury*'.

7.4.7 Dual Alkali Process of FGD

Here, the sulphur dioxide (SO_2) is absorbed by contacting 'Sodium sulphite (Na_2SO_3)' solution and forms 'Sodium bisulphite ($NaHSO_3$)'. When treated with lime, $CaSO_3$ and Na_2SO_3 are formed. Some of the 'Sodium bisulphite ($NaHSO_3$)' reacts with oxygen present in flue gases and forms 'Sodium bisulphate ($NaHSO_4$)'. The level of oxidation during absorption is a function of the scrubber design, oxygen content of flue gas and the operating temperature of scrubber. The spent scrubbing solution contains sodium bisulphite ($NaHSO_3$) and some 'Sodium bisulphate ($NaHSO_4$)' which when reacted with lime / calcium hydroxide [$Ca(OH)_2$] regenerates sodium sulphite while forming calcium sulphite hemi-hydrate ($CaSO_3.0.5\ H_2O$) and mixed sulphite and sulphate of calcium [$CaSO_3.CaSO_4.4H_2O$]. The chemical reactions in the process take place as follows:

$$SO_2 + Na_2SO_3 + H_2O = 2NaHSO_3$$

$$Ca(OH)_2 + 2NaHSO_3 = Na_2SO_3 + CaSO_3 . 0.5\ H_2O + 1.5\ H_2O$$

$$Na_2SO_3 + 0.5\ O_2 = Na_2SO_4 .$$

$$2Ca(OH)_2 + 4NaHSO_3 + Na_2SO_4 = CaSO_3 .CaSO_4 .4H_2O + 3Na_2SO_3 .$$

$$2NaHSO_3 + O_2 = 2NaHSO_4$$

After regeneration of sodium sulphite by lime, the insoluble calcium sulphite/ sulphate salts are separated from the liquor by thickening and filtering. This cake is washed to further recover soluble salts of sodium, however small amount of sodium salts are lost with the cake. The filtrate solution is returned / recycled to the scrubber.

Due to the formation of some sulphate, a part of Na_2SO_3 is lost in the process which is made up by addition of soda ash/ sodium carbonate (Na_2CO_3).

$$(Na_2CO_3) + SO_2 = Na_2SO_3 + CO_2.$$

The mixture of sulphite and sulphate of calcium is disposed off with ash.

High quality gypsum can also be produced if required by introducing an additional oxidation stage for converting calcium sulphite to calcium sulphate according to the chemical reaction:

$$2\ CaSO_3 + O_2 + 4\ H_2O = 2\ [CaSO_4 . 2\ H_2O]\ \textit{i.e. Gypsum}$$

The process efficiency of the system for removing oxides of sulphur may be more than 90%. As the scaling problems in simpler processes like lime stone /lime systems have been solved, the dual alkali process is not being favoured now.

7.4.8 Double Contact Flow Scrubber (DCFS)

In DCFS, the flue gas is introduced from the top of scrubber to contact absorbent liquid which is sprouted upwards at high velocity from special ceramic nozzles installed at the bottom of scrubber. The sprouted liquid reaches the top of the scrubber and then disperses .The falling absorbent liquid after dispersion from top and the sprouted absorbent liquid going up collide and generate fine liquid drops thus forming a dense liquid layer of fine drops in the gas liquid contact space. This produces a surface renewal of liquid drops which together with mixing of gas due to collision of falling liquid and sprouted liquid efficiently promotes SO_x absorption. Higher the sprouted height, higher is the efficiency of SO_x removal and for 10 m height, removal of about 95% is achieved. The flue gas outlet is through a mist eliminator at the side of tower towards the bottom. The sump of the tower is provided with air rotary sparger (ARS) for oxidation as in the lime stone gypsum process. This process also reduces the emission of SPM from the stack. *[Ukawa et al., 1996].*

7.4.9 Non-regenerative FGD System using Magnesium Oxide

The major difference between magnesium and calcium based systems of flue gas desulphurization are:

- Magnesium hydroxide $Mg(OH)_2$ or magnesium oxide (MgO) is used as absorbent for oxides of sulphur (SO_x) instead of calcium hydroxide $Ca(OH)_2$ or quick lime (CaO). Normally aqueous slurry of magnesium hydroxide is used as absorbent.
- The magnesium hydroxide after absorbing sulphur dioxide (SO_2) gets converted to magnesium sulphite ($MgSO_3$) and after absorbing sulphur trioxide (SO_3) is converted to magnesium sulphate ($MgSO_4$). The quantity of sulphur trioxide is very small and therefore sulphate formation is not significant.

- $MgSO_3$ is less thermally stable than $CaSO_3$, therefore, if necessary, the SO_2 can be recovered for use in the manufacture of sulphuric acid (H_2SO_4) or elemental sulphur by further processing.
- The solubility of $MgSO_3$ in water is more than that of $CaSO_3$. Therefore, the solution of $MgSO_3$ gives more alkalinity in scrubbing liquor than $CaSO_3$ under similar conditions. Therefore the scrubbing liquor removes the SO_2 more efficiently than in case of calcium based systems.

Features of the Process are

(a) The flue gases are admitted at the bottom of a packed bed in an absorption tower. $Mg(OH)_2$ slurry is supplied over the top of the bed in the tower. The gases travel up the bed losing, almost, all their SOx content which reacts with the sorbent [$Mg(OH)_2$]. The gases get washed by flushing water sprayed in the hood portion of tower after which they pass through 'Demister' before going to stack. The slurry after absorbing SOx collects at the bottom of the tower. This slurry is partly recirculated (for ensuring full utilization of its absorbing capacity) while the balance is passed through an oxidizing stage where it reacts with oxygen of the air blown through it. $MgSO_3$ in the effluent slurry is converted to $MgSO_4$. Fresh slurry of MgO is continuously supplied. The reaction which takes place in the oxidizing stage is as follows

$$2MgSO_3. + O_2 = 2MgSO_4.*$$

[**In fact,* $MgSO_4$ *has a water of crystallization and it therefore becomes* $2MgSO_4. 7H_2O$].

The effluent from oxidizing stage is filtered, concentrated, and the $MgSO_4$ cake is separated and after drying it may be disposed off as this is a stable and safe material.

(b) The process is simple, requires lesser equipment, smaller space and less maintenance. $MgSO_4$ has high solubility and there is an insignificant problem of scaling.

(c) High desulphurization efficiency can be achieved by this process.

7.4.10 Selective Catalytic Reduction for Removal of both NOx and SOx

In this process, selective catalytic reduction (SCR) reactor for oxides of sulphur is preceded by SCR for reduction of oxides of nitrogen (NO_x), where ammonia (NH_3) is dosed in the flue gas. The ammonia converts the oxides of nitrogen (NO_x) to nitrogen. The flue gas from there is led to a gas to gas heat exchanger for being heated. The heated gas is then led to 'SO_2 converter', where it is catalytically converted to sulphur trioxide (SO_3) in presence of a catalyst. The sulphur trioxide when reacts with water forms sulphuric acid (H_2SO_4). The reactions in the process are as follows:

$$6\,NO_2 + 8NH_3 = 7N_2 + 12H_2O;$$

$$4NO + 4NH_3 + O_2 = 4N_2 + 6H_2O.$$

$$2SO_2 + O_2 = 2SO_3. \text{ (Catalytic conversion);}$$

$$SO_3. + H_2O = H_2SO_4$$

$$2NH_3 + H_2SO_4 = (NH_4)_2SO_4.; \text{ or } 2NH_3 + SO_3 + H_2O = (NH_4)_2SO_4.;$$

Ultimately, sulphur is removed as ammonium sulphate which can be well used as a fertilizer.

7.4.11 Other Non-Regenerative FGD Processes

There are several other FGD systems which are not applicable for power plants. These are briefly mentioned below:

(i) *Activated Carbon Absorption Process.*

In this process, dust free cooled flue gas is passed over a bed of activated charcoal which absorbs SO_X. When the spent charcoal is heated, it gives out SO_2 which can be led to sulphur recovery/H_2SO_4 production unit. This process has some industrial applications.

(ii) *Electron Beam Process*

In this process the flue gases are subjected to NH_3 injection and electron beam irradiation whereby ammonium sulphide and ammonium nitride are formed which get collected along with ash in the dust collector. Both NO_x and SO_x are captured in the process. Such a process has been reported to have been developed by MHI Japan.

(iii) *Compact Membrane Absorber*

TNO Institute or Environmental Energy Technology Apeldoorn Netherlands and other Dutch firms have devised gas liquid contactor which consists of microporous hollow fibres of polypropylene. These fibres separate the flue gas from counter flow absorbent ($NaSO_3$ solution). SO_2 diffuses through fibres and is absorbed by sulphite to form $NaHSO_3$. The process can be made regenerative if $NaHSO_3$ is treated with soda-ash to regenerate Na_2SO_3.

7.5 REGENERATIVE PROCESSES OF DESULPHURIZATION

7.5.1 Magox Process of FGD

In this process, sprayed magnesium hydroxide slurry in the reactor reacts with SO_x in the flue gas to form mainly magnesium sulphite and small quantities of magnesium sulphate. The magnesium sulphite is then calcinated at 800°C – 1000°C to release SO_2 and produce MgO. The SO_2 released after calcination is led to H_2SO_4 or sulphur manufacturing facility while MgO is hydrated to regenerate $Mg(OH)_2$ for recycling. The process is suitable for high sulphur coal. *[This process is somewhat similar to the process described in **para 7.4.9 of this chapter** except that in the Magox process, the sorbent is regenerated and the SO_2 can be obtained in purer form for futher utilization.]*

$$Mg(OH)_2 + SO_2 = MgSO_3 + H_2O; \; Mg(OH)_2 + SO_3 = MgSO_4 + H_2O$$

$$MgSO_3 \text{ when calcinated} = MgO + SO_2.$$

A Simplified flow diagram of the process is shown in **Fig. 7.5.**

7.5.2 Wellman Lord Process of FGD

This is a popular regenerative FGD process. In this process sodium sulphite (Na_2SO_3) solution is sprayed into the flue gas which reacts with SO_2 to form sodium bisulphite (Na HSO_3).

$$[Na_2SO_3 + SO_2 + H_2O = 2Na\,HSO_3]$$

The Sodium sulphite is regenerated by thermal decomposition of sodium bisulphite releasing SO_2 and Na_2SO_3. The SO_2 is sent for further processing to produce H_2SO_4 or sulphur, while sodium sulphite is recycled. Because of high temperature involved in the process of thermal decomposition of sodium bisulphite (Na HSO_3), some sodium sulphate (about 15%) and some sodium thio-sulphate ($Na_2S_2O_3$) are also formed in addition to sodium sulphite. These are separated out and the loss of sulphite is made up by addition of soda ash (Na_2CO_3) or caustic soda (NaOH) which reacts with SO_2 to form sodium sulphite. A simplified flow diagram of the process is shown **in Fig. 7.6.**

$$Na_2CO_3 + SO_2 = Na_2SO_3 + CO_2;$$

$$2\,NaOH + SO_2 = Na_2SO_3 + H_2O$$

*A somewhat similar regenerative process using ammonium sulphite instead of sodium sulphite is used in industries as described in **Para 8.3.3.2, Chapter 8.***

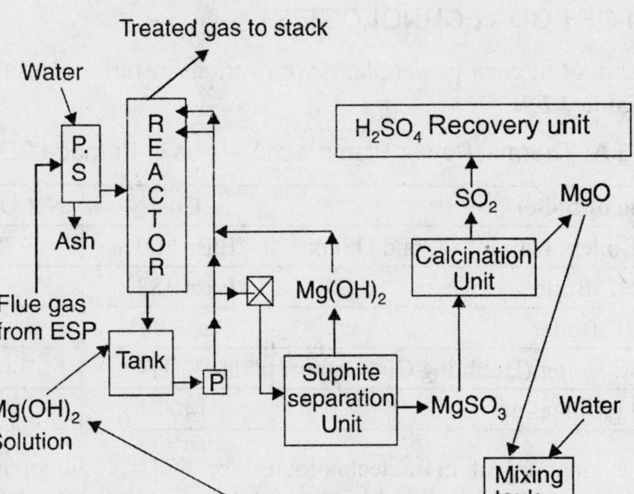

Legend: P–Pump; PS–Pre-Scrubber

Fig. 7.5: Magnesium Oxide Process

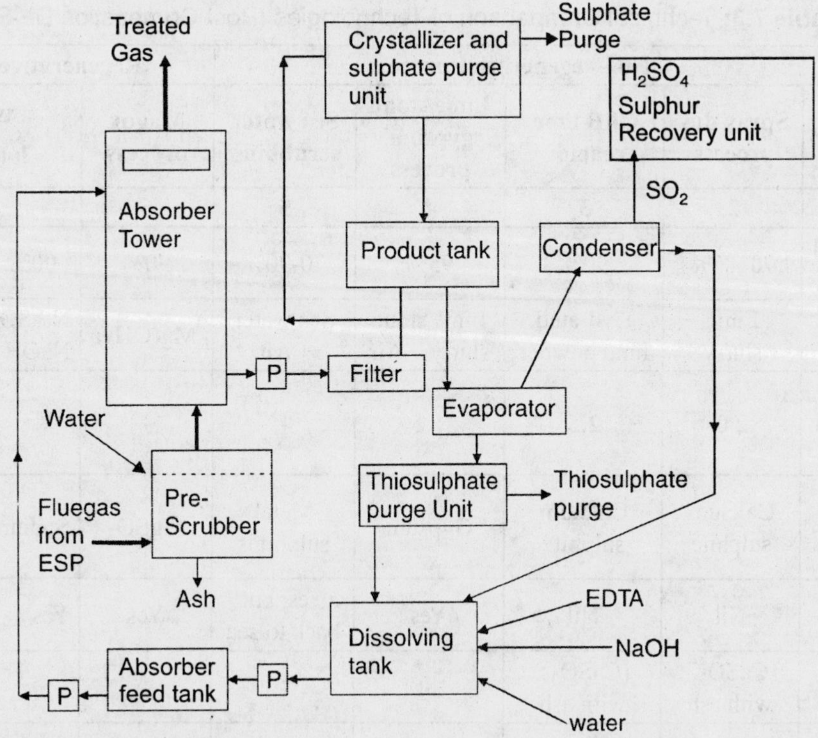

Legend: P–Pump; EDTA–A chemical

Fig. 7.6: Wellman Lord Process

7.6 EVALUATION OF FGD TECHNOLOGIES

The efficiencies and cost of thermal power plants with various in-furnace desulphurization systems are compared below in **Table 7.2 A**

Table 7.2 A: Thermal Power Plants with Various in-Furnace FGD Systems

Plant with type of boiler	Cost	Net Overall Efficiency
Plant with PF Boilers with DNO_x and DSOx	Base 100%	35–37%
Plant with BF BC Boiler	Base 88%	35–37%
Plant with CFBC Boiler	Base 90%	34–36%
Plant with PFBC Boiler (Bubbling Circulating)	110%/114%	34–41%/38–46%
Plant with coal gasification	140%	about 41%

The 'Technical evaluation' of main technologies for FGD is shown in **Table 7·3**, while the 'Comparative financial aspect' is presented in **Table 7·4** for pulverised fuel fired boiler based power plants.

For Indian coal which has a high ash content, low calorific value and low sulphur content, FGD process may be chosen according to the location of plant, degree of desulphurisation needed, space availability for FGD, disposal of liquid effluent and, economic parameters. The spray dryer process, circulating fluidized lime reactor process or lime stone - gypsum processes may be found suitable.

Table 7.3: Technical Comparison of Technologies (Post Combustion De-SO_x)

	Non-regenerative process				Regenerative process	
Items	**Spray dry process**	**CFB lime reactor**	**Lime stone gypsum process**	**Sea water scrubbing**	**Magox process**	**Wellman-lord process**
1	**2**	**3**	**4**	**5**	**6**	**7**
SO_x removal efficiency	70–80%	90%	95%	70-90%	90%	>90%
Reagent	Lime slurry	Hydrated lime powder	Lime stone slurry + Air	Sea water + Air	$Mg(OH_2)$	Na_2SO_3 ; Na_2CO_3; NaOH
Practical Ca : S molar ratio	2.0	2.5	1.2	-	-	-
Reaction product main	Calcium sulphite	Calcium sulphite	Gypsum	Metal sulphites	$MgSO_3$	Sodium bisulphate
Liquid effluent	Nil	Nil	Yes	Yes, but back to sea	Yes	Yes
Waste product	$CaSO_3$, with ash	($CaSO_3$), with ash	-	-	-	-
Regeneration of reagent	-	-	-	-	Yes	Yes

Contd...

Contd...

Items	Non-regenerative process				Regenerative process	
	Spray dry process	CFB lime reactor	Lime stone gypsum process	Sea water scrubbing	Magox process	Wellman-lord process
1	2	3	4	5	6	7
Marketable by-product	-	-	Gypsum	-	H_2SO_4 or sulphur	H_2SO_4; Sulphur; Na_2SO_3 ; Sodium thiosulphate
Make-up fresh water requirement	Yes	Yes	Yes	No fresh water required	Yes	Yes
*Power consump-tion of as % of generation	1–1.5%	1–1.5%	1.5–2.2%	1–1.55	1.8–3.0%	1.8–3.0%
Usage and suitability	For low sulphur coals. Simple and less problems	For low sulphur coals. Less problems and simple	For high sulphur removal efficiency	For costal stations	For high sulphur coal S>3%	For high sulphur coal S>3%. More suitable for oil refineries & power plants >1000MW.

**The figures of power consumption are for Indian coals.*

Table 7.4: Cost Comparison Post Combustion De-SOx

**Costs with H_2SO_4 recovery system*

Items	Spray dry process	CFB lime reactor	Lime stone gypsum	Sea water scrub-bing	Magox process	Well-manlord process
1	2	3	4	5	6	7
Comparative capital cost. Base: Lime stone Process.	65	40	100 (Base)	130	150 to 185%*	160 to 195%*
Comparative capital cost of FGD as % age of capital cost of power project with FGD	14.3%	8.8%	22%	28.5%	27-41%*	28-43%*
Comparative annual fixed +variable cost	75	70	100 (Base)	88	145*	148*
Comparative cost added to electricity as %	7.5%	7%	10%	9%	14.5%	14.8*

REFERENCES

1. Ara, Tokuma et al., "Present situation and develop trend of desulphurization technology", MHI Review, Vol. 28, No. 1 (1991).
2. Council of Power Utilities publication "Need for rationalization of non-coking coal prices," 4/96.
3. Gebel, Karl et al., "FGD technology experience seminar", by NTPC, January 14–16, 1997, New Delhi.
4. Kuroda, Hiroshi, "Recent tech. development in SO_2 and NO_x removal system for flue gases from TPP", Hitachi Rev., Vol. 36, No. 6, 1987.
5. Kant, Rajni, "Control of SO_x emission and clean technology for coal based TDPs", Journal of International Association Elec. Gen., T and D (Afro- Asian), New Delhi, October–December 1993.
6. Kant, Rajni, 'Post combustion SOx mitigation in thermal power stations', Water and Power International Journal , Vol 61. No. 3, July–Sept 2004.
7. MHI Technical review (under new products), Vol. 32, No.1, Feb 1995).
8. Okina, Susumu et. al., 'Development of Automatic calcium carbonate slurry analyzer for optimum control system of desulphurization plant, MHI Technical review vol. 34, No.1, Feb 1997 and 'New products' MHI Technical Review, Vol. 35, No.1, Feb 1998).
9. Radchiffe, P., "FGD economics", EPRI Journal, September 1992.
10. Sood, H.N. et al., "FGD system and installation for coal fired plants", I. Chem Engg. Symposium Series 131, March 1993.
11. Swanekamp, Robert, "Emission control tech. continue to clear Air", Power, Vol. 146, May/June 2000.
12. Takahasi, Shigeki et al. ' Development of Semidry FGD and furnace limestone systems', MHI Technical Review Vol.30, No.1, Feb1993.).
13. Ukawa, Naohiko, et al., "Development of waste water concentration and solidification system for FGD system", MHI Tech. Rev. Vol. 31, No. 1, February 1994.
14. Ukawa, Naohiko, et al., Development DCF Scrubber for FGD MHI Tech Review, Vol. 33, No. 2, June 1996.
15. Vegvoda, Josef et al., Inst, of Chem. Tech. Czech Republic, "FGD experience in Czech Republic", (9) Elec. Power International fall 1999, Vol. 9, No. 3.

QUESTIONS

1. What is the justification for using SO_x removal technologies in India when the Indian coal in general has low sulphur?
2. List out the post combustion SO_x reduction technologies and point out the differences between regenerative and non regenerative processes of SO_x reduction.
3. On which considerations the regenerative processes of SO_x reduction is recommended?
4. Describe the Spray dryer process of flue gas desulphurization. How can you increase the efficiency of this process? How does the use of bag filters improve the efficiency of the process?
5. Describe the 'Limestone Gypsum process of flue gas desulphurization. How can you improve the efficiency of the process?
6. Write short notes on:
   (a) Circulating fluidized bed limestone process of FGD,
   (b) Dual alkali process of FGD,
   (c) Double contact flow scrubber,
   (d) Sea water scrubbing process of FGD.

7. Describe 'Limestone sludge process of FGD'. What modification makes this process suitable for removal of mercury from flue gases?
8. Write a short note on FGD process using Magnesium oxide instead of calcium oxide / limestone. How can you make this process regenerative?
9. Write a short note on: (a) Wellman Lord Process of flue gas desulphurization; (b) Acid dew point.
10. Evaluate the different FGD technologies.

CHAPTER 8

AIR POLLUTANTS FROM INDUSTRIES AND THEIR CONTROL

8.1 CHARACTERISTICS OF INDUSTRIAL EMISSIONS

The characteristics of emission of pollutants differ in different industries. The pollutants emitted by any industry may generally fall within the following pollutants listed below:

Carbon monoxide; Oxides of nitrogen (NOx), Nitrous oxide ((N_2O) – a green house gas; Oxides of sulphur (SOx), Hydrogen sulphide (H_2S); Mercaptans; Fluorine (F), Hydrogen fluoride gas (HF) and other fluorine compounds; Chlorine (Cl), Hydrogen chloride gas (HCl) or vapours, and other Chlorine compounds; Methane (CH_4) ; Hydrocarbons – unburnt gaseous hydrocarbons fuel or liquid fuel vapours and Volatile Organic Compounds (VOC); Ammonia (NH_3); Ozone depleting substances such as Chlorofluorocarbons (CFCs), Hydro Chlorofluorocarbons (HCFCs), etc; Suspended Particulate Matter (SPM), Respirable Suspended Particulate Matter (RSP) / Aerosols and micro particles of toxic metals or their compounds such as Arsenic (As), Lead (Pb), Mercury (Hg), Barium (Ba), Beryllium (Be), Boron (B), Cadmium (Cd), Chromium (Cr), Vanadium (V), Manganese (Mn), Nickel (Ni), Phosphorous (P), Selenium (Se), Zinc (Zn), Iron (Fe), etc.; Biological pollutants such as aero-allergens, bacteria, fungus, virus, etc.; Air borne fumes, vapours and dust of 'Pesticides, insecticide, herbicides'; Nuclides / radioactive gases.

It may be mentioned that the more efficient process design/ more efficient equipments, achievement of higher efficiency in combustion of fuel and other related processes, better operation, maintenance and monitoring would, in general, reduce pollution A summary of the air pollutants emitted by some industries is given in **Table 8.1.**

Table 8.1: Industry-wise Air Pollutants

	Industry	Air Pollutants
1.	Steel industry—Blast furnace	CO, NOx, SOx, NH_3, H_2S, hydrocarbons, SPM/ RSP with solids of metallic compounds of Cr, Fe, Mn, Si, Ni, V, etc.
	Steel mill	NO_x, SO_x, NH_3, SPM/ RSP with solids of metallic compounds of Cr, Fe, Mn, Si, Ni, V, etc.
	Galvanizing and tinning	SO_2 , Cl, acid vapours of HCl, H_2SO_4, metallic vapours of zinc and tin
	Annealing furnace	Mainly SO_2,
2.	Electroplating, battery industry	SPM, SC_x, acid fumes, Pb,

Contd...

	Industry	Air Pollutants
3.	Ceramics, refractories	SPM, RSP, Pb, As, Cr, NO_x, Fluorides, chlorides, dust bearing silica, SOx, CO, hydrocarbons
4.	Stone crushing	SPM, RSP, Silica, etc.
5.	Brick kiln	SPM, CO, NO_x, SO_x, Cl, F, HF, HCl
6.	Lead, Zinc, Tin and Copper smelting and Refining Industry	SOx, NOx, SPM, RSP, odour, oxides and sulphides of Pb, Zn, Sn, Cu, Cd, As,
7.	Rayon industries.	H_2S, Carbon disulphide (CS_2), fumes, fibres, foul odour
8.	Caustic soda industry (using Mercury in Process)	Hg, Cl, HCl,
9.	Leather tanneries	Infectious aerosols, H_2S, **mercaptans**, acidic fumes Cr, Cl, chemical vapours, biological dust, foul odour
10.	Coal/oil fired boiler in industry or thermal power stations	F, Cl, SO_x, NO_x, HF, HCl, CO, SPM, RSP, Cu, Ni, Cr, V, Cd, As, Pb, Hg. (In oil fired boilers, pollutants include hydrocarbons also)
11.	Nitric acid manufacture	NOx, acid fumes
12.	Sulphuric acid manufacture	SOx, acid fumes
13.	Asbestos industry	Asbestos fibres, dust, RSP
14.	Chlor alkali	Hg, Cl, HCl,
15.	Calcium carbide	SPM
16.	Hospital wastes	Infectious aerosols, pathogens, virus
17.	Coke ovens and coal gasification	NH_3, NO_x, SO_x, CO, Cl, F, HCl, HF, hydrocarbons, H_2S, SPM, Pb, V, Cr, Ni, CH_4
18.	Uranium/Radioactive metal mining and processing	Radon gas, SPM containing various radioactive metal micro-particles like that of U, Pu, Th, Pb, Sr, Cs, Ra etc.
19.	Coal mining industry	SPM, CH_4, CO, N_2O, H_2S,
20.	Fertilizer plants	F, SO_x, NOx, NH_3, SPM, RSP, Cd, P, odour, Product dust / powder
21.	Fermentation industries	H_2S, CO, Cl, CH_4, NH_3, chemical fumes (acetic acid, formic acid, alcohol).
22.	Paper and Pulp Industry	SOx, H_2S, Cl, Mercaptans like CH_3SH, $(CH_3)_2SH$, foul odour, SPM
23.	Oil Refineries	SOx, CO, NOx, H_2S, F, NH_3, foul odour, mercaptans, PNA, PCB, Benzene, other hydrocarbons vapours, solvent vapours, SPM, Cr, As, V, Pb, Ni, catalyst,
24.	Petrochemical industry	PNA, PCB, F, Cl, mercaptans, other chemical fumes and vapours,
25.	Aluminum industry	SPM, RSP, Al, F, HF, CO, SO_x, CF_4, C_2F_6
26.	Rubber industry	SO_x, H_2S
27.	Arc furnaces	Ozone, SPM,
28.	Mining of nonferrous metals excluding radioactive metals	SPM, RSP, Cu, Zn, Pb, Hg, Cd, Cr, Fe, As, Co

Contd...

Contd...

	Industry	Air Pollutants
29.	Nuclear Power Plants	Krypton (Kr-85, Kr-88), and other radioactive noble gases; Tritium (H-3), iodine (I-135), xenon (Xe-133, Xe-138) and other nuclides, α, β, γ radiations from radioactive materials and wastes in process
30.	Manufacturers and sectors using pesticides, fungicides. insecticides,	Toxic aerosol vapours, fumes, toxic dust, odour, chlorinated hydrocarbons, etc.
31.	Phosphoric acid manufacture	SPM, F, acid fumes,
32.	Dairy industry	NH_3, CH_4, CO, N_2O, H_2S
33.	Welding shops	Fe, Mn, fluoride, ozone, metal oxides fumes, NO_x
34.	Mica mining	SPM, RSP, dust bearing silica,
35.	Cement industry	Lime and cement dust, RSP
36.	Paints and pigments	Lead, nitrobenzene, aniline, vapours of thinners and solvents, base material, Odour
37.	Glass industry	SPM, RSP, Pb, As, Cr, Fluorides, NOx, SOx,
38.	Storage battery	Lead content in SPM / dust, acid fumes,
39.	Dry cell manufacture	SPM, Mn,
40.	Foundries - cupola	SPM, RSP, SOx, CO, NOx, fumes
41.	Gas turbines	Mainly NOx
42.	Diesel engines drives for industry, and DG sets	SPM, RSP, CO, SOx, NOx, hydrocarbons
43.	Incinerators (all types) or refuse fired boilers	SPM, RSP, CO, SOx, NOx, N_2 O, HCl, dioxins
44.	Agricultural waste fired boilers	SPM, RSP, CO, SOx, NOx, N_2 O
45.	Coal washeries	SPM, RSP, (may be CO, N_2O, NOx, SOx due to spontaneous combustion)

***Note**: The symbols used in the table are mentioned in brackets in the above paragraph*

PNA—Polynuclear Aromatic hydrocarbans

PCB—Polychlorinatied Biphenyl hydrocarbous

8.2 CONTROL OF SUSPENDED PARTICULATE MATTER IN INDUSTRIES

In various industries, the control of emission of Suspended Particulate Matter (SPM) is done by various methods including 'Inertial dust collectors', 'Cyclone', 'Battery cyclones', 'Electrostatic Precipitators (ESP) of dry type, Wet type Electrostatic Precipitator (Wet ESP), 'Bag house/ fiber filters / bag filters', and 'Scrubbers'. Each industry would use suitable, effective and economically viable method of control of SPM. Sometimes several types of dust separators are used in combination. It is important to mention that toxic and harmful solid micro pollutants (Fe, Cu, Cd, V, Pb, Ni, Cr, Zn, Co, As, Sb, Sn, Be, B, Mn, silica, mica, fibers (asbestos, or other materials etc.), which accompany the suspended particulate matter in the exhaust gases are generally removed with particulate matter itself. The application of various types of dust collecting equipments in industries is briefly discussed below.

8.2.1 Inertial Dust Collectors and their Applications

The inertial dust collectors are used in combination with other types of dust collectors or as an integral part of an equipment design.

Baffles and Baffled Chambers

Baffles are extensively used to change the direction of motion of particulate laden gases in many situations. These are a type of inertial dust precipitators and are used as an integral part of some dust precipitating systems.

The baffles break the inertia by impingement and change the direction of motion which results in separation of coarser suspended particles from the gas stream and these particles fall at the bottom of chamber or hopper. The baffle can be any type of solid body or plate in the gas stream so that the gas has to pass around it. The Baffle type separator sometimes may consist of straight or curved strips or plates staggered in several rows. When the gases are passed through it, they are subjected to direction changes in motion many times resulting in separation of grit/ dust / material particles.

Gravity Settling Chambers

The Gravity Settling Chambers are least effective in removing dust. These are used only for some industrial requirements along with other type of dust collectors. The baffles can be used in a gravity chamber also to make them more effective for dust precipitation

The gravity settling chambers work on the principle that the velocity of in-coming gas is drastically reduced due to sudden change in section (flow area) and therefore the dust particle settle out under the force of gravity and fall to the bottom of the chamber. Their efficiency would not exceed 60%. The minimum particle size which can be separated by gravity chambers can be Calcuilated from Stoke's low and is given by the equation:

$$dp = \text{Diameter of particle} = [C.\mu.q/g.B.L.(\rho_s - \rho_g)]^{\frac{1}{2}}$$

Where '*C*' is a constant depending upon units of various quantities

μ = absolute gas viscosity, g = acceleration due to gravity, q = Volumetric Flow rate of gas,

B = Breadth of chamber, L = Length of chamber,

ρ_s = True density of solid in the gas

ρ_g = Density of gas.

(If μ in pounds per foot-sec, *g* in feet per sec per sec, *B* in feet, *L* in feet, and ρ_s and ρ_g in pounds per cubic foot, diameter of particle is in feet then the value of constant '*C*' is 18)

As a rule of thumb, Stoke's law applies if the porticle size dp <100 µm.

Dust Traps

Simple inertial 'Dust trap' based on change in direction of gas is used at many places in power stations and industries. [see ***Fig. 4.1 Chapter 4]***

8.2.2 Industrial Application of Mechanical Precipitators /Battery cyclones/ Cyclones

These equipments for collection of SPM can be used in the following:

- Wood, bark, or bagasse fired boilers and also coal fired stoker type boilers in industries

- Various furnaces in industrial applications
- Brick kilns, lime kilns
- Processing of : 'Cement clinkers', 'Phosphate ore', 'Gypsum', 'Alumina', 'Lime', 'Bauxite', 'Sinter from blast furnace', and various other materials
- Roasters for ore or other materials
- Petroleum refinery catalyst recovery
- Coal drying
- Machining operations; and wood working operations
- Material crushing and grinding operations for any solid material in any industry
- Incinerators

As already mentioned earlier, the cyclones are an integral part of a 'Circulating fluidized bed combustion (CFBC)' boiler. [*Mechanical precipitators ie 'Multiclones/ Battery cyclones' and 'Cyclones' have been described in **Chapter 4***]. In certain situations these equipments may be used in combination with other types of dust collectors.

8.2.3 Industrial Application of Dry and Wet Electrostatic Precipitators

The electrostatic precipitators (ESP) in industrial use are of two types, namely 'Dry type Electrostatic precipitator (ESP)' and 'Wet type Electrostatic precipitator (Wet ESP)*. The dry type ESP has already been described in ***Para 4.3, Chapter 4***, while the Wet ESP is described below. Both dry and wet types of ESP are used in industry according to suitability.

**Wet Electrostatic Precipitator (Wet ESP)*

The 'Wet ESP' is an ESP where water or liquid is used to remove the solid particulate matter instead of rapping of electrodes. (Rapping of electrodes is employed in dry type ESP). In the 'Wet ESP', the gases are cooled to saturation temperature before entering its 'electric field'. The wet dust is easily collected after the precipitator is steamed and the temperature inside it adjusted to 85—90 °C. The collected dust is removed by flushing with hot water/ liquid. Normally the 'Wet ESP' may be vertical- flow, single-compartment, single- stage dust collector. The shell of the wet ESP may be cylindrical made of steel and lined inside with 'acid- resistant bricks' laid over a bed of 'poly-isobutylene' or other material. The collecting electrodes are lead plated and may be shaped as hexagonal honey-combs, while the discharge electrodes may be lead -plated star -shaped cross section. The shape and design may differ from manufacturer to manufacturer. *A general view of a typical 'Wet ESP' is shown in **Fig. 8.5.***

Wet ESPs have found use for cleaning gases of Sulphuric acid mist, selenium and arsenic at sulphuric acid plants after the dust is removed by dry ESP. The oxides of sulphur which are emitted during metallurgical processes of non- ferrous metals are being normally used for manufacture of sulphuric acid. The discharge electrodes are suspended by tie rods from cross pieces fixed to supporting insulators set in boxes. The insulator heating is provided for preventing condensation of acid on the insulators. Very high collection efficiency can be obtained by operating wet ESP at low gas velocities of around 0.5–0.6 m/s.

Wet ESP is seldom used as sole dust separator; but is used in combination with several other types of dust separators as integral to a dust precipitation / collection system. Wet ESPs are good for collection of high resistivity particulate matter, while dry type ESP's have difficulty in collecting such particulate matter.

Application of Dry Type ESP

Dry type ESP may find application in the following industries:

- Cement industry for cleaning of flue gas from cement kilns and for recovery of cement dust from kilns
- In processing of : phosphate ore, gypsum, alumina / bauxite, roasting of ores of non ferrous metals or any other solid material where material dust is likely to be emitted
- Sintering machines (Iron oxide and fluxing dust)
- Cleaning of blast furnace gas intended to be used as fuel
- Open hearth furnaces, basic oxygen furnaces, electric arc furnaces (steel plants) – cleaning of gases
- Coke ovens for cleaning coke oven gas
- Nonferrous metal (zinc, lead, copper, etc.) metallurgical operations— ore roasters, smelters, sinter kilns, converters
- Incinerators
- Elemental phosphate furnaces
- Sulphuric acid production
- Cleaning of various types of gases in chemical industry
- Petroleum cracking and catalyst recovery (Petroleum industry)
- Pulp and Paper manufacturing industry (Wood and bark boilers, lime kiln, and black liquor recovery boiler)
- Carbon black
- Glass manufacturing

As already mentioned, 'Wet ESP's' are used in combination with other types of dust collectors. [For an example of application of 'Wet ESP' refer to ***Para 8.4.5 and Fig. 8.4*** *of this chapter.*

8.2.4 Industrial Application of Fabric Filters

Fabric filters or bag house can be used for collection of suspended particulate matter / dust in several industries such as:

- Processing of gypsum, alumina, phosphate, chalk, lime, titanium dioxide, Manufacture of any powdered material, flour mills. carbon black production
- Fertilizer production
- Pesticide powder manufacture
- Lead furnaces
- Cattle feed factory
- Machining, wood working operations
- Along with 'Spray dry process' of SOx removal
- Metallurgical industries including electric furnaces in steel plants or other industries.
- Cement industry

[For description of bag or fabric filters/ bag house, see ***Para.4.4, Chapter 4].***

8.2.5 Application of Various Types of Scrubbers

8.2.5.1 Wet Scrubbers

Wet scrubbers are widely used for dust removal in various processes where gas flows are low and sufficient water/ liquid effluent for scrubbing is available. The wet scrubbers are efficient for effectively controlling emission of SPM. Wet scrubbing besides removing particulate matter also removes soluble gaseous pollutants, and other toxic micro particles which is an added advantage of this method. The effluent from scrubber would contain the removed suspended solid as well as soluble solids and gases. The safe disposal or use of the waste water from such scrubber needs to be considered so that it may not cause water pollution. There are several types of wet scrubbers as mentioned below:

(a) Wet Cyclone Scrubber

If water spray is introduced in a cyclone collector, the dust collection becomes more efficient, while the spray of water checks the re-entrainment of the dust in the gas stream. These scrubbers are used in various processes (e.g. *used at non-ferrous metallurgical plants).* These may be used in conjunction with other types of scrubbers to remove particulate matter.

(b) *Venturi Scrubber*

The dust/ particulate matter laden gases are passed at a high velocity through a venturi converging cone which terminates in a small length of cylindrical cross section (throat), where the velocity of gases is the maximum. The throat velocity may be more than 60 m/s. The water is introduced ahead of venturi throat. When the water is introduced, it disintegrates into fine droplets. The energy required for atomizing the water is mainly taken from high speed gas stream. High degree of turbulence is conducive to flocculation of particulate matter. The gases then travel through a diverging cone and their velocity falls. The converging cone, throat and diverging cone all form a high speed atomizer. The coarse liquid droplets with entrapped particulate matter from the high speed atomizer are then led to a wet cyclone for removal of dust.

The optimum dimensions may be taken as:

Diameter of throat = D; Length of throat = 0.15 to 0.5 D;

Length of converging conical portion = $3D$;

Diameters at gas inlet may be about 2.3 D;

The diameter at gas outlet from diverging conical portion may be similar to dia at the inlet of converging portion. In case of peripheral water inlet, the location of water inlet is 1.55 D ahead of the throat.

Venturi scrubbers are efficient, but can be used for small gas flows. Efficiency of removal of particulate matter is lower for smaller particles. These can be used for variable and intermittent small gas flows. Its applications may be in 'Paper mill furnaces', small 'Metallurgical furnaces', 'Sulphuric acid plants', etc. The disadvantage in venturi system is the '*Severe erosion* of venturi throat' because of high velocity.

A type of venturi scrubber with peripheral water inlet is shown in **Fig. 8.1a.** *[The venturi scrubber may be set in vertical or horizontal fashion.]*

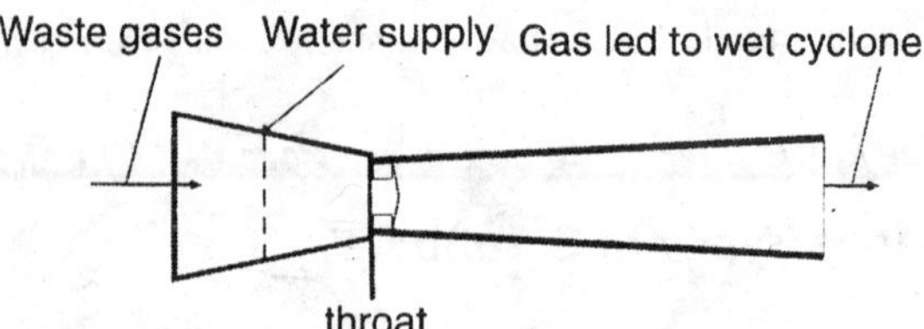

Fig. 8.1(a): Schematic Diagram of Venturi Scrubber with Peripheral Liquid Injection

(c) *Spray Tower*

In the spray towers (empty), the gas travels from bottom to top, while the water is sprayed from top of the tower, the falling fine drops of water remove dust, but some fine dust would still be carried away with the gas. Such spray towers are used as a means of cooling, primary cleaning and separation of the particulate matter and cinder from the gases before final dust collector. These can be used in treating blast furnace gas.

(d) *Packed Tower Scrubber*

These are spray towers with packing material inside it. The packing may be of fiber glass / ceramic or porcelain / or other type, with water flow from top to bottom and gas flow from bottom to top. The cleaning of packing material is often cumbersome and therefore packed tower scrubbers for removal of particulate matter are used only in limited processes.

(e) *Mechanical Scrubbers*

Such scrubbers are provided with mechanical means such as rotating discs for shattering / disintegrating liquid into fine droplets which are thrown radially by centrifugal force. The dust laden gas while passing through scrubbers gets cleaned. These types of scrubbers are not in use because of higher capital cost and operation and maintenance cost.

(f) *Tray type Scrubbers*

Such a scrubber has a cylindrical housing in which metal trays with holes are arranged in vertical series. The holes disperse the gas and create bubbles in the liquid above the plate. The bubbles formed provide the interfacial area. The gas flow is counter current to the liquid flow; the gas flows from bottom to top and the liquid from top to bottom.

Use of Wet Scrubbing

Wet scrubbers can be used for controlling emission of particulate matter/ dust. Also these types of scrubbers serve as 'Chemical scrubbers' when the chemical solution is used instead of water for removing soluble gases/ materials. These scrubbers can be used in the following industries.

- Coal / lignite crushing and handling plants at coal / lignite mines, coal washeries, and industries for dust control.
- Pulverized fuel preparation plants for pulverized fuel fired boilers.
- As a part of some flue gas desulphurization processes of exhaust gases containing oxides of sulphur.
- Kraft paper mills.
- Phosphate, gypsum processing.
- Steel plants at some locations, foundries, and hot coating processes.
- Lead furnaces and pyrite roasting.
- Metallurgical processes for production of non ferrous metals.
- Production of sulphuric acid, phosphoric acid, nitric acid.
- Incinerators.
- Lime kilns, various solid material crushing or handling plants.

Table 8.1A mentions the type of scrubber generally used depending on the range of particle sizes, and the loading.

Table 8.1A: Type of Scrubber Used for Different Size Range and loading of Particulate Matter

Particle diameter range μm	Particle loading	Type of scrubber used
0.3 to 1000 μm	High	Centrifugal separator
0.01 to 100 μm	2-3 grains/ft^3 (4.6-6.9 g/m^3)	Venturi separator
0.01 to 1000 μm	Less than 1 grains/ft^3 (2.3 g/m^3)	Packed tower
0.01 to 1000 μm	1-2 grains/ft^3 (2.3-4.6 g/m^3)	Tray tower scrubber

[one Ounce (avoidupois = 16 drams = 437.5 tory grains; one tory ounce = 480 tory grains = 31.1 gram]

8.2.5.2 Chemical Scrubbing / Absorption system

Scrubbing with appropriate chemicals is often used to remove some polluting gases, vapours, etc. from the exhaust gases. These pollutants are dissolved, absorbed, or oxidised / reacted upon and converted to another materials by these chemicals. When the target pollutant is oxidized, the scrubber is called *oxidising scrubber.* In wet absorption system, there are some key points which are very important for its performance such as:

- Saturation of gas by liquid vapour [the efficiency is directly affected by the degree of saturation of gases at the operating temperature.]
- Gas and solvent distribution across the cross section of reactor / scrubber.
- Proper mixing of gases and sorbent/ solvent

The wet absorption system is common for industrial use. These absorption equipments may be of various types such as stage wise absorption equipment, packed columns or towers, and high energy contactors.

(i) *Stage Type Absorption*

The stage absorption equipment has a group of horizontal metal plates or trays, arranged in a vertical fashion in a cylindrical housing. Each horizontal plate is one stage of absorption. The plates can be bubble type equipment, sieve trays, ballast trays.

In the bubble type equipment, the gas flows through small nozzles set in holes under each bubble cap and goes out through the serrated rims of the caps. Bubbles and foam are created in liquid tray. The bubbles provide interfacial surface for absorption.

The sieve tray type absorptin equipment is similar to bubble type, but there are no caps and the plates have many small holes. When the gas rises through the holes, bubbles are created which provide interfacial surface for absorption.

The ballast type absorption equipment is similar to sieve tray type except that there are floating plugs over the holes in the tray. The gas lifts the plugs and bubbles through the liquid creating intimate contact for absorption. In case when gas is shut, the plugs drop down to plug the holes.

Although the stagewise absorption equipments are very efficient absorption systems, but the pressure drop in them is higher than in packed towers. These absorbers can operate with low liquid flow rates and can handle particulate loading well.

The Stage wise absorbers may achieve the following efficiencies of absorption:

Hydrochloric acid gas—up to 99%; Sulphur dioxide—up to 99%; Soluble VOCs—up to 99%

(ii) *Packed Column or Towers*

The vertical packed towers are common with gas entering bottom and liquid / solvent trickling from top. The packing may be structured or random. The structured packing has more interfacial surfaces per unit volume for absorption, but has higher pressure drop and more problems of plugging than the random packings. These systems have low energy consumption and good efficiency, but have high water/ liquid consumption and cannot handle high particulate loading. The packed tower type absorbers may achieve the following efficiencies of absorption.

Hydrochloric acid gas—up to 99%; Sulphur dioxide –up to 99%;

Nitrogen dioxide—60–80%; Soluble VOCs—up to 99%

(iii) *High energy contactors*

Venturi scrubbers are a type of high energy contactor which has been described earlier. These types of equipment can handle highly reactive gases and high particulate matter loading, but have high energy consumption and high water/ liquid consumption. The high energy contactors may achieve the following efficiencies of absorption.

Hydrochloric acid gas—up to 90%; Sulphur dioxide—up to 90%;

Nitrogen dioxide—up to 50%; Soluble VOCs—up to 70%

Type of scrubber used for different applications is given in **Table 8.1B**.

Table 8.1B: Type of Scrubber Used Based on Application

Application	Source of Gas	Particle size range µm	Type of scrubber
Particulate removal	Iron, coke & silica dust	0.1—10 µm	Venturi scrubber
Acid Fume	Sulphuric acid mist	------------	Venturi with cyclonic spray
Acid and fumes	Titanium dioxide, and hydrogen chloride fumes	0.5—1.0 µm	Venturi scrubber
Cement particulate	Cement dust	0.5—55 µm	Venturi scrubber
Incinerator exhaust	Fly ash, hydrochloric acid, sulphur dioxide	0.05—1.0 µm	Venturi with packed tower
Ammonia removal	Ammonia, ammonium chloride	-----------	Packed tower
Paper industry	Hydrogen sulphide, particulate, sodium carbonate	0.5—50 µm	Venturi with packed tower / tray tower
Chemical industry	Acetic acid fumes, particulate	0.5—10 µm	Tray tower

8.2.5.3 Dry Scrubbers

Dry absorption/ adsorption systems are similar to chemical reactors which use dry bulk chemicals. Dry scrubbing / absorption/ adsorption systems with a suitable powdered material (like activated carbon, alumina, etc.) is sometimes used for purposes of adsorbing / absorbing of vapours, gases, ionic material, or solid compound dust.

In the manufacture of aluminum, 'dry scrubbing by alumina powder' is advantageous as it captures fluorides, Polycyclic Aromatic Hydrocarbons (PAH) / other hydrocarbons, oxides of nitrogen (NO_x), and oxides of sulphur (SO_x). The 'dry scrubber systems with aluminum oxide absorbent' is used for control of emissions from reduction cells and from anode bake ovens. Use of dry scrubbing system with aluminum oxide as adsorbent for on-site baking of anodes has the advantage of capturing all the waste products and making it possible to recycle all the enriched alumina and absorbed material directly into the reduction cells. Dry scrubber systems applied to the pot fumes and to the anode baking furnace result in the capture of 97% of all fluorides from the process. ***[Also see Para 8.4.3.3 ahead]***

In some industrial processes, dry scrubbing/ absorption system may be combined with incineration for controlling emissions of tar and volatile organic compounds (VOCs) and to recover energy. There are three types of dry absorption systems namely: Dry scrubbers, Spray dryers, Fluid bed reactors.

The dry scrubbers can handle hot flue gases and high particulate loadings. Dry sorbent or dry absorbing/ adsorbing material is injected into the gas stream. Mixing of sorbent with the gases is the key to the performance. The absorption system has dust collector at the end.

The spray dryer utilizes a wet solvent into a hot stream of gases where the liquid is evaporated and the solvent is dried which then mixes with the gases to perform its function. It can handle high particulate loading and hot gases. The particulate matter is removed at the end of absorption by cyclones and bag house. ***[Also see Chapter 7, Para 7.4.1.]***

Fluid bed reactors use granulated sorbing material which is fluidized within the reactor to absorb or adsorb the targeted pollutant gas. The sorbent is retained in the bed material. The material going with outlet gases is collected along with the dust by cyclones/ or other type of dust collector and part of it is recirculated. ***[Also see Chapter 7, Para 7.4.3 .]***

Adsorption

The adsorbent has a limited life and it has to be regenerated or replaced. The materials used as adsorbents are: Alumina; Chemical coated alumina; Silicates like aluminium silicate, calcium silicate, and magnesium silicate; Silica gel; Clay; Fuller's earth; Activated carbon; Organic polymers; etc. Adsorption is used for the following purposes in industries.

- Purification of air intake or air in circulation
- Purification of exhaust gases for removal of toxic gases, unwanted components and odours.
- Sorbent recovery from air/ gas leaving a process such as in dry cleaning, spray painting, etc.
- Fractionalization of gases.

Note: *Wet scrubbing is not recommended in aluminum manufacture, since liquid effluent, high in fluorides and hydrocarbons will require extensive treatment and disposal.*

8.2.5.4 Bio-scrubbers / Bioreactors / Biofilters

Although biological treatment is most often practiced for sewerage/ waste water, but the removal of some pollutants from gases by microorganisms by biological treatment is also very advantageous. Hydrocarbons and odorous emissions can be removed by biological treatment in bioscrubbers / biofilters. The biological filters/ scrubbers consume very little energy and the treatment may be carried out at ambient pressure and temperature without creating any NO_x emission. Natural biological degradation and immobilized microorganisms are used in biological treatment / control systems. The biological systems for removal of pollutants may be of the following types:

- Fixed film reactors — Biofilters and trickle bed reactors
- Suspended growth—Bio- scrubbers

The biodegradation reactions by microorganisms oxidize the hydrocarbons (VOCs) into CO_2 and H_2O. The sulphur containing organic compounds on biodegradation may form inorganic acids. (Some sulphur may be consumed by specific species of microorganisms for their growth.).

The bioscrubber is essentially a packed bed in which the microorganisms are present (suspended) in the scrubbing liquid which is called 'suspended growth slurry'. The suspended material is termed as 'floc'. The inert packing of the scrubber serves the purpose of increasing contact surface. The scrubber system has a slurry tank in which activated sludge is kept in suspension by stirring and mixing. Air is also bubbled through in the slurry tank for providing oxygen. The slurry is pumped to scrub the incoming gases in the scrubber. A schematic diagram of bioscrubber is shown in the **Fig. 8.1 b(i)**

As compost mixed with peat, wood chips, bark, leaves, etc. can be a good bio- material containing necessary microorganisms, it can be used as filtering material for removing pollutants. A biofilter containing one or two beds of above material (of less than 1m height) along with a compost slurry mechanical recirculation system can be used. Such a system with a single layer (bed) of biological filtering material is schematically shown in **Figure 8.1 b (ii).** The waste gas enters at the base of the filter and diffuses into biologically active layer attached to the particles of filtering material. Aerobic reaction occurs in the biofilm where the pollutants are metabolized by the microorganisms. The resulting gaseous products go out from the top of the filter. Some water is sprayed on the bed material to provide moist environment for the growth of microorganisms and aerobic conditions are maintained.

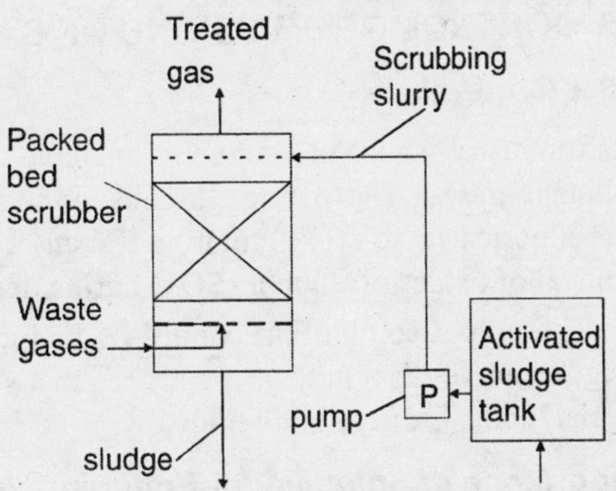

Fig. 8.1 b(i): Packed Bed Bio-Scrubber With Activated Sluge Tank

Three types of microorganisms are generally present in biofilter, namely—fungi, bacteria, and actinomycetes. The actinomycetes are organisms which resemble both the bacteria and fungi in some ways. The growth and activity of microorganisms is dependent on the environment of biofilter ie the oxygen supply, absence of toxic materials, ample inorganic nutrients, optimum moisture conditions, appropriate temperature, near neutral pH range.

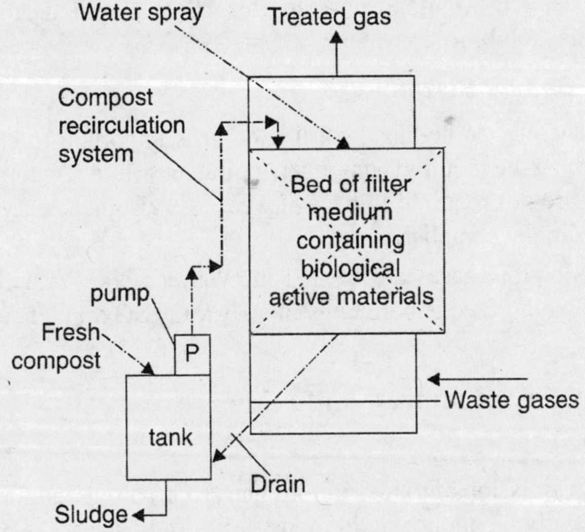

Fig. 8.1 b(ii): P—pump

Biofilter with a Bed of Biologically Active Material and Recirculation System

In a trickle bed reactor, there is a fixed biofilm attached to packing material in the bed and liquid/ water trickles through the bed. The packing elements are larger than in a bioscrubber. The gas to be treated enters at the bottom of the reactor, while treated gas goes out from the top. The effluent liquid is collected in a tank and recirculated / sprayed from the top of the bed. A schematic diagram of trickle-bed filter is shown in **Fig. 8.1 b(iii).**

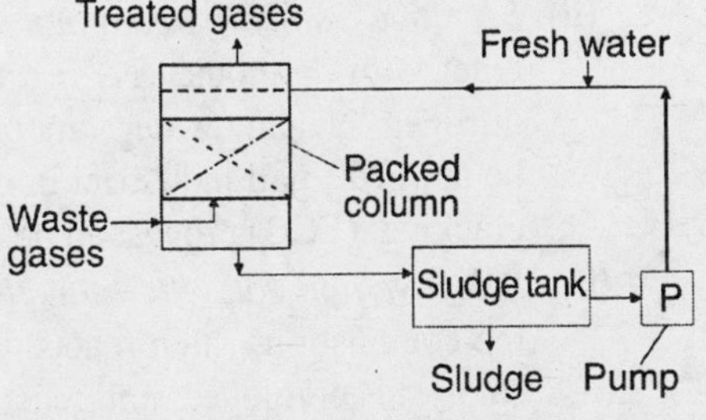

Fig. 8.1 b(iii): Trckle-Bed Bioreactor

8.3 CONTROL OF EMISSIONS OF GASEOUS POLLUTANTS IN INDUSTRIES

8.3.1 General

Each industry would use suitable, effective, and economically viable method for control of various pollutant gases. The acidic gaseous pollutants like hydrochloric acid gas (HCl), chlorine (Cl), hydrofluoric acid gas (HF), fluorine (F), etc. are mostly removed from the exhaust gases along with the removal of oxides of sulphur (SOx) during desulphurization.

The proper combustion control of fuel can bring down the level of CO in the exhaust gases to insignificant level. It may be mentioned that some pollutants in a particular industry may require some special methods of control/ removal.

Importance of Solubility of Pollutants in Liquid Sorbents

Solubility of gases in liquids plays an important part in cleaning the gases from gaseous chemical impurities. The amount of a particular gas that can be dissolved in a liquid depends on:-

The properties of the gas and absorbing liquid (sorbent), the temperature of sorbent liquid and the partial pressure of the gas. Higher the partial pressure of the gas, greater is the amount of the gas which can be dissolved in the sorbent liquid. Lower the temperature of liquid sorbent, the greater is the amount of the gas dissolved in it. Ammonia gas, HCl gas, HF gas, etc. have high solubility in water. The chlorine, etc. have medium solubility in water.

Capture of Polluting Gas by Scrubbing

When the gases are cleaned by scrubbing with liquids, it is important to choose suitable liquid with ample capacity for absorbing the required component pollutant, while remaining un-reactive to the other component gases which are not required to be removed. It is also necessary to have adequate contact between liquid and gases in the scrubber.

If the exhaust gases or flue gases are treated in 'Wet scrubbers' for removal of SPM/ RSP, as mentioned earlier, the soluble gaseous pollutants like HCl, Cl, SOx, HF, F, etc., and soluble vapours / fumes are also removed..

8.3.2 Control of NOx Emission from Industries

For Industrial Boilers:

The methods of control of emission of oxides of nitrogen (NOx) discussed for fossil fuel power stations in ***Chapter 5*** would generally be applicable in industries also. In short, the principle of these methods and their applicability to industries is mentioned below:

(i) *Low NOx burners, Overfire air, and Air- stage firing:* These provisions can also be applied in industrial boilers like boilers in power plants.

(ii) *Selective Catalytic Reduction (SCR)*: In this process, the oxides of nitrogen NO and (ie NOx), react with ammonia (N) in the presence of a flow-through honeycomb catalyst to form nitrogen (N_2) and water vapour (H_2O). This process of NOx removal from exhaust gases can be applied to all industrial boilers, heaters, furnaces, incinerators, as well as 'Fluid catalytic cracking' (FCC) process in petroleum refinery.

(iii) *Selective non catalytic reduction (SNCR):* In this method, ammonia or urea is added during the combustion which reacts with NOx to produce molecular nitrogen and water vapour. No catalyst is employed, and reaction takes place in a narrow high-temperature range. This technology can also work on industrial boilers (fired by fossil fuel or other fuels), fired heaters, furnaces and incinerators, but the efficiency of NOx removal by this is quite low, and therefore is not generally used.

(iv) *Water Scrubbing:* Since NO has a very low solubility in water, the scrubbing with water for 'mainly removing SPM and SO_x can remove NO only very slightly, while it may remove comparatively more soluble constituent of NOx L.e. NO_2 to a better degree.

NOx Reduction Methods specific to Diesel Engines and Gas turbines:

For diesel engines refer to ***Para 13.3.3.4 of Chapter 13*** and for gas turbines, refer to ***Para. 5.7, Chapter 5.***

8.3.3 Control of SOx in Industrial Exhaust Gases

8.3.3.1 SO_x Control Methods Discussed in Chapters 6 and 7—Applicability

Generally the methods of control of emission of SOx discussed for fossil fuel power stations in Chapter 6 and 7 may be applicable in industries. In short, the applicability of these methods for industries is mentioned below:

(i) Water scrubbing is universally applicable for removal of SPM and soluble gases.

(ii) Post combustion FGD (Flue gas de-sulphurization) methods for removal of **SOx** both non-regenerative and regenerative, may be applicable to the exhaust gases from industrial boilers, and furnaces of other industrial processes as in power plants.

(iii) The in-furnace removal of $\mathbf{SO_x}$ may be applicable in some specific cases in the industries.

(iv) In cases of some industries, if the use of 'Low sulphur raw materials' and 'Low sulphur fuels' is economically feasible, the emission of SOx can be lowered to a great extent.

Note: *In the non-regenerative FGD, the sorbent material used for absorption /adsorption of SOx is not recovered while in regenerative FGD process, it is recovered for reuse / recycling.*
Non-regenerative FGD processes include Spray dry process, Lime stone sludge process, Lime / lime stone gypsum process, sea water scrubbing, Dual alkali process, Double contact flow scrubber, and Selective catalytic reduction for NOx, SOx, etc.
Regenerative processes include Wellman Lord Process, Magox process, and Activated charcoal process, etc. During the recovery of sorbent, SO_2 is recovered which is either converted to sulphuric acid or elemental sulphur while the recovered sorbent is recycled to the process.

8.3.3.2 De-SO_x Processes based on Absorption by Ammonium Sulphite or Zinc Oxide

There are some other chemical processes which are used sometimes for 'Capturing the sulphur dioxide' from industrial processes for utilizing it for the manufacture of 'Sulphuric acid' or 'Elemental sulphur'. These processes achieve high recovery rates of . Three such processes of 'SO_2 capture' are described below:

(i) ***Ammonia Cyclic Regenerative Process:***

This process uses Ammonium sulphite [$(NH_4)_2SO_3$] solution as sorbent for from the 'de-dusted' exhaust gases (the particulate matter/dust is to be removed from exhaust gases before gases are sent to SO_2 absorption tower).

The $(NH_4)_2 SO_3$ solution is profusely sprayed from the top of the absorption tower, while the de-dusted exhaust gases flow from bottom to top. Two or more absorption towers can be used in series for completing the SO_2 absorption. After absorbing the SO_2, the ammonium sulphite in the solution is converted to ammonium bisulphite ($NH_4 HSO_3$).

The gases after passing through the absorption tower are led to a wet scrubber to remove any remaining SO_2 and other gases/ chemicals soluble in water. The absorption reaction of SO_2 would be as follows:

$$(NH_4)_2SO_3 + SO_2 + H_2O = 2NH_4HSO_3.$$

On boiling the ammonium bisulphite, the '*pure* SO_2 ' is evolved which can be used for manufacture of 'sulphuric acid' or 'elemental sulphur'.

$$2NH_4HSO_3 = (NH_4)_2SO_3 + SO_2 + H_2O$$

Ammonium sulphite vapour is separated out by condensing and is then recycled, while the pure SO_2 is sent for utilization. This method is useful in cases of removal of SO_2 from exhaust gases from metallurgical processes. A schematic diagram of the ammonia cyclic method of enriching SO_2 is shown in **Fig. 8.1c**.

(ii) ***Ammonia Non-Cyclic Non- Regenerative Process***

In this process also, the SO_2 is also absorbed by ammonium sulphite [$(NH_4)_2SO_3$] solution according to the reaction:

$$(NH_4)_2SO_3 + SO_2 + H_2O = 2NH_4HSO_3.$$

The resultant ammonium bisulphite [NH_4HSO_3] may be divided into two parts. One part is treated with acid say sulphuric acid, phosphoric acid or nitric acid to recover sulphur dioxide for utilization as follows:

$$2NH_4HSO_3 + H_2SO_4 = (NH_4)_2SO_4 + 2SO_2 + 2H_2O \text{ [No heating required]}$$

The ammonium sulphate [$(NH_4)_2SO_4$] can be used as fertilizer. The remaining part of Ammonium bisulphite [NH_4HSO_3] is neutralized with Ammonium hydroxide (NH_4OH) i.e. ammonia solution for converting it into $(NH_4)_2SO_3$ for recycling in the process:

$$NH_4HSO_3 + NH_4OH = (NH_4)_2SO_3 + H_2O$$

Since only a part of the sorbent is returned to the process, it is a non regenerative process.

(iii) ***Zinc Oxide Regenerative Process:***

This process is also a regenerative process and uses zinc oxide (ZnO) slurry for absorbing SO_2 from de-dusted exhaust gases from metallurgical processes. The zinc oxide slurry is sprayed from top of the absorption tower, while the gases flow from bottom to top. On absorbing SO_2, the zinc oxide is converted to zinc sulphite crystals (having water of crystallization). The Zinc sulphite crystals are separated by filtering and are sent to furnace for heating where it gives out the SO_2 and is again converted

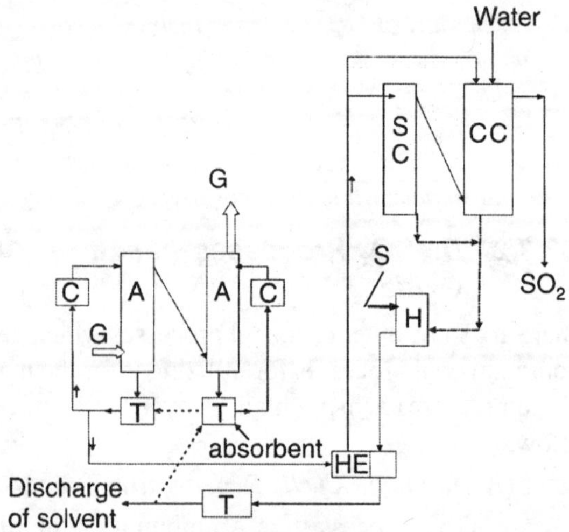

Fig. 8.1 c: Ammonia Cyclic Method of Enriching Sulphur Dioxide

to 'zinc oxide'. The products of reaction in furnace are condensed to separate out zinc oxide, which is sent for recycling and the SO_2 is led for utilization. The reaction in the process is as follows:

$$SO_2 + \text{Zinc oxide slurry} = ZnSO_3.nH_2O \text{ (zinc sulphite crystals)} + \text{water}$$

$$ZnSO_3.nH_2O = SO_2 + ZnO + nH_2O \text{ (water vapour)}$$

8.3.4 Removal of Hydrogen Chloride Gas from Industrial Exhaust Gases

Waste gases from a number of industries may carry HCl (hydrogen chloride gas or hydrochloric acid vapours). When the 'HCl content'* in gases is small and other soluble gases are present, this acidic gas can be eliminated from the exhaust gases by:

- Cold water scrubbing – The resultant effluent should be neutralized by any alkali say lime which is cheapest one.
- Scrubbing by milk of Lime [$Ca(OH)_2$ suspension in water]– Venturi scrubbers can be used.
- Scrubbing by soluble alkalis like soda ash [Na_2CO_3] solution in water, or caustic soda solution in water
- Scrubbing by Magnesium hydroxide [$Mg(OH)_2$] suspension in water

***Note:** *If the content of HCl in exhaust gases is high, hydrochloric acid of 27 % strength can be obtained. The vapour pressure of HCl over a 27% hydrochloric acid is about 14.2 mm Hg at a temperature of 40 . It may be interesting to note that if the HCl content of gases at atmospheric pressure is more than 1.87 % (i.e. 14.2 × 100 / 760), 27 % hydrochloric acid can be produced provided the liquor temperature does not increase beyond 40°C. The dissolution of HCl in water gives rise to evolution of heat and the heat must be extracted during successive absorption for production of hydrochloric acid of required strength. It is also important that if the intention is to produce hydrochloric acid from these gases, the gases must be cleaned of dust and other impurities before going in for production of hydrochloric acid from it. In case, the gases also contain oxides of sulphur, the production of hydrochloric acid may be very difficult as both acid gases are soluble in water.*

8.3.5 Chlorine Removal from Industrial Exhaust Gases

The chlorine (Cl_2) is removed from the exhaust gases by absorption by milk of lime* [$Ca(OH)_2$] or by milk of magnesium hydroxide [$Mg(OH)_2$]. Reactions of Chlorine with milk of lime or milk of magnesium hydroxide would be as mentioned below:

$$2Cl_2 + 2Ca(OH)_2 = Ca(OCl)_2 + CaCl_2 + 2\ H_2O$$

$$2Cl_2 + 2Mg(OH)_2. = Mg(OCl)_2. + MgCl_2 + 2H_2O$$

These reactions hold when there is excess hydroxide in the absorbing liquor, but if all the hydroxide is spent, chlorine would react with water to form hydrochloric acid and hypochlorous acid. The reaction is reversible and hence the absorption is adversely affected. Therefore there should always be excess amount of hydroxide present in the absorption liquor.

$$Cl_2 + H_2O = HCl + HOCl\ ;\ HCl + HOCl = Cl_2 + H_2O$$

The hypochlorites of calcium or magnesium are present as pollutants and should be removed from effluents by interaction with iron fillings or decomposed by heating with catalyzer like copper/nickel salts.

Packed scrubbers or venturi scrubbers can be used for removal of chlorine by milk of lime. As chlorine and acidic gases are very corrosive, anti-corrosive treatment is given to all equipments handling chlorine or any acidic gas.

***Note: Bleaching powder**

Following reactions are of interest in connection with the removal of chlorine by treatment with lime and subsequent formation of bleaching powder.

$$2Cl_2 + 2Ca(OH)_2 \quad = \quad Ca(OCl)_2 + CaCl_2 + 2\,H_2O$$

The calcium hypochlorite [$Ca(OCl)_2$] if treated with hydrochloric acid (HCl) gives out chlorine in purer form which can be used for making various products

$$Ca(OCl)_2 + 4HCl \quad = \quad CaCl_2 + 2H_2O + 2Cl_2$$

The calcium chloride [$CaCl_2$] with excess of calcium hydroxide [$Ca(OH)_2$] can form basic calcium chloride [$CaCl_2.Ca(OH)_2.H_2O$]

Bleaching powder is a complex of calcium hypochlorite and basic calcium chloride according to Bunn, Clark and Clifford' view (1935)

$$2Cl_2 + 3Ca(OH)_2 \quad = \quad \underbrace{Ca(OCl)_2 + CaCl_2.Ca(OH)_2.H_2O}_{\text{Bleaching powder}} + H_2O$$

8.3.6 Hydrogen Fluoride Removal from Industrial Exhaust Gases

The gases from Aluminum- reduction cells and Cryolite (*cryolite is* $Na_2\,AlF_6$) plants contain hydrogen fluoride (HF) gas. This gas can be removed from de-dusted gases by:

- Dry scrubbing by Alumina [Dry scrubbing is a prevalent method in case of aluminum manufacture (see Para 8.4.3 ahead].
- Scrubbing by water with the formation of 'Hydrofluoric acid'.
- Chemical scrubbing by alkali solution of soda ash or caustic soda with the formation of 'Sodium fluoride (NaF).

In case of water scrubbing, the silica deposit on the packing of scrubbers may pose difficulty in the use of scrubbers and in collection of hydrofluoric acid. The silica is formed due to the presence of $H_2\,Si\,F_6$ and H_2O which react to form silica. This difficulty may be overcome by increasing the density of spraying and using mechanical absorbers.

The sodium fluoride obtained by chemical scrubbing with alkali solution may be used in the manufacture of pure 'Cryolite (sodium aluminum fluoride $Na_2\,AlF_6$)'. The scrubbing with alkali solution is very efficient and only one stage spraying is necessary.

It may be mentioned that the water absorption is used only when production of hydrofluoric acid is required and would be worthwhile if the content of other water soluble impurities is not very significant. The absorption by water is only partial and three or four (3 or 4) successive stages of absorption (scrubbing/collecting towers) would be necessary. The first stage scrubbing of gases is done by strongest acid (HF) and it yields the finished acid. The last (4th) scrubbing of gases is done by water plus some weakest acid (HF). The liquor from 4th scrubber is sent for scrubbing in 3rd scrubber; from there liquor is sent to 2nd stage scrubber and from there the liquor goes to first stage scrubbing. The liquor is cooled at every stage because the hydrogen fluoride dissolves in water with evolution of heat. A schematic diagram for removal of gaseous hydrogen fluoride from metallurgical gases to obtain hydrofluoric acid is shown in **Fig. 8.2.]**

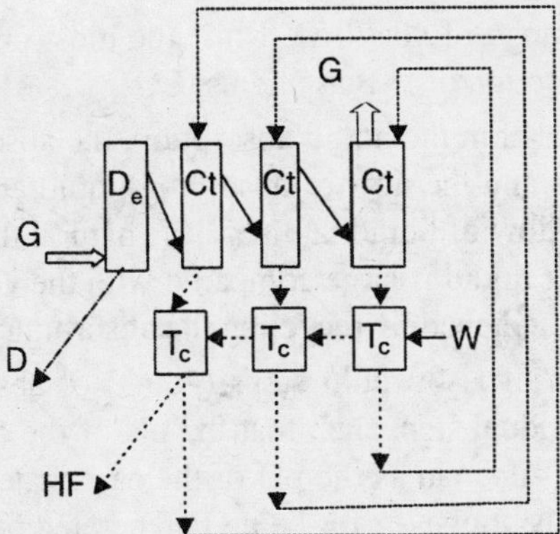

Ct–Scrubbing/ collecting tower; G–Gas; T_c–Tank with cooling coil; D–Dirty Condensate; W–Water; HF– Strong Hydrofluoric acid D_c–Dust removal and cooling without direct contact with liquid

Fig. 8.2: Schematic Diagram for Removal of HF from Gases to Obtain Hydrofluoric Acid

8.3.7 Removal of Hydrogen Sulphide, Carbon disulphide, Carbonyl sulphide Mercaptans, etc. from Gases

Hydrogen sulphide (H_2S), carbon disulphide (CS_2), carbonyl sulphide (COS), mercaptans, etc. are sometimes required to be removed from natural gas, synthesis gas, refinery tail gas, and flue gas from boilers /furnaces / industrial process, etc.

The hydrogen sulphide can be combusted in air to form SO_2 which then can be removed by methods already described. The H_2S can also be absorbed by certain media. Various absorption media are used for this purpose.

8.3.7.1 Removal of Hydrogen Sulphide by Sodium Carbonate Solution

Mostly the hydrogen sulphide (H_2S) and CO_2 occur simultaneously in exhaust gases. Both of these gases are soluble in sodium carbonate solution. The selectivity of H_2S for absorption compared to CO_2 depends on the ratio of H_2S content to CO_2 content of the gases being treated. At higher pH value ie pH higher than 12, the absorption of CO_2 is higher.

In an effort to remedy the above inconvenience in the absorption of H_2S from gases containing both H_2S and CO_2, methods for selective absorption of H_2S have been developed. For instance, efforts have been made to selectively absorb H_2S in solutions of strong oxidizing agents, such as potassium permanganate, sodium dichromate or ferric salts. The other selective methods use alkaline solutions, such as sodium carbonate or potassium carbonate solutions, the operational conditions during the absorption being carefully adjusted. By using a carbonate solution, such as a sodium carbonate solution, (instead of a hydroxide solution, such as a sodium hydroxide solution), the selectivity for the absorption of H_2S against CO_2 can be augmented to about 30-50%.

According to *Swedish Patent Application 9300533–8*,(which was filed on 18 February 1993), for higher*absorption of H_2S as against the absorption of CO_2, the pH range of 10–11 is suitable. The most preferred range is 10.2 to 10.8 and the pH is adjusted continuously by addition of sodium hydroxide* (NaOH) and not sodium carbonate. In the pH range of 10.2 to 10.8, optimum absorption of H_2S is obtained. The temperature of the liquid used in the absorption is not very critical, but preferably it should be between 20°C and 80°C; above 80°C the absorption of H_2S may decrease. More preferably the temperature should be in the range of 40–70°C, while the most preferred range of temperature is 60-70°C for a better selective absorption of H_2S as against CO_2.

To optimize the selectivity to H_2S in the liquid absorption, the absorption should be carried out in such a manner that the flow of gas and the flow of absorbing liquid are counter-current, and the flow of gas should be turbulent and the flow of liquid laminar. It is natural that the separation of H_2S would be better if the volume of absorbing liquid is large compared with the volume of gas containing H_2S. A higher ratio of liquid to gas can be achieved by recycling the absorbing liquid.

Instead of one step absorption of H_2S, multi steps (2or 3) of absorption have the advantage of shortening the length of each individual step, such that the pH of the carbonate solution does not fall below range in the individual step. As soon as the pH of the carbonate solution at the end of the step falls to about 10.0–10.2, the liquid is drawn off for being regenerated by means of NaOH and then it is recycled.

It has also been found that the removal H_2S in the above type of process can be made more efficient, while drastically reducing the chemical consumption, if the method is divided into two stages, namely a first stage, in which the major part of the H_2S is removed in accordance with the method described above ie in 2 or 3 steps, and a second stage, in which the remaining H_2S (a minor portion) is essentially removed by combustion to SO_2, which is then absorbed in an alkaline solution in a wet scrubber.

As the selectivity of H_2S compared to CO_2 depends on the ratio of H_2S content to CO_2 content of the gases being treated, the absorption of H_2S as compared to CO_2.would go on decreasing in successive stages after the first stage. So it is better to remove the last remaining minor portion of H_2S by combustion to SO_2 which then can be absorbed.

Note: *The following reaction forms sodium carbonate from sodium hydroxide and carbon dioxide.*

$$2NaOH + CO_2. = Na_2CO_3. + H_2O$$

Application of the Invention in 'Paper Mill Industry'* for Absorbing H_2S

The emissions from paper and pulp mills have H_2S which needs to be removed. The invented method of removing H_2S by Na_2CO_3 can be used for removing H_2S in the sulphate pulp and paper industry. When combusting '*Black liquor*', the '*Green liquor*' may be used as the carbonate-containing alkaline solution. This constitutes a particular aspect of the invention. *Green liquor* is an aqueous solution that usually contains about '1.3–1.4 mole of Na_2CO_3 per litre', 'about 0.4–0.6 mole of Na_2S per litre', and 'about 0.3-0.5 mole of NaOH per litre'.

The 'White liquor' is an aqueous solution that contains NaOH, Na_2S and Na_2CO_3, usually in the proportions of about 2.3–2.6 mole of NaOH per litre, about 0.4–0.6 mole of Na_2S per litre, and about 0.25–0.4 mole of Na_2CO_3 per litre. Both green liquor and the white liquor are the intermediate effluents in the paper and pulp industry. The white liquor may be used in the first stage like the hydroxide for regenerating the carbonate-containing alkaline solution

Since both green liquor and white liquor contain carbonate and hydroxide, they may be used either separately or together, as the carbonate-containing solution for regenerating the carbonate-containing alkaline solution.

In the second stage, the combustion of H_2S contained in the gas can be done, say in a lime sludge re-burning kiln or a bark burning boiler. If the heat content of H_2S containing gas is too low, an auxiliary fuel, such as oil or natural gas, may be used in the combustion to produce satisfactory oxidation of the H_2S to SO_2. After the combustion of H_2S containing gas in the second stage, SO_2 formed is removed by absorption in an alkali- containing solution in a wet scrubber.

The absorbing alkaline solution is recycled to a large extent. The pH of the alkaline solution in the SO_2 scrubber is preferably in the range of about 6–8, and most preferred in the range of about 7–7.5. At these pH values, the absorption of carbon dioxide is negligible, and the selectivity to the SO_2. separation consequently is 100% *[Ref: Document Type and No.: WIPO Patent Application- WO/1995/007750, Kind Code: A1].*

***Note on Sulphate (Kraft) Process of Paper and Pulp industry**

The 'Sulphate process' of pulp making (also called Kraft process) uses sodium sulphate, sodium hydroxide, and sodium sulphide. 'Black liquor' is a waste liquid obtained after pulping the raw material in a "digester" with chemicals. The black liquor contains residual caustic soda, sodium sulphide, sodium carbonate, sodium sulphate, silica, dissolved lignin, etc. The cellulosic fibre separated from black liquor in the separator is washed with water and the concentrated wash water is mixed with black liquor and sent for recovery of chemicals.

Black liquor is concentrated by solar evaporation or otherwise and then incinerated with the addition of sodium sulphate. The organics like lignin are burnt out and the smelt is dissolved in water, yielding green coloured liquid called "Green liquor". Lime is added to 'Green liquor' resulting in precipitation of (mostly) calcium carbonate as "Lime mud", and the effluent, white in colour, is called "White liquor" which is recycled to "Digester" (where the pulp is made by digesting wooden chips by chemicals). The white liquor contains the recovered chemicals.

The washed pulp fibre is further washed with water and then bleached with chlorine. The pulp fibre is further bleached by caustic and then again by 'Hypochlorite'. The pulp fibre is then ready for processing for paper making. It is therefore quite obvious that the wash water would have caustic waste in the brown coloured waste water and chlorination / hypochlorite waste in light coloured waste, these waste liquids are in addition to the black liquor mentioned previously.

8.3.7.2 Removal of Hydrogen Sulphide by Amines

It is well-known that H_2S and CO_2 can be removed from gases by using amino-alcohols, such as monoethanol- amine and diethanolamine. Absorption may, for instance, be used for producing H_2S in pure form, which optionally is further processed to sulphur.

The chemical solvents of choice are aminoalcohols which are weak basic compounds that react with certain acid gases, forming weak chemical bonds. These bonds are easily broken upon mild heating. The strength of the alkanol-amine's bond and the corresponding heat needed to break the bond depends on the number of organic hydrocarbon groups attached to the nitrogen atom. Depending on the number of such groups, alkanol-amines are classified as either primary, secondary, or tertiary.

Alkanolamines (aminoalcohols) are used primarily to absorb CO_2 and H_2S . Carbon monoxide and nitrous oxides do not chemically react with alkanolamines. Sulphur dioxide and nitrogen dioxide form very strong chemical bonds with alkanol-amines. The heat required to break these bonds would decompose the alkanol-amines themselves. They, therefore, must be removed from the gases to be treated before treatment by alkanolamines.

8.3.7.3 Amines Used Commercially for Absorption of H_2S

All types of amines have at least one hydroxyl group and one amino group. In general, the hydroxyl group reduces the vapour pressure and increases the solubility, while the amino group provides alkalinity in water solution to cause absorption of H_2S. Higher pressure and lower temperature favours H_2S absorption.

The amines used commercially for the absorption of H_2S are:

- Mono-ethanol-amine (MEA)----- Primary Amine
- Di-ethanol-amine (DEA)------ Secondary Amine
- Di-propanol-amine (DIPA))------ Secondary Amine
- Methyl-di-ethanol-amine (MDEA)----- Tertiary Amine

MEA is effective at removing virtually all H_2S and CO_2, but requires a large quantity of heat to regenerate, i.e. for breaking the chemical bonds formed. DEA is used when the specification allows for some CO_2 to be left in the treated gas. DEA does not form non-regenerative degradation products with carbonyl sulfide, which makes it a suitable choice for treating refinery gases. MDEA is a weak base that reacts much faster with H_2S than with CO_2, making it particularly selective under the proper design conditions. It is used selectively on higher-pressure gas streams [*20.6 bar (g) and above*] for deep H_2S removal with only moderate CO_2 removal. It is not used for lower pressure gas streams. It is very corrosive. MDEA can also be used non-selectively for bulk removal of CO_2, from gas streams, but bulk removal requires the aid of absorption enhancing additives. The Di-propanol-amine (DIPA) is specially used for gases where the H_2S is mixed up with *Sulfolane i.e. Tetra-hydro-thiophen dioxide and water*, carbonyl sulphide, carbon disulphide and mercaptans. It is suitable for treatment of oil refinery gases.

8.3.7.4 *Removal of Carbonyl sulphide, Carbon disulphide* and *Mercaptans*

Carbonyl sulphide (COS) and carbon disulphide (CS_2) react with primary amines, *forming non-regenerative decomposition products*, except in the case of Di-glycol-amine (DGA). The carbonyl sulphide and carbon disulphide react with *secondary amines* (not significantly with other amines) forming compounds which can be regenerated.

Mercaptans are such weak acid gases that they are not readily removed by any of the alkanolamines. Mercaptans are often removed by treatment with caustic soda.

8.3.8 Ammonia Removal

The removal of ammonia is easily done by scrubbing with water. It normally gets removed along with dust during water scrubbing. If gases contain acidic gases (like hydrogen chloride, oxides of sulphur, hydrogen fluoride, etc.) ammonia in the gases normally gets neutralized by acidic gases in presence of moisture and forms ammonium salts. In case of excess ammonia in the gases, it would form ammonium hydroxide on water scrubbing which if economical, can be recovered for use by stripping. Another process of removal of ammonia from gases is Chemisorption of ammonia.

Ammonia Air Stripping

In stripping process, the pH value and temperature are adjusted so that higher percentage of free ammonia is present in water. [*Free ammonia is in equilibrium with Ammonium ions depending upon the pH value and temperature*]. The ammonia gas can then move from water to air and thus be removed by air stripping in a packed tower.

The internal packings inside stripping column distributes the water into thin layers to create a large surface area. The air is blown from bottom of the column to top. The ammonia is stripped from the falling water droplets into the air.

Chemisorption of Ammonia

The ammonia laden gases are treated with a dilute sulphuric acid solution which then produces ammonium sulphate. The ammonium sulphate can be used as soil fertilizer.

8.3.9 Removal of Dioxins from Combustion Exhaust Gases

Dioxin is commonly used to refer to: Dioxins and dioxin-like compounds, *which are a diverse range of chemical compounds which are known to exhibit "dioxin-like" toxicity. Chemically a dioxin is a* heterocyclic 6-membered ring, *where 2 carbon atoms have been substituted by oxygen atoms:* 1,2-Dioxin, 1,4-Dioxin. The sources and ill effects have already been dealt with in ***Para 2.7, Chapter 2.***

8.3.9.1 Dioxin Removal by High Temperature Combustion

It is known that Dioxins decompose at temperatures above 700°C. If the waste is incinerated at high temperatures and the exhaust gases are cooled quickly and filtered, the dioxins are avoided from the incinerator exhaust. Some municipalities adopt this method, but it is costly.

8.3.9.2 Dioxin Removal by Injection of Activated Carbon

Dioxins exhausted from the stack of municipal refuse incinerators get almost decomposed at the part burning at elevated temperature of exhaust gas. But they recombine at the process of heat utilization facility/ heat exchanger because they are not decomposed completely. [Gas to air heat exchanger is adopted generally as a method to recover heat in semi-continuous incinerator and mechanical batch incinerator.]

The dioxins are reduced by adopting two stage combustion technique and latest combustion controls, but still some dioxins remain in the emission at the bag filter outlet. For overcoming these problems, an adsorption removal technique with powdered activated carbon is adopted as a method that reduces dioxins. Removal of these dioxins can be done by injection of fine powder of activated carbon before the bag filters. Dioxins at gas-side are adsorbed by the activated carbon, and the carbon along with dioxins is retained in the solid deposit layer at the surface of filtering fabric of filter bags. The dust along with carbon and dioxins is collected and removed. The system is suitable to be adopted for municipal waste incinerators, various industrial waste incinerators, and melting furnaces, etc.

8.3.9.3 Removal of Dioxins using Catalysts

The exhaust gases from the incinerator are first led to heat exchanger* to recover heat and then led to a wet dust collector** to remove dust, sulphuric acid mist, hydrogen chloride (HCl), mercury (Hg) and others, which have serious influence to the life of the catalyst tower. The gases from wet dust collector are then passed through a reheater which reheats the gases to a temperature >170°C. (The preferred range of temperature for performance and longer life of catalyst is 230 to 270°C.) The gases are then led to 'Catalyst tower' where the NOx and Dioxins decompose.

Notes:

- The heat exchanger may be a 'Gas to gas heat exchanger', 'Gas to air heat exchanger' or any other type of heat exchanger.
- A mixing scrubber can be used as the wet dust collector. With this dust, sulphuric acid mist, hydrogen chloride (HCl), mercury (Hg) and others, which have serious influence on the life of the catalyst in 'Catalyst tower', may efficiently be removed.

Mercury Removal Provision

In the above method of exhaust gas treatments, it is possible to incorporate addition of an oxidizer with injection of an oxidizing agent in the exhaust gases which would oxidize the mercury and other pollutants. The oxidized mercury and the dust, etc. are removed in the dust collector when the gases after oxidation pass through it.

In case of use of wet dust collection, it is possible to install a solid/ liquid separator which separates slurry discharged from the wet dust collector into solid and liquid phases. A mercury adsorbing tower is incorporated for adsorbing mercury from the separated liquid. With this provision, mercury contained in the combustion exhaust gas can be adsorbed and recovered/. ***(Ref: 1).***

8.3.9.4 Removal of Dioxins by Water – Low Cost Method

A research team led by Tateo Usui, professor of chemical reaction engineering at Osaka University, announced on October 24, 2003, that they had developed simple dioxin-removal equipment in collaboration with Sumikin, Kansai Industries Ltd., Japan.

They conducted experiments in which exhaust fumes were blown into water, and obtained good results of reducing the amount of dioxins in the fumes by more than 95 percent. In addition, they established a simple method to remove more than 95 % of dioxins in the effluent water. The New Energy and Industrial Technology Development Organization (NEDO) and the Japan Science and Technology Agency (JST) have invested in this project for commercialization of this process.

8.4 AIR POLLUTION AND ITS CONTROL IN METALLURGY OF NON-FERROUS METALS

8.4.1 General Process of Nonferrous Metallurgy

The production of nonferrous metals involves:- (i) Crushing, (ii) Water washing, (iii) Dense medium separation so that the rich ore can be separated from unwanted materials, (iv) Froth floatation for beneficiation / concentration of ore, (v) Roasting*of ore, (vi) Smelting** and (vii) Refining.

The production processes of nonferrous metals may pollute air by SPM including dust of metals or their compounds, gaseous pollutants like oxides of sulphur (mostly SO_2), CO etc.

Notes:

* 1. *Roasting of ore is a process in which the ore, either mixed with some suitable substance or alone is simply heated below its melting point. The heating may be in a current of air. During this process some volatile impurities may be converted into their oxides or go out as volatile substances. Some metallic compounds may be converted into some other compounds either by thermal decomposition or by oxidation by air or by interaction with the suitable substances added in the process.*

** 2. *Smelting is any process of melting or fusion, for recovering metal from their ore. Here the ore concentrate is heated after mixing with flux material like lime stone, silica, iron oxide, etc. The infusible impurities (Gangue) present in the ore get converted to fusible substances (Slag) which can easily be removed. The flux material is chosen according to the nature of gangue. For an acidic gangue like silica, a basic flux like lime is used and the silica is converted to calcium silicate which is fusible. The specific density of slag should be less than that of metal to be extracted so that the slag floats on the molten metal. If iron oxide is present as impurity, it will also form iron silicate like calcium silicate as a part of slag and get removed.*

$$CaO + SiO_2 = CaSiO_3 \text{ (slag)}; FeO + SiO_2 = FeSiO_3 \text{ (slag)}$$

Air Blast Roasting

The roasting of concentrates containing metal sulphide is done with a current of air to convert the metal sulphides into metal oxides. Copper pyrites, Zinc bled (zinc sulphide), Galena (lead sulphide) ores are subjected to air blast roasting by which the sulphide is converted to oxide and a small amount of sulphate.

$$2MS + 3O_2 = 2MO + 2\,SO_2 \text{. [Here 'M' represents metal]}$$

Chlorinating Roasting

Chlorinating roasting as in case of Argentite [the sulphide ores of Silver (Ag_2S) with some antimony and copper sulphides], the ore concentrates are mixed with common salt. The silver sulphide is converted to silver chloride.

$$Ag_2S + 2NaCl = 2Ag\,Cl + Na_2S$$

Calcinating

In case of carbonate ores, the crushed material is calcinated to convert carbonates to oxides, with evolution of CO_2.

$$MCO_3 = MO + CO_2.$$ [Here 'M' represents metal]

Metal Production by Reduction

The resulting material after roasting or calcinating of ores which contain metal oxides is then smelted together with coke where the metal oxides are reduced to metals and CO is evolved. The oxygen supply should be so regulated that CO_2 is not formed, otherwise the reaction may take place in reverse direction resulting in slowdown in reduction of oxides.

$MO + C = M + CO$ [Here 'M' represents metal]

The CO evolved is not wasted, but used as fuel for heating in the process. The metals obtained are then refined by various processes like liquefaction, distillation, vacuum arc refining, electrolytic process or other methods.

There can be auto reduction by heating in absence of air. The partially roasted sulphide ore which contains a mixture of metal oxide and metal sulphide is heated without air:

$2MO + MS = 3M + SO_2$. [Here 'M' represents metal]

Note: *Roasting / Chlorinating / Calcination processes, etc. described above do not apply to 'Aluminum production' for which the para No. 8.4.3 ahead may be referred.*

Energy Recovery and Conservation

At the time of process design and furnace design, heat conservation should always be considered. The use of low-mass refractories may be preferred. Recovery of energy / heat before or after the treatment (for removal / control of pollutants) should be incorporated in the process where practicable. The heat energy available from processes can be utilized in the following ways:.

1. Production of steam in waste heat boilers which can be used in process or for electricity production.
2. The use of the heat of reaction to smelt or roast concentrates or melt scrap metals in a converter.
3. The use of hot process gases to dry the feed materials.
4. Pre-heating of a furnace charge using the energy content of furnace gases or hot gases from another source.
5. Pre-heating of combustion air or heating of leach liquors.
6. The use of combustible contents in some raw materials as a fuel, provided that VOCs and dioxins are not emitted.
7. Use of emitted combustible gases like carbon monoxide, methane, etc. as gaseous fuel.

8.4.2 Air Pollution Control in Metallurgical Processes

Range of raw materials available to various metallurgical industries is wide and the process choice is governed by the raw materials. The furnace should be designed for the raw materials to be used. For better control of pollution, sealed and semi sealed furnaces should be used, but there may be exceptions. Blending of the raw materials, process control, management and the collection of fumes also have major influence on pollution control.

Hierarchy in the Choice of Some Processes

- Thermal or mechanical pre-treatment of secondary material to minimize organic contamination of the feed.
- Use of sealed furnaces or other sealed process units to prevent fugitive emissions, allow heat recovery and allow the collection of process gases for other uses. (The CO can be used as a fuel and SO_2 can be used for making useful by-products).
- Use of semi-sealed furnaces where sealed furnaces are not available or there are some difficulties in their use.
- Minimization of material transfers between various processes to minimize pollution and energy losses.
- Launders to be used in preference to ladles for molten materials, if such transfers are unavoidable. (In some cases the restriction of techniques to those that avoid molten material transfers may prevent the recovery of some secondary materials that would then enter the waste stream. In these cases the use of secondary or tertiary fume collection is appropriate so that these materials can be recovered.)
- Providing hooding and ductwork to capture fumes arising from hot metal, matte* or slag transfers and tapping. (* matte—a product of smelting of sulphide ore)
- Furnace or reactor enclosures to be used, if necessary, to prevent release of fume losses into the atmosphere.
- The furnace to be fully enclosed and ventilation air drawn off by extraction fans to a suitable treatment and discharge system where primary extraction and enclosure are likely to be ineffective.
- Maximum use of the energy content of sulphidic concentrates.

Dust Capture and Removal

The dust collected by dust collection equipments are re- circulated in the process to recover the metal from it and to reduce pollution of air by these toxic metals. In case the quantity of metal in the dust is very insignificant, the same can be disposed off in environmentally safe way or utilized.

Overall, for dust and associated metal removal, fabric filters (after heat recovery/gas cooling) can provide the best performance provided modern wear resistant suitable fabrics are used, and continuous monitoring is used to detect failures. Modern filter fabrics offer significant improvements in performance, reliability and longer life and therefore offer cost savings in the medium term. The systems like bag burst detection systems and on-line filter cleaning methods can be incorporated even in existing installations during maintenance.

For sticky or abrasive dusts, scrubbers or properly designed wet electrostatic precipitators (specifically for the process) can be effective if provided.

Note: *The use of fabric filter may, sometimes, require prior removal of hot particles to prevent fires. Cyclones and heat exchanger/ cooler may be used before fabric filter.*

Sulphur Capture, Removal and Recovery

Sulphur capture is an important requirement when sulphur containing ores or concentrates are roasted or smelted. The SO_2 produced by the process is collected and sold out or can be converted to sulphur, gypsum, or sulphuric acid. The process choice depends on the existence of local markets for the product. The recovery of SO_2 generated in case of metallurgical processes of lead, mercury, copper, etc. is profitable. A major part of SO_2 recovered may be used for producing the chosen products and remaining

SO_2 can be removed by absorbents like calcium hydroxide (Lime) forming solid products like calcium sulphite / calcium sulphate which may be removed by dust collection equipments.

It is also possible to use Ammonia cyclic regenerative process as already described in ***Para. 8.3.3.2*** of this chapter for absorption and recovery of SO_2 in pure form which can be used for manufacture of 'Sulphuric Acid' or 'Elemental Sulphur'.

The production of sulphuric acid in a double contact sulphuric acid plant with a minimum of four passes, or in a single contact plant with gypsum production from the tail gas and using a good catalyst, are considered to be the best available technologies. Plant configuration will depend on the concentration of SO_2 produced in the roasting or smelting stage.
Gas treatment for the smelting or incineration stage should include a SO_2 removal stage. Removal of Doxins may also be sometimes necessary.

Note: *Alternatively, the SO_2 can also be absorbed during roasting by sorbents like 'Sodium sesquicarbonate' if added in the charge along with the ore.*

The 'Trona mineral ore' found in nature consists of about 70-95 % of crystals of 'Sodium sesquicarbonate' which is a complex salt of sodium carbonate and sodium bicarbonate (Na_2CO_3. $NaHCO_3$. $2H_2O$). Crushed 'Trona ore' can be used instead of 'Sodium sesquicarbonate' in the charge along with the ore to be roasted.

Control of Dioxins

The presence of dioxins or their formation during a process needs to be considered for many of the pyro-metallurgical processes used to produce non-ferrous metals. These techniques may be used in combination. Some non-ferrous metals are reported to catalyse the synthesis of dioxins. It is sometimes necessary to have a clean gas prior to further treatments / processing. For achieving better control over dioxins, it may be necessary to incorporate the following:

(i) Quality control of scrap inputs depending on the process used, use of the correct feed material for the particular furnace or process, selection and sorting to prevent the addition of material that is contaminated with organic matter to minimise the potential for dioxin formation.

(ii) Use of optimum combustion conditions and the use of oxygen injection in the upper part of a furnace to ensure complete combustion of furnace gases.

(iii) Use of correctly designed and operated afterburners and rapid quenching of the hot gases to < 250°C.

(iv) Absorption of dioxins by activated carbon in a fixed bed or moving bed reactor or by injection into the gas stream, and removal as filter dust efficiently.

(v) Use of a catalytic oxidation stage or fabric filters with catalytic coating.

(vi) The treatment of collected dusts in high temperature furnaces to destroy dioxins and to recover metals.

Preferred Emission Control Methods at Various Stages of Metallurgical Processes:

(i) ***Material handling/ Grinding/ Drying***

Dust and metals — Use of suitable handling method including dust proofing of various equipments, dust confinement, dust extraction, ventilation, and collection of exhaust dust laden air/ gases and its treatment via cyclones/ multiclones or if necessary via fabric filters or passing through a scrubber.

(ii) ***Sintering, Roasting, Smelting, Converting, and Refining***

Dust and metals in exhaust gas — Heat recovery, dust collection via cyclone and fabric filter. (with metal dust recycling)

VOC and Dioxins — After-burner oxidation, adsorbing by activated carbon or other adsorbent

CO — After-burner oxidation

SO_2 — Wet Scrubber, utilisation for sulphuric acid manufacture.

(iii) ***Slag treatment***

Dust — Gas / fumes collection, cooling, collection by fabric filter

SO_2 — Wet Scrubbing

CO — After-burner oxidation

(iv) ***Leaching and Chemical refining***

Chlorine — Gas collection and reuse, wet chemical scrubber

(v) ***Carbonyl Refining***

CO — Use of sealed process, recovery and reuse

Hydrogen — After-burner oxidation

Dust — Dust removal by fabric filter from tail gas.

(vi) ***Solvent Extraction***

VOC — Containment of vapours, collection of vapours/ fumes and recovery by solvent, carbon absorption if necessary.

(vii) ***Thermal Refining***

Dust — Gas collection and use of fabric filter

SO_2 — Wet scrubbing.

(viii) ***Molten salt Electrolysis***

Fluoride, Chloride — Gas collection, alumina scrubber and use of fabric filter.

(ix) ***Electrode Baking and Graphitization***

Dust, SO_2*, Fluoride, Poly Aromatic Hydrocarbons (PAH), and Tar* — Gas collection, condensation / cooling, use of ESP, after burner oxidation, use of alumina scrubber and fabric filter.

(x) ***High Temperature Reduction***

Hydrogen — Use of sealed process, collection and reuse.

(xi) ***Melting and casting***

Dust — Gas collection and use of fabric filter

VOC, Dioxins — After-burner activated carbon injection.

(xii) ***Metal powder products***

Dust and Metal — Gas collection and use of fabric filter (with metal dust recycling.)

Abatement Technology wise Reported Emissions by European Commission:

'European commission on integrated prevention and control, Joint research centre, Institute of prospective technological studies (Sevelli), European IPPC bureau' has published a reference document on best available technologies in nonferrous metal industries in December 2001. According to it, the reported abatement technology-wise emissions are given in **Table 8.2**.

Table 8.2: Technology-wise Reported Emissions

Abatement technology	**Component of Emission**	**Minimum emission**	**Maximum emission**	**Remarks**
Fabric filter, hot ESP, and Cyclone	Dust (metals depend on composition)	1mg/ Nm^3	100 mg/ Nm^3	*Specific emission per t. of metal produced 10–80 g/t **Specific emission per t. of metal produced 5–10 µg/t
Activated Carbon filter	Total carbon	< 20mg/ Nm^3 .	---------	
After burner including temp. quench for dioxins	Total carbon Dioxins(TEQ) PAH(EPA) HCN	<2 mg/ Nm^3 <0.1 ng/ Nm^3 <1 µg/ Nm^3 <0.1 mg/ Nm^3	100 mg/ Nm^3 * 5ng/ Nm^3 ** 2500 µg/ Nm^3 10 mg/ Nm^3	
By chlorine recovery	Chlorine	<5 mg/ Nm^3		
Oxidizing Scrubber	NO_x	-----------<100 mg/ Nm^3		
Optimised combustion and low NOx burners	NO_x	10 mg/ Nm^3 — 500 mg/ Nm^3		
Wet or Semi dry Scrubber	SO_2 . Hydrocarbons Chlorine	<50 mg/ Nm^3 <10 mgC/ Nm^3 <2 mg/ Nm^3	250 mg/ Nm^3 200 mgC/ Nm^3 -------	
Alumina scrubber	Dust Hydrocarbon PAH(EPA)	<1 mg/ Nm^3 <1 mgC/ Nm^3 <20 µg/ Nm^3	20 mg/ Nm^3 50 mgC/ Nm^3 2000 µg/ Nm^3	
Cooler, ESP, Lime or Activated carbon adsorption & fabric filter	PAH (EPA) Hydrocarbon	0.1 mg/ Nm^3 20 mgC/ Nm^3	6 mg/ Nm^3 200 mgC/ Nm^3	
Sulphuric acid Plant— Double contact process	Conversion of SO_2 to SO_3 Balance SO_2 in gases	99.3 % 0.7 %	99.7 % 0.3 %	
Sulphuric acid Plant— Single contact process	SO_2 Conversion to SO_3 Balance SO_2 in gases	95 % 5 %	99.1 % 0.9 %	
EPA—Methods of measurement as per EPA of US *PAH—Polycyclic Aromatic Hydrocarbons; (mg C/Nm^3 — milligram carbon per normal cubic meter)*				

Control of Emissions Associated with the Use of Certain Reagents

Several specific reagents are used in chemical treatment of solutions of metals or in various metallurgical processes. **Table 8.3** gives some examples of reagents/ gaseous pollutants produced due to their use and suggested treatment methods for gases.

Table 8.3: Treatment Methods for Pollutants Due to Use of Reagents in Processes

Process /reagent used	Pollutant in exhaust gases	Treatment method used
1. Solvents	VOC, Odour	Containment, Condensation, Absorption / Adsorption by Activated carbon,
2. Sulphuric acid, sulphur in fuel, or raw materials	Sulphur dioxide	Wet / Semidry scrubbing, Utilization in manufacture of sulphuric acid
3. Aqua Regia	NO, Cl, NOx	Caustic Scrubbing System
4. Chlrine, HCl,	Chlorine	Caustic Scrubbing System
5. Nitric Acid	NOx	Oxidize, and absorb, recycle; Scrubbing
6. Sodium or Potassium Cyanides	HCN	Oxidize by Hydrogen per oxide or Hypo-chlorite
7. NH_3 orAmmonium com-pounds	Ammonia	Recovery of Ammonia / scrubbing
8. Ammonium Chloride	Aerosol	Recovery by sublimation, Wet scrubber system
9. Hydrazine	N_2H_4 (Carcinogen)	Wet Scrubbing, Absorption / adsorption by activated carbon
10. Formic acid	Formaldehyde	Caustic scrubbing system
11. Sodium chlorate/HCl,	Chlorine oxides (explosion)	Control of process end point
12. Refining Zn/Pb – use of As /Sb oxide	Arsine / Stibine	Permanganate Scrubbing
13. Pitch, etc.	Tars and PAH	Afterburner/oxidation, condenser and ESP or dry absorber.

8.4.3 Aluminum Production and Air Pollution Control

8.4.3.1 Production Process of Aluminum

Primary aluminum is produced by the electrolytic reduction of pure alumina (Al_2O_3). The purified alumina is dissolved in a molten bath of fluoride compounds (electrolyte), and an electric current is passed through the bath, causing the alumina to dissociate to form liquid aluminum and oxygen. The oxygen reacts with carbon in the electrode to produce CO_2 and CO. Molten aluminum collects at the bottom of the individual cells or pots and is removed under vacuum into tapping crucibles.

For obtaining pure alumina, 'Bauxite ore' is crushed, beneficiated, dried and finely crushed and then reacted with caustic soda at high temperature and pressure in a reactor

$$Al_2O_3.\ n\,(H_2O) + 2NaOH \ = \ 2Na\,AlO_2. + (1+n)\,H_2O.$$

Aluminum hydroxide ($Al(OH)_3$) is precipitated from sodium aluminate ($NaAlO_2$) by cooling and flash evaporation. The steam produced is recycled to the process for heating the slurry:-

$$NaAlO_2. + 2H_2O. = Al(OH)_3 + 2\ NaOH$$

The aluminum hydroxide is precipitated at 120 °C. The effluent is then clarified / filtered. Caustic soda (NaOH) solution is recovered. The precipitate is then Calcinated at 1000°C in a kiln or fluidized bed furnace to produce *pure alumina* (Al_2O_3)

The Alumina (Al_2O_3) is mixed with cryolite (Na_3AlF_6— melting point 1006°C) to keep mobility during reaction. This, then, is placed in cavities of cathode blocks and anode, of carbon. Alumina is fed intermittently. Some quantity of calcium fluoride or aluminum fluoride is fed which reduces the temperatures of the molten bath.

$$2Al_2O_3. + 3\ C = 4Al + 3\ CO_2\ .$$

The aluminum (Al) is thus produced. The cryolite and other fluorides added are the source of fluorides in the emission. There may be other impurities which appear in emissions. NOx may form during high temperature reactions. A flow diagram of production process of Aluminum is given **in Fig. 8.3 A.**

8.4.3.2 Air Pollutants in Aluminum Plant

At the bauxite production facilities, bauxite dust is emitted to the atmosphere from crushers, dryers and materials-handling equipment, through vehicular movement, and from blasting. Although the bauxite dust is not hazardous, it can be a nuisance if containment systems are not in place, especially on the crushers, dryers and handling equipment. Other air emissions could include nitrogen oxides (NOx), SO_2 and other products of combustion from the bauxite dryers / kilns.

In the aluminum smelting plant, air emissions include: (i) Alumina dust, *red mud* dust* from handling facilities, dust and kilns from calciners/ kilns (ii) Coke dust from coke handling; (iii) Gaseous and particulate fluorides, SO_2, CO_2, CO and various dusts from the electrolytic reduction cells; (iv) Tar vapour, gaseous and particulate fluorides, and carbon particulates from the baking furnace; (v) Coke dust, tars, and polycyclic aromatic hydrocarbons (PAHs) from anode-forming plant / anode production plant; (vi) Carbon dust from the rodding room; (vii) Fluxing emissions and oxides of carbon (CO and CO_2) from smelting, casting, and finishing; (viii) Sodium salts/ caustic aerosols from cooling towers; (ix) Products of combustion such as SO_2 and nitrogen oxides from boilers, calciners, mobile equipment, and kilns.

The electrolytic reduction cells (pot line) are the major source of air emissions, with the gaseous and particulate fluorides being of prime concern. The electrolysis also results in emissions of carbon tetrafluoride (CF_4) and carbon hexafluoride (C_2F_6), which are strong greenhouse gases. Emission values that have been reported for uncontrolled gases from smelters are 20–80 kg per ton (kg/t) of product for particulates, 6–12 kg/t for hydrogen fluoride, and 6–10 kg/t for solid fluoride particles. Corresponding concentrations are 200–800 mg/ m^3; 60–120 mg/m^3; and 60–100 mg/m^3.

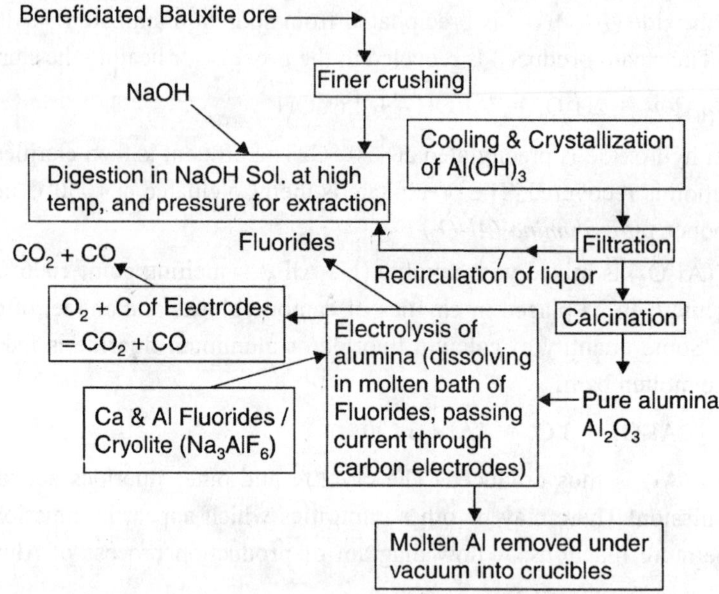

Fig. 8.3 A: Flow Diagram of Aluminum Production Process

Secondary Aluminum Production (Aluminum from Scrap)

Atmospheric emissions from secondary aluminum melting (from scrap) include hydrogen chloride and fluorine compounds. Demagging (which refers to removal of magnesium from the smelt or molten scrap of Aluminum) may lead to emissions of chlorine, hexachloroethane, chlorinated benzenes, and dioxins and furans. Chlorinated compounds may also result from the melting of aluminum scrap that is coated with plastic. Solid wastes from the production of secondary aluminum include particulates, pot lining refractory material, and salt slag. Processing of this slag may emit air pollutants. Particulate emissions, possibly containing heavy metals, are also associated with secondary aluminum production ie production from scrap.

***Note on red mud:** The red mud is a byproduct of Aluminum production by 'Bayer's process'. It is an insoluble matter when the bauxite ore is digested by caustic soda solution at elevated temperature and pressure. The composition of red mud varies and range may be as follows:*

Fe_2O_3----*30–60%;* Al_2O_3.----10–20%; SiO_2 ----*3–50%;*

CaO---2–8%; Na_2O ---- *2–10% ;* TiO_2—*Trace to 25% ;*

Utilisation of 'Red mud' has been a subject of major scientific research. An astonishing high number of potential uses have been reported in the literature or patents. For the decade 1967–1977 alone, 193 published papers and patents have been cited. Various uses include:-

(i) Metallurgical ones (iron and steel production, and recovery of Titanium, alumina and alkali materials and other minor constituents);
(ii) Production of building materials (constructional brick, light weight aggregates, bricks roofing and flooring tiles, cements, etc.);
(iii) Use in Catalysis, ceramics (pottery, sanitary ware, special tiles and glasses, glazes, ferrites, etc.);
(iv) Other miscellaneous direct uses (in waste treatment, as a filler, as a fertilizer, etc.).

8.4.3.3 Prevention and Control of Air Pollutants in Aluminum Plants

In the smelter, computer controls and point feeding of aluminum oxide to the centerline of the cell help reduce emissions, including emissions of 'Organic fluorides' such as CF_4, which can be held at less than 0.1 kg/ton aluminum. Energy consumption is typically 14 megawatt hours per ton (MWh/t) of aluminum, with prebake technology.

At bauxite facilities, the dust emission from major sources like dryers is controlled with electrostatic precipitators (ESPs) or baghouse. Dust removal efficiencies of 99% are achievable. Dust from conveyors and material transfer points are controlled by hoods and enclosures, and dust extraction system. Dust from truck movement can be minimized by providing metalled road surfaces and by ensuring that vehicles do not drop material as they travel. Dusting from stockpiled material can be minimized by the use of water sprays or by enclosure in a building. In case of 'Calciners', the alumina dust losses are controlled by ESPs, while the SO_2 and NOx emissions are reduced to acceptable levels by contact with the alumina.

In the smelter, primary emissions from the reduction cells are controlled by collection and treatment using '*dry sorbent injection*', while the 'fabric filters' or ESP's are used for controlling particulate matter. Primary emissions comprise 97.5% of total cell emissions; the balance consists of secondary emissions that escape into the pot room and leave the building through roof ventilators. Secondary emissions result from the periodic replacement of anodes and other operations; the fumes escape when the cell hood panels are temporarily removed.

Use of Dry Scrubbing

The '*dry scrubber systems*' with aluminum oxide absorbent is used for control of emissions from reduction cells. When anodes are baked on site, the same system of '*dry scrubbing system*' is used. The system has the advantage of being free from waste products and all enriched alumina and absorbed material are recycled directly to the process. The system results in the capture of 97% of all fluorides from the process. '*Dry scrubbing*' may be combined with incineration for controlling emissions of tar and Volatile Organic Compounds (VOCs) and to recover energy.

Note: *Wet scrubbing for control of primary emissions is not recommended since a large volume of toxic liquid effluent, high in fluorides and hydrocarbons will require treatment and disposal.*

Aluminum Production from Scrap— Air Pollution Control

For secondary aluminum production from scrap, the principal treatment technology downstream of the melting furnace is dry sorbent injection using lime, followed by fabric filters. Other waste gases from processing should be filtered as well. Waste gases from aluminum scrap pretreatment that contain organic compounds of concern may be treated by post combustion treatment. The scrap material should be thoroughly sorted out as far as possible to remove the unwanted material.

World Bank Guide Lines for Air Pollution limits in Aluminum Smelting

All maximum levels should be reckoned for at least 95% of the time that the plant or unit is operating to be calculated as a proportion of annual operating hours.

Parameter Maximum value:

Particulate matter -- 30 mg/ N m^3.; Hydrogen fluoride --1 mg/ N m^3.

Total fluoride -- 2 mg/ N m^3.; VOCs -- 20 micro gram/ N m^3.

Modern smelters using good industrial practices are able to achieve the following in terms of pollutant loads per tonne of aluminum on an annualized basis:

Hydrogen fluoride—0.2–0.4 kg/t; Total fluoride—0.3–0.6 kg/t; Particulates—1 kg/t; Sulphur dioxide—1 kg/t; Nitrogen oxides—0.5 kg/t. *(CF_4 emissions should be less than 0.1 kg/t).*

8.4.4 Lead Smelting and Air Pollution Control

8.4.4.1 Indirect Smelting Process for Lead

The ore is concentrated by various methods including floatation process as indicated in para. 8.4.1 of this chapter. Limestone can be mixed with the ore concentrate for roasting for absorption of SO_2 and also to act as fluxing material. Coke is used as fuel. The mix is spread on a moving grate, and air is blown through at a temperature of 1400°C . Coke and sulphur in the ore burn to provide heat for the roasting process. Sulphur is converted to SO_2. Roasting fuses the remaining ingredients into a brittle product called sinter, which consists of oxides of lead, zinc, iron, and silicon along with lime, metallic lead, and some remaining sulphur. This material is broken into lumps as it is discharged from the moving grate. The prefluxed, lumpy sinter is then loaded into the top of a heated blast furnace, along with the coke as fuel. A blast of air is admitted to the lower part of the furnace to aid combustion of the coke and carbon monoxide is produced. At a temperature of about 1200°C , the CO reacts with the metallic oxides of lead and various other metals producing CO_2 and molten metal. The non-metallic wastes form a slag with the fluxing materials present in the furnace.

On completing the reduction process, the furnace is tapped and the lead (Pb) is drawn off to flow into kettles or molds which is called 'bullion' which contains 95 to 99 % lead together with dissolved metallic and nonmetallic impurities (oxide and sulphide).

The bullion is maintained at a temperature just above its melting point, say about 330°C. At this temperature, the solubility of copper in lead is very low, so that the copper content segregates and forms a scum, or dross, on the surface of the bath in the drossing kettle. After this is skimmed off, more copper and other impurities are brought to the surface by stirring / agitating it with submerged air lances. These impurities are also skimmed off and the remaining base bullion is refined to yield lead of commercial quality

The sintering and roasting of lead concentrates can be carried out in 'down- shaft sintering' process (where the gases are sucked through a bed of burden materials) or by 'up-shaft sintering' process (where the blast of gases is forced through a bed of materials from bottom to top. In both the down shaft and up shaft processes (for lead), the dust collected by dust collecting equipments should preferably be recycled for further extraction of metals from it which would in effect lower the pollution by heavy metals.

Pollution Control in Down- shaft Sintering

In the 'down- shaft sintering', the gases from all suction boxes are mixed; their SO_2 content is not that high (0.5 to 1.5 %), They are de-dusted by passing first through 'Cyclone', then through 'Empty Scrubbing tower', then through an 'Electrostatic Precipitator (ESP)' and finally again wet scrubbed before being exhausted to the atmosphere. Alternatively, in place of ESP, a fabric filter may be used. At the end of the 'Fabric filter' the gases may require wet scrubbing* for removal of any acid gases or any remaining solid matter. *[*There is a possibility of escape of dust from fabric filters only in case of filter bag failure].*

ESP has to face the problem of high resistivity because the lead oxide (PbO), lead sulphide (PbS), zinc oxide (ZnO) which are present in the gases have high resistance. It is therefore necessary to cool the gases and humidify them. For this reason, empty scrubbing tower before the ESP is put in the process where the oxides of sulphur and other soluble gases/ matter get partly removed and the gases are cooled and humidified.

The '*Wet type ESP*' can also be used replacing both the 'Empty spray tower' and 'ESP (dry type)'. It is also possible to avoid the use of ESP / wet type ESP and fabric filter altogether with an alternative scheme such as:

The shaft gases are first passed through a 'Cyclone', then through 'Empty Scrubbing tower', then through 'Venturi tube scrubber', and then finally through 'Wet cyclone ie spray cyclonic eliminator'. The 'Liquid effluent' from all types of the 'Scrubbers' which dissolves acidic gases from the gas stream must be '*neutralized*' and the solids which contain toxic metals including lead, arsenic, selenium, etc. must be removed before disposal. Because of concern for higher water pollution, this scheme is not recommended.

Pollution Control in Up-shaft Sintering

In 'up- shaft sintering' process, the gases are rich in oxides of sulphur (more than 5%). These gases rich in SO_2 are first passed through a 'Cyclone' or an 'Intertial dust collector' (say baffled chamber), then through an 'Empty Scrubbing Tower', and then through an 'ESP' (dry type ESP) and then led for sulphuric acid manufacturing unit. The sulphuric acid plant manufactures sulphuric acid from these gases of sintering plant rich in SO_2. The other gases which are lean in content of oxides of sulphur can be treated like those in the case of down-shaft sintering gases.

It is possible to enrich the liquid effluent of the scrubbers by recycling and sending it to sulphuric acid plant for using up the dissolved sulphur dioxide gases in the effluent and then recovering it for use. This would eliminate acid pollution from the effluent. Alternatively, the liquid effluent should be neutralized before disposal.

The flow diagram for cleaning of gases from up-shaft sintering machine for lead is shown in **Fig. 8.3 B** and flow diagram for dust removal systems for lead smelter shaft furnace gases is shown in **Fig. 8.3 C.**

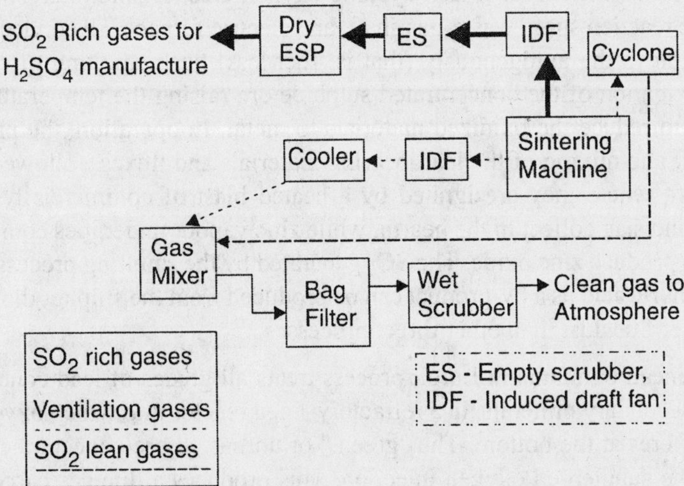

Fig. 8.3 B: Flow Diagram for Cleaning Gases from up-Draft Sintering Machine

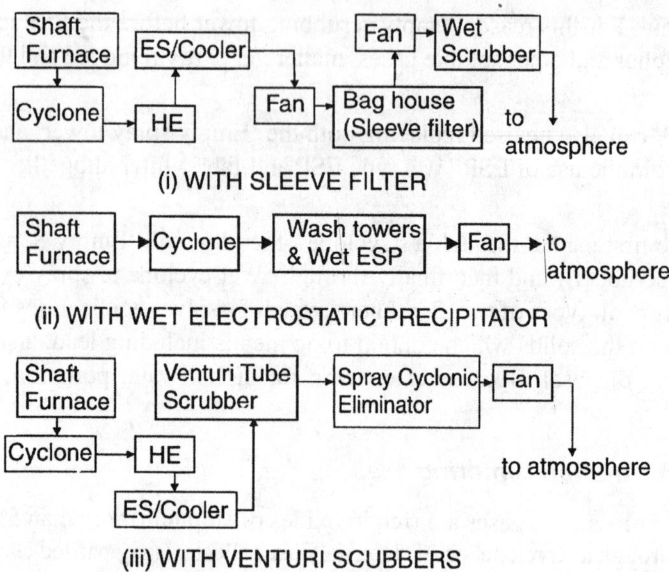

HE–Heat Exchanger;
ES/Cooler–Empty Scrubber/Cooler

Fig. 8.3 C: Flow Diagram for Dust Removal Systems for Lead Smelter shaft Furnace Gases

8.4.4.2 Direct Smelting Process for Lead

The direct smelting processes require neither the sintering of ore concentrates nor the use of metallurgical coke. They produce lower volumes of gases and dust. In general, direct smelting can be divided into two categories: (1) submerged smelting, as in the QSL (Queneau-Schuhmann-Lurgi) and Isasmelt processes, in which the refining reactions occur in a liquid (i.e., molten metal, matte, or slag), and (2) suspension smelting, as in the KIVCET process, in which the reactions occur between gases and solids.

KIVCET is a Russian acronym for "Flash-cyclone-oxygen-electric-smelting." A three-part KIVCET furnace comprises the reaction shaft, waste-gas shaft, and electric furnace, all connected with a common settling hearth. It employs the autogenously (that is, fuel-less) flash smelting of raw materials, with the heat-producing oxidation of the concentrated sulphide ore raising the temperature to 1,300° -1400 °C, which is enough to reduce the oxidized materials to metal. In operation, the process involves the proportioning, drying, and mixing of the lead-bearing materials and fluxes, followed by their injection into the reaction shaft, where they are ignited by a heated blast of commercially pure oxygen. The smelted lead bullion and slag collect in the hearth, while zinc vapour undergoes combustion with CO in the electric furnace to produce zinc oxide. The SO_2 generated by the smelting process is tapped from the waste shaft. The Sulphuric acid as a by-product can be produced from the sulphur dioxide. The KIVCET process produces lesser flue dust than other direct processes.

The QSL, or Queneau-Schuhmann-Lurgi, process treats all grades of lead concentrates, including chemically complex secondary minerals, in a refractory-lined reactor into which oxygen and natural gas are blown through tuyeres at the bottom. The "green," or unroasted ore concentrates are first oxidized in a molten bath by the submerged oxygen injection. This produces a flue gas carrying oxides of lead and zinc as well as a slag containing 80 percent of the zinc from the charge. The injection of natural gas produces CO which reduces the metal oxides in the gases. The concentrates employed in the QSL process are not briquetted or dried before being fed to the reactor. Moisture content is held around 7 - 8 percent in order to minimize dusting.

In the Isasmelt process, a gas or air lance is brought in through the top of a furnace and its tip submerged in the sulphide concentrate. A blast from the lance produces a turbulent bath in which the concentrates are oxidized to produce a high-lead slag which is tapped continuously and transferred to a second furnace, where it is reduced with coal. Crude lead and slag are tapped continuously from the second furnace and separated for further refining. In the direct smelting process, the scheme for air pollution control would not be much different than that for indirect smelting.

8.4.5 Zinc Smelting and Air Pollution Control

The process gases from Zinc sulphide ore roasting would have dust and $\mathbf{SO_2}$. The $\mathbf{SO_2}$ in the gases may be to the extent of 9 to 14 % and dust may be 60 to 130 g/ (NTP). The dust may contain small amounts of oxides / sulphates / compounds of lead (Pb), copper (Cu), cadmium (Cd), arsenic (As), selenium (Se) and large amount of oxides / sulphate / compounds of zinc (Zn). Even some metals may be present if they are volatile. The dust collected by dust collecting equipments is recycled for extracting the metals and reduce pollution.

The gases from zinc roasting / smelting are first treated in cyclones / multi cyclones and then in ESP of dry type, and then led for the manufacture of sulphuric acid (H_2SO_4).

The production of H_2SO_4 requires purified oxides of sulphur without any dust and compounds of arsenic and selenium (which are poisons for vanadium catalyst used in the production). The gases from 'dry type ESP' are passed through 1st stage washing tower (sprayed with 30–40% sulphuric acid), then through 2nd stage washing tower (sprayed with 15–20 % sulphuric acid) and then through '*Wet type ESP*' from where they are led to a humidifying tower. The humidified gases are then passed through a second stage 'Wet type ESP' before the gases are sent to contact chamber for converting the $\boldsymbol{SO_2}$ into Sulphur trioxide (SO_3) for the manufacture of H_2SO_4. All the acid effluent from washing towers is cleaned of solid contents and recycled.

A flow sheet of cleaning and removal of arsenic and selenium compounds from the zinc roaster gases rich in $\boldsymbol{SO_2}$ for manufacture of H_2SO_4 is shown in **Fig. 8.4.** *A general view of a typical tubular wet type ESP is shown in* ***Fig. 8.5.***

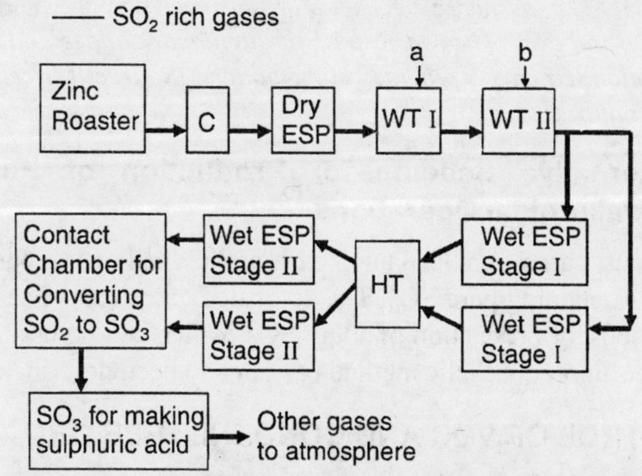

C—Cyclone; a—Wash down by 30 to 40% H2SO4; b—Wash down by 15 to 20 % H_2SO_4; HT—Humidifying Tower; WT Wash Tower

Fig. 8.4: Flow Sheet for Cleaning and Removal of Arsenic & Selenium Compounds from the Zinc Roaster Gases rich in SO_2 for the Manufacture of Sulphuric Acid

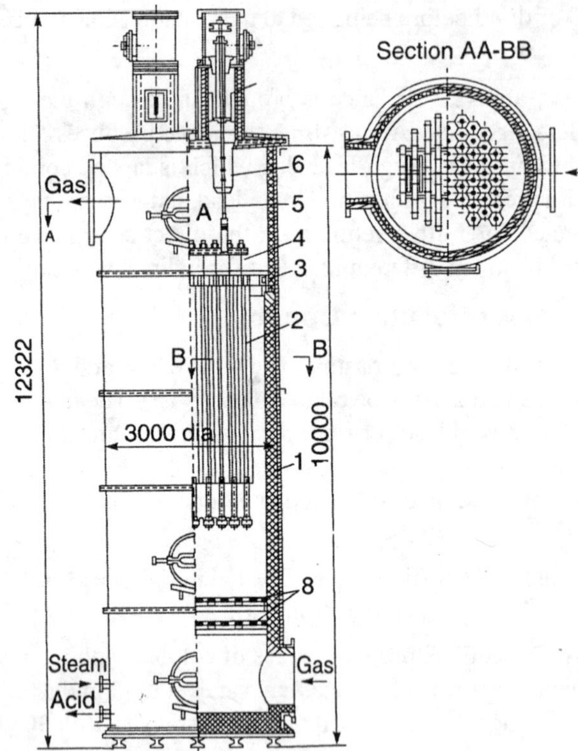

Fig. 8.5: General View of a Typical Tubular Wet ESP

Note: *The top of shell houses a lead plated steel grid 3, from which hang lead collecting electrodes 2 shaped as hexagonal honey combs. Lead plated discharge electrodes 4 of star shaped cross section, secured at the top of a frame. The discharge electrodes are suspended by the tie rods 5 from cross pieces, fixed to support insulators set in boxes. The rods 5 pass through quartz pipes and carry protective charge disks 6. Double gas distribution screen 8 is provided at the bottom. (**Source**: Miscellaneous Literature of old Russian Power stations in India).*

8.4.6 Alternative Scheme for Production of Pure SO_2 from Lead and Zinc Metallurgical Operations

As the exhaust gases from lead and zinc metallurgical operations are rich in SO_2, they can be used as a resource for obtaining pure SO_2. The de-dusted gases can be sent for absorption of SO_2 by 'Ammonia cyclic method' for production of 'Pure SO_2' which can be used for production of H_2SO_4 or elemental sulphur. The ammonia cyclic method has already been described **in Para 8.3.3.2 of this chapter.**

8.5 CONTROL OF VOC AND CONDENSABLE ORGANIC COMPOUNDS

The first requirement for control of Volatile Organic Compounds (VOC) and condensable organic compounds is their containment and their collection from the source because, if they spread in air, it may not be possible to mitigate them. In all industrial processes, fumes of any VOC should be confined by closed enclosures in the form of hoods or otherwise. These may then be sucked by a fan and sent for suitable treatment. Various types of treatments are available for these VOC's such as:

- Condensing them at low temperatures including 'cryogenic condensing' depending on the condensation temperature of the organic compound.— Condensing technique is best applied for removing VOC's / condensable hydrocarbons when their concentration is higher than 5000 ppmv. With high concentration of VOC pollutants, condensers are frequently employed as preliminary 'air pollution control device' prior to other devices/ methods. In condensers, one or more volatile components of vapour mixture can be separated from the remaining vapours. The condensation ie phase change from vapour to liquid can be achieved by lowering the temperature or increasing the pressure. In most cases refrigeration technique is employed to obtain the low temperatures required for acceptable removal. The surface condensers of shell and tube type allow for direct recovery of the VOC's from the gas stream, and the coolant does not get polluted.
- Adsorption by materials like activated carbon
- Flaring
- Thermal oxidation with the help of some auxiliary fuel in a furnace / kiln with heat exchangers for recovery of heat energy — The thermal destruction / oxidation may take place at temperature of about 800°C. Oxidation temperature can be lowered by a catalyst and then it is called catalytic thermal oxidation / catalytic incineration / catalytic oxidation.
- Absorption scrubbing by some suitable solutions / solvents depending on the solubility of the organic compound in a particular sprayed liquid.
- Vapour recovery — The recovery of the organic compound may also be possible from the adsorbent material or the solution/ liquid adsorbent (in which the organic compound has been adsorbed) either by simple heating or treatment with some chemicals in a particular case.
- Removal of VOCs by bio-scrubbing / biofilters as discussed in Para 8.2.5.4 of this chapter above — The removal of VOC's by bio- filtration or bio-scrubbing is dependent on the biodegradability of the particular VOC. The bio-degradability of some compounds like alcohols, aldehydes, ketones, ethers, esters, organic acids, amines, thiols, etc. is very high. Slowly bio-degradable VOCs include hydrocarbons, phenols, methylene chloride, etc., while the VOCs which degrade very very slowly include halogenated hydrocarbons, Polyaromatic Hydrocarbons (PAH), carbon disulphide, etc. [**Also refer to Para 8.3.9 of this chapter** regarding removal of dioxins].

Thermal oxidation for VOCs

The thermal oxidation is primarily used for control of emissions that require destruction of a pollutant like volatile hydrocarbons (VOCs), and hazardous substances/ gases. Where halogenated hydrocarbons are involved, great care is required as the emissions from thermal oxidation process would contain halogens/ halogen oxides/ halogen acids in addition to water and carbon dioxide. These halogen compounds should be absorbed in alkali solution. In the thermal oxidation process, the pollutant is destroyed by reacting with oxygen at high temperature. In order to lower the reaction temperature catalysts are used.

8.6 CONTROL OF AIR POLLUTION FROM CERAMIC INDUSTRY

The ceramic industry manufactures:

(a) Refractory items for high temperature application,

(b) Acid proof and alkali proof items for protection against chemicals,

(c) Porcelain materials for various purposes — electrical insulators, white wares, porcelain enamel, etc.

(d) Abrasives,

(e) Structural clay products,
(f) Special purpose ceramics.

There is a wide range of ceramic products requiring different composition of raw materials and glazing materials. They require definite maximum temperature and firing / cooling procedure and time schedule varying with the manufacturing process followed and the end product required.

At the firing stage, the air pollution control is very important as most of the air polluting emission can emanate from the kiln. There are several types of kilns in use in the industry such as 'down draft kiln', 'tank furnace', 'vertical shaft kiln', 'shuttle kiln' 'chamber kiln', 'top hat kiln', and 'tunnel kiln'. The down draft kiln is the highest polluting type. The tank furnace has a high pollution level, but lesser than down draft type. Different types of kilns are used for the manufacture of different products. Even for the same product, different types of kilns may be used in different factories.

Pollutants in Emission from Stacks

The emission from the stack of the kiln would comprise of:

- Particulate matter including respirable suspended particulates, silica, compounds of lead, arsenic, chromium, etc.
- Oxides of Sulphur (SOx), and oxides of nitrogen (NOx),
- Fluorides and chlorides
- CO, hydrocarbons [and CO_2]

Sources of Pollutants

Other sources of emissions in ceramic industry apart from kiln are:

- Spray glazing—Emission of aerosols of spray material,
- Mechanical finishing operations —Suspended particulate matter (SPM) including respirable suspended particulates (RSP)
- Basic raw material and its processing operations like crushing, lime calcination.—SPM emission including RSP with objectionable solids,
- Heat recovery operations – SPM, and gaseous pollutants,
- Handling, transportation, drying, etc.—Fugitive emissions including RSP

Emission Control Measures in Ceramic Industry

Emission prevention and control measures include:

- Confining the fugitive and particulate matter emission by dust suppression, dust containment, and dust extraction systems.
- Isolating primary dust generating equipments/ areas.
- The exhaust from the hooded conveyers or any other closed dust generating areas like crusher house, transfer points in conveyer system led to a wet scrubber.
- Some other methods for reducing SPM like regular cleaning and wetting of ground within premises, construction of wind breaking walls, green belt planting along the periphery.
- Use of hydro-jet for cleaning the plant premises.
- Provision of dust collecting equipments like Multiclone / ESP / Bag filters / Scrubber, etc. for flue gases from furnaces before stack,
- Provision of 'Low NO_x, burners', over fire air and means to create air turbulence, grit re-firing in case of use of coal/ lignite fuel, proper combustion control, etc. for furnaces. Provision of sorbent injection for reduction of NO_x, SO_x, etc.

- Provision of equipments for removal of balance soluble gaseous pollutants by wet scrubbing, but after thorough removal of dust.
- Above all, the proper operation; process optimization; proper equipment maintenance, overhauling and renewals
- Some changes of raw materials and their sources, change of fuel used, modification in process, etc. with a view to reduce pollution (For example: use of raw material without /or with low fluoride and chloride content for enamel and glazing; use of gas as fuel instead of solid or liquid fuel; etc.).

8.7 CONTROL OF AIR POLLUTION FROM STONE CRUSHING UNITS

The main air pollutant from the stone crushing units is particulate matter. The measurement of 'Suspended Particulate Matter (SPM)' should be conducted at least twice a month all the 12 months of the year but a daily routine measurement is better. The measurement is done at various locations and but the contribution value of SPM at a distance of 40 meters from a controlled isolated unit as well as from a unit located in a cluster should be less than the limits given under the 'Indian Environment protection rules1986' (which is 600 mg/ Nm^3). The following measures may be taken to control the dust.

(i) Dust containment system would be adopted, the conveyers should be hooded and crusher house should be closed and isolated. The closed space should be provided with induced draft and the exhaust of dust laden air is sent to a wet scrubber.

(ii) Dust suppression measure should be taken.

(iii) Enclosures should be made for primary dust generating equipments and areas. The dust laden air should be exhausted and led to a water scrubber to remove the dust.

(iv) Construction of wind breaking walls should be carried out.

(v) The roads inside premises should be metalled.

(vi) Regular cleaning wetting of the ground within the premises should be a practice. Hydro-jet should be used for cleaning of plant and premises.

(vii) Growing of a thick green belt along the periphery is a must.

8.8 COAL MINING AND COAL WASHERIES— AIR POLLUTION CONTROL

The following practices should be adopted for coal handling or coal processing plants at coal mines / coal washeries:

- Enclosures should be provided for coal crushers/ pulverizers / crusher house and coal transfer towers. The coal conveyors be provided with hoods. The induced draft be provided for these enclosures and the exhaust of the enclosure be led to wet scrubber.
- The water preferably mixed with wetting agents be sprayed at all strategic points including coal transfer points, coal conveyers, coal loading and unloading / tippling points, during stacking, etc.
- Water sprinkling system be provided for all coal heaps, coal stacks and around coal crushers. The sprinklers be provided with fine atomizers.
- The storage bunkers, hoppers, chutes should be provided with erosion resistant and impact resistant lining to check frequent leakages of dust
- Emission from 'Vehicles and Equipments' used in the mining area, coal handling areas, and washery areas should be controlled within the limits given in the 'Standards'.
- Spontaneous combustion* of coal in the coal stacks should be affectively monitored and controlled. The temperature monitoring of coal stacks should be done. The coal should be consolidated layer

by layer during stacking as the air pockets in the coal stacks are the potential source of spontaneous combustion of coal. The pollutants emitted during this process may include carbon monoxide, nitrous oxide, NO_x, sulphur dioxide and hydrocarbons which go out in atmosphere and it is not feasible to eliminate them.

- There is emission of coal gas / methane from mine coal bed which makes matters more complex in case of mine fire. The coal bed methane may be tapped as a source of energy.
- All precautions required against fires in the mining and washery areas should be taken and the fire-fighting equipments should be kept in good working conditions, for which fire-fighting drills should be carried out.
- The areas and roads in and around the coal handling plant, coal washery should be pucca / asphalted / concreted as far as possible.
- Thick and wide 'Green belts' should be provided around such plants.
- The over-burden stacked in the mining area should be covered with good soil and tree plantation should be done on it and greenery should be developed. Wherever possible, the over burden may be filled in the mined areas from where coal has been recovered.
- Proper drainage of areas should be constructed to avoid water logging.
- The drain water may be conserved by creating settling tank/ pond. The water from the pond can be used for various services.

***Note on Spontaneous Combustion of Coal / Lignite**

Spontaneous combustion is phenomenon associated with coal and lignite. When such a solid fuel in 'fuel stacks' is slowly oxidized by the atmospheric oxygen present in air pockets of the fuel stack, the temperature slowly builds up as the natural cooling at the particular point may be inadequate. When the temperature goes up to ignition point, the fire would appear spontaneously. The higher the volatile content of solid fuel, the higher is the chance of spontaneous combustion. The high volatile coals / medium volatile coals and lignite (which normally have higher volatile content) exhibit this phenomenon more often. ***[Also see Para 20.3, Chapter 20]***

8.9 NEW AIR FILTERING TECHNOLOGY FOR AIR BORNE VERY FINE PARTICLES

Scientists at Washington University in St. Louis have invented a new filtering technology to protect from inhalation of fine particulate, allergens, virus, aerosol, etc. (Research team was led by Professor Pritam Biswas, head of energy, environmental and chemical engineering at the University.) The device is named as SXC unit (or SXC-ESP).

It works by placing a charge on air borne particles and then using an electrical field to trap the particles. It also irradiates biological air borne pollutants to inactivate them. Thus, the device, if used, can protect the entire respiratory system against air borne fine and very fine particles/ dust, aerosols, pollens, allergens, virus, etc. (which otherwise may lead to various diseases including asthma, allergic reactions, infections, etc.) This technology may also be applicable to any limited indoors space any where including factories, hospitals, residences, control rooms, etc.

8.10 AIR POLLUTION CONTROL IN IRON AND STEEL PRODUCTION

Iron and steel plants have coke making plant, ore sintering plant, iron making plant (Blast furnace), steel making plant (Basic oxygen furnace, Open-hearth furnace, Electric arc furnace), iron and steel foundries, etc.

The coke oven gases consist of particulate matter, hydrogen, methane, ethane, carbon monoxide, carbon dioxide, ethylene, propylene, butylene, acetylene, hydrogen sulphide, ammonia, nitrogen, oxygen, tar vapours, Poly Aromatic Hydrocarbons (PAH), etc. The gases are used as fuel after cleaning of particulate matter from them by ESP, Venturi scrubbers, cyclones, etc. and after the recovery of useful co-products like benzene, xylene, toluene, tar acids, naphthalene, creosote oil, coal tar pitch, etc. The particulate matter collected from dust separators should be examined for iron content so that the iron content is not wasted away and corrective measures may be taken in the process.

The ore sintering (mixing the powdered ore with coke and then firing) eliminates the undesirable elements/matter and produces a physically and chemically stable product suitable for charging in blast furnace. The air pollutants in the gases from this process would be particulate matter, carbon monoxide, sulphur dioxide, chlorides, fluorides, ammonia, hydrocarbons, arsenic, etc. These gases are cleaned by cyclones and ESP / Bag house and wet scrubbers..

The Blast furnace is charged with sintered ore along with coke and lime. The iron oxide is reduced to iron metal. The iron and slag are tapped from the blast furnace. The iron is cast into ballets. The slag which contains the impurities is used to produce slag wool thermal insulation material. The blast furnace gases contain lot of combustible gases like carbon monoxide, hydrocarbons, etc. These gases can be used as fuel after cleaning of particulate matter by cyclone /ESP / Bag house and wet scrubber. The blast furnace gases can be used to produce power.

The steel is produced by eliminating the excess carbon and undesirable impurities from iron. Many processes are in use for producing steel like 'Basic oxygen furnace, Open-hearth furnace, and Electric arc furnace'. The particulate matter in the emissions from these processes may contain iron oxide dust. The particulate matter is removed by Cyclones, ESP / Bag house and Wet scrubber. The emissions also have sulphur dioxide, and other soluble gaseous pollutants, which get removed by wet scrubbing.

8.11 AIR POLLUTION CONTROL METHODS USED IN SOME OTHER INDUSTRIES

Emissions and their control in some industries are mentioned in **Table 8.4**

Table 8.4: Control Methods Used in Some Industries

Product Manufactured	**Air Pollutant Emissions**	**Control methods used**
Normal super phosphate	Si F_4, HF Particulate matter	Cyclonic / Venturi scrubber Wet scrubber / Bag house
Diammonium phosphate	Gaseous fluorides and Ammonia	Cyclonic / Venturi scrubber with 30% Phosphoric acid
Triple superphosphate	Si F_4, HF	Cyclonic / Venturi scrubber
Triple super phosphate granular	Si F_4, HF Particulate matter	Packed scrubber / Venturi scrubber
Lime	Particulate matter	Cyclones and wet scrubber
Sodium carbonate by Solvay process	Particulate matter	Wet scrubber
Sodium hydroxide (Caustic soda)	Chlorine, Mercury	Alkaline scrubbers, Chemical scrubbing and adsorbers**

Phosphoric acid	Particulate matter, Fluorides, Phosphoric acid vapours, H_2S	Bag house, Alkaline scrubbers, Mist eliminators
Arsenic as by product of nonferrous metal smelters	Arsenic trioxide	Baghouse or ESP
Cadmium as by-product of non-ferrous metal smelters	Cadmium and cadmium oxide	Baghouse
Zirconium, Titanium Vanadium, Tungsten, Molybdenum,	Chlorine, Chlorides, Silicon chloride Ammonia	Wet Scrubbers Conversion of ammonia to ammonium sulphate fertilizer
Coke ovens	Particulate matter Hydrocarbons,	Aspiration to draw pollutants into oven, Venturi scrubbers, ESP.
** [*Also see **Chapter 10** for mercury removal*]		

REFERENCES

1. Japanese Patents
   Patent document 1: Japanese Patent Publication 2001-272023 gazette,
   Patent document 2: Japanese Patent Publication 2003-117343 gazette,
   Patent document 3: Japanese Patent Publication 2000-185217 gazette,
   Patent document 4: Japanese Patent Publication 2003-24746 gazette,
   Patent document 5: Japanese Patent Publication Heisei 7-75720 gazette

2a. Swedish Patent Application 9300533-8, (filed on 18 February 1993),

2b. WIPO Patent Application- WO/1995/007750, Kind Code: A1

3. Reference document on Best available techniques in non ferrous metals Dec. 2001, European commi sion, Joint research centre, Institute of Prospective Technological studies, Seville, Technology for sustainable development, European IPPC Bureau Ins.Poll. Prevention & control.
4. Pollution Prevention And Abatement Handbook, 262 Project Guidelines: Industry Sector Guidelines, World Bank Group Effective July 1998.
5. Rao, B. Padma S., 'Prevention and control of air pollution from ceramic industry', Chemical engineering world, vol. XXXII, No. 7, JULY 1997.
6. Hindustan Times Dated 27-2-201, Lucknow.
7. Sulzer Technical Review 3 / 2014 'From Wastewater Treatment to Sweet Pineapples'.

QUESTIONS

1. Enumerate the industries which use mechanical dust precipitators for control of SPM.
2. Write a short note on Wet Electrostatic Precipitator (Wet ESP).
3. Enumerate the industries which use dry type Electrostatic Precipitator (ESP) for control of SPM.
4. Enumerate the industries which use Bag filters/ Fabric filters for collection of dust.
5. Write short note on 'Wet scrubbers' using water as medium for removal of gaseous pollutants from industrial exhaust..
6. Describe various types of chemical scrubbing/ absorption system by liquid chemicals for removal of gaseous pollutants from industrial exhaust. Discuss the applicability of liquid chemical absorption systems.
7. Write a short note on 'Dry scrubbing' and its applicability in industries.

8. Describe removal of air pollutants by biological methods like bio-filtration, bio- scrubbing from gases.
9. Briefly discuss the removal of 'Volatile organic compounds' by various methods including condensation, adsorption / absorption, thermal oxidation, catalytic decomposition, and bio-scrubbing.
10. Discuss the ammonium sulphite based and zinc oxide based SO_x recovery system from exhaust gases from various processes in industries.
11. Write a short note on removal of dioxins from industrial exhaust gases.
12. Discuss the removal of hydrogen sulphide by sodium carbonate solution.
13. Discuss the hydrogen sulphide removal by amines.
14. Write short note on removal hydrogen fluoride from exhaust gases and production of hydrofluoric acid.
15. Write short notes on recovery of the following air pollutants from industrial exhaust gases: (a) Chlorine (b) Hydrogen chloride (c) Fluorides.
16. Discuss the utilization of sulphur dioxide emitted in metallurgical operations of lead, zinc, etc. for production of sulphuric acid.
17. How are the particulate matter and SOx controlled/ removed in case of metallurgical operations of lead and zinc.
18. Write short notes on (a) Roasting (b) Smelting in case of metallurgy of nonferrous metals like lead, zinc, etc.
19. Write a short note on aluminum production from alumina ore and the control of pollutants evolved in the process.
20. Enumerate the anti-pollution measures required to be taken in:
    (a) Ceramic industry. (b) Stone crushing industry.
21. Write a short note on pollution control measure to be taken in coal mining industry and coal washeries. What is 'Spontaneous combustion of coal'? Why is it necessary to prevent it from the point of view of pollution apart from loss of coal calorific value and fire?

CHAPTER 9

AIR POLLUTION CONTROL IN PETROLEUM REFINERIES

9.1 SOURCES OF AIR POLLUTANTS IN PETROLEUM REFINERIES

There are many equipments and processes in a petroleum refinery which are potential sources of air pollutants. Air emission sources in petroleum refineries generally include:

- ***Petroleum cracking units, and fractionators along with coking units*** (these units produce various products from the crude oil)
- ***Sulphur recovery unit***—(which produces elemental sulphur/ or if required sulphuric acid)
- ***Boilers, IC Engines, and Gas Turbines.***

  Boilers are used to produce steam required for processes or to raise the temperature of feed materials to meet requirements of reaction or distillation. Boilers are typically fired with residual oil, distillate oil, refinery gas, or natural gas. The flue gases from the regenerator (of catalyst) are quite rich in carbon monoxide (CO) content and therefore may be used as a fuel. Reciprocating engines are used to drive gas compressors and are usually fueled by natural gas or refinery gas. Gas turbines can fire a variety of fuels and are used as cogeneration units in combined cycles to produce electricity. The 'Waste heat recovery boilers' can feed steam to steam-turbine and also fulfill the need of process steam.
- ***Process Heaters.***

  Process heaters are used to indirectly heat process fluids to a moderate processing temperature. Fuel is combusted in an insulated chamber and the heat is transferred to either tubes or a cylindrical vessel containing the process fluid. Process heaters use natural gas, refinery fuel gas, fuel oil or residual oil as fuel.
- ***Process Vents***

  Process vents having low emissions include those used for venting from processes like separation, conversion, and treatment. The process vents having higher emissions are mainly from three specific processes namely: 'Catalytic cracking unit's catalyst regeneration', 'catalytic reformer's catalyst re-generation', and 'sulphur recovery plant'.
- ***Flares.***

  Flares are commonly used for the disposal of waste gases mainly during start-up, shutdown and emer-gencies.
- ***Blowdown Systems.***

  Blowdown systems consisting of piping / valves / surge vessels, etc. are used for the removal of liquids and vapours from various process units going under shut down for maintenance and repairs. The blow down is also done during emergencies for preventing dangerous conditions from occurring. During the blow down of such liquids or vapours, some concentrated air emissions may take place. To mitigate such highly concentrated air emissions, the blow down system is connected to a refinery flare unit for safe disposal.

- ***Leakages from Process Equipment***

  Emissions occur from process equipment through various components like pumps, valves, and flange leakages and pressure relief valve discharge. Emission also occurs during leakages due to seal failures, valve seat passing, component failure, process upsets, etc. The leakages may be of liquid, vapour or gaseous.

- ***Waste Water Treatment.***

  Many component equipments of the waste water collection and treatment system are open to atmosphere and allow waste water containing organic impurities to come in contact with ambient air. The emission of air pollutants may be from drains and collectors, oil-water separators, flotation systems, and basins and ponds of wastewater collection and treatment systems.

- ***Cooling Towers***

  A large quantity of cooling water is used for cooling in 'Refining processes'. The 'Cooling towers' are used to transfer heat from the cooling water to the atmosphere. The water which comes to cooling tower for being cooled may contain hydrocarbons from leaking equipment. The hydrocarbon vapours / Volatile Organic Compounds (VOC) may pollute air.

- ***Storage Vessels***

  Storage vessels for various liquids / oils / effluents / products used throughout the refining process have either fixed roof or floating roof. The escape of vapour / VOC from such storage vessels may create air pollution.

- ***Loading Operations***

  Loading losses occur as organic vapours in empty cargo tanks are displaced to the atmosphere by the liquid being loaded into the tanks. (Crude oil may be transported from production centres to a refinery by marine tankers and barges, rail tank wagons, tanker trucks, and pipelines. Petroleum products are sent out similarly). **Figure 9.1** shows simplified flow diagram of typical crude oil refining process

9.2 MAIN EMISSIONS FROM REFINERIES

Main emissions of air pollutants from refineries are oxides of sulphur (SO_x), oxides of nitrogen (NO_x), hydrocarbons/ Volatile Organic Compounds (VOC), and suspended particulate matter (SPM). The Fluid Catalytic Cracking Units (FCCU) and coking units are the largest sources of SO_x, SPM and also big sources of NO_x and VOC. Boilers and process heaters are also a large source of SO_x, NO_x, hydrocarbons/ VOC and SPM. The internal combustion oil engines (IC oil engines) are important sources of NO_x, SO_x, hydrocarbons/ VOC and SPM, while gas fired Gas turbines and gas- engines are sources of NO_x. The 'Sulphur recovery units' is a large source of SO_x. Flaring is large source of SO_x. The equipment leakage is one of the big sources of VOC. The waste water treatment, storage tanks, cooling towers and loading operations are important sources of VOC. Leakage of exhaust gases which contain many air pollutants may occur from flanges, dampers, fans, and other places.

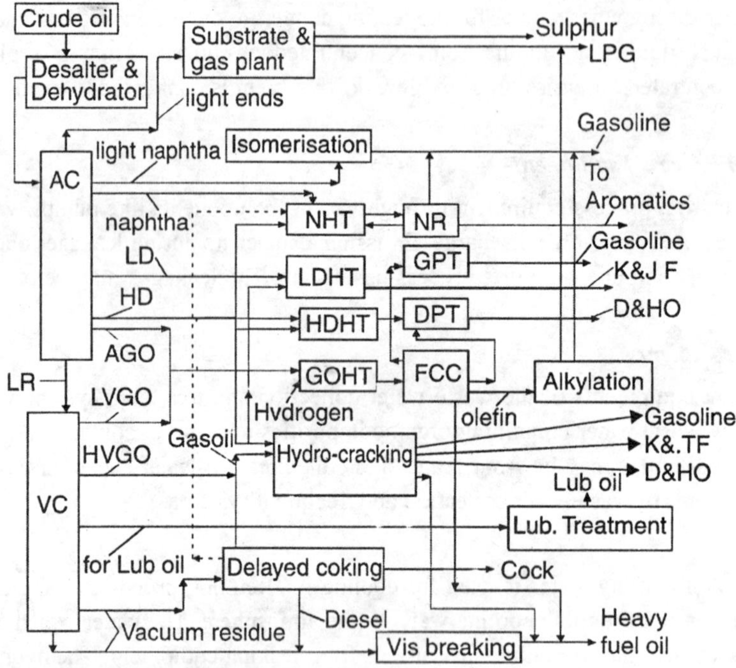

HD—Heavy Distillate, LD—Light Distillate, AC—Atmospheric Column, AGO—Atmospheric Gas Oil, LVGO—Light Vacuum Gas Oil, NHT—Naphtha Hydro Treatment, HVGO—Heavy Vacuum Gas Oil, NR—Naphtha Reforming, LDHT—Light Distillate Hydro Treatment, HDHT—Heavy Distillate Hydro Treatment, GOHT—Gas Oil Hydro Treatment, K&JF—Kerosine & Jet Fuel, FCC—Fluid Catalytic Cracking, GPT—Gasoline Post Treatment, DPT—Diesel Post Treatment, LR—Long Residue, D& HO—Diesel & Heat Oil

Fig. 9.1: Simplified Flow Diagram of A Typical Crude Oil Refining

9.3 PETROLEUM CRACKING PROCESSES

9.3.1 Fluid Catalytic Cracking Unit (FCCU) in Petroleum Refinery

In the 'Fluid catalytic cracking (FCC)' process, the 'heavy petroleum oil feed' is vaporized upon contact with hot Zeolite based 'Catalyst' at a temperature of 665°C to 760°C in the transfer line (pipe sloped upward called 'Riser' where feed is sprayed on the catalyst) leading to 'Reactor'. Cracking of large molecules of 'heavy oil feed' takes place on the 'Surface and in the pores' of the 'Catalyst' during the 'Transfer to reactor' and also in the 'Reactor' itself. The desired vaporized products on catalytic cracking after passing upward through the 'Reactor bed' flow out of the reactor. The mixture of catalyst fines and hydrocarbons are passed through a 'Cyclone system' for separation of solids and the hydrocarbons are then sent for separation into various fractions (like fuel gas, LPG, gasoline, naphtha, jet oil, diesel oil and heavy oils) to the 'Fractionator distillation column' downstream.

During the trip up the 'Riser', the cracking catalyst is "Spent" by reactions which deposit coke on the catalyst and greatly reduce its activity and selectivity. The 'Spent catalyst' is disengaged from the cracked hydrocarbon vapors and needs to be regenerated before it can be recycled.

Petroleum coke is an undesired product of cracking. It is a black, condensed hydrocarbon deficient in hydrogen and contains carbon, hydrogen, sulphur, nitrogen, metallic compounds, etc. The sulphur, nitrogen, and metallic compounds contained in crude oil tend to accumulate in the heavier liquid fractions and coke produced during refining. Metallic impurities include nickel and Vanadium. The 'Nickel' enhances coke formation, while the 'Vanadium' acts as catalyst to oxidize 'SO_2' to 'SO_3' during combustion of sulphur compounds. The sulphur and nitrogen in the coke are in the form of large, heavy organic molecules.

As the deposit of coke along with accompanying metallic impurities on the 'Spent catalyst' interferes with the desired cracking reactions, it has to be continuously removed and cleaned / regenerated in the 'Regenerator'. The 'Spent catalyst' from the 'Reactor' constantly moves to a fluidized bed 'Regenerator' from where after regeneration the 'Regenerated catalyst' goes to 'Stripping section' where steam is injected to strip the hydrocarbon vapours from the catalyst particles. From stripping section, the 'Regenerated catalyst' goes to 'Transfer line' and meets the 'heavy oil feed' to 'Reactor'.

In the regenerator, the coke is burned off to clean the catalyst for its return to the reactor. Compressed air is fed to the regenerator for its combustion. The heat generated by the combustion reaction vaporizes the feed oil and supplies the energy necessary for the endothermic cracking reactions taking place in the reactor vessel plus whatever heat losses occur.

The LPG from FCC unit is an important source of C_3-C_4 olefins and *isobutane* for manufacture of polymers like polypropylene. A typical simplified flow diagram for 'Fluid catalytic cracking unit' is shown **in Fig. 9.2.**

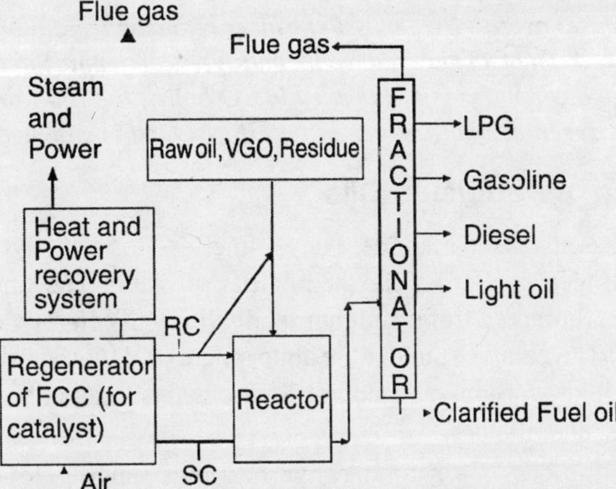

S–Spent Catalyst, RC–Regenerated Catalyst, VGO–Vacuum Gas Oil

Fig. 9.2: Typical Simplified Flow Diagram of a Fluid Catalytic Cracking (FCC) Unit

Notes:

(i) *Catalyst for FCC: Powdered catalyst is employed in 'Fluid catalytic cracking' process (FCC). Previously fluid cracking employed a low activity alumina catalyst and a reactor where the catalyst particles were suspended in a rising flow of feed hydrocarbons in a fluidized bed. But nowadays active zeolite based catalyst is used. Still these Alumina-catalyzed cracking systems are used in school and university laboratories for experiments. The alumina catalyst is crushed pumice stones, containing mainly aluminium oxide and silica into small porous pieces.*

(ii) *Alkanes: Alkanes are compounds like methane, ethane, propane, butane, pentane, etc.* [CH_4, C_2H_6, C_3H_8, C_4H_{10}, C_5H_{12}]

(iii) Alkenes: Alkenes are compounds like ethylene, propylene, butylene, pentylene, etc. [C_2H_4, C_3H_6, C_4H_8, C_5H_{10}].

(iv) *Alkynes: Alkynes are compounds like acetylene or ethyne, allylene or propyne, crotonylene or butyne, etc. [C_2H_2, C_3H_4, C_4H_6].*

Air Pollutants Formed in the Regenerator

Air pollutants are formed in the regenerator chemically during the combustion process of impurities along with coke entrained with the catalyst particles entering the regenerator. The pollutants in the exhaust gases from regenerator include CO, hydrocarbons, catalyst particles i.e. suspended particulate matter (SPM), SO_2 and SO_3 and oxides of nitrogen (NO_x). The NO_x from the regenerator is predominantly in the form of nitric oxide (NO). Most of the NO_x is fuel NO_x as the temperatures in the regenerator is not high enough to produce much of the thermal NO_x (thermal formation of NO_x due to high temperature). The organic nitrogen compounds in the coke break down into such species as ammonia (NH_3), hydrogen cyanide (HCN), nitrous oxide N_2O, nitric oxide (NO), and molecular nitrogen (N_2).

Note: *Low-NO_x burners are available for fired regenerator, but they depress only the thermal formation of NO_x and not the NO_x formation due nitrogen in the fuel (called fuel NO_x). The NO_x emission may increase due to poor mixing of oxygen in the regenerator. The higher the nitrogen in the 'Crude oil feed' to catalytic cracker, the higher may be the overall NO_x formation in the process. NO_x from the regenerator may also be affected by higher local oxygen (O_2) concentrations there, caused by poor mixing. Increasing the amount of oxygen in the regenerator, decreases both the carbon in the regenerated catalyst and CO formation, but may make the remaining carbon richer in nitrogen. The CO depresses the formation of NO_x. Therefore, higher excess oxygen increases overall NO_x emission. Under reducing conditions when oxygen supply is less, more CO would be produced. The CO may reduce the nitrogen compounds to reduction products like HCN (hydrocyanide) and NH_3 and when this gas is burned in CO boiler, the HCN and NH_3 may form more NO_x. Under both high and low oxygen combustion conditions, the NO_x can be higher.*

9.3.2 Hydro-cracking of Petroleum Oils

'Hydro-cracking' process is also used in modern day refineries. It is a catalytic cracking process assisted by the presence of 'Hydrogen gas' at an elevated partial pressure. The function of 'Hydrogen' is the purification of the hydrocarbon feed from 'Sulphur' and 'Nitrogen'. In 'Hydro-cracking' a bi-functional catalyst is normally used. The catalysts, under the atmosphere of 'Hydrogen', are capable of rearranging and breaking chains of hydrocarbons and adding hydrogen to aromatic compounds and olefins for conversion to napthalenes and alkanes.

The products of this process are saturated hydrocarbons ranging from ethane, LPG, to heavier hydrocarbons consisting mostly of iso-paraffins with low sulphur content. However, the main products produced from hydro cracking are 'Jet fuel' and 'Diesel', along with some LPG, kerosene, and high octane gasoline fractions. The reaction conditions i.e. temperature, and pressure, catalyst activity would be different for obtaining different products. This process is used in India, Europe and Asia because of high demand for diesel and kerosene. In the US, fluid catalytic cracking is more common because the higher demand for gasoline.

9.3.3 Steam Cracking of Petroleum Oils

Steam cracking is a petrochemical process in which saturated hydrocarbons are broken down into smaller, often unsaturated hydrocarbons. It is a principal industrial method for producing the *lighter alkenes / olefins, including ethylene and propylene.*

In steam cracking, a gaseous or liquid hydrocarbon feed like 'Naphtha', LPG, ethane is mixed with steam and briefly heated in a furnace in absence of oxygen. Typically, the reaction temperature is high (about 850°C), but the reaction time is very small (the residence time is in ms (milliseconds) and velocities are more than that of sound. On reaching the cracking temperature, the gas is quickly quenched to stop the reaction by quenching with oil in header or by a heat exchanger. The products produced in the process would depend on the 'feed composition', amount of steam present in reaction, cracking temperature and furnace residence time.

The process also results in slow deposition of carbon (coke) on the reactor walls, degrading the reactor efficiency. The steam cracking furnace can usually run only for a few months at a time between 'De-cokings ie removal of coke'. The furnace is to be isolated from the process for removal of coke. The hard coke layer is then converted to CO and CO_2 by heating in the atmosphere of steam or a steam/air mixture. After this reaction is complete, the furnace can be returned to service.

9.3.4 Thermal Cracking of Petroleum Oils

Thermal cracking process operates at 400°C–460°C and was used by many refineries previously, but now new type of high pressure thermal cracking is used. Modern high-pressure thermal cracking operates at absolute pressures of about 7,000 kPa. It may be noted that 'Light hydrogen-rich products' are formed at the expense of heavier molecules which get depleted of hydrogen. The reaction is known as 'Homolytic fission'. The products are 'Alkenes' from which polymers can be produced.

9.3.5 Visbreaking

Visbreaking is also a type of thermal cracking used in refineries to 'reduce the viscosity of oil'. Visbreaking units typically convert about 15 % of the feedstock to naphtha and diesel oils and produce a lower-viscosity residual fuel. [Thermal cracking units provide more severe processing and often convert as much as 50 to 60 % of the incoming feed to naphtha and light diesel oils]. The uses of visbreaking are:

(i) Reduce the viscosity of: (a) residue from 'vacuum distillation of crude oil', (b) residue from 'hydro-skimming operations', (c) natural bitumen from tar sands, (d) high viscosity crude oils.

(ii) Reduce the amount of residual fuel oil produced by a refinery. (Residual fuel oil is generally regarded as a low value product).

(iii) Increase the proportion of middle distillates in the refinery output.

The visbreaking can be of two types, namely coil/furnace visbreaking and soaker visbreaking. A typical flow diagram of Vis breaking is shown **in Fig. 9.3.**

Coil/ Furnace Visbreaking

In this the cracking occurs in the furnace tubes (or "coils") when the feed is heated in furnace. The liquid on exit from the furnace is quenched (to stop the cracking reactions) by a heat exchange with the virgin material being fed to the furnace. The extent of the cracking reaction is controlled by regulation of the speed of flow of the oil through the furnace tubes. The quenched oil is then sent to a fractionator where the products of the cracking (gas, LPG, gasoline, gas oil and

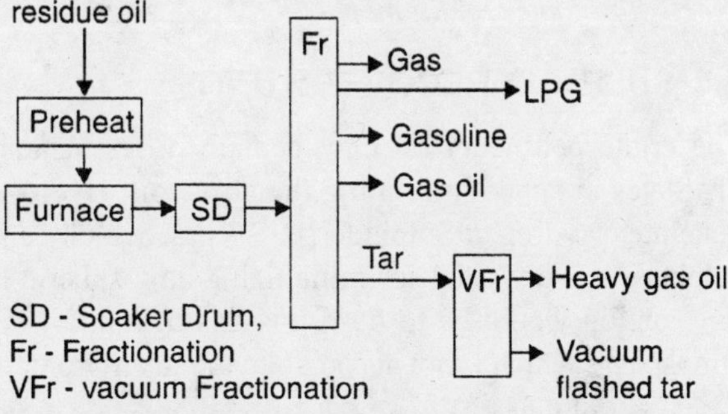

Fig. 9.3: Schematic Diagram of a Visbreaker Unit

tar) are separated and recovered. The coils of furnace require frequent de-coking as coke deposits on coils during cracking reactions.

Soaker Visbreaking

In soaker visbreaking, most of the cracking occurs not in the furnace, but in a drum (soaker drum) located after the furnace, where the oil is held at high temperature for required period for allowing the cracking to occur and then it is quenched. The oil is then sent to a Fractionator for separation of various products. The coke deposits on furnace coils in this case is insignificant, while there is coke deposits in drum, but is not very serious as the temperatures in the drum are not that high. The de-coking can be done periodically.

Further cracking of Tar

Visbreaker tar can be further refined by feeding it to a 'Vacuum fractionator'. Here additional heavy gas oil may be recovered and routed either to catalytic cracking, hydrocracking or thermal cracking units in the refinery. The vacuum-flashed tar (sometimes referred to as pitch) is then routed to fuel oil blending. In few refinery locations, visbreaker tar is routed to a delayed coker for the production of certain specialist cokes such as anode coke or needle coke.

Note: *Visbreaking preferentially cracks aliphatic compounds which have relatively low sulphur contents, low density and high viscosity and the effect of their removal can be clearly seen in the change in quality between feed and product. A too severe cracking in a visbreaker will lead to the asphaltene colloid becoming metastable(excited state of material). Subsequent addition of a diluent to manufacture a finished fuel oil can cause the colloid to break down, precipitating asphaltenes as sludge. It has been observed that a paraffinic diluent is more likely to cause precipitation than an aromatic one. Stability of fuel oil is assessed using a number of proprietary tests (for example "P" value and SHF tests).*

P-Peptizaein value, xylene or toluene equivalout

SHF- sedimentation by hot filtration

Example of Reduction of Viscosity by Visbreaker

(a) *Feed stock to visbreaker: ---Viscosity at 100°C, 930 centistokes, Density 1.020 kg/L , sulphur 4% by weight.*

(b) *After passing through visbreaker, a part of the feed is converted to naphtha or diesel oil while the characteristics of residual oil may be as follows:---*

Viscosity at 100°C, 115 centistokes; Density 1.048 kg/L; Sulphur 4.7% by weight.

[There will be considerable variation from visbreaker to visbreaker as no two will operate under exactly the same conditions].

9.4 SULPHUR RECOVERY UNIT

The crude petroleum oil may contain different amounts of sulphur. Some sulphur is passed on to distillates obtained in refining the crude oil. Hydro-treating/ hydro-cracking various distillates would generate 'Hydrogen Sulphide (H_2S)' gas, which can be converted to useful elemental sulphur (S) in sulphur recovery plant for minimizing atmospheric pollution. In the absence of sulphur recovery, the only option would be to burn this gas in refinery furnaces, releasing huge amounts of SO_2 into the atmosphere which is not acceptable due to environmental concerns.

For 'Sulphur recovery', H_2S is recovered from various refinery gaseous streams by scrubbing with an 'Amine' solution and converted to 'Elemental sulphur' by controlled oxidation. One third of this H_2S

is burnt to SO_2 to obtain H_2S/ SO_2 stoichiometric ratio of 2:1.

The recovery of elemental sulphur from the acid gas containing H_2S by '*Claus process*' would involve the following chemical reactions:

$$2H_2S + 3O_2 = 2SO_2 + 2H_2O;$$

$$2H_2S + SO_2 = 3S + 2H_2O$$

$$\text{Or simply: } 2H_2S + O_2 = 2S + 2H_2O$$

The reaction of SO_2 and H_2S to obtain elemental sulphur would occur at about 1390°C (2500 F). At high temperature, conversion up to 70% into sulphur may be obtained in the *Claus reaction* at the above stoichmetric ratio.

Alternatively, if SO_2 can be available as such from other processes in the refinery itself, then there is no need to burn a part of H_2S to obtain SO_2. In order to obtain 'Pure SO_2' for the above reaction, the SO_2 formed as a result of burning high-sulphur fuel oil in refinery processes is absorbed by scrubbing the flue gases with aqueous ammonia or ammonium sulphide forming ammonium bisulphate:

$$2NH_4OH + SO_2 = (NH_4)_2SO_3 + H_2O;$$

$$(NH_4)_2SO_3 + H_2O + SO_2 = 2NH_4HSO_3 \text{ (ammonium bisulphate)}$$

The pure SO_2 is recovered from the ammonium bisulphate sorbent in purer form by heating

$$2NH_4HSO_3 = SO_2 \text{ (recovered)} + 2H_2O + (NH_4)_2SO_3$$

9.5 AIR POLLUTION CONTROL MEASURES FOR REFINERIES

9.5.1 Control Options for Various Pollutants

The main sources of particulate matter are 'Fluid Catalytic Cracking (FCC) regenerators' and 'Coking operations'. The particulates from coke handling equipments can be reduced by enclosing of equipments. The particulate matter of the flue gases from 'Fluid catalytic cracking unit' can be removed by wet scrubbing or other methods.

Fugitive emissions (leaks) from storage tanks, sewers, process units, seals, valves, flanges, and other fittings would contain 'Volatile Organic Compounds (VOCs)'. By providing floating roofs on all open tanks and a vapour recovery system on tanks which already have roofs, VOC emissions from oil tanks can be minimized. The VOC detectors can be used to pin point place of leakage for the repair at the earliest.

The main source of oxides of nitrogen (NOx) and oxides of sulphur (SOx) are the fuel oil heaters and regenerators of FCC. The 'NOx emissions of FCC' may go up when:

(a) The catalyst contains more combustion promoter
(b) The oxygen in the flue gas goes up.
(c) The regenerator temperatures are higher.
(d) The feed nitrogen contents are higher.

It may be noted that 'combustion promoter' is generally a noble-metal material that accelerates the reaction between CO and O_2 to form CO_2. Because of removal of CO, the combustion promoter inhibits the following reactions:

$$2\,NO + 2\,CO = N_2 + 2\,CO_2$$

$$2\,NO_2 + 4\,CO = N_2 + 4\,CO_2$$

A major portion of the sulphur in the feed to an FCC unit ends up in coke on the catalyst. SO_x are formed in the regenerator when the coke is burned away. If sulphur in the feed to FCC decreases SO_x would naturally decrease.

Summary of Emission Control Methods Employed

The general methods of pollution control employed in refineries are:

(i) For emission of NO_x from boilers*:

- 'Low NO_x burners', 'Over fire air', and 'Air stage firing'.
- Selective Non- catalytic Reduction (SNCR)
- Selective Catalytic Reduction (SCR).

(ii) For emission of SOx from boilers*:

- Wet scrubbing
- Use of 'SO_x adsorption materials'
- Use of ammonia cyclic method **(see Para 8.3.3.2, Chapter 8)**.
- Use of natural gas as fuel instead of solid or liquid fuel.

(iii) For emission of suspended particulate matter (SPM) from boilers* :

Wet Scrubbing

Electrostatic precipitation by ESP

Use of natural gas as fuel instead of solid or liquid fuel.

*[*Note: The emission from heaters in which fuel is combusted would be controlled in a similar manner].*

(iv) For emission of SPM from FCC units:

Cyclones /multiclones,

Electrostatic precipitator and

Wet scrubbing.

(v) For emission of NOx from catalytic cracking units (FCCU):

- Selective Catalytic Reduction (SCR)** which can achieve reduction in NO_x emission of about 90%.
- Selective Non-Catalytic Reduction (SNCR), Thermal De-NO_x system, which can achieve reduction in thermal NO_x emission of only up to 60% (maximum)
- Use of NO_x combustion promoters and specifically engineered catalyst to decrease NO_x which can achieve NO_x red*uction of about 45%.*

Notes: ***Excess ammonia over that required to participate in the SCR reactions would appear in the SCR effluent. The ammonical effluent is called as 'Ammonia slip'. Efficiency of removal of NO_x depends on the amount of catalyst; the NH3-to-NO_x ratio; and the local distribution of ammonia & NO_x (ie the mixing of ammonia and NO_x), and flow across the SCR inlet.*

For FCC, if the 'SCR for NO_x removal' is located upstream of 'SO_x removal system', there is a risk of fouling of catalyst by 'ammonium sulphate []' and/ or 'ammonium bisulphate [H]'. In case, SPM is not removed before 'SCR for NO_x removal' there is a risk of plugging and the scouring action of entrained particulate matter. The risk can be greatly minimized by efficiently removing SPM. If water scrubbing is used in the process, reheating would be necessary (by use of gas to gas heat exchanger). For SCR (Selective catalytic cracking process) ***refer to Para 5.4.1, Chapter 5, and Paras 13.3.3.4 and 13.6.5, Chapter 13.***

(vi) For emission of SO_x from catalytic cracking units (FCCU):

- Water scrubbing (Water scrubbing is a favoured method as it removes SPM also).

- Can be controlled by absorption (including ammonia cyclic method) or catalyst adsorption. The recovered pure SO_2 can be converted to any useful product.
- Desulphurization of gases.

(vii) For emission of hydrogen sulphide (H2S) from 'Hydro-cracking Units':

The H_2S can be removed by absorption by amine after removal of SPM from the gas. The scrubbing by amine solution is done in a packed tower. The amine solution is regenerated and the H_2S is recovered and sent to the sulphur recovery plant.

Recovery of Energy from Flue Gases of FCC Unit

The energy carried away by waste flue gases (in the form of pressure and temperature of these gases) should be recovered to affect overall cost saving and consequent reduction in emission of greenhouse gases. For this, there can be two methods:

1. Providing a Waste Heat Recovery Steam Generator (WHRSG) to generate steam for use.
2. Providing a 'Hot gas expander' which is a single stage 'Gas turbine' to drive an 'Air compressor' which provides compressed air to 'Regenerator'. Alternatively, the hot gas expander (single stage gas turbine) may drive an electric generator to generate power.

9.5.2 Various Alternative Methods for SOx Control in Refineries

Apart from the methods of desulphurization mentioned previously, following methods of SO_x control are sometimes used:

9.5.2.1 *Carbon Adsorption Developed by Lurgi*

In this process, hot flue gas first goes through a cyclone and a dust collector for particulate removal. After this, the gas is cooled with water and sent to an adsorption tower packed with activated carbon where the SO_x is adsorbed on activated carbon. Water is sprayed into the tower intermittently to remove the adsorbed gas as a weak aqueous acid which is led to gas cooler where it scrubs remaining SOx from the gas and the clean gas goes out from stack. Cooler discharge can be sold as dilute sulphuric acid.

9.5.2.2 Reinluft Carbon Adsorption Process

In this process, the SO_x adsorbent is a slowly moving bed of activated carbon. This activated carbon is made from petroleum coke and activated by heating under vacuum at 1100°F (593°C). The adsorbed SOx on activated carbon can be recovered from it for use.

9.5.2.3 *Catalytic Oxidation Process (Monsanto Process)*

The particulates from hot gases are removed by cyclone and ESP. The particulate free gases are then passed over a fixed bed of catalytic agent (Vanadium pentaoxide V_2O_5) to catalyze SO_2 to SO_3 in a converter. The gases from converter are then led through a series of heat exchangers and then finally through a packed tower where they contact recycled dilute sulphuric acid. The gases after the absorption tower go for removal of acid mist by 'Wet ESP' or by any other method. Some dilute acid is recycled to the absorption tower. [Also see ***Para 8.4.5 Chapter 8***].

9.5.2.4 Absorption of SO_2 by Manganese Dioxide (MnO_2)

The manganese dioxide may be used as an absorbent for SO_2 from gases and then reacted by ammonia. Ammonium sulphate is obtained which is a fertilizer and MnO_2 can be recovered. Such a process is used by Mitshibishi Heavy Industry (MHI).

9.5.2.5 Ammonia Cyclic Method

This process is described in ***Para 8.3.3.2, Chapter 8***.

9.5.2.6 Use of SO_x Transfer Additives in FCC Regenerator

The proprietary transfer additives developed by Davison Chemicals when added react with SOx in regenerator of FCC and form sulphated additives which during circulation in reactor section get reduced to H_2S which is recovered by amine solution and sent to sulphur recovery plant. The balance SOx in the gas stream can be treated by any of the simple removal methods like wet scrubbing.

9.6 FLARING AND VENTING OF GASES AT REFINERIES AND DRILLING SITES

The necessity of flaring (burning) of gases arises both in refineries and on-land drilling site/ off-shore drilling sites. In order to control environment damage, the following practices should be followed.

- The flaring should be done through a stack and not at ground level in refinery.
- Cold venting of gases should never be resorted to and all gases should be flared through a stack if their release is required.
- In case of any elevated flaring the stack height would not be less than 30 m. The stack height should always be such that the maximum ground level concentration of pollutants should never exceed the ambient air quality standards.
- Burning of effluent in pits would never be carried out.

Ground Flaring

In case of oil/ gas drilling sites (both on-shore and off- shore), ground flaring may be permitted in special cases only where no bad effect is expected on crop production in adjoining area due to flaring. In order to minimize un-favourable effects of ground flares (when ground flare is needed), the Flare–pit should be constructed of RCC, surrounded by a permanent wall made of refractory bricks of minimum 5 m height to reduce glaring (heat radiation at ' Group gathering station', 'Oil collecting station', and 'Group collection station' of the oil/ gas drilling sites). A green belt of 100 m width should be developed around the flaring area after the refractory wall. In case the development of requisite green belt is not feasible, enclosed ground level flare system can be adopted only in exceptional cases, but it should be designed with a requisite enclosure height to reduce the Ground Level Concentration (GLC) up to or below the requirement in the standards.

9.7 EMISSIONS FROM PETROCHEMICAL PROCESSES

The petro-chemical processes which use air oxidation type reactions emit large, continuous amounts of gaseous emissions to atmosphere, but most of the petrochemical processes are enclosed processes and therefore there is only a small amount of fugitive emission.

The emissions from petrochemical processes may consist of some of the following pollutants: Carbon monoxide, methane, ethane, ethylene compounds, methanol, organic acids and anhydrides, aldehydes, ethers, hydrogen cyanide, hydrogen sulphide, hydrogen chloride, sulphur dioxide, etc.

The soluble gases/vapours like formaldehyde; methanol, sulphur dioxide, etc. can be removed by wet scrubbing. Hydrogen sulphide and sulphur dioxide may be recovered and sent to sulphur recovery plant. The organic emissions may be removed by 'Catalytic after- burner'. In case of carbon black manufacture, the calorific value of exhaust gases may be suitable for their use as fuel, and these may be used as fuel in a 'Waste heat recovery boiler'.

REFERENCES

1. *James H. Gary and Glenn E. Handwerk (1984). Petroleum Refining Technology and Economics (2nd Edition ed.). Marcel Dekker, Inc..*
2. *Robert E. Maples (2000). Petroleum Refinery Process Economics (2nd Edition ed.). Pennwell Books.*
3. *James G. Speight (2006). The Chemistry and Technology of Petroleum (4th Edition ed.). CRC Press.*
4. *C.T. Baird (1989), Guide to Petroleum Product Blending, HPI Consultants, Inc. HPI website.*
5. *Conroy, C.F and D. H. Linden, "Successful Application of Stand-alone FCCU Expander / Generator Sets," International Symposium on Turbomachinery, combined-Cycle Technologies and Cogeneration – IGTI-Vol.1. Greenhouse Gas.*

QUESTIONS

1. Enumerate the sources of different pollutants in an oil refinery.
2. Enumerate the methods of removal of various pollutants in the oil refinery and describe the method of utilization of H_2S and SO_x for recovery of elemental sulphur.
3. Write a short note on pollution control in 'Fluid Catalytic Cracking unit' of an oil refinery.
4. Write short notes on: (a) Hydrocracking, (b) Vis breaking, (c) Steam cracking in an oil refinery.
5. Write short notes on: (a) Hot gas expander, (b) Ground Flaring, (c) Pollutants evolved in petrochemicals.

CHAPTER 10

CONTROL OF MERCURY EMISSION FROM THERMAL POWER STATIONS AND INDUSTRIES

10.1 AIR POLLUTION BY MERCURY

The mercury and its compounds are highly poisonous and toxic. The sources and ill effects of mercury pollution have already been mentioned in ***Para.2.15.5, Chapter 2.*** Mercury is released in gaseous form when coal is burned. Moreover there are some industrial processes which emit elemental mercury and dust of solid mercury compounds in atmosphere. It falls on earth and enters the food chain. The mercury (Hg) emission is a real threat from coal fired boilers of power stations and from some industries which emit mercury and its compounds in environment. The mercury is volatile and elemental mercury directly pollutes the air. Also mercury / its compound converted to solid phase are emitted as RSP (respirable particulate matter). The air pollution on this account would cause lots of health problems directly and also by entering the food chain. The water from wet scrubbers / Flue Gas Desulphurization (FGD) / ash handling water may also sometimes contain some mercury; this may also pose some problems if not monitored.

As an example, it may be mentioned that 'Caustic soda' manufacture from old process of mercury cells is severely polluting industry. There are three processes of caustic soda manufacture:

- Mercury cell Process— the oldest technology employed which has been largely closed down around the world because of its mercury pollution potential and lower efficiency.
- Diaphragm cell— the dominant technology.
- Membrane cell— preferred technology for new plants. (Now used in Australia for its smaller plants)

The caustic soda manufacture uses electricity and common salt. In manufacture, for every 80 tonne of caustic soda, 71 tonne of chlorine and 2 tonne of hydrogen are co-produced. Each tonne of caustic soda requires around 2200 kWh of electricity and approximately 1.5 tonne of common salt (commercial quality).

According to the rules of government of India, it is mandatory that new caustic soda manufacturing plants should be based on 'Membrane Cell Technology' and not on 'Mercury Cell Technology'. These rules have expedited the eventual shift from mercury cell technology to membrane cell technology. While in 1960s and 1970s, most of the companies were using mercury or diaphragm cell technology, but now a days in India, majority of companies use diaphragm cell or membrane cell technologies, while some companies still use mercury cells technology.

The mercury can exist in flue gases in three states, namely elemental mercury, mercurous or monovalent mercury and divalent mercury. Mercury also binds with other chemicals to form inorganic compounds like '*Mercuric chloride*' and organic compounds like '*Methyl mercury*'. The species (form) of mercury present is critical to the mercury removal technology to be used at a plant. The presence of sulphur, chloride and other elements can bind with mercury, which may block its absorption by '*Carbon*'. There are now sustained efforts to reduce mercury emission. New ways to trap mercury going up in air are being discovered. *The Environment Protection Authority of USA announced 'Clean Air Mercury Rules (CAMR)' in 2005.*

Mercury removal is a complex process and its removal efficiency is dependent on large number of factors like:

- Type, constituents, and analysis (details) of fuel specific to a plant
- Form of mercury present in flue gas
- Types of burners and firing system
- Boiler operating conditions, like excess air, furnace temperature, turbulence in furnace, combustion conditions, unburnt carbon in ash, etc.
- Design and type of dust / suspended particulate matter removal system, temperature and residence time of flue gases in particulate control device, etc.
- Design of FGD System

10.2 TECHNOLOGIES FOR REDUCTION IN MERCURY EMISSION

10.2.1 Types of Mercury Removal Technologies

The basic mercury removal technologies are:

- Physical cleaning of coal prior to combustion
- Converting mercury in combustion gases to solid phase and removal with ash or dust.
- Adsorption of mercury on adsorbents, (say, activated carbon, unburnt carbon in fly ash, etc.)
- Converting mercury into water soluble compounds which get removed in wet scrubbing / wet FGD, Excitation of elemental mercury in flue gases by UV light so that it combines with sulphur dioxide / oxygen / other materials present in flue gases and removal with ash or dust..
- Electro-catalytic oxidation (ECO) and removal in 'WET ESP* (Wet electrostatic precipitator)' [* For wet ESP refer to **Para 8.2.3 and 8.4.5 Chapter 8]**

10.2.2 Role of Various Factors and Equipments in Mercury Removal

10.2.2.1 Role of Type and Quality of Coal in Mercury Removal

When coal is beneficiated / washed, the calorific value of coal increases and the ash content and the impurities like heavy metals including *mercury* are reduced. The un-beneficiated and low calorific coal on combustion pollute more and have higher *mercury* in their emission. Generally, therefore, it is better to use clean coal. It has been observed that the quality of coal affects the removal of mercury by various dust removal equipments. The removal of mercury is higher in case of bituminous coal than in cases of low grade coals like sub- bituminous coal and lignite.

10.2.2.2 Reduction of Mercury Emission by Dust Removal System

The fly ash may help in removal of mercury, but it depends on the characteristics of fly ash, its temperature and contact time. The solid phase mercury is removed along with the removal of SPM by the dust control system like scrubber / fabric filter. The Bag filters have the highest capacity to remove mercury, while in a cold side ESP, the mercury removal is much lower than in a Bag filter. In a '*Hot side ESP*'*, the removal of mercury is insignificant. A low 'NOx combustion system' along with 'Bag Filter' is a good combination for removal of mercury.

It may be mentioned that 'Wet scrubber' can remove 90% of the oxidized mercury, but it cannot remove elemental mercury because its solubility in water is low. However strict watch would be necessary over pollution of water by mercury in case of wet scrubbing. *[*For hot side ESP, refer to notes under **Para 4.3.3 Chapter 4**]*

10.2.2.3 Role of Chlorine Content of Coal in Mercury Removal

The *chlorine content* of coal being combusted in the furnace increases the oxidation of elemental mercury (Hg) to solid phase i.e. 'oxidized mercury compounds' which are solids. Higher the chlorine content, higher is the reduction in mercury emission in air.

10.2.2.4 Role of Carbon Particles in Mercury Removal

The *unburnt carbon* in fly ash adsorbs elemental mercury and gets removed in the particulate removal system. ***[Ref: 2]***. The injection of '*Activated Carbon*' in flue gas after an Air Preheater and primary dust collecting equipment like ESP before the 'Bag filters' adsorbs the elemental mercury. ***[Ref: 4]***.

But it has been found that Sulphur trioxide (SO_3) gets over the carbon before the 'Mercury' and therefore limits the capacity of 'Activated Carbon' in adsorbing elemental mercury. It is therefore better to remove SO_3 by desulphurization through 'Spray Dry Absorber' (which injects 'Lime' in flue gas system) before injection of 'Activated Carbon' in flue gases for removal of mercury.

10.2.2.5 Role of Spray Dry Absorber Type FGD System in Mercury Removal

It has been observed that Spray Dry absorption (SDA) type 'Flue Gas Desulphurization (FGD) process' not only removes oxides of sulphur (SO_x), but also helps mercury removal from flue gases. The removal of 'Elemental mercury' also depends on the type of coal. When the 'Spray dryer absorber' is employed in plants fired by 'bituminous coal', it removes both the 'elemental mercury' and 'oxidized mercury', but when employed in plants fired with 'low grade coals', it removes only 'oxidized mercury' ***[Ref: 5]***.

10.2.2.6 Role of Wet FGD System in Mercury Removal

There is a significant reduction in mercury in flue gas after it has passed through the treatment in wet FGD system. The wet FGD though more effective in reducing mercury emission in bituminous coal fired units, are also significantly effective for sub-bituminous coal and lignite.

Passing of flue gases into the lime stone slurry through a sparger tube in a tower called *Jet Bubbling Reactor*, removes *mercury* in addition to oxides of sulphur. This FGD process was developed by *Chiyoda Corp., Yokohama, Japan* and the reactor is named as CT -121' ***[Refer Para.7.4 6. 2, Chapter. 7.]***

10.2.2.7 Role of Selective Catalytic Reduction (SCR) in Mercury Removal

It has been found that SCR Systems convert 'elemental mercury' to 'oxidized mercury' which then can be removed by 'wet scrubber', but it works better in 'bituminous coal' fired plants than in 'sub-bituminous coal' fired plants. A wet scrubber along with 'Selective catalytic reduction' is a good combination for removal of mercury. ***[Ref: 5]***.

It may, however, be noted that the effluents from a scrubber should be sent to a settling tank and a part of the decanted water may be recycled. The effluent released to the environment should not have mercury content more than the permitted limit. The solids collected in the settling tank also require monitoring of mercury contents and all precautions are required in their disposal.

10.3 CHEMICAL INJECTION FOR REDUCTION OF MERCURY EMISSION FROM COAL FIRED BOILERS

As already mentioned, the chlorine content of coal plays a positive role in reduction of mercury emission. Therefore in order to increase the chlorine content of coal, calcium chloride ($CaCl_2$) in required amount is injected / dosed in the coal before the 'coal preparation plant / pulverizing plant'. The pulverized coal gets well mixed with calcium chloride before it is fired in the furnace. Lime dosed in spray dryer absorber is employed to absorb the oxides of sulphur by lime. Activated carbon is dosed in the flue gases between 'Spray Dryer Absorber (SDA)' and bag filter for adsorbing mercury. The flue gases after passing through the 'spray dryer absorber (SDA)' are led to 'Bag filter' where the SPM is collected. *[The desulphurization of gases is done prior to dosing of activated carbon for reasons already mentioned in the previous para.]*

Schematic diagram of chemical injection for reduction of mercury emission from coal fired boilers is shown **in Fig. 10.1.** [Also refer ***Para. 8.3.9.3, Chapter 8,*** for removal of mercury along with dioxins.]

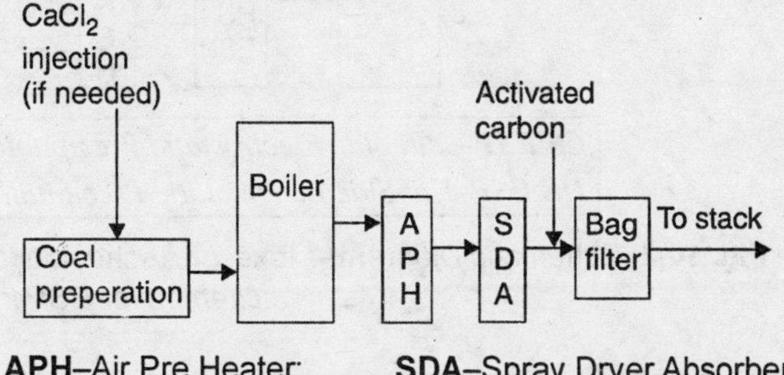

Fig. 10.1: Chemical Injection for Reduction of Mercury Emission from Coal Fired Boilers (***Ref:*** 5)

10.4 TOXECON PROCESS OF REDUCTION OF MERCURY EMISSION

TOXECON Technology was initially formulated by US National Energy Technology Laboratory (NETL), and has been patented by US Electric Power Research Institute. Its application is claimed to reduce mercury emission from coal fired stations significantly and also remove SPM easily. The technology was demonstrated at 'WE Energies Michigan Presque Isle' power plant. A schematic diagram of the TOXECON technology for mercury removal is shown with two alternatives for location of ESP in **Fig. 10.2.**

It is mainly based on adsorption of mercury by activated carbon from gases whose bulk of SPM has been removed in advance by ESP or other primary dust collecting equipment. The activated carbon is injected in the gas stream after the ESP ahead of bag filter. The bag filter provides an excellent contact between 'mercury in gas and activated carbon' resulting in efficient removal of mercury. The residual ash escaping from primary dust collecting equipment or ESP together with activated carbon which has already absorbed the mercury is collected in bag filter. The ash from primary dust collecting equipment / ESP can be used as usual without any danger of mercury pollution, while the solids collected in bag filters which contain absorbed mercury, should be suitably disposed off taking all due precautions *[Ref: 4].*

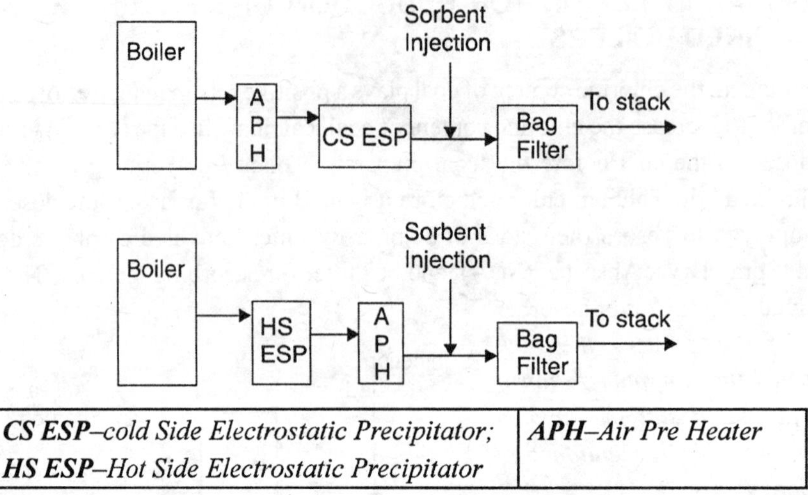

Fig. 10.2: Schematic Diagram – Toxecon Technology for Mercury Removal [*with two alternative schemes for ESP location*]

10.5 ELECTRO-CATALYTIC OXIDATION FOR REMOVAL OF MERCURY

The Electro-catalytic Oxidation (ECO) technology was developed by Powerspan Corp., New Durham, NH, US. It is a multi – pollutant treatment technology and can reduce oxides of nitrogen (NO_x), sulphur dioxide (SO_2), fine particulate matter and mercury. This technology consists of three stages:

(i) 'Gas Reactor' called 'Dielectric barrier discharge reactor', which oxidizes the gaseous pollutants. The nitric oxide (NO) is oxidized to nitrogen dioxide and nitric acid, and a small amount of SO_2 is oxidized to SO_3 / sulphuric acid and mercury is oxidized to mercuric oxide.

(ii) Ammonia (NH_3) scrubbing removes unconverted SO_2 and nitrogen dioxide (NO_2). SO_2 in presence of moisture forms ammonium sulphite which is disposed along with the ash. The NO and NO_2 are reduced to nitrogen.

(iii) A 'Wet ESP' captures acid aerosols, fine particulate matter and *oxidized mercury.* Water from wet ESP is treated and a part of this treated water is recirculated and balance is discharged. Solids/ brine sludge from effluent treatment plant is treated for removal of mercury before disposal. The chemical reactions in the process are **[Ref: 6 &7]** :

$$2NO + O_2 = 2\,NO_2\,;\ 2SO_2 + O_2 = 2SO_3;\ 2Hg + O_2 = 2HgO$$

$$6NO_2 + 8NH_3 = 7N_2 + 12H_2O;\ 4NO + 4NH_3 = 4N_2 + 6H_2O.$$

$$SO_3 + H_2O = H_2SO_4;\ 2NH_3 + H_2SO_4 = (NH_4)_2SO_4;$$

$$2NH_3 + SO_3 + H_2O = (NH_4)_2SO_4;\ 2NH_4OH + SO_2 = (NH_4)_2SO_3 + H_2O$$

A schematic diagram of the system is shown in **Fig. 10.3.** The *escape of mercury* in the brine sludge in a single power failure can be as much as 10 times the loss of mercury in the sludge produced in a day under normal running of the plant. The mercury carryover in sludge/effluent causes serious pollution of water and land directly and in addition it may cause air pollution indirectly. Any air/ fumes from mercury cell room would carry elemental mercury/ mercury compounds which can cause direct air pollution.

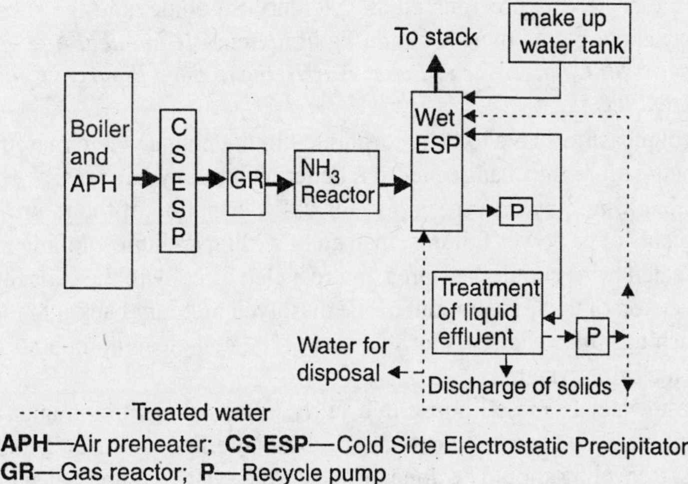

APH—Air preheater; **CS ESP**—Cold Side Electrostatic Precipitator
GR—Gas reactor; **P**—Recycle pump

Fig. 10.3: Electro Catalytic Oxidation for Removal of Mercury

10.6 CONVERSION OF ELEMENTAL MERCURY TO SOLID COMPOUND BY ULTRA VIOLET (UV) LIGHT

It has also been found that when flue gases are exposed to UV light, the elemental mercury (Hg) gets excited and reacts with other compounds present in the flue gas and it is thus converted to solids which are easily removable. The excited mercury reacts with SO_2 and oxygen to form mercuric sulphate and mercuric oxide, both of which can be removed easily. The UV light is applied in the 'Particulate matter collector like bag filter and Multiclone' or the wet scrubber. This method is attractive for softer varieties of coal with low calorific values *[Ref:3]*.

10.7 MITIGATING MERCURY POLLUTION FROM CAUSTIC SODA INDUSTRY BY CHANGE OF TECHNOLOGY

10.7.1 Drawbacks of Mercury Cell Technology of Caustic Soda Production

In the production of caustic soda based on mercury cell technology, the mercury is not supposed to get consumed, but it gets entrapped in the circulating brine solution and all the streams of products and bye products. This leads to contamination of the water, wastewater, air and solid wastes generated from the production activities. The quantity of mercury involved in the production is very large, its leakage, spillage and even evaporation of the spilled mercury are common.

A limit of 0.01 mg/L has been prescribed for the levels of mercury in the effluent along with a limitation of a maximum of 10 cum effluent per tonne of product. But in practice, the limits are not maintained. The control measures in attending to leakages / spillage / washings of cell room floor are required to be strictly maintained. Attention needs to be given to all the mercury cell-based chlor-alkali plants, especially, the manner in which the return brine from the electrolysis cells is treated soon after the power failures. The mercury concentration in the return brine after power failures may go up to even 200 mg/L as against a concentration of about 7 mg/L in the brine in the normal running of the plant. Generally, 'Sodium Sulphide' is added to the brine after power failures for avoiding the chlorine nuisance, but this addition of 'Sodium Sulphide' results into a bigger nuisance as the high level of mercury present in the brine gets precipitated as 'Mercury Sulphide' and the subsequent brine sludge from the clarifier may contain upto over 1000 mg/kg of mercury ***[Chemical Age of India 35 (9):1984].***

Precautionary Measures till Changeover to Better NaOH Technology [Source: Central pollution control board, India, 23rd Nov. 2000.]

The following control measures have to be incorporated in the plants where mercury cell technology is used for the time being before the changeover to a better technology happens:

- Provide a reliable 'Stand-by Power supply' for de-chlorination of the return brine in the existing mercury cell plants after power failures, instead of adding sodium sulphide to eliminate chlorine. This de-chlorinated brine should be stored and recycled directly to the cells in a controlled manner to ensure conversion of the high amount of the dissolved mercury back into the elemental mercury
- Collect the washings immediately after the start of the cell cleaning operation and recycle of the washings into the brine system.
- Collect the wastewater from cell house in a fairly large sedimentation tank and provide a pump with nipple and hose pipe arrangements in the discharge line and use of this collected water for floor washings. Recycling should be done into the brine system as and when possible with the help of same pumping arrangements. (Chemical Age of India 37 (11) : 1986). Also, the floor should be preferably swept and the wet washings should be avoided as far as possible.
- The solid wastes resulting from the Hg cell room should be heated in a closed system and the mercury should be recovered through condensation (indirect cooling).
- The hydrogen gas should be treated at source with the help of activated carbon adsorption technique. This will eliminate the involvement of the mercury in the HCI production system. The hydrogen produced can find so many industrial uses also.
- The production of caustic soda should be optimized on the basis of demand. The chlorine production and demand should also be balanced. The caustic soda or lime slurry can be used for removal of chlorine on the basis of the use of resulting 'Calcium hypochlorite' or 'Sodium hypochlorite'. [Also see **Para 8.3.5 of Chapter 8**, removal or nuetralization of chlorine.]
- Each plant authorities should study the level of excess chlorine which can be maintained in the circulating brine of the production system as the presence of chlorine in brine avoids precipitation of mercury into the brine sludge. The mercury should be recovered from the brine sludge before disposal which should be made in a secured landfill with proper arrangements for the collection and recycle of the leachate.
- The mercury streams with mercury should be segregated at the source itself in the plant and recycled into the brine system. This will result into minimization of mercury input load to the final treatment system, and the steps like precipitation, ion- exchange etc. can be decided by individual plant authorities depending upon the level of mercury control that can be achieved at source.

10.7.2 Changeover to Membrane Cell Technology in Caustic Soda Production

The mercury cell technology of production of caustic soda is fraught with environmental pollution and therefore in most countries, this technology has been or is being abandoned and alternative non polluting technologies have been / are being used. The membrane cell technology is supposed to be the best. As already mentioned, the setting up of new plants or expansion of caustic soda units based on mercury cell process has already been banned by the *Government of India* and the existing mercury cell based plants are also in the process of switching over to the membrane process.

Membrane Cell Process

Chlorine is normally produced by passing an electric current through a solution of brine which is common salt (NaCl) dissolved in water. The Caustic Soda ie Sodium hydroxide (NaOH) and Hydrogen (H_2) are obtained.

$$NaCl = Na + Cl \text{ (by electrolysis)}$$

$$2Na + H_2O = 2NaOH + H_2 \text{ .(by reaction of sodium with water)}$$

All these three products (namely caustic soda, hydrogen and chlorine) are highly reactive, and technologies have been developed to separate them and keep them apart. Hydrogen is a combustible gas used in various processes including the production of hydrogen peroxide and ammonia as well as the removal of sulphur from petroleum derivatives.

In the membrane cell process, the anode and the cathode are separated by an ion-exchange membrane. Only sodium ions and a little water pass through the membrane. The brine is de-chlorinated and re-circulated. Solid salt is usually added to re-saturate the brine. After purification by precipitation-filtration, the brine is further purified with an ion exchanger.

Table 10.1: Mercury Removal Capability of Various Air Pollution Control System [for bituminous coal fired plants] **[Ref: 5]**.

Se. No.	Air pollution controls used	Baseline Hg removal	Mercury control used	Hg removal Expected
1	Cold side ESP	10–30%	+activated carbon injection	50–70%
2	ESP + Bag filter	--------	+activated carbon injection and bag filter with pulse jet control (TOXECON)	Upto 90 %
3	Bag filter with pulse jet control	50–70%	+activated carbon injection	70–90%
4	Spray dry absorber and ESP	30–50%	+activated carbon injection	50–70%
5	Spray dry absorber & Hot Side ESP	-----------	+activated carbon injection and bag filter with pulse jet control	Upto 90 %
6	Spray dry absorber & bag filter with pulse jet control	50–70%	+Activated carbon injection	70–90%
7	Cold side ESP and wet FGD	50–60%	+Injection of other sorbents	60–90%
8	Cold side ESP and wet FGD	----------	+Injection of other sorbents and Activated carbon	80–90%
9	SCR, cold side ESP and wet FGD	70–90%	+Injection of other sorbents / (other than activated carbon)	80–90%
10	SCR, cold side ESP and wet FGD	--------	+Injection of other sorbents and activated carbon	90 %

The caustic solution leaves the cell with about 30% concentration and at a later stage in the process, is usually concentrated to 50%. The chlorine gas contains some oxygen and is purified by liquefaction and evaporation. The consumption of electrical energy is low and the amount of steam needed for concentration of the caustic is relatively small (< 1 tonne / tonne of caustic soda).

10.8 COMPARATIVE ROLE OF VARIOUS AIR POLLUTION CONTROL DEVICES/ SYSTEMS IN MERCURY REMOVAL

Various air pollution control devices / systems are used in practice to reduce various types of pollutants in the emissions. These devices also incidentally reduce mercury emissions. A comparison of mercury removal capability of these systems in bituminous coal fired plants is shown in **Table 10.1,** while the mercury removal performance of various equipments in cases of different coals/ lignite is shown in **Table 10.2.**

Table 10.2: Mercury Removal Performance of Various Equipments for different Coals / Lignite *(Source of data: Ref: 1)*

	Approximate % tage of Mercury removal		
	Bituminous coal	**Sub-bituminous coal**	**Lignite**
1. Cold side ESP	Up to 40 %	About 2 %	About 3 %
2. Hot side ESP	About 2 %	About 1 %	---
3. Fabric / Bag Filter	Up tp 80 %	Up to 60 %	About 2 %
4. Spray Dryer + Bag filter	Up to 80 %	About 3 %	About 1 %
5. Wet Flue gas desulphurization	Up to 58 %	15 %	15 %

From the above table is may be seen that the use of specifie technologies for minimizing mercury emossion is much more essential in case of sub-bituminous cool and lignite.

10.9 DUST REMOVAL FROM GASES OF 'MERCURY ORE ROASTER'

Mercury is obtained from its chief ore, 'Cinnabar' mercury sulphide (HgS). The cinnabar is mainly found in Spain, Italy, Austria, and California (US).

$$HgS + O_2 = Hg + SO_2 .$$

Most of SO_2 can also be absorbed by adding 'Sodium Sesquicarbonate' along with the ore. The sodium sesquicarbonate is a complex salt of sodium carbonate (Na_2CO_3) and sodium bicarbonate ($NaHCO_3.2H_2O$). Naturally found Trona ore which consists of about 70–95 % of crystals of 'Sodium Sesquicarbonate' can be crushed and used instead of 'Sodium Sesquicarbonate' in the charge along with the mercury ore for roasting. The roaster gases would contain some remaining part of SO_2 which would be removed after the mercury is condensed. There is evolution of dust in roasting of fine mercury ores during production of mercury. A typical flow diagram of dust collection from gases from 'Fluid bed roasting' is shown in **Fig. 10.3.**

The gases from roasting furnace are divided in two parallel paths; each path consisting of a cyclone, Multiclone and a fan. The fans convey the gases to upstream of a multi stage horizontal flow dry type ESP's. The ESPs are maintained under slight vacuum. The gases at the outlet of ESPs are led to condensers for condensing 'Mercury vapour' to obtain mercury. The condensed mercury is purified by vacuum distillation. The Boiling Point of Mercury is 357 °C and its vapour is colourless. At about 350 °C ie just below the boiling point, mercury slowly combines with oxygen to give mercuric oxide HgO. The exhaust gases from mercury condensers may have SO_2 and traces of mercury.

Before releasing the exhaust gases to the atmosphere, SO_2 and traces of mercury should be removed. Both SO_2 and traces of mercury from the exhaust gases (from condenser) can be removed by methods already discussed. *[One method of removal is by 'Wet scrubbing', but the liquid effluent from the scrubber should be neutralized and suitably treated for removal of mercury and before release to the environment].*

Additional mercury specific removal (>99%) can be achieved using e.g. the Boliden Norzink technique where mercury is adsorbed in a solution of calomel (mercurous chloride) in a packed tower type scrubber as described in next paragraph.

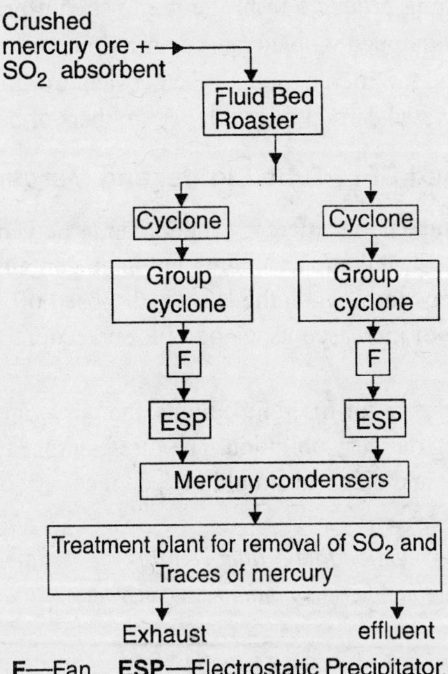

Fig. 10.4: Typical Flow Diagram for Dust Removal From Gases of Fluid Bed Roaster of Mercury Ore

10.10 BOLIDEN-NORZINK PROCESS OF RECOVERY OF MERCURY IN METALLURGICAL PROCESSES

In this process, the de-dusted gas containing mercury vapour is scrubbed in a packed bed tower with a solution containing $HgCl_2$. This reacts with the metallic mercury in the gas and precipitates it as calomel (Hg_2Cl_2):

$$HgCl_2 + Hg = Hg_2Cl_2$$

Hg_2Cl_2 is removed from the circulating scrubbing solution and partly regenerated by chlorine gas to $HgCl_2$, which is then recycled to the washing stage. The mercury product bleed is either used for mercury production or stored.

A part of the precipitate, the calomel, is electrolyzed to give metallic mercury and chlorine gas which is reacted with the rest of the calomel to produce the reagent, $HgCl_2$, used in the scrubbing. Efficient removal of mercury from gases and recovery of mercury is the main environmental advantage of the above processes. [*An alternative process uses sodium thiocyanate as the scrubbing medium and mercury is precipitated as the sulphide. This can be recovered in a primary mercury roaster.*]

10.11 MERCURY POLLUTION FROM WASTE MATERIALS AND CONTROLS

10.11.1 Waste Incineration

Since all mercury emitted to air from waste incineration originates from mercury in the waste, the most important primary measure is to reduce the input of mercury to the incinerator. This can be accomplished by sorting out mercury containing products from the waste stream prior to incineration. Most of the modern waste incinerators are equipped with advanced emission control for different pollutants. Some of these devices can be optimized for 'mercury control' e.g. by controlling temperature. Mercury specific control can be achieved by activated carbon injection or scrubbers or both.

10.11.2 Mercury from used CFL, Tube Lights, and Mercury Vapour Lamps

Mercury is also released when tube lights, mercury vapour lamps and compact fluorescent lamps (CFL) are disposed off after their use. Each compact fluorescent lamp contains about 5mg of mercury which comes out in chemical or gaseous form when the lamp is disposed off. This type of mercury pollution has gained importance because of increased usage of CFL on account of their high efficiency and low energy consumption.

A collection and disposal system for fused tube lights, mercury vapour lamps and CFL is needed to be evolved to check the resulting mercury pollution. The fused lamps / tube lights should be collected at central place and their breaking and disposal should include recovery of mercury by any of the known methods.

Important note: *It is important to mention that if light emitting diodes lamps (LED lamps) are used, we can save more power and at the same time there is no danger of mercury pollution.*

REFERENCES

1. Chemical Age of India 3(a): 1984; 37 (ii): 1986.
2. Connie Senior, 'Mercury Control — Two pollutants for the price of one', Power, Vol.149, No. 5, June 2005].
3. Global Monitor, Power Vol. 147, No. 8, Oct 2003 p.14.
   [Source Central pollution control board 23rd Nov. 2000].
4. Global Monitor, Power, Vol. 154, no 3, March 2010, p.11].
5. Jin-Gang Yu, Bao-Yu Yue, Xiong-Wei Wu, Qi Liu, Fei-Peng Jiao, Xin-Yu, Jiang, Xiao- Qing Chen, 'Removal of mercury by adsorption: a review', Enviromental Science and Pollution Research, March 2016, Volume 23, Issue 6, pp 5056–5076.
6. Kennedy Maize 'Technology could deliver 90% Hg reduction from coal', Power, Vol.153, No.7, July 2009, pp.52–53.

7. Dr Pelter, Robert, 'Mercury removal standards are coming. Where is the technology?', Power, Vol.147, No.4, May 2003, p.40–48.
8. Special report, pp.36-38, Power, Vol.146, No.3, May/ June 2002.

QUESTIONS

1. Write a short note on air pollution by mercury.
2. Enumerate different factors which affect the mercury removal from flue gases/ exhaust gases.
3. What are the roles of different flue gas types of dust collection systems and desulphurization systems as regards mercury removal from flue gases/ exhaust gases from boilers?
4. Describe the mercury removal process from flue gases of boilers by (a) Chemical injection, (b) Toxecon Process.
5. Describe mercury removal from boiler flue gases by Electro catalytic oxidation method.
6. State the draw backs of mercury cell technology of caustic soda production as regards pollution by mercury.
7. Enumerate the precautions to be taken for reducing pollution by mercury when the mercury cell technology is used for caustic soda production.
8. Write short note on 'Membrane cell technology for caustic soda production' pointing out the advantages of the process.
9. Describe a dust removal system for mercury ore roaster.
10. Write short note on mercury production by chemical process so that the pollution by mercury could be reduced to minimum.
11. Suggest ways to reduce pollution from waste material and its incineration.

CHAPTER 11

AIR POLLUTION BY NUCLIDES / RADIOACTIVITY: SAFETY AND CONTROL

11.1 RADIOACTIVITY

11.1.1 Radioactive Decay

Radioactive decay is the process by which an unstable atom loses energy by emitting ionizing particles (ionizing radiations). The emission is spontaneous, in that the atom decays without any physical interaction with another particle from outside the atom. Usually, radioactive decay happens due to a process confined to the nucleus of the unstable atom, but, on occasions (as with the different processes of electron capture and internal conversion), an inner electron of the radioactive atom is also necessary to the process.

As for types of radioactive radiation, it was found that an electric or magnetic field could split such emissions into three types of beams. For lack of better terms, the rays were given the alphabetic names 'Alpha', 'Beta', and 'Gamma'. While alpha decay was seen only in heavier elements (atomic number 52, tellurium, and greater), other two types of decay were seen in all the radioactire elements.

In analyzing the nature of the decay products, it was obvious from the direction of electromagnetic forces produced upon the radiations by external magnetic and electric fields that alpha rays carried a positive charge, beta rays carried a negative charge, and gamma rays were neutral. From the magnitude of deflection, it was found that alpha particles were much more massive than beta particles. Passing alpha particles through a very thin glass window and trapping them in a discharge tube allowed researchers to study the emission spectrum of the resulting gas, and ultimately proved that alpha particles were Helium nuclei. Other experiments showed the similarity between classical beta radiation and cathode rays: They are both streams of electrons. Likewise, gamma radiation and X-rays* were found to be similar high-energy electromagnetic radiation.

Radio-nuclides can undergo a number of different reactions. Radioactive decay results in a reduction of summed rest mass, once the released energy (the disintegration energy) has escaped in some way. Although decay energy is sometimes defined as associated with the difference between the mass of the parent nuclide products and the mass of the decay products, this is true only of 'Rest mass' measurements, where some energy has been removed from the product system. The decay energy always carries mass with it, wherever it appears (see mass in Special Relativity Theory) according to the formula $E = mc^2$. The decay energy is initially released as the energy of 'emitted photons' plus the 'Kinetic energy' of 'Emitted particles' (that is, particles that have 'Rest mass'). If these particles come to thermal equilibrium with their surroundings and photons are absorbed, then the decay energy is transformed to thermal energy, which retains its mass.

Decay energy therefore remains associated with a certain measure of mass of the decay system invariant mass. The energy of photons, kinetic energy of emitted particles, and, later, the thermal energy of the surrounding matter, all contribute to calculations of invariant mass of systems. Thus, while the sum of rest masses of particles is not conserved in radioactive decay, the system mass and system invariant mass (and also the system total energy) is conserved throughout any decay process.

11.1.2 Formation of Clusters of Ions of Radionuclides

When a radioactive nuclide decays, and decay products are formed, the loss of electron from it may form a positive ion which can attract liquid and even solid material and thus forming clusters of atoms or particles in the submicron region ranging from 0.001 to 0.01 μm.

Air is permanently ionized by radiation from natural radioactivity of air, and by cosmic radiation which consists mostly of positively charged particles. In the free atmosphere, the rate of production of small ions is in balance with the rate of neutralization by recombination, and the rate of attachment to condensation nuclei. Condensation nuclei are mostly the sub-micrometer particles in the range of 0.005 to 0.01 μm.

In air containing water vapour, the positive ions are mostly hydrated protons, p+ or H+ $(H_2O)n$, where n may be any number between 1 and 8. The negative ions are mostly hydrated O- or OH-. The formation of clusters of water molecules around ions is very rapid, but the clusters do not grow beyond about 0.001 μm (1 nm) diameter and remain as small ions until they become attached to condensation nuclei. They then become large ions which are mostly in the accumulation mode. The intermediate ions in the size range of 0.001 to 0.010 μm (1–10 nm) also exist. If sulphate or nitrate vapours are present in air, their molecules will dissolve in the water clusters and these will then grow into the intermediate ion range.

A possible mechanism for formation of radiolytic nuclei is radiolysis of water, leading to formation of H_2O_2 which then .oxidizes traces of SO_2 to give H_2SO_4 Addition of O_3 to the air also increases nuclei production, whereas addition of NO, a well-known radical scavenger, inhibits it.

The radio nuclides / ions in the atmosphere exist in two forms: (1) as "unattached clusters" with a diffusion equivalent diameter ranging from 0.5 to 5 nm and (2) as "aerosols-attached clusters" with particle diameters varying between 5 and 3000 nm.

Beryllium-7

It is known that small quantity of Beryllium (Be) is present in automobile smoke with possible traces of Beryllium-7 (Be-7) which is a relatively short-lived nuclide (half-life 53.3 days). Be-7 is also naturally occurring radionuclide of cosmic origin which is probably formed by spallation processes of light nuclei of atmosphere, such as carbon, nitrogen and oxygen, when they absorb protons and even neutrons of the primary component of cosmic rays.

Once Be-7 (inorganic ion) is formed in the troposphere, it rapidly associates primarily with submicron-sized aerosol particles. Beryllium-7 in these fine aerosols may subsequently enter the marine as well as the terrestrial environment and vegetation via wet or dry deposition. Following this deposition, Be-7 will tend to associate with particulate matter (particle-reactive element).

Beryllium-7 as well as other natural radio nuclides like Pb-212, Pb-210, Na-22, S-35, P-32, and P-33 participate in the formation and growth of the accumulation mode aerosols (0.07–2.0 μm diameter), which is a major reservoir of pollutants in the atmosphere. Following its production by gas-phase nuclear transformation, this isotope condenses on the aerosol population growing by condensation of non-radioactive species, e.g. $H_2 SO_4$ or organics. Refer to **Para 2.10.5, Chapter 2** for effects of beryllium pollution.

11.1.3 *Net Radioactivity Concentration on Earth's Surface*

The net radioactivity on earth surface would be different from place to place and therefore human beings and animals are subjected to varying quantity of environmental radiation depending upon the location.

11.2 COSMIC AND TERRESTRIAL RADIATION

Certain amount of air pollution by radioactive gases and nuclides is of natural origin.

Cosmic rays are predominantly high energy protons originating from essentially interstellar protons. Cosmic ray radiation consists of 85 % protons, 14 % alpha particles and about 1 % nuclei of atomic number between 4 and 26. These particles are highly penetrating and possess extremely high energy. Cosmogenic radionuclides like H-3, Be-7, C-14 and Na-22, produced by cosmic ray interaction of the heavy charged particles coming from outer space, with the atmosphere, contribute to the natural radiation exposures as internal exposure through inhalation.

Annual effective doses from cosmic ray radiation around the world are estimated to range from 0.26 to 2.00 milli Sv per year. On the average, at sea level the cosmic radiation is about 0.30 mSv/yr, but it increases with increase in altitude. At high altitude it can be double or triple.

Terrestrial radiation comes mainly from the primordial (existing from beginning ie original) radio nucleides on earth or its atmosphere like K40, Rb87, radioactive decay series of U238, and Th232. These radiations may be alpha, beta and gamma. Because gamma rays have the highest penetrating power (although their ionizing power is lower than both beta and alfa rays), the natural environmental gamma radiation has the maximum impact. The terrestrial radiation decreases as the altitude increases. The presence of radioactive substances in rocks and soil give rise to external as well as inhalation and ingestion exposures. Levels of terrestrial radiation differ from place to place around the world as the concentrations of these nuclides.

Cosmic rays and ionizing radiation from natural radio-active substances in soil and water produce back ground radiation. These may be from potassium – 40 (k40) or other radio-active substances in living tissues, cosmic rays, radio nuclides in rocks and soil, etc. Natural waters have been found to contain 'Alpha activity' mainly derived from rocks and minerals and 'Beta activity' mainly due to potassium isotope. Natural environmental radiation affecting human beings may be 'Cosmic radiation component and terrestrial radiation component'.

Notes:

(1) Gamma rays and X rays: These are high energy radiations having characteristics as given in the Table 11.1 A:

Table 11.1A: Characteristics of Gamma rays and X-rays

Class	Frequency	Wave length	Photon Energy
Gamma rays	300 – 30 EHz	1pm – 10 pm	1.24 Mev – 124 kev
HX (hard X-rays)	30 – 3 EHz	10 – 100 pm	124 – 12.4 kev
SX (soft X-rays)	3 EHz – 300 PHz	100 pm – 1 nm	12.4 – 1.24 kev

(2) Solar Electromagnetic Radiations and their Effects— *Solar radiations may be a source of various radiations including ionizing radiations. The details about solar radiations and their effects are given below.*

(i) ***Frequencies and Wavelengths of Solar Electromagnetic Radiations:*** *Besides visible light and heat, there are other types of radiations from sun which affect the life on earth. Different radiations from sun have different frequencies, wavelength and photon energies. Different radiations from sun are summarised in the* ***Table 11.1 B:***

Table 11.1B: Characteristics of Various Radiations from Sun

Class of Solar Electromagnetic Radiations	Frequency	Wavelength
Gamma	300 – 30 EHz	1pm – 10 pm
HX (hard X-rays)	30 – 3 EHz	10 – 100 pm
SX (soft X-rays)	3 EHz – 300 PHz	100 pm – 1 nm
EUV (extreme ultraviolet)	300 – 30 PHz	10 – 100 nm
NUV (near ultraviolet)	30 – 3 PHz	100 nm – 1μm
Visible light		0.39μm to 0.76 μm
	3PHz – 300 THz	1 – 10 μm
NIR (near infrared)	<300 – 30 THz	10 – 100 μm
MIR (mid infrared)	30 – 3 THz	100 μm – 1mm
FIR (far infrared)	3THz – 300 GHz	1 – 10 mm
Radio waves		
EFH (extremely high frequency)	300 – 30 GHz	1mm – 1cm
SHF (super high frequency)	30 – 3 GHz	1 – 10 cm
UHF (ultra high frequency)	3 GHz to 300 MHz	10 cm – 1m
VHF (very high frequency)	300 – 30 MHz	1 – 10 m
HF (high frequency)	30 – 3MHz	10 – 100 m
MF (medium frequency)	3MHz – 300 KHz	100 m – 1km
LF (low frequency)	300 – 30 KHz	1 – 10 km
VLF (very low frequency)	30 – 3 KHz	10 – 100 km
VF / ULF (voice frequency or ultra low frequency)	3KHz – 300 Hz	100 – 1Mm
SLF (super low frequency)	300 – 30 Hz	1 – 10 Mm
ELF (extremely low frequency)	30 – 3 Hz	10 – 100 Mm

A popular unit of wavelength is angstrom

one angstrom = 10^{-8} *cm, one fermi (fm)* = 1×10^{-15} *meter*

one picometer (pm) = 1×10^{-12} *meter*

one Exa hertz (EHz) = 1×10^{18} *Hz*

one Tetra hertz (THz) = 1×10^{12} *Hz*

one Giga hertz (GHz) = 1×10^{9} *Hz*

(ii) ***Interactions of Electromagnetic Radiations with Matter:*** *Different radiations interact with matter differently as evident from examples shown in the following Table. Some of the interac-*

tions of these radiations may affect life on earth. Some radiations in nature may be extremely damaging, but the damage would depend upon the intensity of the radiation. Examples of Main Interactions with Matter are given in ***Table 11.1 C****.*

Table 11.1C: Interaction of Electromagnetic Radiation with Matter

Radio	Main interactions with matter
Radio Waves	Collecltve oscillation of charge carriers in bulk material (plasma oscillation). An example world be the oscillation of electrons in an antenna.
Microwave through far infrared radiations	Plasma oscillations, molecular rotation
Near infrared radiations	Molecular vibration, plasma oscillations (in metals only)
Visible radiations	Molecular electron excitation (including pigment molecules found in the human retina), plasma oscillations (in metals only)
*Ultraviolet (UV) radiations	Excitation of molecular and atomic valence electrons, including ejection of electrons (photoelectric effect). Also **see Para 20.1 of Chapter 20.**
X-rays	Excitation and ejection of core atomic electrons, 'Compton scattering' (for low atomic numbers).
Gamma rays	Energetic ejection of core electrons in heavy elements, Compton scattering (for all atomic numbers), excitation of atomic nuclei including dissociation of nuclei.
High-energy Gamma rays	Creation of particle-antiparticle pairs.[A single photon at very high energies can create a shower of high energy particles and antiparticales upon interaction with matter].

(iii) Thermal Effect of Electromagnetic Radiation: *The basic structure of matter involves charged particles bound together in many different ways. When electromagnetic radiation is incident on matter, it causes the charged particles to oscillate and gain energy. This energy may immediately re-radiate and appear as scattered, reflected, or transmitted radiation or it may dissipate into other microscopic motions within the matter /may come to, thermal equilibrium and manifest itself as thermal energy in the material as heating. Any type of electromagnetic energy can be transformed into thermal energy on interaction with matter when it is absorbed.*

(iv) Ionizing Effect: *There are some exceptions as high energy photons of ionizing radiation of far ultra violet/ X-ray/ gamma radiation range may create effects like 'Fluorescence', 'Harmonic generation', 'Photochemical reactions', 'Photovoltaic effect', etc. (but can also create heating effect ultimately). Ionizing electromagnetic radiation creates high-speed electrons in a material and breaks chemical bonds, but after these electrons collide many times with other atoms in the material eventually most of the energy is downgraded to thermal energy; this whole process happens in a tiny fraction of a second. This process makes ionizing radiation far more dangerous per unit of energy than non-ionizing radiation. This also applies to the ultraviolet (UV) spectrum, even though almost all of it is not ionizing, because UV can damage molecules due to electronic excitation which is far greater than their heating effects.*

(v) ***Biological Effects of Electromagnetic radiation***—*The effects of electromagnetic radiation upon living cells, including those in humans, depend upon the power and the frequency of the radiation. From radio waves to visible light (ie low frequency radiations) the best-understood effects are those due to radiation power alone, acting through the effect of simple heating when the radiation is absorbed by the cell. For these thermal effects, the frequency of the radiation is important only as it affects radiation penetration into the organism as for example microwaves penetrate better than infrared. But the 'World Health Organization' has classified radio frequency electromagnetic radiation as a possible 'Group 2b carcinogen'.*

There are 'Complex biological effects' of the non-thermal electromagnetic fields. At higher frequencies (visible and beyond), the effects of individual photons of the radiation begin to become important, as these now have enough energy individually directly or indirectly to damage biological molecules. All frequencies of UV radiation have been classed as 'Group 1 carcinogens' by the World Health Organization. (WHO) Ultraviolet radiation from sun exposure is the primary cause of skin cancer.

Thus, at UV frequencies and higher (and probably somewhat also in the visible range), electromagnetic radiation does far more damage to biological systems than simple heating predicts. This is most obvious in the 'Far or extreme ultraviolet', and also X-ray and gamma radiation, are referred to as 'Ionizing radiation' due to the ability of photons of this radiation to produce ions and 'Free radicals' in materials (including living tissue). Such radiation can produce severe damage to life although they produce little heating effect. These radiations therefore are considered as a separate group than other electromagnetic radiations. **(Source: Wikipedia free encyclopaedia).**

11.3 RADON AND URANIUM IN ROCKS

Radon is present in air in very low concentration in a natural environment. This is a naturally occurring radio-active gas found in enriched quantities in ground water, natural gas, and soil. It can enter the living environment of human being through the use of ground water, natural gas burning, and building materials. Uranium concentrations are higher in rocks of granite composition. Most of these rocks have moderate amounts of uranium (5 to 25 ppm) and moderate amounts of solid 'Radon' (500 to 2000 pC_i/L or 18500 to 74000 Bq /m^3). Radon 222 is derived from the radioactive decay of U238. Radon is soluble in water, but its solublity is temperature sensitive. The partial pressure of 'Radon' in air is very low and it is readily degassed from water if exposed to air. Radon is not ionized in solution. Uranium, Thorium and radioactive Potassium are always present in natural rocks. (The decay of Radon–222 forms Radon-220 which is named as Thoron).

11.4 OTHER SOURCES OF RADIATION

Apart from radioactivity derived from rocks, minerals, natural sources, it may also come from:

(a) Fall out from 'Nuclear weapon tests', and 'Nuclear explosions'.

(b) Radionuclides from materials used in industry and wastes from industries, including the following:-

- Exploration, mining and milling of Uranium and Thorium.
- Uranium enrichment plants.
- Fuel fabrication for nuclear power stations.
- Nuclear fuel cycle/ operation of 'Nuclear power stations'
- Other peaceful uses of radioactivity.
- Management of high radiation level wastes.

11.5 RADIATION INVOLVED IN NUCLEAR POWER PRODUCTION

Gaseous effluents may be from dissociation of coolant / liquid wastes. The liquid wastes are removed to hold up tank to permit decay of short-lived isotopes. After the decay of short lived isotopes, the gaseous effluents formed from dissociation of liquid effluents, coolants or otherwise are monitored and diluted with air and discharged through a tall stack when meteorological conditions are suitable for dispersion high into the atmosphere. The stack gases from nuclear station may be the source of Krypton (Kr) –85 and Iodine (I) –135. There may also be Krypton– 88, Tritium, Xenon (Xe)–133, Xenon–135, and Xenon–138 in the stack gases.

The liquid wastes from nuclear power stations may contain Cobalt–58, Strontium–89, Strontium–90, Iodine–131, Cesium (Cs)–134, Cesium–137, Barium–140, Cerium (Ce)–144, etc.

The radiation sources may be some solid wastes like scrap materials or discarded objects which may have become contaminated with radioactive matter. The spent fuel elements are the most hazardous material from point of radioactivity which needs proper processing and disposal.

Types of emission products from Nuclear Reactors of different types are shown in **Table 11.2.**

Table 11.2: Types of Emission products from Nuclear Reactors

Types of Reactors	Emission Products
1. Pressurized water reactor	(i) Long lived noble gases (ii) Liquid fission products, (iii) Tritium liquid
2. Heavy water moderated reactors	(i) Long-lived noble gases, (ii) Short-lived noble gases, (iii) Tritium gas (iv) Fission products (v) Tritium liquid
3. Boiling water Reactor	(i) Long lived noble gases (ii) Short lived noble gases (iii) Liquid fission products (iv) Tritium liquid
4. Gas cooled Graphite reactor	(i) Short lived noble gases (ii) Long lived noble gases (iii) Liquid fission products (iv) Tritium liquid.

Major source of radioactive waste from reactor systems is the coolant which gets contaminated with activation products, radioactive corrosion products and fission products. In case of water cooled reactors, coolant is continuously cleaned by on-line filters and ion exchange beds in the clean-up systems. Liquid effluents consist of regenerant solutions from ion exchangers, leakages from the reactors etc. Spent ion exchange resins and adsorption materials, and reactor components and equipment rejected during repairs originate the radioactive solid waste.

Bulk of radio-activity generated in the reactors however, remains confined in the fuel. The radioactivity with the above mentioned waste streams accounts for only a very small fraction of total radioactivity contained in the fuel.

Fuel reprocessing of spent fuel from reactor is done in reprocessing plant for separation and recovery of uranium and plutonium from fission products. This generates some highly radioactive liquid waste and a significant quantity of 'low level' and 'medium level' radioactive liquid

Types of Radioactive Wastes in Nuclear Fuel Cycle:

(i) Mining of Uranium ore – Radioactive dust, Radon gas, radioactive drainage from mines, etc.

(ii) Milling of ore – Radium and other radioactive elements in liquid wastes, etc.

(iii) Fuel elements fabrication – Solid waste as metal scrap, low level liquid waste, etc.

(iv) Reactor operation – Radioactive gases, liquid waste, spent fuel, heavy radio nuclides (Actinides, etc.)

(v) Fuel Reprocessing – Highly radioactive solid, liquid and gaseous wastes, etc.

Radioactive Substances Present

Nuclear reactor may be the source of Tritium (H-3), Carbon (C) –14, Xenon (Xe) –133, Xe –135, Xe –138, Krypton (Kr) –85, Kr –87, Kr –88, Iodine (I) –131, Plutonium (Pu) –238, Pu –239, Strontium (Sr) –89, Sr –90, Cesium (Cs) –134, Cesium (Cs) –137, Cobalt (Co) –58, Co –60, Cerium (Ce) –144.

The nuclear fuel reprocessing may be the source of Tritium, Xe –133, Xe –135, Xe –138, Kr –85, Kr –87, Kr –88, Iodine (I) –129, I –131, Pu –238, Pu –239, Uranium (U), Thorium (Th), Sr –89, Sr –90, Co –58, Co –60, Cerium (Ce) –144, Cesium(Cs) –134, Cesium (Cs) –137.

The Nuclear fuel fabrication facilities may be the source of Uranium, Thorium, Radium (Ra) –226, Polonium (Po) -210, Yttrium (Y)-90, Y-91, Zr-95, Niobium (Nb)-95, Technetium (Tc)-99, Promethium (Pm)- 147, Sr-90, Cesium (Cs)-137, Cerium (Ce)-144, Barium (Ba)-137, Praseodymium (Pr)-144, etc.

11.6 RADIOACTIVITY FROM 'FOSSIL-FUEL-FIRED POWER PLANTS'

There is also a small radioactivity involved in coal combustion which increases the background radiation in the environment. Coal has insignificantly low concentration of radio nuclides, but when burnt in boilers, the ash, particularly the fly ash, gets enriched in some radio nuclides. A typical coal may contain about 2 ppm of Th -232, about 1.1 ppm of Uranium isotopes (Radium–226, Thorium–232, and Radon) which are released with fly ash emission. Since the electric generation from coal is much more than from nuclear power stations, the environmental concern about the pollution by radio nuclides from coal fired power stations assumes importance.

It has been reported that fly ash samples taken from three thermal power stations namely Talcher, Angul, Dhamanjodi in Odisha (India) had radioactivity content. The values of radiation found were as follows:

Name of Power Station	Gross Alpha radiation Bq/kg	Gross Beta Bq/ kg
Talcher in Odisha	310	1090.9
Angul in Odisha	282.4	1174.8
Dhamanjodi in Odisha	285.7	944.6

Typical values of Ra–226 per unit energy generated in atmospheric emission of radioactivity were estimated as:

(i) Talcher power plant — 1020 MBq/GW/year

(ii) Angul power station — 824 MBq/GW/year

(iii) Dhamanjodi power station — 884 MBq/GW/year

The ash from other power stations which use coal from other sources may have pollution by same or different radio nuclides and therefore a watch on radio isotopes and radioactivity created by fly ash should be kept **[Ref: 16]**.

In the vicinity of two large oil-shale-fired power plants in northeast Estonia*, depth-dependent activity concentrations of natural radio nuclides in soil were reported. In the surface soil considerably different higher concentrations of 'K-40, Ra-226 and Th-232' were found than in deeper soil layers depending upon the relative concentrations of radio nuclides in fly-ash and in deep soil layers.

***Note:** It has been reported that in the surface soil considerably higher concentrations of 40K, 226Ra and 232Th were found than in soil layers in the vicinity of two oil- shale- fired power stations in North eastern Estonia. The activity concentration of natural radionuclides in soil were determined by gamma spectrometry.*

It was observed that the increase or decrease in level of radionuclides for different sampling sites was dependent on the relative concentrations of radionuclides in fly-ash (of oil shale filter ash) and in deep soil layers.

The concentration ratio of Radium-226 and Thorium-232 in the oil-shale filter and the mean ratio of Radium-226 and Thorium-232 deposited on to the ground were almost similar. The atmospheric deposition rates of fly-ash radionuclides on to the ground were estimated and compared to other relevant published data and were found in agreement

11.7 HALF-LIFE OF RADIOACTIVE ISOTOPES

It is known that the radioactive emission decreases with time at a rate which is proportional to the intensity (I) at any instant. The intensity is proportional to the number of radioactive atoms (N) present at the instant.

i.e. (-) $dI/dt = \lambda I$ where λ is decay constant

This can also be written as (-) $dN/dt = \lambda N$

The minus sign denotes that the intensity and the number of radioactive atoms decrease with time. Integration of the above equation between the limits 0 to time t would yield the following relation:

In $\left(\frac{N_t}{N_0}\right) = -\lambda t$ where N_0 is the number of the atoms present at any arbitrary zero time.

or $\left(\frac{N_t}{N_0}\right) = \exp(-\lambda t)$

$e = 2.7183$; $\log_{10} e = 0.4343$

Taking log on both sides to the base 10,

$$\log_{10}\left(\frac{N_t}{N_0}\right) = \lambda t \log_{10} e = -0.4343\ \lambda t$$

If $\left(\frac{N_t}{N_0}\right) = \frac{1}{2}$, the amount of radioactive element at time t – (N_t) is reduced to half of N_0 (the original amount). This time is the half-life of a radioactive substance and is denoted by Γ. Therefore,

$$\log_{10}\left(\frac{1}{2}\right) = -0.4343\ \lambda\Gamma;$$

or $\Gamma = (0.693 / \lambda)$

***Note:** Since the amount of and hence the activity of a radio-element is reduced to one half of its original value in the time Γ, the fraction time n Γ will be $(\frac{1}{2})^n$. Although this fraction may become very small, but it can theoretically never be zero. After 10 intervals of half life the activity falls to $(\frac{1}{2})^{10}$, which is about 0.1% of original amount so that the remaining activity is negligible in comparison to the initial value. The half life of some radioactive isotopes is mentioned in **Table 11.3**.*

Table 11.3: Half-Life of Certain Radioactive Isotopes

Radio Active Isotopes	Half-Life
1. Carbon – 14 (C-14)	5568 years

2. Phosphorus – 32 (P-32)	14.3 days
3. Iodine – 131 (I-131)	8 days
4. Strontium – 90 (Sr-90)	28 years
5. Cesium – 137 (Ce-137)	30 years
6. Plutonium – 238 (Pu-238)	86 years
7. Plutonium – 239 (Pu-239)	24,000 years
8. Plutonium – 240 (Pu-240)	6,600 years
9. Plutonium –241 (Pu-241)	13 years
10. Cobalt – 60 (Co-60)	5.3 years
11. Iron – 59 (Fe-59)	44 days
12. Sodium – 24 (Na-24)	14.8 hours
13. Potassium – 40 (K-40)	1.8×10^9 years
14. Thorium – 234 (Th-234)	24 days
15. Thorium – 232 (Th-232)	1.4×10^{10} years
16. Thorium – 230 (Th-230)	8.3×10^4 years
17. Barium – 140 (Ba-140)	12.8 days
18. Uranium – 233 (U-233)	1.6×10^5 years
19. Uranium – 235 (U-235)	7×10^8 years
20. Uranium – 238 (U-238)	4.6×10^9 years
21. Radon – 222 (Rn-222)	3.82 days
22. Strontium – 89 (Sr-89)	52 days
23. Thorium – 231 (Th-231)	24.6 years
24. Uranium – 234 (U-234)	2.4×10^5 years
25. Radium – 228 (Ra-228)	6.7 years
26. Radium – 226 (Ra-226_	1.58×10^3 years
27. Tritium – (H-3)	12 years
28. Xenon – 133 (Xe-133)	5.3 days
29. Cerium – 144 (Ce)	290 days
30. Yttrium – 90 (Y-90)	62 hours
31. Yttrium – 91 (Y-91)	57 days
32. Zirconium – 95 (Zr-95)	65 days
33. Technetium – 99 (Tc-99)	One million years
34. Promethium – 147 (Pm-147)	4 years
35. Niobium – 95 (Nb-95)	35 days
36. Praseodymium – 144 (Pr-144)	17.5 minutes
37. Barium – 136	2.6 minutes
38. Beryllium – 7	53.3 days

11.8 ILL EFFECTS OF RADIOACTIVE METERIAL

Radioactive materials generally cause gene mutation, ionization of body fluids, chromosomal mutations, cancers, destruction of body cell tissue and they adversely affect the reproductive system. When mother is exposed to radiation during pregnancy, it causes severe mental retardation and leukemia in infants. Radioactive metals like non-radioactive heavy metals are also nephrotoxic and damage kidneys.

Radio-active substances can enter an organism along with food and water, through inhalation and through skin. Food and water can be contaminated by radio nuclides. Inhalation of radio nuclides along with air is very dangerous.

Radium and Radon

Radium, which is a daughter product of Uranium, readily acts as proxy for calcium in the bone structures. $Radium^{226}$ is known to cause carcinomas of para-nasal sinus and the mastoid* air cells of internal ear. Its absorption in body especially through lungs is high, but elimination from body is very slow and it is a bone seeker. Radon (Rn^{222}) is a decay product of Radium and causes cancer, particularly in lungs owing to its radioactive emission of alpha particles and gamma rays causing carcinogenic damage to lung cells mostly through inhalation. Radon decays into Po^{218}, Pb^{214}, Bi^{214}, Po^{214} which have a combined half-life of less than 50 minutes.

These solid charged particles are adsorbed readily on dust particles and some of the ions remain unattached. The dust particles and also the unattached ions when inhaled tend to lodge in lungs whose tissues are then subjected to relatively rapid delivery of radioactive disintiegration during combined short half-life. (*Mastoid bone – temporal bone behind the ear).

Radon

Regarding the sources of Radon, refer para 11.3 of this chapter. The risk for lungs can increase if the concentration of radon in living space is more than 4 pCi /L as estimated by EPA (Environment Protection Agency) of U.S.A. It has also been suggested that carcinomas in head/brain are related to decay of Ra^{226} to Rn^{222} and accumulation within head cavities of Rn^{222} gas and its progeny.

Natural Uranium

Natural Uranium is also a bone seeker, but its effects on health are more related to it being a heavy metal. It is specifically Nephro-toxic (causing kidney damage). The difficulty in demonstrating that natural U is radiotoxic is probably related to the fact that the level of natural Uranium provides a very low dose of alpha radiation to the bone and bone marrow. However, the daughter products of Uranium like Radium etc. provide more evidence of damage.

Ionization and Penetration of Radiation

Radiation emitted by radio isotopes can destroy body cell tissue by producing ions when radiation passes through it. Alpha (α) radiation produces highest numbers of ion pairs (30 to 100,000 ion pairs per CM of dry air. The degree of ionization depends on mass, charge, velocity of primary beam and collision angles. Formation of a single ion pair in air at S.T.P requires an average expenditure of 32.5 eV of energy. Alpha (α) radiation is however least penetrating. Gamma (γ) radiation is highly penetrating more than 'Beta (β)' and 'Alpha (α)', but produces very few ion pairs much less than 'Beta (β)' and 'Alpha (α)'. Beta (β) radiations produce much less ion pairs than 'Alpha (α)' radiation (say 100 ion pairs per CM of dry air). Beta (β) radiation is less penetrating than 'Gamma (γ)' radiation, but more penetrating than 'Alpha (α)' radiation.

Susceptibility of Body to Radiations

Certain cells of body are much more susceptible to radiation than others. The increasing sensitivity order is somewhat like the following.

- Nerve cells
- Bone cells
- Muscle cells
- Connective tissue cells
- Endothelial cells (lining the close cavities of body such as the heart and blood vessels)
- Cells of tubules of the kidneys
- Bile duct cells
- Alveolar cells (which absorb oxygen and release CO_2)
- Basal cells (cells of gonads*, bone marrow, skin, alimentary canal)
- White blood cells formed in bone marrow (Granulocytes)
- White blood cells formed in spleen and lymph nodes (Lymphocytes)

(* Gonads—Organ that produces sex cells)

Higher the metabolic rate in a cell, lower is its resistance to radiation. Cells involved in reproduction process are more susceptible to damage, promoting mutations by radiation. Permanent damage to cell is caused by formations of free radicals during ionization. Free radicals promote formation of abnormal linkages that result in mutations and usually death of cell.

There is possibility of health hazards in children exposed to radiation during certain periods of pregnancy of mother. This includes likelihood of Leukemia (Blood Cancer) and severe mental retardation. Cell instead of being killed may sometimes gets modified on exposure to radiation and ultimately develops into a cancer. The probability of occurrence of cancer is higher for higher doses. The severity of any cancer depends on the dose. If a stem cell whose function is to transmit genetic information is damaged owing to radiation exposure, hereditary mutation of various types may develop in the descendants of the exposed person. Effects of low doses however cannot be determined unequivocally.

Amongst the radio isotopes related to nuclear reactors, fuel processing, and fuel bundle fabrication, I –131 is known to adversely affect thyroid, Po – 210 the kidneys, Co – 60 the liver, U – 238 the lungs and kidney, Pu –239 the lungs and bones, and Tritium the whole body, while the U – 233, Ra – 226, Sr – 89, and Sr -- 90, are bone seekers.

11.9 SAFETY FROM NUCLEAR RADIATIONS

11.9.1 Radiation Exposure

The term radiation means energy propagation in the form of electromagnetic waves or photons or in the form of sub-atomic particles. Ionizing radiation has radio action of sufficiently high energy to cause (in the medium through which it passes) the production of pairs of ions, i.e of atoms or group of atoms that have either lost or gained one or more electrons to become positively or negatively charged, and the corresponding complementary electrons. If the medium in which ion pairs are produced is a biological material (more specifically cellular material), biological ill effects would be produced.

The definition of exposure to ionizing radiation is based on the amount of ionization in the air. The term radiation (absorbed) dose generally means the amount of energy which is absorbed from ionizing radiation by a unit mass of material. This quantity is expressed as energy per unit mass, in Joules per kilogram, which takes the special name gray (Gy) [1 Gy = 1,000 milli-gray (m Gy)]. For purpose of protection from radiation, the adsorbed dose is weighted to take account of the effectiveness of different radiation types and the radiosensitivity of various organs and tissues.

The dose potential is a quantity that takes into account the characteristics of a particular type of radiation and its potential for causing damage to body cells and tissues.

The term 'Effective dose equivalent', which is measured in Sievert (Sv), expresses the effects of irradiation of whole body of man.

The conversion from Gy to Sv is possible by multiplying the absorbed dose by the relative biological effectiveness (RBE) also known as quality factor of the radiation concerned. For natural Gamma rays, the RBE has a value of unity (1.0). For photons in intermediate energy range, one mGy is approximately equal to one mSv i.e. (one mGy ≅ one mSv).

Where radiation source can be controlled, a system of dose limitation can be applied for safety. In case, where the source of exposure cannot be controlled, the system of dose limitation cannot be applied. A 'prescribed radiation dose' means the 'absorbed radiation dose limit recommended by ICRP (International Commission of Radiological Protection)'.

The average external dose as per "United Nations Scientific Committee on Effects of Atomic Radiation (UNSCEAR)" is 800 m Sv per year and the internal dose is 1500 m Sv per year. Inhalation of radon daughters constitutes nearly 2/3 of internal dose. In India authorties have done extensive surveys of the natural background radiation. The Monazite rich sand along the coast of Kerala and Tamil Nadu have the highest external dose rates of natural background radiation.

11.9.2 Standards for Radiation Exposure

Maximum exposure limits of Gamma Radiation as recommended by International Commission of Radiological Protection (ICRP) are given in **Table 11.4**.

Table 11.4: Maximum Limits of Exposure to Gamma Radiation

Occupational Workers	Public
2.0 R/year 20.0 m Sv / year 1.0 mR / hour	0.10 R* / year 1. 0 m Sv / year 11. 5 μR / hour (above background)

[* *R (Roentgen)* = 2.58×10^{-4} *C/kg (coulombs per kg)]; [mR = milli Roentgen*]

The term low radiation dose is used to mean a radiation dose lower than designated level; sometimes it is also informally used to mean a low dose rate per unit time. In specialized radiobiological forums, low radiation dose (and dose rate) refers to exposures for which it is very unlikely that more than one event of energy adsorption from radiation will occur in the critical parts of a cell (and damage it) within the time during which repair mechanism in the cell can operate. Thus, UNSCEAR (United Nations Scientific Committee on Effects of Atomic Radiation) concluded that low radiation dose refers to a total dose of less than 200 m Sv and dose rate below 0.1 m SV per minute (which in fact is very high dose rate).

For the general public low radiation doses are deemed to correspond to levels similar to those from, for instance, natural background exposure or some very common radiation exposures such as those arising during air travel. Natural background exposure varies widely around the world. Some "normal" and "elevated" values of annual dose rates are as follows: for cosmic ray, 0.38 mSv (2.0 mSv), for terrestrial radiation 0.43 mSv (4.3 mSv) and for exposure to Radon, 1.2 mSv (10 mSv), leading to an average total of around 2.4 mSv per annum (elevated doses are given in brackets).

Note: 1. Units of ' Dose equivalent'

SI Unit Sievert (Sv) or milli Sievert (mSv)

Older Unit Rem (1 Sv = 100 rem)

2. Units of 'Radioactivity exposer'

SI Unit Coulomb / kg

Older Unit Roentgen (R) / kg of air

These dose rate levels of a few m Sv per annum are expected to deliver, during a lifetime, dose of around 100 m Sv, which are of the order of magnitude of low radiation doses designated by UNSCEAR.

On the basis of above radiation doses, the International Atomic Energy Agency (IAEA) in its revision of the 'Basic Safety Standard (BSS)', recommended 200 becquerels of $Radon^{222}$ per m^3 as the action level for dwellings and 1,000 becquerels per m^3 for workplace. The safe Radon concentration is defined in terms of 'Working Level (WL)' and 'Action Level (AL)'. WL corresponds to any combination of Radon daughters in air which on ultimate decay yields 20.8 micro Joule of potential alpha energy per cubic metre of air. 'Action level' relates to the effective dose in the range of 3 to 10 m Sv in a year. The corresponding Radon concentration (F = 0.4) is 200 – 600 Bq/m^3 in air (i.e. 5.4 to 16 p Ci/L) assuming an annual occupancy of 7,000 hours in dwellings. The action level in working place corresponding to annual occupancy of 2,000 hours is higher at 5,000 – 15,000 Bq/m^3 (13.5 to 50 p Ci/L); (ICRP – 65, 1993).

11.9.3 Control of Radiations

Factors that control the amount or dose of radiation from a source are:

(i) Time of Exposure
(ii) Distance of source from body
(iii) Shielding between source and body

A combination of above factors is used in practice.

Different Ways of Shielding from Different Radiations:

- Alpha radiations can be stopped even by a paper.
- Beta radiations—can be stopped by low density plastic, wood, water, acrylic glass, etc.
- Gamma radiations and X-ray—very difficult to be stopped. Materials like water, concrete, lead, polyethylene, paraffine wax, can shield to some extent.

Gamma Ray Intensity Reduced to 50% by:

(a) 6 cm of concrete
(b) 1 cm of lead
(c) 150 meters of air
(d) 9 cm packed soil
(e) 0.2 cm of depleted Uranium

Radioactive Waste

The radioactive wastes are classified in three categories namely 'Low level waste', 'Intermediate level waste' and 'High level waste'.

- Low level waste up to 10-4 Micro Ci /mL*
- Intermediate Level Waste 10-4 Micro Ci / mL to 10 Micro Ci /mL.
- High level Wastes More than 10 Micro Ci /m

The 'Low level wastes' can be discharged in environment under controlled condition, while the 'Intermediate level wastes' need treatment before disposal in environment. The 'High level wastes' require very careful waste management with measures like concentration, containment, storage (delay and decay) till the activity levels come down to desirable values for release into environment. *[One micro Curie = 37,000 Becquerel ie 1μCi = 37,000 Bq; 1Bq = 2.703×10^{-11} Curie]

11.10 PROTECTION FROM RADIATION IN NUCLEAR POWER PLANTS

11.10.1 Basic Design Protection from Radiation at Nuclear Power Plants

There are three basic safety barriers in the design of nuclear power plants:

(i) **Cladding of Nuclear Fuel:** The nuclear fuel rods consist of uranium pellets contained in zirconium container / cladding. This nuclear fuel cladding should be kept intact as this is the first barrier of fission product release. The radioactive fission process occurs inside the fuel cladding and it would remain contained if the cladding does not have crack or holes/openings.

(ii) **Cooling System of Reactor Core:** The second barrier for radiation is the reactor cooling system which consists of an elaborate piping system self contained loop of coolant to reactor core. The coolant is contained within the loop and not exposed elsewhere in the plant, but if the fuel cladding has even a small crack / leak, the radioactive fission products would leak into the reactor coolant system and may create problems in case of slight exposure of coolant and therefore the coolant system must remain intact to act as a second barrier for preventing the release of radiation.

The cooling system of a reactor is one of most important and crucial system and its electrical power supply needs to be maintained by several alternatives to keep it running under all circumstances. The failure of cooling system* causes overheating of core and accumulation of hydrogen leading to hydrogen or steam explosions and also damage / melting of reactor core. This damage to equipment is accompanied by contamination of whole site. The failure of one reactor may also lead to failures of equipments at other reactors units.

During the overheating of core, the spent fuel / fuel may burn / melt and release radioactive elements like Cesium-137 and Strontium- 90 which have long life (even half life of Cesium- 137 is 30 years and half life of Strontium- 90 is 28 years) in the atmosphere.

The light-water reactor of Fukushima Daiichi power plant (Japan) was struck by tsunami sea water flooding in March 2011, and its cooling system went out of order and the reactor core went on heating resulting in explosions (due to accumulation of hydrogen in the reactor because of overheating) and melting of reactor core. Fukushima reactors had 8 hours battery reserves for cooling, yet the fuel / core melted in three reactors. Most reactors may have even lesser reserve power and can get into trouble as much longer blackout can occur.

(iii) **Containment Structure:** The third barrier for fission product release is the containment structure which is a thick concrete structure housing the radioactive core as well as the reactor coolant system and other components vital to the nuclear steam supply systems. The containment structure is designed to prevent release of radiation to the atmosphere even if the first two barriers fail because of some leakage. The containment structure is designed to withstand the high pressures and temperatures that may occur in an accident / failure within the fuel or reactor coolant system.

Some General Points for Consideration for Nuclear Installations:

(a) Clustering of Nuclear Reactors

Overheated fuel risks hydrogen or steam explosions which damage equipment and contaminate the whole site therefore clustering many reactors together (to save money) can make failure at one reactor cascade to the rest.

(b) Spent Fuel Disposal

During the life span period of reactor, another problem is of accumulation of spent fuel which contains radioactive elements like plutonium (half life of Pu-238 is 86 years, Pu-239 is 24,000 years, Pu-240 is 6600 years) has been a big problem and as it has to be safely stored in special containers and places till it is finally disposed. This has been a big challenge for nuclear power stations

(c) Uniquely Unforgiving Nature of Nuclear Power

No mistake in design, operation, and maintenance is permitted in a nuclear installation. Besides this, as per Swedish Nobel physicist Hannes Alfvén, 'No act of God can be permitted'. Human mistakes have created its half-century history of a few calamities, a steady stream of worrying incidents, and many near-misses.

America has been lucky so far. The Three Mile Island power plant had a hydrogen explosion accident in 1979, but the radioactivity could be contained by the containment dome structure. If the dome had not been built with double-strength as it was under an airport landing path, it may not have withstood the hydrogen explosion. In 2002, Ohio's Davis-Besse reactor fault was luckily caught just before its massive pressure-vessel lid rusted through.

The smarter choices of utilizing renewable sources of energy are sweeping the global energy market. Half the world's new generating capacity in 2008 and 2009 was renewable. In 2010, renewable energy got a huge boost while nuclear energy development was scanty.

11.10.2 Precautions during Visit to a Nuclear Installation:

(i) At each important location, signs are painted or marked on sign boards for guidance and safety which must be carefully seen and taken cognizance of.

(ii) A time limit is to be observed for visit or for working in a radiologically controlled area to control exposure to radiations. Radiation areas are provided with international radiation symbols.

(iii) One is required to have: 'Dosimeter', 'Radiation monitor', foot covering, hand covering, hard hats, coveralls and scrubs or modesty clothes, when entering a radiologically controlled area. [*The foot and hands should be covered, overall should be worn and hard hat / helmet should be put on. The shoes are also to be covered and bare shoes are not allowed.*]

(iv) Only those objects which are really needed should be carried into radiologically controlled area. If the objects or clothes become contaminated with air borne radioactive particles / dust, these objects / clothes have to be left there itself.

(v) In order to prevent spreading radioactive dust and contaminating oneself during a visit to a radiologically controlled area, no object should be touched, rubbing against walls, railings and equipment should be avoided, ones face, mouth, and eyes should not be touched with hand (which are covered with gloves).

(vi) While getting out of radiologically controlled area, a person has to take following precautions in order to protect oneself and not to spread radioactive particles or dust in environment:

- The gloves, overall and foot covering, should be removed without touching their outer surface with bare hands.
- Before leaving the radiologically controlled area, monitoring of small articles and radiation count on ones body should be done. Nobody should step back into the radiologically controlled area with bare shoes.
- While getting out of radiologically controlled area, one has to pass over a sticky pad so that the radiological dust is removed from the bottom of the shoes. One should be careful not to fall down when passing over the sticky pads.

11.10.3 Radiation Protection during Accidental Release of Radioactivity in Environment

General

The nuclear technology has an obvious probability of hazard of release of radioactivity in the environment when a failure of 'equipment, system, control and instrumentation, protective gear, etc.' occurs. The potential danger from an accident at a nuclear power plant is exposure to radiation which can come from the release of radioactive material from the plant into the environment. In case of a major hazard, the radioactive particles and gases may rise like a plume or a cloud and spread radioactivity in the vicinity. The radioactive particles may deposit on various places, objects, soil, water, and vegetation. The radioactive particles may get ingested in body of animals / humans through air inhalation, drinking water, and food chain. As already discussed, radiation has a cumulative effect and therefore longer a person is exposed to radiation, greater would be the ill-effects.

Planning Emergency Zones

The nuclear power authority / company and government should amply publicize the information material regarding the ways to protect oneself from radiation in a 80 km radius from the nuclear power plant. The general public should also be advised to obtain such information material and keep it ready at hand.

A 16 km radius zone is the primary emergency zone around the nuclear plant where Sirens should be located to be heard within a 16 km radius from the plant (where the public could be harmed by direct radiation exposure). All emergency planning by government and nuclear plant authority to deal with radioactivity leakage should be done for the area including shifting of population of the area to another location in case of major nuclear hazard. The plan should be firmly in place and the responsibilities of each action should be well defined and facilities should be available.

The next emergency zone is within 80 km radius of the nuclear power plant where radioactive materials could contaminate water supply, food crops, milk, animals, cattles, etc. during major accidental emergencies due leakage of radioactivity from the plant into environment. The plan to deal with such emergencies to protect the public should also be in place with the government and nuclear power authority / company.

Minimizing Radiation Exposure

As the exposure to radiation decreases with the distance between the source of radiation and oneself, sometimes the evacuation of population becomes necessary in case of major hazard. Shielding by heavy and dense material between the source of radiation and oneself is important for protection. Remaining indoor during radiation leakage would be better to limit the exposure. Most of the radioactivity may lose its strength with the time period and so the above precautions may be necessary for a definite period only.

Some Guidelines for Public for Nuclear Plant Emergencies:

(i) All doors and windows are to be kept closed and air should be recirculated, while the air conditioner, ventilation fans, furnace, and any other air intake should be closed.

(ii) Food, water, etc. should be kept covered in containers or in refrigerator.

(iii) If possible, it is better to go in a basement or underground area.

(iv) Unless it is absolutely necessary, the telephone should not be used.

(v)) A battery-powered radio and television should be kept and attention should be paid for specific instructions.

(vi) Even when one is told to evacuate, keep the windows, air vents closed. The instruction given should be strictly followed.

(vii) If one is already exposed to radiation, the clothes, shoes, etc. should be changed after immediately taking bath and the exposed/ dirty clothes or other material should be sealed in a plastic bag and kept out of the way. The food not covered previously should be washed and put into containers or refrigerator.

(viii) After the emergency, if any exposure related symptom is suspected, immediate medical attention should be sought.

11.10.4 Survival during an Attack on Nuclear Plant

Nowadays, there is always a threat to national security by terrorists. Advance home work and planning for such emergency by plant authorities and the government is required even with the increased security provided at the plant. Training of staff, providing protection guidelines and necessary protection equipments for dealing with such a emergency is the first step in the direction.

In the event of an attack, a blast might destroy the immediate surroundings of a plant, but the main concern for damage is the potential for nuclear fallout and residual radiation that could rain down on surrounding areas for weeks following an attack. The following advice may be observed for survival of oneself in case of attack on a nuclear power plant.

- Obtain information about: (i) potential radiation threat; (ii) type of radiation and how it can affect the health; (iii) period for which threat is likely to persist; (iv) contamination of water, food, milk, other supplies in the one's residential area.
- If one lives near a nuclear facility, construct an 'underground shelter' in advance with a thick brick / concrete barrier all around where one can live for weeks. If the advance construction could not be done, one should at least dig out a small area underground so as to cover the body. If this also could not be done one should at least remain indoor with all doors and windows closed and recirculated air.
- Stock essential goods like non-perishable food items, water, batteries, blankets and first aid supplies in the house or shelter before nuclear fallout / or before the radiation contamination increased beyond upper limits.

- Consume only stocked uncontaminated food and water
- It is better to shift out about 80 km from the nuclear plant site in case of attack for the time things normalize.
- Try to get updated by radio news and follow guide lines.

11.10.5 Nuclear Disaster due to Failure of Reactor Cooling System

(Japan's Nuclear Disaster in Fukushima March 2011 as a result of tsunami)

An earthquake of magnitude 8.9 occurred at about 24 km below sea level at 2:46 PM Japan local time [11:16 AM (IST)] which was centered off the coast of Honshu (Japan's largest island), Japan, triggering off a ferocious tsunami sea water waves slamming Japan's eastern coast. It caused large scale devastation, casualties, fires, and above all a very severe nuclear disaster. Tokyo Electric Power Company's Fukushima Daiichi Nuclear power plant's reactor (light-water reactor) cooling system failed due to sea water flooding the plant (on account of Tsunami). The cores of the reactors started heating and radiation leakage started. In spite of best efforts to restore the cooling system and using unconventional means of cooling the cores of reactors by pouring sea water, there was no success and the reactor cores continued to get heated. Explosions occurred due to accumulation of Hydrogen and reactor cores badly damaged / melted. Air borne radiations, many times of the safe limits, emanated from the reactors. The radioactive water accumulated and halted repair work. Plutonium was found in soil. Flooding of reactors with nitrogen to halt hydrogen explosions was met with partial success. Due to large accumulation of radioactive water it had to be released into sea.

Radioactive Iodine(I-131) thousands of times higher than the permissible safety limits, was discovered in the sea water near Fukushima (as reported by Times of India dated 31st March 2011, the radioactive iodine I-131 was 3355 times the safety limits). Further the Times of India dated 6th April 2011 reported that: (i) radioactive iodine I-131 in sea water was found 200,000 Becquerel per cm (5 million times more than the legal limits), (ii) I-131 in small fish was found 4000 Becquerel per kg of fish (permitted value 2000 Becquerel of I-131), (iii) radioactive Cesium- 137 found in small fish – 526 Becquerel per kg of fish. It may be noted that Cesium- 137 has a half-life of 30 years while Iodine -131 has a half-life of only 8 days and therefore the Cesium-137 is much more of a serious threat. More than six million people had to be shifted from Fukushima. Even milk, water, vegetables, etc. got contaminated by radioactive elements up to Tokyo (250 km). The radiation travelled long distances and scare was created in four continents.

The Japan's Fukushima reactors were of very old design (1960s design) and it probably did not have safety margin to contain 90 % core melt downs. Every light-water reactor if deprived of power for auxiliaries and cooling system can melt down. As already mentioned in **Para.11.10.1 above,** Fukushima power plant had 8 hours battery reserves for cooling, yet the fuel rods melted in three reactors.

11.10.6 Chernobyl Nuclear Accident, Ukraine, in April, 1986

The Chernobyl nuclear accident occurred on Saturday, April 26, 1986, at 1:23:58 AM local time at the V.I. Lenin Memorial Chernobyl Nuclear Power Station located in Ukraine, near the town of Pripyat, which had been built to house power station employees and their families. The power station was in a wooded, marshy area near the Ukraine-Belarus border, approximately 18 km northwest of the city of Chernobyl and 100 km north of Kiev, the capital of Ukraine.

The Chernobyl Nuclear Power Station had four nuclear reactors, each capable of producing 1000 MW of electric power. At the time of the accident, the four reactors produced about 10 percent of the electricity used in Ukraine. The first of the four reactors was commissioned in 1977, and Reactor No. 4 began producing power in 1983. When the accident occurred in 1986, two other nuclear reactors were under construction.

On April 26, 1986, the operating crew planned to test whether the Reactor No. 4 turbines could produce enough energy to keep the coolant pumps running until the emergency diesel generator was activated in case of an external power loss. During the test, power surged unexpectedly, causing an explosion and driving temperatures in the reactor to more than 2,000°C melting the fuel rods, igniting the reactor's graphite covering, and releasing a cloud of radiation into the atmosphere.

Causes of the Chernobyl Nuclear Accident

The precise causes of the accident are still uncertain, but it is generally believed that the series of incidents that led to the explosion, fire and nuclear meltdown at Chernobyl was caused by a combination of reactor design flaws and operation error.

Loss of Life and Health Effects from the Chernobyl Nuclear Accident

By mid-2005, fewer than 60 deaths could be linked directly to Chernobyl (mostly the deaths included workers exposed to massive radiation during the accident or children who developed thyroid cancer). Estimates of the eventual death toll from Chernobyl varied widely. A 2005 report by the Chernobyl Forum—eight U.N. organizations—estimated the accident eventually would cause about 4,000 deaths. Greenpeace places the figure at 93,000 deaths, based on information from the Belarus National Academy of Sciences.

The Belarus National Academy of Sciences estimates that 270,000 people in the region around the accident site would develop cancer as a result of Chernobyl radiation and that 93,000 of those cases are likely to be fatal.

Another report by the Center for Independent Environmental Assessment of the Russian Academy of Sciences found a dramatic increase in mortality since 1990—60,000 deaths in Russia and an estimated 140,000 deaths in Ukraine and Belarus—probably due to Chernobyl radiation.

However, the biggest challenge facing communities still coping with the fallout of Chernobyl is the psychological damage to 5 million people in Belarus, Ukraine and Russia. People have been led to think of themselves as victims over the years, and are therefore more apt to take a passive approach toward their future rather than developing a system of self-sufficiency.

Seventy percent of the radioactive fallout from Chernobyl landed in Belarus, affecting more than 3,600 towns and villages, and 2.5 million people. The radiation contaminated soil, which in turn continues to contaminate crops that people rely on for food. Many regions in Russia, Belarus and Ukraine are likely to be contaminated for decades.

Radioactive fallout carried by the wind was later found in sheep in the UK, on clothing worn by people throughout Europe, and in rain in the United States.

After the accident, Soviet authorities resettled more than 350,000 people outside the worst areas, including all 50,000 people from nearby Pripyat, but millions of people continue to live in contaminated areas.

In many villages, up to 60 percent of the population is made up of pensioners. In most of these villages, the number of people able to work is two or three times lower than normal. After the accident, Reactor No. 4 was sealed, but the Ukranian government allowed the other three reactors to keep operating because the country needed the power they provided.

Reactor No. 2 was shut down after a fire damaged it in 1991, and Reactor No. 1 was decommissioned in 1996. In November 2000, the Ukranian president shut down Reactor No. 3 in an official ceremony that finally closed the Chernobyl facility. But Reactor No. 4, which was damaged in the 1986 explosion and fire, is still full of radioactive material encased inside a concrete barrier, called a sarcophagus, that is aging badly and needs to be replaced. Water leaking into the reactor carries radioactive material throughout the facility and threatens to seep into the groundwater. The radioactivity in the damaged reactor would need to be contained for 100,000 years to ensure safety. That is a challenge not only for today, but for many generations to come.

11.11 REMOVAL OF RADIOACTIVE SUBSTANCES FROM WASTE WATER

Radioactive substances if not removed from wastes can cause serious sources of radiation and therefore, the wastes should be treated suitably or should be safely stored / disposed.

11.11.1 *Methods for Removal of Radioactive Substances*

Radioactive substances can be removed from water by various treatments as discussed below:

(i) Settling

Suspended radioactive contaminants can be removed by settling. The time of settling of short lived isotope is related to the half-life of particular contaminant.

(ii) Adsorption

Suspended radioactive particles are adsorbed by micro organisms of the biological film on slow filters. Common sand filters retain only a part of radioactive contaminants as the adsorbing capacity of sand is low. The percentage retention of radioactive substances on filtering media on the average may be as follows.

(a) Quartz Sand – 72–89%;

(b) Activated Alumina– 94%;

(c) Activated Carbon– 92%;

(d) Charcoal – 86%; (e) Glauconite – 83%

Some clay minerals like Bentonite are also effective in adsorbing radioactivity if concentration is higher.

Crude vermiculite (belongs to zeolite group of minerals), which is complex silicate of magnesium (Mg), iron (Fe), and alminium (Al), is very effective as 'zeolite filter media' for removal of radioactivity from waste water.

(iii) Coagulation

If radioactive substances are finely dispersed in water, coagulation may be an appropriate treatment. Dose of coagulant should be experimentally determined. For better coagulation the water can be alkalized and concentration of element (radioactive) in question is increased by addition of its non-radioactive isotope in order to promote de-activation of wastewater by coagulation.

Aluminum sulphate, ferric sulphate, ferric chloride, phosphates (Na_3PO_4 and KH_2PO_4), lime with activated sodium silicate, poly electrolytes etc. can be used as coagulants. Coagulation is more effective if the wastewater is turbid and therefore 1 to 4 g/L of clay may be added and pH value may be increased to about 11. After about 100 minutes of contact, the radioactivity of contaminated water may decrease by even 90%. Coagulation may not however be effective for removal of certain Isotopes.

(iv) Addition of Metallic Dust

Metals like Iron, Aluminum, Zinc, Copper etc absorb the products of radioactive decay. If the radioactivity contaminated water is filtered through a bed of metal chips (depth of bed 0.5 to 0.8 m) the radioactivity can decrease by 80 – 85%. The filter may use powdered Iron. A dose of 1000 mg/L of Iron with a contact time of 90 minutes may be effective and may be tried.

(v) Lime Soda Treatment

The lime soda process of water softening can remove some radio Isotopes like Sr – 89, Ba – 140, La – 140, Cd – 115. The reagent doses would be higher (than that for only water softening) in case of removal of radioactive Isotopes.

(vi) Phyto Remediation

Phyto remediation by some plants like sun flower plants has been used for treating radioactivity of water. Phyto Tech Mon Mouth-Junction, New Jersey tested the power of sun flower plants at abandoned uranium factory, Astabula, Ohio (USA) and recorded a reduction of uranium concentration in water from 200 micro gm per litre to 20 micro-gm per litre. Also at the Chernobyl site test in 1996 showed that in 10 days sun flower plants soaked up 95 % of radioactive strontium and cesium in a pool near the stricken reactor. Phytoremediation technology is very promising and is likely to be used more in future **[Ref: 4]**.

(vii) Removal by Proprietary Chemicals

Some proprietary chemicals have been developed as additives for precipitating the radioactive elements like Uranium. Potassium Ferrate $K_2F_eO_4$ is the key ingredient of this additive formulation which can be obtained from Analytical Development Crop., Colorado Springs (US), licensed for this by Los Alamos National Laboratory, Los Alamos (N.M.) **[Ref: 2]**.

(viii) Membrane Process

Reverse Osmosis can remove both radioactive and non radioactive solids from contaminated laundry water of Nuclear Power stations. Electrodialysis can also remove the radioactive components/solids. For both reverse osmosis and electrodialysis, it is necessary that colloidal matter is removed first **[Ref: 15]**.

(ix) Distillation

Although this is very effective method of removing radioactivity from water, but it is very expensive and cannot be used on large scale.

(x) Ion Exchange with Regenerative Columns

Special synthetic resin like strongly basic INDION 103 cross linked polystyrene resin is suitable for recovery of Uranium from leach liquors. The resin Indion 103 and other nuclear grade resins are manufactured in India by Ion Exchange India Ltd. **[Ref: 8]**.

11.11.2 Concentration of Radioactive Liquid Waste

Low and medium radioactive liquid wastes from different sources can be collected and transported to treatment facility by means of permanent piping systems. If in some cases, where volumes are small, specially designed portable tanks can be used for transportation.

Evaporation is used for concentrating the liquid waste. Depending upon the volumes involved and climatic conditions, different evaporation methods can be chosen. For low volume and radioactive liquids, steam evaporation is preferred, while for large volumes involving low radioactivity natural evaporation is desirable. Non-boiling solar pans can be adopted for very high volume reduction when climate has high ambient temperature low humidity and high wind velocity **[Ref: 1, 14]**.

11.11.3 Immobilization of Radioactive Wastes

In a nuclear facility or industry, uranium mining and nuclear power station, there would be a number of waste materials having some radioactivity. These would include sludges from chemical treatments, spent ion exchange materials, spent adsorbents, concentrates from evaporator as well as concentrated clean up solutions, etc. and all these waste materials would need further processing before their disposal. In order that the migration of radioactivity is retarded, these waste concentrates require conditioning for immobilization in a solid form. The media in which the radioactive pollutants are immobilized are commonly referred to as "Matrices". Cement and cement composites are generally employed for this because of their low cost and simplicity.

High radioactive waste liquids generated during reprocessing of spent nuclear fuel are of major concern as they contain more than 99% of total radioactivity handled in the entire nuclear fuel cycle. The decay heat is large enough and therefore these are self-heating/boiling in nature besides being source of intense nuclear radiation. These wastes are stored in high integrity stainless steel tanks located underground vaults with adequate provision for constant cooling. They are generally filled with remotely operated sampling and transfer system. Such a storage facility is only considered as an interim arrangement prior to processing and ultimate disposal. Final disposal / long term storage requires the conversion of these wastes into a stable solid form having properties of high thermal conductivity, chemical durability, radiation stability, mechanical strength, etc. **[Ref: 1, 14]**.

11.11.4 Management of Solid Waste at Nuclear Facility

Radioactivity can result from improperly disposed radioactive solid wastes of nuclear facility which are produced in many different ways. Radioactive solids with relatively high radioactivity are to be collected in containers which can provide safety to personnel and avoid dispersion of contaminants into environment. These containers are to be stored underground in suitable structures. As regards solid wastes with low radioactivity, they would not present much problem. These can be segregated into combustible and noncombustibles. The combustibles can be incinerated in specially designed incinerator with gas cleaning system. Noncombustibles can be compressed and put into containers which are stored underground in suitable structures. The sites are selected for underground storage after a detailed study considering geological formation, nature of soil, ground water level, hydrology of area, proximity of population etc. Safe handling and transportation facilities must be available for the management of solid wastes having radioactivity. Long term storage of high level radioactive wastes is possible in suitable geological formations which have been stable physically and chemically. Waste management scientists of BARC (India) have been investigating and studying different geological formation available in India (***Ref: 1, 14***)

REFERENCES

1. Balu, K (BARC), 'An overview of nuclear waste management practices in India, April 1996, Geological Survey of India, Lucknow.
2. Chemical Engineering, April 1990, (McGraw hill).

3. Diwivedi, K.K., A.M.D. GSI Hyderabad, "Environmental Studies related to Atomic mineral exploration and exploitation." Symp. Earth science in Env. Assessand Management Lucknow GSI Apr. 1996, Vol. – 2. Sep. Pub. No. 48.
4. Environews 4/1998, ISEB, NBRI, Lucknow, India.
5. Genutis, Laura, eHow Contributor 'How to prevent radiation in a nuclear power plant', updated: May 12, 2010.
6. Global Security: Nuclear Power Plants: Vulnerability to Terrorist Attack; Captain Dave's Survival Center: The Nuclear Threat. Read more: How to Survive a Nuclear Power Plant Attack | eHow.com http://www.ehow.com/how_7586712_survive-nuclear-power-plant-attack.html#ixzz1HOItCkEK. Read more: How to Prevent Radiation in a Nuclear Power Plant | eHow.com; http://www.ehow.com/how_6510471_prevent-radiation-nuclear-power-plant.html#ixzz1HOH5hB00
7. Gurdeep and Harish Goel, Advanced Inorganic chemistry, chapter 9 Nuclear & Radio Chemistry, Publishing House Meerut.
8. Ion Exchange India Limited Bombay's Brochure Indion-Ion exchange resins- properties and application- Data.
9. Mahoney, Jessica, eHow Contributor, 'How to Survive a Nuclear Power Plant Attack' updated: December 01, 2010.
10. Mays et al., "Cancer risk from the life time intake of Ra and U isotopes"Health physics 48, 1985.
11. Radio Biology of Humans and Animals. MIR Publishers Moscow.
12. Radiological Protection. 'A summary Handbook of I.C.R.P Publications and Recommendations' D.R.D.O Publication by Nagarthnam. A. 1994.
13. Satoskar et al., "Pharmacology and phara cotherapeutics 11th Edition Bombay Popular Prakashan Bombay. 1998.
14. Sinha, D.K. et al., (Atomic Mineral Div. Nagpur), "Uranium & Radon concentration in ground waters of Decan Trap country and environmental hazard in Kelori-Nainpur Area, Seoni- Mandla Dist. Madhya Pradesh,. Symp Earth Sciences in Environmental Assessment and Management G.S.I, Lucknow, April 1996, Sp. Pub. 48(2) 1997.
15. Stnna, 'Westinghouse membrane system' in Sourirajan Ed, reverse Osmosis Synthetic membranes, National research council of Canada, Ottawa, 1977.
16. Vijayan, V. and Das, S., 'Studies on Radio activity of fly ash from Thermal power Plants', (VIII – 10). Inst. Of Physics, Bhubaneshwar. International Conf, Fly Ash Disposal and Utilization CBIP. New Delhi.
17. Voskoboinik, David, Nuclear Power, People's Publishing House New Delhi.
18. West, Larry, 'About.com Guide Chernobyl Nuclear Accident, Ukraine'.
19. Wikipedia free encyclopedia.

QUESTIONS

1. How does the radioactivity affect the environment air/ water/ land/ food?
2. What are the sources of 'Cosmic radiation' and of 'Terrestrial radiation'?
3. Write short note on sources of radiation on account of human activity (i.e. other than cosmic and terrestrial sources).
4. Define the half life of radioisotope?
5. Write short notes on: (a) Radioactive substances released in the environment in nuclear power generation, (b) Radioactivity in the coal ash from thermal power station.
6. Differentiate between alpha, beta, and gamma rays.

7. Enumerate the radiation likely to be present in in emission or effluent from:
   (a) Stack emission of a nuclear power station;
   (b) Fly ash from a thermal power plant;
   (c) Liquid wastes from nuclear power plant;
   (d) Wastes from Nuclear fuel fabrication facility;
   (e) Nuclear fuel reprocessing facility
8. Write a note on health problems created by nuclear radiation. How does the susceptibility of different body parts vary for nuclear radiation?
9. Write a note on radiation exposure. What are the international standards for the same?
10. What are the basic design protections provided against radiations from a nuclear reactor.
11. Write a note on the planning and management for a possible accidental release of radioactivity in environment. What steps or precautions are to be taken to survive during an attack on nuclear facility?
12. How did the tsunami cause nuclear disaster in Japan at and around Fukushima nuclear plant in March 2011?
13. Write a note on Chernobyl nuclear accident, Ukraine in April 1986.
14. Describe how to remove radioactive substances from water.
15. Describe how to manage the radioactive wastes at a nuclear facility.

CHAPTER 12

AIR POLLUTION BY MUNICIPAL SOLID WASTES, BIOMEDICAL WASTES, E-WASTES AND THEIR CONTROL

12.1 URBAN SOLID WASTES AND HOSPITAL WASTES

12.1.1 Municipal Solid Wastes

Municipal solid wastes and biomedical wastes are big sources of pollution of environment including pollution of air, water, and land. The solid waste collection, handling, sorting, and disposal require a lot of care which is lacking in developing and under-developed countries. The Government of India / Central Pollution Board have laid down 'Municipal Solid waste (Management and Handling) Rules, 2000' for control of pollution by these wastes, but real progress in the direction of their implementation is yet to be made.

There are six stages of solid wastes from generation to disposal and at each stage careful management is needed:

(a) Waste collection; (b) Waste segregation; (c) Storage of waste;

(d) Transportation (e) Processing of wastes; (f) Disposal of wastes

Types of Municipal Solid Wastes

There are various types of municipal solid wastes which include: household wastes; wastes from slum areas; wastes from squatter areas; wastes from animal slaughterhouses / meat and fish market; wastes from hotels/ restaurants; biomedical wastes from other than organized hospitals: remains of animals who have died due to accidents, disease, or natural death; plant wastes from road side trees, green belts, parks, etc.; metallic scrap; plastics; e- waste; industrial wastes; etc.

Air Pollution by Municipal Solid Wastes

Air pollution from municipal solid wastes may arise from all stages ie generation, collection, handling, segregation, and disposal of waste and hence proper management, handling, disposal, and monitoring is required. The type of air pollution from the waste would include the following:

- Bacterial and Viral pollution — By decay of organic wastes
- Dust / Suspended Particulate Matter(SPM) pollution which may include particulates of 'Toxic metals like lead, chromium, mercury, cadmium, selenium, cobalt, aluminum, arsenic, etc.'; 'Pesticide / insecticide / biocides'; 'Cyanides' —During handling, Spraying / fogging for malaria control, Burning/ combustion of wastes.

- Halogens like chlorine, fluorine and their gaseous compounds [including ozone depleting substances, and greenhouse gases] — During burning / combustion of wastes.
- Sulphur dioxide (SO_2) — During burning / combustion of wastes
- Hydrogen sulphide (H_2S), mercaptans — By bacterial decay of wastes
- Ammonia (NH_3) — By bacterial decay of wastes
- Carbon monoxide (CO) — By bacterial decay, slow oxidation, or by burning/combustion of wastes
- Methane (CH_4) -- [a greenhouse gas] — By bacterial decay of organic wastes in burial, stacking, compost making, or landfills by wastes
- Nitrous oxide (N_2O) -- [a greenhouse gas] — By bacterial decay and slow oxidation of organic wastes in landfills, in compost making, or by burning of wastes in open
- Other oxides of nitrogen (NO_x) — Oxidation/ burning of waste in open, or during combustion of wastes in incinerators/ boilers.

Besides polluting gases, carbon dioxide and other greenhouse gases (already named above) are also emitted due to bacterial decay, burning /combustion of wastes.

12.1.2 Biomedical Wastes

The Government of India / Central Pollution Board have laid down 'Biomedical waste (management and handling) Rules, 1998' for control of pollution from these wastes. Under the rules, the biomedical waste has been divided into ten (10) categories. Biomedical Waste (BMW) means any waste, which is generated during the diagnosis, treatment or immunization of human beings or animals or in research activities pertaining thereto or in the production or testing of biological items / materials, and including categories mentioned in Schedule I of BMW (Management and Handling) (second Amendment) Rules, 2000 by Ministry of Environment and Forests Notification.

Categories of Biomedical Wastes (Under 'schedule 1' of the biomedical waste management and handling rules, 1998.) with up to date amendments.

(i) ***Category 1.*** Human anatomical wastes ie human tissues, organs,or body parts cut out in operations during treatment.

(ii) ***Category 2.*** Animal waste ie animal tissues, organs, body parts, carcasses, bleeding parts, blood/ body fluid, experimental animals used in research , waste generated by veterinary hospitals, or discharge from hospitals, etc.

(iii) ***Category 3.*** Microbiology and biotechnology waste like waste from lab cultures, stocks or specimen of microorganisms live or attenuated vaccines, human and animal cell culture used in research and infectious agents from research and industrial labs, wastes from production of biological toxins, dishes and devices used for transfer of culture.

(iv) ***Category 4.*** 'Waste Sharp' ie wastes like syringes, needles, scalpels, blades, glass slides, etc. (Used or unused).

(v) ***Category 5.*** 'Discarded and Cytotoxic* drugs' ie contaminated, outdated, discarded drugs. *(*Cytotoxic means 'toxic to body cells')*

(vi) ***Category 6.*** 'Soiled Wastes' ie contaminated items (including cotton, dressing, plaster casts, linens, bedding, etc.) with blood / body fluids, etc.

(vii) ***Category 7.*** Solid wastes other than mentioned above (i) to (v) ie disposable tubing, catheters, intravenous tubing, etc.

(viii) *Category 8.-* Liquid wastes ie waste generated from washing, cleaning, house-keeping, or disinfecting of pathological laboratories/ clinics / hospitals.

(ix) *Category.9-* Incinerator ash after incineration of biomedical wastes.

(x) *Category.10-* Chemical waste ie chemicals used in production of biological items, and chemicals used in disinfection like insecticides, etc.

Air Pollution by Biomedical Wastes

The air pollution from Biomedical wastes is of very serious nature as it can spread many diseases very quickly by polluting air, water, land, and can infect not only the persons handling or touching it, but also the general public by air borne bacteria and virus. The other types of pollutants, as mentioned above in **Para 12.1.1** may also be emitted, but the severity of infection is much more.

If the biomedical waste is mixed up with municipal solid waste, the whole of the waste becomes infective very quickly, therefore biomedical waste should never be mixed with other municipal wastes. E-waste is another category of waste which often gets mixed up with municipal waste and biomedical waste. It needs to be segregated and recycled in an oderly and safe manner in designated industries as the same may create environmental pollution*.

In India, E-waste is covered in Schedule 3 of "The Hazardous Wastes (Management and Handling) Rules, 2003". Under Schedule 3, e-waste is defined as "Waste Electrical and Electronic Equipment including all components, sub-assemblies and their fractions except batteries falling under these rules". "Guidelines for Environmentally Sound Management of e-waste" formulated by the Ministry of Environment and Forest, Government of India, in the year 2008 followed the same definition.

According to the very recent "the e-waste (Management and Handling) Rules, 2011", 'electrical and electronic equipment' means equipment which is dependent on electric currents or electro-magnetic fields to be fully functional and 'e-waste' means waste electrical and electronic equipment, whole or in part or rejects from their manufacturing and repair process, which are intended to be discarded.

Realizing the growing concern over e-waste, the Government of India has been supporting several initiatives. The Central Pollution Control Board (CPCB) prepared "Guidelines for Environmentally Sound Management of E-waste" in May, 2008 and "the e-waste (Management and Handling) Rules, 2011". As per these rules, the responsibilities of producer, collection centres, consumers, dismantler, and recycler/ reprocesser are given in the **Appendix 12.**

The major approach to control the pollution by e-waste is to reduce the concentration of hazardous chemicals / materials/ elements through recovery and recycle. The e-waste can act as 'Secondary resource' for raw materials for important industries. The process would comprise of dismantling, removal of easily accessible parts containing valuable substances, segregation of ferrous and non- ferrous metals, sending ferrous materials to steel industry and non-ferrous metals to non-ferrous metallurgical industries and treatment and disposal of dangerous materials. It may be noted that silicon used in electronic and solar PV cells may be produced from silicon e-waste.

12.2 MUNICIPAL SOLID WASTES COLLECTION, SEGREGATION, STORAGE AND TRANSPORTATION

Collection

The collection of wastes in cities remains a challenging task. The procedures are well documented in the rules which provide for house to house and place to place collection of garbage transferring the same to big garbage bin of the area by closed carts or vehicles, and evacuation of garbage bins by mechanical means. During collection the 'Biomedical wastes' from doctor's clinics or other places should never be mixed up with general solid waste, but should be collected separately. Similarly, 'Industrial wastes' as well as 'e-wastes' should be collected separately.

Segregation

The segregation of waste should be organized. The non- bio-degradable waste and recyclable wastes should be segregated from the biodegradable wastes. Most of the segregation may be possible at collection stage itself. Public awareness program regarding segregation may be of some help.

Storage of Wastes and Transportation

Adequate storage facility should be created at convenient places and most of the handling and transportation should be done mechanically in a safe way without littering. The persons engaged on the job should be made to wear the personal protective devices for their own protection. Closed carts, vessels, vehicles, etc. should be used. Separate bins or containers should be used for bio-degradable wastes (Green colour), recyclable wastes (White colour), and other wastes (black colour). The vehicles and mechanical handling equipment used themselves should be efficient and should be least polluting.

12.3 PROCESSING AND DISPOSAL OF MUNICIPAL SOLID WASTES

12.3.1 Conversion of Waste to Compost

The biodegradable waste should preferably be converted to compost bio-manure. The Vermicomposting technique (in which worms, mainly earth worms are used to form bio-manure) can also be used.

Biofertilizers are the natural fertilizers which are microbial inoculants of bacteria, algae, fungi generally combined with nitrogenous matter of plant and animal / human waste. Biofertilizers may have nitrogen fixers like (Rhizobium, Azotobacter for legume crop, Azolla for low land paddy, Acetobacter for sugar cane, and Azoctobacter and Agospirellium for non-legume crop), Phosphate Solubilizing Bacteria (PSB), and 'Mycorrhizae fungi'. Composting is a process to convert waste to bio manure / fertilizer which may have the above mentioned useful microorganisms. The bio- fertilizers augment the availability of nutrients to plants.

It has been well established that the use of chemical fertilizers deteriorates the soil in the long run. Each succeeding year, the soil would need increased quantity of the chemical fertilizer to maintain the same productivity level if chemical fertilizers are used. Moreover, the manufacturing of chemical fertilizers is a very energy intensive process. Increased energy production requires burning of more fuel. The production of energy as well as chemical fertilizer process adds to air pollution, water pollution, greenhouse gases and global warming. But with the exclusive use of compost/ bio fertilizers, we reduce these bad effects on environment.

12.3.2 Composting Process

Municipal solid wastes and sludge from municipal sewage treatment can be composted into organic fertilizer which is a good soil conditioner and nutrient for plants/agricultural crops. In compost manure formation, biological decomposition of organic solid waste takes place in presence of moisture. The excess of moisture inhibits transfer of oxygen and forces the process to become anaerobic*, while excess ventilation lowers the moisture content and inhibits the biological decomposition.

Oxygen is necessary for the microbiological activity particularly for aerobic decomposition producing carbon dioxide, water and humus substances. Both oxygen and moisture are necessary for the process. Too much excess air can dry out the compost and lowers the temperatures.

Composting is a biological decomposition of organic solid waste to a relatively stable end product. The decomposition is brought about by bacteria and fungi. At the start of process, the temperature is less and mesophilic bacteria are active, but due to bacterial action the temperature rises (say 45–60°C) and thermophilic bacteria become active. Normally the fungi which are active may be 'Geothrichium', 'Penicillium', 'Rhizopus Aspergillus' and 'Mucor', but dominance and the presence of various fungi would depend on nature and composition of organic matter.

Due to bacterial action organic acids are produced in the initial stage causing a fall of pH value, but due to mixing and aeration which is done with the aid of machinery (during composting process) would improve the pH value. When composting is done without the aid of machinery (manual systems) and where mixing and aeration is not possible, the pH value is corrected by addition of lime or other buffers.

Mechanical composting systems have mixing, and aerating mechanisms for accelerating the composting process. Stable fertilizer can be produced in about 15 to 30 days time which in manual systems takes months.

It is well known that carbohydrates and trace elements are also necessary for microbial action. Required ratio of Carbon, Nitrogen and Phosphorus (C. N. P.) is approximately 100:5:1. An addition of resource of nitrogen and phosphorous to the 'process of formation of compost' accelerates it and also adds to the value of the fertilizer. The animal wastes and agricultural wastes also form a good material for making compost manure.

***Note:** It may be noted that during composting some anaerobic decomposition takes place forming methane which goes out to atmosphere (methane is more powerful greenhouse gas than carbon dioxide). However, a major portion of carbon remains sequestered in the soil and overall conversion of waste to compost and its use as fertilizer to replace the chemical fertilizer is environmentally friendly.*

Safety from Pathogenic Pollution in Composting

From point of view of pollution and spreading of diseases, it may be mentioned that most of the pathogens (disease causing microorganisms) except a few may not be able to survive at 60 °C reached in the composting longer than half an hour. There are antagonistic processes caused by antibiotic inhibitors organism and fungi which may kill most of the pathogens. Composting is therefore thought to be safe for producing manure/ biofertilizers.

12.3.3 Conversion of Municipal Waste to Various Types of Fuels

Liquid Fuels

The biodegradable waste can also be converted to renewable liquid fuels such as 'Alcohol' and 'Bio-diesel'.

Pyrolysis of Waste

Thermal decomposition in absence of oxygen of plastic or other organic waste materials (called Pyrolysis) can be done for obtaining useful fuels like gaseous fuel, liquid fuel or coke. The process almost eliminates the formation of 'Dioxins' because of absence of oxygen.

Gaseous Fuel

Methane can be generated not only by animal dung, human excreta, septic tank sludge but also by other biodegradable wastes like municipal solid waste / garbage. The garbage is shredded and materials unsuitable for fermentation in it are separated out. Night soil and septic tank sludge after being subjected to biological 'Denitrification', is mixed with the shredded biodegradable materials and subjected to 'Fermentation and Methanization' for forming Methane in a biological reactor. The methane is separated and sent for utilization while the resulting sludge is dewatered and led for composting to make useable biofertilizers / bio manure for farming. A system for solid waste conversion to methane is shown in the flow diagram **Fig. 12.1.**

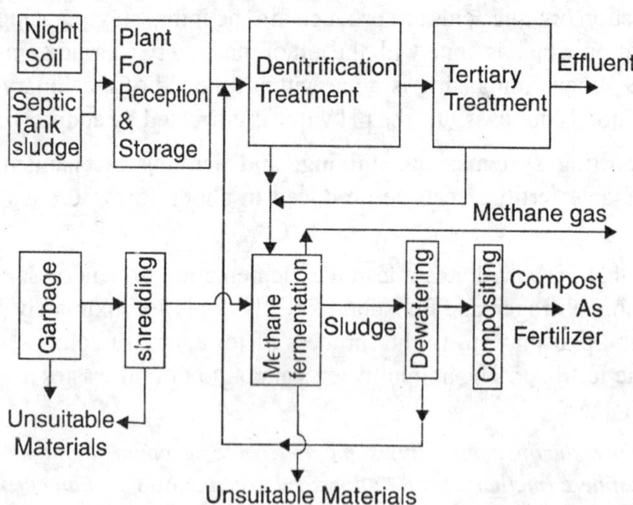

Fig. 12.1: Flow Diagram for Methane Fermentation–Organic Waste

A similar system was developed by MHI Japan and the fermentation process used by them has two key equipments – (i) Mix separator and (ii) Twin reactor. The mix separator, in which the sludge and garbage are fed; it separates the materials unsuitable for fermentation, adjusts the moisture content and temperature and homogenizes the charge converting it in the form of slurry for feeding to the twin reactor for methane generation. The twin reactor has a pre-chamber located inside it which prevents short circuits of charge and ensures effective fermentation. The mixed liquid effluent (slurry) from the pre-chamber is pumped and dispersed at the top of the twin reactor. Mechanical stirring along with gas stirring blowers are jointly used to ensure shredding of scum material. The MHI process can degrade quite a significant content of organic matter to produce biogas which can be utilized as energy source.

The gas produced by Anaerobic Digester may typically have 50% to 80% Methane, and 20% to 50% carbon dioxide depending upon the feed stock and completeness of digestion process. The heat content of this biogas may be about 60% of Natural gas, say 600 BTU per cubic foot. The remaining solids from anaerobic digester are nutrient rich which can be used as fertilizer or soil amender._

Solid waste anaerobic digesters can form a part of system of generation of power or heat or both. There are environmental as well as economic advantages of such a system. There can be centralized anaerobic digester plants which can treat solid organic waste, sewage sludge, etc. The bio-degradable wastes can be converted to 'Biogas' or to hydrogen. **[Ref: 10, 11]**.

12 3.4 Use of Combustible Municipal Solid Waste as Fuel

Waste combustion is particularly popular in countries such as Japan where land is a scarce resource. Denmark and Sweden have been leaders in using the energy generated from incineration of wastes for more than a century; the energy is used as heat or for producing electricity. A number of other European countries like Luxembourg, the Netherlands, Germany and France also use energy of waste incineration.

The garbage and plant waste should not be burnt in open to reduce its volume instead it can be compressed. The combustible waste should be segregated, shredded, dried, and then used in a waste fired boiler to, raise steam for any use including power generation. As an example, the following system followed by MHI, Japan, seems attractive for reducing air pollution along with other types of pollution from municipal solid waste and also to generate some energy.

Waste Incineration with Thermal Cracking, Ash Melting and Electricity Production

Incineration of municipal solid waste having very significant content of combustible matters may side by side generate some electricity by utilizing calorific value of the waste. However incineration of solid waste, if done, should keep in view that the emission of air pollutants is minimized and reduced below the permitted values and also that the leachate from it should not cause water pollution. The ash can be converted to slag and the emission control measures should be taken. MHI Japan has developed a system of incineration of municipal solid waste with direct ash melting, thermal cracking of waste and generation of electricity.

In this system the waste is first shredded and dried on drying feeder by high temperature (hot) sand from FBC (fluidized bed combustion) boiler i.e. char combustion chamber. The temperature maintained on the feeder is 200 – 250°C and the water vapour and chlorine content of waste is ejected and homogenized. The waste along with sand is fed into thermal cracking gasifying furnace and the hot sand from fluidized bed combustion boiler is also further added for supplying heat for thermal cracking and a temperature of 400 – 450°C is maintained in the gasifying furnace. Two products are formed here namely cracked gas and char. The char so produced has a very little chlorine content. The char along with sand is fed to the Fluidized Bed Combustion boiler (FBC boiler) where the char burns off to give out heat. The cracked gas (as a heat source) is fed to a melting furnace of vertical 'Swirling type' where ash separated by bag filters from the flue gases of FBC boiler is melted to form slag. The non-combustibles like iron, aluminum and other metals are discharged from thermal cracking furnace in non-oxidized state. The sand and small size particles discharged from thermal cracking furnace and some quantity from FBC boiler is fed back to FBC boiler's combustion chamber. The flue gas from melting furnace is led to FBC boiler for its heat utilization. A simplified flow diagram of the system is shown in **Fig. 12.2.**

Air Pollution Control in the above MHI System

The flue gases from FBC boiler are finally passed through gas cleaning equipment comprising:

(i) Dust removing bag filter — for the removal of suspended particulate matter;

(ii) Reacting bag filter — for removal of acid gases; (Slaked lime is dosed in the flue gas before the inlet to reactive bag filter).

(iii) Selective Catalytic Reactor (SCR) — for removal of NO_x before going to stack.

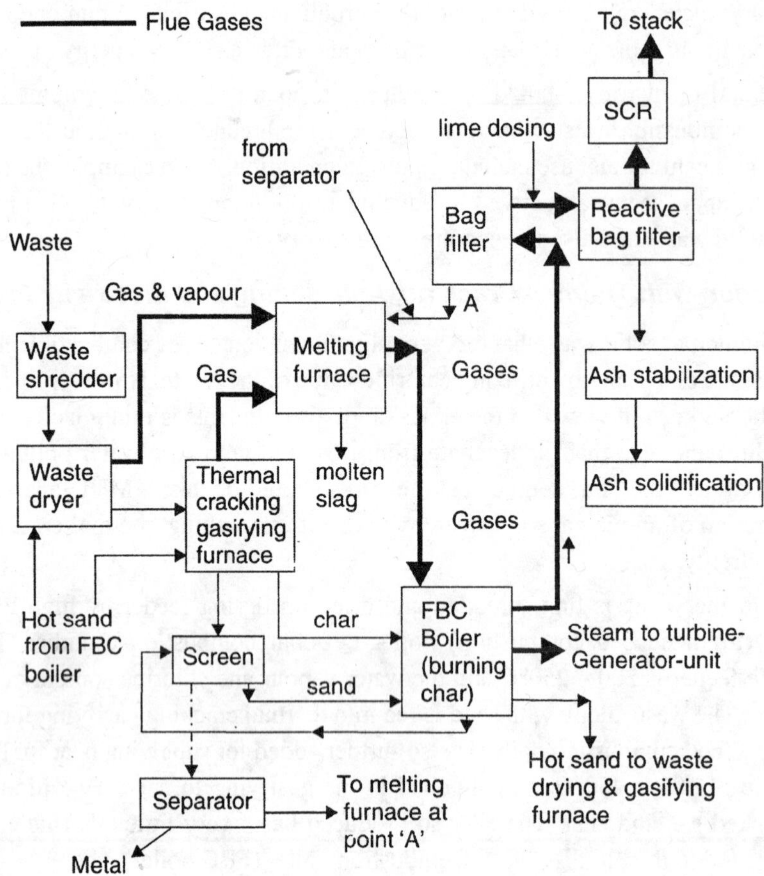

Fig. 12.2: Flow Diagram– Municipal Solid Waste Incineration with Thermal Cracking, Ash Melting, and Power Generation

The emissions of oxides of sulphur and nitrogen (SO_x and NO_x), hydrochloric gas and dioxins are all very well controlled in the above system much below the permitted levels.

The ash from 'Dust removing bag filter' is led to a melting furnace to form slag while the ash from 'Reactive bag filter' is stabilized and solidified. The process avoids water pollution by leachate and reduces air pollution due to fine dust. The gaseous pollutants like NO_x, SO_x, HCl, and VOC are reduced by bag filter and reactive bag filter.

The steam raised in the FBC boiler is generally used for driving 'Steam turbine generator set' for electricity generation. The standard steam water cycle is used.

12.3.5 Land Filling by Municipal Solid Wastes

The municipal solid waste can be filled in specially prepared trenches after separating out non-biodegradable materials. In course of time, say after 6 months or more the waste gets converted to bio-manure, which can be used in agricultural fields. The biogas consisting of mostly methane and carbon dioxide can be tapped as landfill gas.

The municipal garbage is often used as a landfill material for filling uneven land. From landfill areas, the taping of biogas / landfill-gas is feasible. The collection of bio gas from landfill would reduce air pollution / emission of methane which is both a pollutant and a potent greenhouse gas.

The garbage dumps should never be made near river / lake / ponds, etc. to avoid water pollution. Such dumps should be located away from population as there is always a danger of spread of air borne pathogens / virus near such dumps. The dumps should be covered with earth and should not be kept exposed to atmosphere.

The landfill gas collection system including control system at land fill site should be installed. This would minimize 'Odour generation', and prevent 'Off-site migration' of gases. This would also protect 'Vegetation planted' on the rehabilitated landfill site besides reducing greenhouse gases and obtaining biogas which is an extra source of energy generation.

Ambient Air Quality Monitoring at Landfill Sites

The concentration of methane gas emitted at landfill site should not exceed 25% of the lower explosive limit. The land fill gas from collection facility at land-fill site should be utilized. If it is not possible, the gas should be flared, but in any case, it should not be allowed to escape to atmosphere. The ambient air quality at landfill site should be monitored to meet the following standards.

Standards/ Limits in India for Pollutants in Ambient Air at Landfill sites:

(i) Sulphur dioxide: 120 μg/m^3. (24 hr. average)

(ii) Suspended particulate matter: 500 μg/m^3. (24 hr. average)

(iii) Methane: Not exceeding 25 % of the 'Lower explosive limit' which is equivalent to 650 mg/m^3.

(iv) Ammonia daily average (sample duration 24 hours): 4 mg/m^3.

(v) CO: 2 mg/m^3 (1 hr average) and 1 mg/m^3 (8 hr. average).

According to regulations, the monitoring should be carried out:

- At least six times a year for cities with population >50 lakhs,
- At least four times a year for cities with population between 10 and 50 lakhs,
- At least two times a year for cities with population between 1 and 10 lakhs.

12.3.6 Incineration of Hazardous Sludges and Wastes

Energy saving technology for sewage sludge incineration is also in use. Mitsubishi Heavy Industries Ltd (MHI) Japan installed a system at Northern Sewage Treatment Center of Yokohama Municipal Sewage Bureau which has been operating since 1989. The system used has the following special features:

(i) Fluidised bed type incinerator. Heat of flue gases of incinerator is used for preheating both the air (used for combustion) as well as oil. This heated oil is used as heating medium in an indirect heat dryer of hollow disc type for pre-drying of de-watered sludge.

(ii) Heat recovery system employs shell type heat exchangers which are suitable for gases with high dust density.

(iii) The gases/vapours from sludge/waste dryer are scrubbed with water and then led to an incinerator. Portion of digester gases produced in sewage treatment plant supplement the fuel requirement of incinerator.

(iv) Flue gas cleaning system has cyclone separator, electrostatic precipitator and an absorption tower for gaseous pollutants including VOCs.

Biogas Utilization

The heat energy of biogas can be utilized for:

- Combined heat and power plants

- Industrial use as fuel
- Domestic use as fuel
- Maintaining the temperature of anaerobic digester
- Producing hot water.
- Space heating.
- Sludge drying.
- Industrial use as fuel
- Domestic use as fuel

Belt dryers are commonly used for drying bio-solids and utilizing heat from reciprocating engines. The combined heat and power systems at times may be supplemented by natural gas in case of insufficient biogas generation. Such combined heat and power systems are operating at many places in the world such as:

(i) Waste treatment plant of East-bay Municipal utility District Oakland, California, USA (with reciprocating engines).

(ii) Lewiston waste water treatment plant, Lewiston, near Niagara falls, in upstate NewYork (with combustion turbines).

(iii) Encina (California) water pollution control facility (with reciprocating engines).

(iv) Santa Barbara's El Estero California waste water treatment plant (Molten carbonate fuel cells)

12.4 TREATMENT OPTIONS FOR DIFFERENT BIOMEDICAL WASTES

The biomedical waste of different categories is mentioned in **Para 12.1.2.** The treatment options for different categories of biomedical waste (as per schedule I of biomedical waste management and handling rules 1998) are indicated in **Table 12.1.**

Table 12.1: Treatment Option for Different Categories of Biomedical Wastes

Category	Content of the category	Treatment options for different categories of biomedical wastes
Cat.1	Human anatomical wastes	Incineration / deep burial
Cat.2	Animal wastes	Incineration / deep burial
Cat.3	Microbiological and biotechnology waste	Local autoclaving / microwaving and incineration
Cat.4	Waste Sharp	Disinfection (chemical treatment / autoclaving / microwaving) and then mutilation and shredding
Cat.5	Discarded medicines and Cytotoxic* drugs	Incineration/ destruction and drug disposal in secured landfills
Cat.6	Soiled waste	Incineration/ autoclaving/ microwaving
Cat.7	Solid wastes	Disinfection by chemical treatment / autoclaving / microwaving and then mutilation and shredding
Cat.8	Liquid waste	Disinfection by chemical treatment including neutralization and discharge into drains
Cat.9	Incineration ash	Disposal in municipal landfills
Cat.10	Chemical waste (both liquid and solid)	Chemical treatment including neutralization after which 'Liquids' are to be disposed into drains and 'Solids' in secured landfills

The colour coding and type of container for different treatment option exercised for different categories of waste are specified in schedule II of the biomedical management and handling rules, 1998; these are indicated in the **Table 12.2.**

Table 12.2: Colour Coding, Type of Containers, for Different Treatment Options

Waste category	Colour Coding for particular treatment	Type of containers
Cat. 1, Cat.2, Cat. 3, Cat. 6	Yellow colour for incineration / deep burial	Plastic bags
Cat. 3, Cat. 6, Cat. 7	Red colour for autoclaving,/ microwaving / chemical treatment	Disinfected container / Plastic bags
Cat. 4, Cat. 7	Blue / White translucent colour for autoclaving / microwaving/ chemical treatment and then destruction / shredding	Plastic / puncture proof containers
Cat. 5, Cat. 9, Cat.10 (Solids)	Black colour for disposal in secured land fills.	Plastic bags

Notes:

(i) *The colour code is given according to the treatment option chosen.*

(ii) *The plastic container / bags for biomedical waste should not be made of chlorinated plastic (PVC).*

(iii) *If a category 3 waste is disinfected locally, it may not be put in containers or bags.*

(iv) *The level on the biomedical waste container is given in accordance with the schedule III of 'Biomedical waste management and control rules, 1998' where two separate levels are allotted - one for bio-hazardous waste and other for 'Cytotoxic* waste'. **[* Cytotoxic means toxic to body cells].***

12.5 EMISSION STANDARDS IN INDIA FOR INCINERATORS

The 'Schedule V of Biomedical waste management and handling rules 1998', and also 'Schedule IV of Municipal solid waste management and handling rules, 2000' specify that:

The combustion efficiency shall be at least 99.00 %, which would be computed by the formula:

Combustion Efficiency = $100 \times [\%CO_2 / (\%CO_2 + \%CO)]$

The temperature of the primary chamber of incinerator shall be within the temperature range 750–850°C while the residence time in the secondary chamber of incinerator at a temperature of 1000–1100°C shall be at least one second with minimum 3% oxygen in flue gases of the stack.

The revised (2009) Emission standards specified for such incinerators are:

(i) Particulate Matter—50 mg / Nm^3 with correction for 11% O_2 in flue gases

(ii) Oxides of nitrogen (NO_x)–450 mg/Nm^3 with correction for 11% CO_2 in flue gases

(iii) Hydrogen chloride (HCl)–50 mg/Nm^3 with correction for 11% O_2 in flue gases

(iv) SO_2_200 mg/mg^3

(v) CO–100 mg/Nm^3

(vi) Dioxins & Furans

For existhing units—0.2 TEQ/Nm^3

For New Unit—0.1 TEQ/Nm^3

(vii) The volatile organic compounds in ash— to be limited to 0.01%

(viii) The Toxic metals in ash— to be limited to within the permissible values under the 'Hazardous waste management and handling rules, 1989, India.'

(ix) Only low sulphur fuel like LDO / LSHS shall be used.

(x) The minimum stack height should not be less than 30 meters..

12.6 INCINERATION OF BIOMEDICAL WASTES

Incineration is widely used technique for biomedical wastes. The wastes which are to be incinerated should not be chemically treated by any chlorinated disinfectant. All incinerator systems should be provided with pollution control devices so that the pollution limits imposed by pollution control rules are not crossed at any time during operation. Incineration has particularly strong benefits for the treatment of certain types of wastes such as biomedical wastes and certain hazardous wastes where pathogens and toxins are to be destroyed by high temperature. Incineration may also be required in case of chemical multi-product plants with diverse toxic or very toxic wastewater streams, which cannot be routed to a conventional wastewater treatment plant.

The burning of medical waste generated by hospitals may create problems as biomedical waste incinerators emit toxic air pollutants and toxic ash residues. They may be major source of dioxins in the environment. The 'International Agency for Research on Cancer', an arm of 'WHO', acknowledged the cancer-causing potential of dioxins and classified them as 'Carcinogen' for humans'.

Prevention of Pollution

In order to avoid production of dioxins, no bags made from Polyvinylchloride (PVC) or any other plastic containing chlorine or any chemical /compound with chlorine content should ever be introduced in incinerator. The chlorinated pesticides /insecticides should not be used for disinfection so that they may not find their way to an incinerator.

Red colour bags should not be incinerated as red colour contains cadmium, which causes toxic emissions. If mercury-containing items are put into a red bag for infectious waste and sent to an incinerator or other waste treatment technology, mercury which is a potent neurotoxin will contaminate the environment. Mercury containing items should be treated separately.

12.7 NON- BURN TECHNOLOGIES FOR STERLIZING WASTES

Because of the concerns for pollution by incineration, there is worldwide awareness to rethink about the options for non-incineration technologies. As per 'Health Care without Harm' which is an international coalition of 470 organizations in 52 countries, the non-incineration treatment technologies are being developed; some non- incineration technologies are still essentially prototypes, while others, such as autoclave technology, have been used for many years.

12.7.1 Autoclaving

Autoclave is a type of steam sterilizer. The barriers to the penetration of steam have to be removed to properly sterilize the waste by moist steam. There should be an intimate contact of steam and waste. The plastic bag in which the waste is packed is a barrier for steam penetration. The air in the wastes is also a barrier and therefore evacuating air from autoclave chamber is advantageous. There are some modified designs of autoclaves. Models have been developed with various modifications such as vacuum system, shredding system, deodorization, parameter control and monitoring systems, etc.

It may be mentioned that certain types of wastes should not be treated in an autoclave such as anatomical and pathological wastes, radioactive waste of any type, organic solvents, laboratory chemicals, and chemotherapy waste.

Some Important Operational Points for Autoclaves

Dedicated autoclave should be used for the purpose of disinfecting and treatment of biomedical waste.

(a) When operating a gravity flow autoclave, medical waste should be subjected to:
- A temperature of not less than 121°C and pressure of >1 atm (15 psi) for an autoclave residence time of not less than 60 minutes; or
- A temperature of not less than 135°C and a pressure of >2 atm (31 psi) for an autoclave residence time of >45 minutes; or
- A temperature of not less than 149°C and a pressure of >3.5 atm (52 psi) for an autoclave residence time of >30 minutes.

(b) When operating a vacuum autoclave, medical waste should be subjected to a minimum of one pre-vacuum pulse to purge the autoclave of all air. the waste should be subjected to the following:
- A temperature of not less than 121°C and pressure > I atm (15 psi) for an autoclave residence time of >45 minutes.; or
- A temperature of not less than 135°C and a pressure >2 atm (31 psi) for a residence time of >30 minutes.

(c) Medical waste would only be considered treated when the required temperature, pressure, and residence time indicated above were reached during autoclaving. If for any reason these were not reached, the process has to be repeated properly.

(d) Each autoclave should be provided with computerized recording devices, monitoring system effectively keeping watch over the process. The record should show the date, time, load identification which was treated, residence time, temperature, pressure, and graphic representation of process for each treatment cycle.

(e) Each auto clave should be able to completely and consistently kill bacteria and other pathogens at the maximum design capacity. For validating this, a biological indicator used is 'Spores of Bacillus stearothermophillus' in the form of vials or spore strips with at least 1×10^4 spores per milliliter. The test should be performed on regular basis.

(f) To ensure that the waste load has been properly treated, a chemical indicator strips / tape (which changes its colour when the required temperature is reached) is put on the waste package at several locations. This would indicate that required temperature was reached. It is a very simple routine test.

Examples of installation of Autoclaves in India

Some examples where 'Autoclaves' were installed instead of incinerators are mentioned below:

(i) Sanjay Gandhi Mangol Puri, New Delhi *System*: Dedicated Waste Autoclave (Tuttnatuer 66190EP - Israel); Installed: September 1997; Supplier: Omron Medical, New Delhi

(ii) Escorts Heart Hospital, New Delhi *System*: Dedicated Waste Autoclave and Shredder, 360 kg./h (Tuttnatuer 5590EP - Israel); Installed: April 1999; Supplier: Omron Medical, New Delhi

(iii) Tata Memorial Hospital, Mumbai *System*: Hydroclave (advanced autoclave using indirect steam sterilisation technology and a rotary flow), (Hydroclave H-25 - Canada); Installation: June 1999; Supplier : Consolidated Waste Management India (P) Ltd., Mumbai.

12.7.2 Microwaving

It should be ensured that microwave treatment of waste completely and consistently kills bacteria and other pathogens. This is to be validated by testing the microwaving equipment and process by approved biological indicator at maximum design capacity of each microwave unit. The approved '*Biological indicator*' for microwave is '*Bacillus Subtilis spores*' which is used in the form of vials or spores strips with at least 1×10^4 spores per milliliter.

[*Warning: It is very important to note that mircowave treatment is not to be used for Cytotoxic (toxic to body cells), or hazardous, or radioactive wastes, or contaminated animal carcasses, contaminated body parts, or large metal items].*

12.7.3 Deep Burial

Burial Site: The location for deep burial site is authorized by prescribed authority under rules. The institution has to maintain a record and map of all pits used for deep burial. The deep burial pits should be distant from habitation. The sites are chosen so as to ensure that no contamination of ground water or surface water occurs and the sites would not be prone to soil erosion or flooding. The deep burial site should be relatively impermeable and no shallow well would be located in its vicinity.

Burial Pit/ Trench: A burial pit or trench should be dug about 2 meters deep. It should be half filled with waste, and then covered with a layer of 50 cm lime over which rest of the space is filled with soil. It is to be ensured that animals do not have any access to the burial sits. Covers of galvanized iron / wire mesh may be used. On each occasion, when waste is added to the pit, a layer of 10cm of soil shall be added to cover the wastes. The burial should be performed under close and specially dedicated supervision.

12.8 DESTRUCTION TIME AND TEMPERATURE FOR SOME PATHOGENS IN WASTES

Some pathogens may be associated with solid wastes which may pollute the environment. These organisms get destroyed at and beyond the temperature and time indicted in **Table 12.3.**

Table 12.3: Destruction Temperature & Time for Some Pathogens in Solid Wastes

Organisms	Time and Temperature for destruction
1. *S. typhosa*	No growth beyond 46°C, death in 30 minutes at 55–60°C and 20 minutes at 60°C, destroyed in a short time in compost environment
2. *Salmonella sp.*	In 1 hour at 55°C and in 15–20 minutes at 60°C.
3. *Shigella sp.*	In 1 hour at 55°C.
4. *E. coli*	In 1 hour at 55°C & in 15–20 minutes at 60°C.
5. *E. histolytica cysts*	In few minutes at 45°C. and in few seconds at 55°C.
6. *Taenia saginata*	In a few minutes at 55°C .
7. *Trichinella spiralis larvae*	Quickly killed at 55°C, instantly at 60°C
8. *Br.abortus or Br.suis*	In 3 minutes at 62–63°C and in 1 hour at 55°C.
9. *Micrococcus pyogenes var. aureus*	In 10 minutes at 54°C.
10. *Streptococus pyogenes*	In 10 minutes at 54°C .
11. *Mycobactercum tuberculosis var. hominis*	15–20 min. at 66°C /momentary heating at 67°C.

12. *Corynebacterium diptheriae*	In 45 minutes at 55°C.
13. *Necator americanus*	In 50 minutes at 45°C
14. *A.lumbricoides eggs*	In 1 hour at 50°C.

12.9 USES OF SOME IMPORTANT INDUSTRIAL WASTES

Utilization of industrial wastes is better than their disposal which may pollute the environment. Some industrial wastes can be employed usefully as indicated in **Table 12.4.**

Table 12.4: Uses of Some Important Industrial Solid Wastes

Waste	Areas of application
Fly ash	(i) Direct mixing in cement up to 10% (ii) Raw material in Ordinary Portland Cement (OPC) manufacture, Manufacture of oil well cement (iii) Making sintered fly ash light-weight aggregates (iv) Cement / silicate bonded fly ash/clay binding bricks and insulating bricks (v) Cellular concrete bricks and blocks, lime and cement fly ash concrete (vi) Precast fly ash concrete building units (vii) Structural fill for roads, construction on sites, land reclamation, etc. (viii) As filler in mines, in bituminous concrete (ix) As water reducer in concrete and sulphate resisting concrete
Blast furnace Slags	(i) Manufacture of slag cement, super sulphated cement, metallurgical cement; (ii) Non-portland cement (iii) Making expansive cement, oil well, coloured cement and high early-strength cement (iv) In refractory and in ceramic industry (v) As a structural fill (air-cooled slag) (vi) As aggregate in concrete (vii) Production of slag wool which is used as heat insulation
Ferroalloy and other metallurgical slags	(i) As structural fill (ii) In making pozzolona metallurgical cement
Byproduct gypsum	(i) In making of gypsum plaster, plaster boards and slotted tiles (ii) As set controller in the manufacture of portland cement (iii) In the manufacture of expensive or non-shrinking cement, super sulphated and anhydrite cement (iv) As mineralizer (v) Simultaneous manufacture of cement and sulphuric acid
Chromium sludge Lime sludge from paper and sugar industries	(i) As raw material component for coloured (ii) cement manufacture (iii) Manufacture of lime pozzolana bricks/binders (iv) Recycling in parent industry (v) Manufacture of building lime/ masonry cement

*Red mud from aluminum production	(i) As a corrective material (ii) As a binder (iii) Making construction blocks (iv) As a cellular concrete additive for coloured composition for concrete (v) Making heavy clay products and red mud bricks (vi) In the formation of aggregate (vii) In making floor and all tiles (viii) Red mud polymer door
Lignin from Pulp and paper mills	Lignin can be used for manufacture of a number of useful chemicals like vanillin, vanillic, catechols, dimethyl sulphide, dimethylsulphide, etc.
Baggasse and press mud from Sugar Industry	(i) Bagasse which is used for making boards, card boards, other type thick paper or as fuel (ii) Press mud (obtained after cane juice clarification) which is used for manufacturing of biomanure by biocomposting
* See note on Red mud in **Para 8.4.3.2 Chapter 8**	

12.10 MANAGEMENT AND CONTROL OF E-WASTES

E-Waste is another category of waste which often gets mixed up with municipal waste and biomedical waste. It needs to be segregated and recycled in an orderly and safe manner in designated industries as the same may create environmental pollution.

In India, E-waste is covered in Schedule 3 of "The Hazardous Wastes (Management and Handling) Rules, 2003". Under Schedule 3, e-waste is defined as "Waste Electrical and Electronic Equipment including all components, sub-assemblies and their fractions except batteries falling under these rules". "Guidelines for Environmentally Sound Management of e-waste" formulated by the Ministry of Environment and Forest, Government of India, in the year 2008 followed the same definition.

According to "the e-waste (Management and Handling) Rules, 2011", 'electrical and electronic equipment' means equipment which is dependent on electric currents or electro-magnetic fields to be fully functional and 'e-waste' means waste electrical and electronic equipment, whole or in part or rejects from their manufacturing and repair process, which are intended to be discarded.

Realizing the growing concern over e-waste, the Government of India has been supporting several initiatives. The Central Pollution Control Board (CPCB) prepared "Guidelines for Environmentally Sound Management of E-waste" in May, 2008 and "the e-waste (Management and Handling) Rules, 2011".

As per '*e-Waste Management and Handling Rules, 2011*' the responsibilities of producer, collection centres, consumers, dismantler, and recycler/ reprocessor are given in the **Appendix 12.**

The major approach to control the pollution by e-waste is to reduce the the concentration of hazardous chemicals / materials/ elements through recovery and recycle. The e-waste can act as 'Secondary resource' for raw materials for important industries. The process would comprise of dismantling, removal of easily accessible parts containing valuable substances, segregation of ferrous and non- ferrous metals, sending ferrous materials to steel industry and non-ferrous metals to non-ferrous metallurgical indsustries, and treatment and disposal of dangerous materials. It may be noted that silicon used in electronic and solar PV cells may be produced from silicon e-waste.

Environmental Pollution by E-Waste

Mercury leaches when certain electronic devices, such as circuit breakers are destroyed and so also, Polychlorinated Biphenyls (PCBs) leach out from condensers. When brominated flame retardant plastic or cadmium containing plastics are landfilled, both polybrominated diphenyl ethers (PBDE) and cadmium may leach into the soil and groundwater. It has been found that significant amounts of lead ion are dissolved from broken lead containing glass, such as the cone glass of cathode ray tubes, get mixed with acidic waters and are a common occurrence in landfills.

Uncontrolled fires may arise at landfills. When exposed to fire, metals and other chemical substances, such as the extremely toxic dioxins and furans (TCDD—tetrachloro dibenzo—dioxin, PCDDs—Polychlorinated Dibenzodioxins, PBDDs—Polybrominated dibenzo—dioxin and PCDFs—poly chlorinated dibenzo furans) from halogenated flame retardant products can be emitted. The most dangerous form of burning e-waste is the open-air burning of plastics insulated cables/ wires for recovering copper and other metals. The toxic fall-out from open-air burning affects the local environment and broader global air currents, depositing highly toxic byproducts in many places throughout the world. Incineration of E-waste possesses another threat. It can emit toxic fumes and gases, thereby polluting the surrounding air.

REFERENCES

1. Arora, D.R., Sterilization and Disinfection. Textbook of Microbiology. New Delhi: CBS Publishers; 2003. pp. 61-4.
2. Biomedical Waste (Management and Handling) Rules 1998, India [with up to date ammendments].
3. Brannen, L., Cedar, R., Cellarius, D., Davis, S.C., Harvie, J.P., and Christensen, J.H., Medical waste treatment technologies: Evaluating non-incineration alternatives. Washington, DC Health Care Without Harm; 2000.
4. Doug Hinrichs et al. "Using biogas to fuel distributed generation and combined heat & power plants" Nov. Dec. 2005 'Power' (McGraw Hill).
5. Forbes, B.A., Sahm, D.F., Weissfeld, A.S., Diagnostic Microbiology. 11 th ed. St. Louis, Mo. Bailey and Scott's; 2002. p. 19.
6. Gautam, V., Thapar, R., and Sharma, M. Biomedical waste management: Incineration vs. Environmental safety. Indian J. Med. Microbiology. 2010;28:191-2.-6
7. MHI Technical Review, Vol 28, No1, Feb1991.
8. Mitsubishi Heavy Industry (MHI) LTD Japan. Technical Review. Vol.36 No.2, June 1999, under new products p. 53.
9. MOEF guidelines for e-Waste 2008 and E- Waste Management and Handling Rules 2011 Government of India.
10. Municipal Solid Waste (Management & Handling) Rules 2000. (EPA, India).
11. Schulze, K. L., "Continous Thermophilic Composting", Compost Science Vol.3 No.1 (Spring 1962) pp. 22-23.
12. Sharma, M, and Gautam V., Biomedical Waste. In: Singh KK, Juwarkar A, Singh AK, Tomar A, (Editors). Air, Water and Soil Pollution . Delhi Kalyani Publishers; 2007. pp. 396-423.
13. Wiley, John S., 'Refuse and Refuse – sludge composting', Jr. Boston Soc. Civil Engrs. Reprint Jan. 1962 .
14. Yoshinori Terasawa et al. "Commercialization of MSW incineration system with direct ash

melting by Thermal Cracking for high efficient generation of Electricity." MHI Technical Review Vol. 38 No. 2 Jan. 2001 p. 8

QUESTIONS

1. Enumerate the types of wastes included in municipal solid wastes. How these waste create air pollution and greenhouse gases.
2. Enumerate the categories of biomedical wastes. How these waste create air pollution.
3. Write a short note on conversion of municipal wastes to compost and what are its advantages?
4. Write a short note on the use of solid organic municipal waste as a source of energy. Enumerate advantages of converting the municipal wastes to gaseous fuel.
5. How can you control the air pollution from land filling by municipal solid waste?
6. Write a note on incineration of hazardous sludges and wastes.
7. Enumerate treatment options for different categories of biomedical wastes.
8. Write a short note on incineration of biomedical wastes and air pollution **control** in the process.
9. Write short notes on nonburn technologies for disinfection of biomedical wastes.
10. How would you make autoclaving an effective way to disinfect the bio-medical wastes?
11. How does e-waste create environmental problems? What measure should be adopted for dealing with the e-wastes?

ANNEXURE 12

Responsibilities of the Producer of Electrical and Electronic Equipments

The producer of the electrical and electronic equipments listed in Schedule I shall be responsible for:–

- Collection of e-waste generated during the manufacture of electrical and electronic equipment and channelizing the same for recycling or disposal;
- Collection of e-waste generated from the 'end of life' of their products in line with the principle of 'Extended Producer Responsibility' (EPR), and to ensure that such e-wastes are channelized to registered dismantler or recycler.
- Producer shall, as necessary, ensure collection and channelization by authorizing collection agencies;
- Setting up collection centers or take back system either individually or collectively;
- Financing, and organizing a system to meet the costs involved in the environmentally sound management of e-waste generated from the 'end of life' of its own products and historical waste available on the date from which these rules come into force. Such financing system shall be transparent. The producer may choose to establish such financial system either individually or by joining a collective scheme.
- Providing contact details such as address, telephone numbers/helpline number and e-mail of authorized collection centers to consumer(s) or bulk consumer(s) so as to facilitate return of used electrical and electronic equipment.
- Creating awareness through publications, advertisements, posters, or by any other means of communication and information booklets accompanying the equipment, with regard to the following:
- Information on hazardous constituents as specified in sub-rule 1 of rule 13 in electrical and electronic equipment;
- Information on hazards of improper handling, accidental breakage, damage and/or improper recycling of e-waste;
- Instructions for handling the equipment after its use, along with the Do's and Don'ts;
- Affixing a visible, legible and indelible symbol given below on the products or information booklets to prevent e-waste from being dropped in garbage bins containing waste destined for disposal;
- Obtaining an authorization from the concerned State Pollution Control Board or Pollution Control Committee in accordance with the procedures under rule 9;
- Maintaining records in Form 2 of the e-waste handled and such records should be available for scrutiny by the State Pollution Control Board or the Committee concerned
- Filing annual returns in Form 3, to the concerned State Pollution Control Board or Pollution Control Committee, on or before the 30th day of June following the financial year to which that return relates.

RESPONSIBILITIES OF COLLECTION CENTERS FOR E-WASTES

Collection centre(s) shall -

- Obtain an authorization in accordance with the procedures prescribed under Rule - 9 from the State Pollution Control Board or Pollution Control Committee concerned as the case may be and provide details such as address, telephone numbers/helpline number, e-mail, etc. of such collection centre(s) to the general public.
- Ensure that the e-waste collected by them is stored in a secured manner till these are sent to registered dismantler(s) or recycler(s) as the case may be;

- Ensure that no damage is caused to the environment during storage and transportation of e-waste;
- File annual returns in Form 3, to the concerned State Pollution Control Board or Pollution Control Committee concerned on or before the 30th day of June following the financial year to which that return relates.
- Maintain records of the e-waste handled in Form 2 and such records should be available for scrutiny by the concerned State Pollution Control Board or Pollution Control Committee.

Responsibilities of consumer or bulk consumer of electrical and electronic equipment

- Consumers or bulk consumers of electrical and electronic equipment listed in Schedule I shall ensure that e-waste generated by them is channelized to authorized collection centers or registered dismantlers or recyclers or is returned to the pick up or take back services provided by the producers; and
- Bulk consumers shall maintain records of e-waste generated by them in Form 2 and make such records available for scrutiny by the State Pollution Control Board or the Pollution Control Committee.

Responsibilities of dismantler of items of e-Waste

Every dismantler shall,-

- Obtain authorization and registration from the State Pollution Control Board in accordance with the procedure under the rule - 9 and 11;
- Ensure that no damage is caused to the environment during storage and transportation of e-waste; ensure that the dismantling processes do not have any adverse effect on the health and the environment;
- Ensure that the facility and dismantling processes are in accordance with the standards or guidelines published by the Central Pollution Control Board from time to time;
- Ensure that dismantled e-wastes are segregated and sent to the registered recycling facilities for recovery of materials; ensure that non-recyclable / non-recoverable components are sent to authorized treatment storage and disposal facilities;
- File a return in Form 3, to the State Pollution Control Board or the Pollution Control Committee as the case may be, on or before 30th June following the financial year to which that return relates;
- Not process any e-waste for recovery or refining of materials, unless he is registered with State Pollution Control Board as the recycler for refining and recovery of materials.

Responsibilities of recycler/ re-processor of e-Waste

Every recycler shall obtain authorization and registration from the State Pollution Control Board in accordance with the procedure under the rule - 9 and 11;

- Ensure that the facility and recycling processes are in accordance with the standards laid down in the guidelines published by the Central Pollution Control Board from time to time;
- Make available all records to the Central or State Pollution Control Board or Pollution Control Committee of Union Territories for inspection;
- Ensure that residue generated thereof is disposed of in a hazardous waste treatment storage disposal facility;
- File a return in Form 3, to the State Pollution Control Board or the Pollution Control Committee as the case may be, on or before 30th June following the financial year to which that return relates.

CHAPTER 13

CONTROL OF AIR POLLUTION FROM MOBILE SOURCES AND STANDARDS APPLICABLE

13.1 BACKGROUND

There has been an alarming growth in transport sector all over the world and emission from this sector has increased adding to the environmental pollution. The transport sector is a big source of air pollutants. The transport sector includes mobile sources like cars, trucks, off- vehicles, engine driven boats / ships, diesel engine / diesel electric trains, aeroplanes, etc. Technological advancements in designs of all types of transports along with better upkeep and maintenance practices are essential for reduction of emissions. It may be mentioned that the transportation services also add to emission of noise and greenhouse gases besides the air pollutants.

Vehicular traffic is a major cause of air pollution, specially, in cities. It is much more complex to deal with emission from vehicles than emission from stationary sources. The presence of carbon monoxide, hydrocarbons and carbon particles in the smoke indicate incomplete combustion of fuel, and their increase would lower the efficiency and raise fuel consumption besides causing air pollution. Reducing the number of road vehicles by providing cheap transport with higher carrying capacities like buses and metro trains can reduce the air pollution. This would also reduce congestion on roads, number of road accidents and loss of working hours.

13.2 GENERAL MEASURES FOR EMISSION CONTROL FOR INLAND TRANSPORTATION SYSTEMS

Pollutants like particulate matter, oxides of nitrogen (nitric oxide & nitrogen dioxide), hydrocarbons, carbon monoxide and oxides of sulphur (due to the presence of sulphur in some of the fuels used) are present in the exhaust from land based transports which include road vehicles, trains, in- land river transportation system and off-road mobile machines (like cranes, earthmoving machines, tractors, etc.). The emission of hydrocarbons may consist of Volatile Organic Compounds (VOCs)/ Poly-aromatic Hydrocarbons (PAHs) /Nitro- PAHs, Poly-chlorinated Biphenyls (PCBs)/ benzene/ benzo(a)pyrene, etc. which are very toxic even in very small quantities. The following general measures may go a long way in reduction of overall emission from land based transport sector and also enhance public convenience

Project Planning and Implementation

- Augmentation of capacity of 'Railway tracks' and 'Railway system' in the country along with modern safety system.
- Laying new railway tracks for achieving higher train movements and covering those areas by railways which have not been hitherto covered. [This includes doubling the tracks]
- Electrification of railway tracks.

- Augmentation of electric supply system including electricity generation, transmission, and transformation capacities for railways.
- Development of 'Mass rapid transport system' for metro-cities and other big cities.
- Development of Inland River transport system equipped with anti pollution measures.
- Improvement in road infrastructure which would involve widening of roads and improving their condition, constructing new express ways, and national high ways, and also construction of new roads for convenient connection with rural areas of the country.
- Enhanced, efficient and convenient public road transport system instead of individual vehicles.
- Technological changes in vehicles and railway system including their prime movers to increase efficiency and to curb fuel consumption and emission of pollutants.
- Adopting modern technologies in railways and vehicle design with a view to reduce fuel consumption and suppression of emission of pollutants.
- Stoppage of production and use of leaded petrol / gasoline.
- Changeover to fuels and technology which can reduce fuel consumption and emission of pollutants. Changeover to less polluting fuel such as Compressed Natural Gas (CNG), in place of diesel oil after making necessary modification in the vehicles. Change over to hydrogen* as fuel can completely avoid pollution from the engines and efforts in this direction are constantly being made.
- Aggressive promotion of production and use of sulphur free diesel oil.
- Introducing electric drive for vehicles and utilisation of electricity generated through 'fuel cells' and / or 'renewable energy sources like solar or wind' as far as possible for their battery charging.
- Development of Solar powered Transports. {The efforts are going on world over to develop solar powered vehicles. Solar panels are mounted on the vehicle to generate electricity needed for the purpose. Such vehicles would have power electronics and batteries, electrical motor, controller, etc. in place of engine. The future for such non-polluting and non-carbon dioxide emitting solar-powered vehicles is very bright. Adopting solar powered vehicles would immensely benefit the environment}.
- Plantation on all vacant railway land along the railway tracks, and plantation along all roads.

Note: *Hydrogen As Fuel For Vehicles To Reduce Emission — It is well known that hydrogen when used as fuel does not emit any pollutant. Hydrogen can be used as fuel for transport vehicles. In fuel cell powered vehicles, the internal combustion engines are replaced by 'Fuel cell stacks' as prime source of power. Blending hydrogen (up to 30%) in Compressed Natural Gas (CNG) to have on- hand experience was taken up as a 'project' in select vehicles (buses, cars, three- wheelers) by Ministry of Transport through Society of Indian Automobile Manufacturers (SIAM) and Indian Oil Company Limited.*

Examples of Fuel Cell Powered Vehicles:

As already mentioned earlier, the use of hydrogen as fuel via fuel cells pushes up the efficiency to around 54 %. In Brno, Czech Republic, a vehicle 'Mercedes-Benz O530 Citaro' powered by hydrogen fuel cells is being used. Several Automobile manufactures have committed to develop vehicles using hydrogen..

In India, Banaras Hindu University (BHU), Varanasi, developed motor cycles for demonstration which are fuelled by hydrogen stored in tanks as metal hydride.

UK started a fuel cell pilot program in January 2004; the program ran two Fuel cell buses. Western Australia's Department of Planning and Infrastructure has been operating fuel cell buses on trial in Perth as part of its sustainable transport. The Hydrogen Expedition is developing a hydrogen fuel cell-powered ship and using it to circumnavigate the globe, as a way to demonstrate the capability of hydrogen fuel cells.

The 'Hydrogen based fuel cell vehicles' are being developed by some automobile manufactures who have tied up with manufacturers of fuel cells. As an example **Table 13.1a** gives the name of company, name of vehicle, and state/type of hydrogen fuel used. The fuel cell based vehicles may take time to get commercialized because of the cost factor.

Table 13.1a: Examples of Hydrogen Fuel Cell Powered Vehicles

Name of company	Name of vehicle	Type of hydrogen fuel
Daimler- Chrysler	NECR 4	Liquid hydrogen
Daimler- Chrysler	NECR 4a	Compressed hydrogen
Daimler- Chrysler	NECR 4 and Commander Jeep	Menthol (which may be obtained by converting hydrogen to menthol)
Daimler- Chrysler	Chrysler Natrium	Hydrogen stored in metal hydrides / Sodium boro-hydride
Toyota	FCHV (Kluger V)	Compressed hydrogen
Honda	FCX	Compressed hydrogen
General Motors	Hydro Gen 3	Compressed hydrogen
Ford	Focus FCV (Hybrid)	Compressed hydrogen

Maintenance and operational measures

- Careful operation of transports observing all safety and pollution norms and instructions of manufacturers.
- Routine maintenance along with diagnostic check up and thorough inspection of vehicles and railway system including track, locomotive, passenger bogies, wagons, all electrical and mechanical components/ equipments /and systems, all electronics, safety equipment/ devices and systems etc.
- Well planned overhauling of vehicles and railway system using good quality spares and components and ensuring a good 'Works monitoring'.
- Ensuring application of energy and material conservation measures.
- Ensuring improvement in traffic management on roads so that the 'Traffic jams' do not occur and vehicles can move with reasonable speed. (The traffic jams and low speeds of vehicles cause higher fuel consumption and higher amount of carbon emissions and pollutants.)
- Proper maintenance of roads,, clearing 'road- side encroachments' for smooth movement of traffic.
- Travelling at moderate and steady speed and maintaining proper air pressure in the vehicle tyres. (Such measures are also helpful in reducing the fuel consumption).
- Checking adulteration of fuel.
- Utilization of suitable percentage of biofuel for the engines.
- Observing that there is a complete combustion of injected fuel
- Observing that the lubricating oil/ engine oil consumption does not increase.
- Ensuring NO_x emission control by: Lower temperature of intake air; and Retarded combustion; etc.

- Ensuring particulate emission control by: High injection pressure; Fine fuel atomization; Intensive air motion; High excess air; Low consumption of lubricating oil, etc. Also provide 'Oxygen sensors'

To materialize the above works, it would require various positive policy initiatives and planning (both technical as well as financial) by the government, railway administration, public works departments (in-charge of roads), and related public and private authorities in-charge of transport system.

Economic Eco-friendly Measures regarding Vehicles

The following economic eco-friendly measures if taken can lower the pollution from vehicular traffic.

(i) As the emission from old vehicles is very high and their efficiency is lower compared to new vehicles, the regulations can require for the phasing out of old vehicles after a certain time limit and giving consumer enough incentive to change his old vehicle with a new one by giving credits at lower interest rates.

(ii) Provision of credits at lower interest rates to 'Fuel refiner' for producing clean fuel.

(iii) Discouragement for low efficiency and for exceeding emission limits for engines by way of penal action and fines.

(iv) Blocking production of lower efficiency vehicle/ engines and tightening the emission regulations and their implementation.

(v) Provision of subsidies to buyers for buying clean vehicles.

(vi) Independent authority may test the vehicles at manufacture's works.

(vii) Emission testing should be implemented strictly

(viii) Recycling of solvents (which are used during maintenance/ replacement of lubricating oil of transports) after refining.

(ix) Old lubricating oil taken out from vehicle may be utilized for purposes other than lubrication.

Mass Transportation for Reduction of Air Pollution

The fuel efficiency of new modern vehicles is much higher than previously manufactured vehicles (like cars, buses, trucks), but considering their carrying capacity in terms of passengers and material, the pollutants including greenhouse gases generated by them per passenger kilometre or per tonne kilometre is much more than that of the mass transport system like trains / railways.

Similarly buses/ trucks would generate lesser pollutants including greenhouse gases per passenger kilometres or per tonne kilometre than the individual cars /small carriers. It is therefore necessary to develop a convenient public mass transport system. It may also be kept in view that the fuel economy and environment pollution by a vehicle also depends on condition of roads apart from its own condition and efficiency.

Electric trains/ metro trains can be run faster to conveniently clear the traffic in cities. Also, the electricity generated for 'electrified railways system' would normally come from big power stations which may generate net lower or same pollution per passenger kilometre or per tonne kilometre than the diesel engine/ diesel-electric engine run trains depending on the mix of electricity generation by hydro, nuclear, thermal, and renewable sources. It is better to electrify the railway tracks. The reduction of vehicular transport because of providing metro train services would definitely reduce pollution in the crowded areas of cities.

It is therefore advisable to introduce more and more railway transportation and make it more convenient and 'Service efficient' so that the public uses railways both for passenger as well as goods transport in preference to using cars and trucks.

13.3 POLLUTION CONTROL IN AUTOMOBILES

13.3.1 Causes of Pollution by Automobiles

There is a large contribution of air pollution in cities from old and inefficient vehicles. The majority of the automobiles are driven by internal combustion engines with gasoline or diesel fuels which are mixtures of various hydrocarbons (compounds of hydrogen and carbon).

If a pure liquid hydrocarbon fuel were mixed with the perfect amount of oxygen and completely burned, the only by-products would be carbon dioxide (CO_2), and water/water vapour, but in practice, the combustion of a pure hydrocarbon fuel in a typical engine (as opposed to ideal combustion situation) is more like the following:

'Pure liquid hydrocarbon fuel' + Air = CO + CO_2 + NO_x + Water vapour + Unburnt carbon in particulate matter + hydrocarbons + balance oxygen and nitrogen.

The exhaust from a typical gasoline engine would be similar as working with pure liquid hydrocarbon with a few variations because some impurities or additions (like lead compounds) which add to the toxicity of exhaust.

The source of formation of oxides of nitrogen (NO_x) is the 'combustion air which has both nitrogen and oxygen in it' and also the nitrogen in fuel (gasoline and diesel oil) which may be present as an impurity. As the hydrocarbons are not completely combusted in the process, they get exhausted through the tail pipe along with carbon monoxide (CO), carbon dioxide CO_2, Particulate Matter (PM) and water vapour or steam. The diesel oil has some amount of sulphur in it and therefore its combustion gases would contain sulphur dioxide also in addition to PM, NO_x, CO, CO_2, steam, etc.

The CO and hydrocarbon emission are lower in diesel engine, but smoke/ PM is quite high as compared to gasoline engine. The emission of NO_x (which does not include nitrous oxide) would be high in case of diesel engine because of higher temperature during combustion as compared to gasoline engine. It may be pointed out that higher the temperature, higher would be NO_x emission, but the PM would be lower and efficiency would be higher. High speed diesel oil contains about 0.05 % of sulphur and hence diesel exhaust contains sulphur dioxide, while gasoline does not contain sulphur and there is no sulphur dioxide in its combustion gases.

Apart from pollutants in 'Exhaust' from gasoline engines, there is emission of pollutants from crank case / blow by (which may consist of mainly un-burnt or partially burnt hydrocarbons), and evaporations of fuel from fuel system. With the increase of wear on piston rings and on the piston, the 'Blow-by' increases. The evaporative emission can take place while filling petrol in the car (when the nozzle is little away from inlet passage, or the auto shut off on tank full does not properly work). When the car is running, its hot engine exhaust system vaporizes the fuel; and the evaporation continues after the car is turned off, until the engine is cooled down (this is called a "hot soak"). This means that evaporative emissions do not stop even after turning off the car engine and the evaporative emissions may take place even from parked cars.

The exhaust from automobile may also have 'Aldehydes, acetic acid, beryllium, cadmium (which is used in production of tetraethyl lead gasoline), barium (which is sometimes added as smoke suppression additive in diesel fuel), phosphorous (which is sometimes added to fuel as corrosion inhibitor), etc.'

The hydrocarbon content of flue gases during braking is highest and therefore, frequent braking and accelerating is wasteful. In cities, there are a large number of automobiles and together they emit large quantities of carbon dioxide (CO_2), carbon monoxide (CO), and other pollutants like hydrocarbons / organic compounds, NO_X, etc. Under certain conditions of atmosphere, temperature and sunlight the hydrocarbons, organic compounds, and NO_X take part in photo-chemical reaction to create 'Photo chemical smog' which have 'Ozone', and some quantities of 'Peroxyacyl Nitrate (PAN)', 'Aldehydes', etc. These products are very harmful to agriculture / vegetation, and human and animal health, specially eyes and respiratory system organs. (Also see **Chapter 2, Para 2.14**).

The modern technology of vehicles and the engines (both gasoline and diesel fuelled) is very advanced and new vehicles when tested are found to abide by the latest emission Standards / Norms regarding emission levels. The crank case ventilation system and vapour recovery system is used in modern vehicles which recycles the 'Blow-by' gases and vapours to engine intake. Oxygen sensors are also used.

The use of catalytic converters/ catalytic reactors is being done nowadays to remove pollutants. They break up NO_X into N and O before they leave the car. Emphasis is being laid now on cleaner petrol with oxidizing agents (converting CO to CO_2) and a new generation of Low Emission Vehicles (LEVs). The dual fuel cars which can use petrol and the Compressed Natural Gas (CNG) which is less polluting are now on the market, but their price range is higher. Prohibition of use of leaded petrol or completely banning its production is an effective measure taken in most countries against the emission of lead compounds in the smoke.

Reduction of sulphur in the fuel of internal combustion engines can reduce sulphur dioxide in exhaust which would in turn lower air pollution and also corrosion of engine parts thereby improving overall performance and life of engine. The fuel refiners are therefore being urged to reduce sulphur in fuel produced by them for internal combustion engines. There are three ways to control emission of hydrocarbons and carbon monoxide

(i) Promoting more complete combustion of fuel
(ii) Reintroducing hydrocarbons emission back into engine for combustion
(iii) Providing 'Catalytic converter' at the exhaust end where further oxidation of products of combustion takes place and the objectionable contents of exhaust gases can be minimized.

It may be pointed out that fuel consumption not only depends on engine, but also on all working parts / components of the vehicle / lubrication / cooling / speed / operation / maintenance and even on some non-working parts.

13.3.2 SI ENGINE-DRIVEN AUTOMOBILES—POLLUTION CONTROL

The following methods of pollution control are used in modern day spark Ignition engine driven vehicles.

Crank Case Ventilation System

The vapours produced in the crankcase during the normal combustion process are redirected into the air/fuel intake system by the 'Crank case ventilation system' to be burned during combustion. The Positive Crankcase Ventilation valve (PCV valve) controls the quantity of vapours admitted to the air/ fuel mixture intake system so that engine performance is not adversely affected.

On idle running, the air/fuel mixture is very critical and therefore very small quantity of these vapours is admitted into the intake system. More vapours are allowed into intake system by PCV valve at higher engine speeds as the air / fuel mixture is then not so critical and the pressures in the engine are higher.

Fuel Evaporative Control System

Gasoline evaporates quite easily. About 20% of all hydrocarbon (HC) emissions from the automobile are from the gasoline tank. Previously these evaporative emissions were vented into the atmosphere, but now a fuel evaporative control system has been introduced which traps and stores the 'evaporative emissions' from the gas tank and carburetor. Charcoal is used as an adsorbent for trapping the fuel vapours which adhere to the charcoal, until the engine is started. When engine starts, the engine vacuum can be used to draw the vapours from the adsorbent medium into the engine, so that they can be burned along with the fuel/air mixture. This system has a redesigned gasoline tank with sealed cap and having space for vapours to collect so that the vapours can be vented to the charcoal adsorbent put in a small container. A purge valve is used to control the vapour flow into the engine. The purge valve is operated by engine vacuum. The charcoal container is provided with a filter which when clogged is required to be replaced.

Air Injection System with Diverter Valve

If air is injected into the exhaust manifold through a valve (called 'diverter valve'), the unburnt matter in the hot exhaust gases (comprising mainly of hydrocarbons, carbon monoxide and carbon particles) would get ignited / oxidized producing carbon dioxide and water vapour/ steam and also waste heat not producing any power. To control the excessive heating / explosion / popping in the exhaust manifold under the conditions of high unburnt matter in the exhaust gases (such as under deceleration), the 'diverter valve' diverts the injection air from entering the exhaust manifold.

Catalytic Converter

The catalytic converter is located in the exhaust system ahead of the muffler. The converter has pellets or a honeycomb made of platinum or palladium inside it. The platinum / palladium material acts as a catalyst to speed up a chemical oxidation of hydrocarbons and carbon monoxide in the exhaust as they pass over it forming carbon dioxide and water vapour.

The dirtier the exhaust, the more heat is developed and if the temperature rise is very high, the converter can be seen to glow from excessive heat and in such cases, it may destroy itself. It is therefore necessary that the vehicle is maintained properly so that the exhaust is not so dirty.

The leaded fuel (which has been phased out in most countries) would put a coating on the platinum or palladium and render the converter ineffective.

Fuel System Design Modification

Carburetor is not used nowadays and instead of this a 'Multi-port fuel injection system' which is more efficient is used. The electronic engine management for accurately regulating fuel supply to cylinders by sensing various engine parameters is very helpful in obtaining a good performance of the engine. The 'turbo-charged (TC)' engines are better. After treatment of the exhaust gases using catalytic converter, exhaust gas recycling may be introduced.

Some good technologies like 'Lean burn technology' with lean air / fuel ratio, high turbulence in combustion chamber', and 'Variable Valve Activation (VVA) providing improved charge control' are emerging. Use of improved materials for various components of engines like piston, piston rings, valves, blades of turbochargers, etc. is being done to improve performance of the engine.

Exhaust Gas Recirculation (EGR)

The production of oxides of nitrogen (NO_X) rises with higher combustion temperature. Lowering the combustion temperature can reduce the NO_x in exhaust. Recirculation / recycling of some portion of exhaust gases into the intake system can dilute the air / fuel mixture and lower the combustion chamber temperature. This is achieved by Exhaust Gas Recirculation valve (EGR valve).

The exhaust gas recirculation together with EGR valve is very effective method of controlling oxides of nitrogen, but it adversely affects engine performance and therefore the amount of exhaust entering the intake system has to be carefully monitored and controlled which may be accomplished through a series of electrical and vacuum switches and the vehicle computer. The system should not allow exhaust gas recirculation at cold condition of engine or at a time when the engine needs full power.

13.3.3 CI Engine /Diesel Engine-Driven Vehicles—Pollution Control

The Compression Ignition engines (CI engines or Diesel engines) are more efficient than the conventional spark ignition engines from the point of view of thermodynamics, but they have large negative environmental potential also because the higher air to fuel ratio in these engines provides a perfect environment of relatively high temperatures and abundant oxygen for the production of NO_x and in addition, direct injection of fuel (as used here) into the combustion chamber creates rich fuel pockets which can cause the formation of particulate matter (soot).

13.3.3.1 *Design Modifications*

The reduction in emissions and higher efficiency can be brought about by some design modifications which include the following:

- Improvement in fuel injection system and use of higher injection pressure Common rail system unit injections' instead of 'multi-cylinder fuel injection pumps'
- Electronically controlled injection system' to provide 'variable injection timing with good dynamic response to engine load, speed, and temperature
- Improved cylinder head design, inlet port and re-entrant combustion chambers'
- 4-Valve system to improve volumetric efficiency and provide better mixing of fuel and air'
- Turbo-charged engines and 'Turbo-charged after-cooled engines' to provide higher specific power, better fuel economy, with lesser emissions'
- With the change in materials of construction of engine and crankcase by strong and lighter materials like 'Special Aluminium alloys (instead of cast iron or steel)' and 'Thermally sprayed special coating' of engine cylinder or liner, it is possible to design vehicle engine which have lighter weight, and lower friction where by reducing fuel consumption and carbon dioxide emission.

The Selective Catalytic Reduction (SCR) technology can be introduced in vehicles also for NO_x reduction [Similar technology is in use for stationary sources of NO_x emission]. Provision of 'Diesel Oxidation Catalyst (DOC)' can eliminate the carbon monoxide (CO) and hydrocarbons (HC). After-treatment, 'Particulate traps', 'Catalytic converters' and 'Selected Catalytic Reactors (SCR for De-NO_x)', if provided, would reduce pollution. Diesel particulate filter (DPF) cleans exhaust gas by forcing the gas to flow through the filter.

13.3.3.2 *Diesel Particulate Filters*

Diesel particulate filters i.e., particulate filters at the exhaust end of diesel engines are highly effective in the elimination of Particulate Matter (PM10) or soot from diesel exhaust. Variety of filter coatings and filter designs are used depending on the engine application and duty cycle. The following types of diesel particulate filter technologies are in the market.

(i) Cordierite Wall Flow Filters

The most common filter is made of 'Cordierite' (a ceramic material that is also used as catalytic converter supports / core). Cordierite filters provide excellent filtration efficiency, are (relatively) inexpensive, and have thermal properties that make packaging them simple for installation in the vehicle. The major drawback is that cordierite has a relatively low melting point (about 1200°C) and cordierite substrates are known to melt down during filter regeneration. This may happen if the filter has become loaded more heavily than usual, particularly with passive systems, but this problem is rare with active systems, unless there is a system breakdown.

Cordierite filter cores look like catalytic converter cores that have had alternate channels plugged - the plugs force the exhaust gas flow through the wall and the particulate collects on the inlet face.

(ii) Silicon Carbide Wall Flow Filters

The second most popular filter material is silicon carbide (SiC). It has a higher melting point (2700°C) than 'Cordierite' ; however it is not as stable thermally. Small silicon carbide cores are made of single pieces, while larger cores are made in segments, which are separated by special cement so that heat expansion of the core will be taken up by the cement, and not the package. The SiC cores are usually more expensive than cordierite cores, however they are manufactured in similar sizes, and one can often be used to replace the other. Silicon carbide filter cores also look like catalytic converter cores that have had alternate channels plugged ; the plugs force the exhaust gas flow through the wall and the particulate collects on the inlet face.

The characteristics of the wall flow diesel particulate filter substrate are as follows:

(a) Broad band filtration (diameters of the filtered particles are 0.2–150 μm);

(b) High filtration efficiency (can be up to 95%);

(c) Good refractory;

(d) Good mechanical properties;

(e) High melting point.

(iii) Ceramic Fiber Filters

Fibrous ceramic filters are made from several different types of ceramic fibers that are mixed together to form a porous media. This media can be formed into almost any shape and can be customized to suit various applications. The porosity can be controlled in order to produce 'High flow and lower efficiency' or 'Lower volume filtration and high efficiency'. Fibrous filters have an advantage over wall flow design of producing lower back pressure. Ceramic filters remove almost completely the carbon particulates, including fine particulates of less than 100 nanometers (nm) diameter with an efficiency of >95% in mass and >99% in number of particles over a wide range of engine operating conditions.

(iv) Metal Fiber Flow through Filters

Some cores are made from metal fibers—generally the fibers are "woven" into a monolith. Such cores have the advantage that an electrical current can be passed through the monolith to heat the core for regeneration purposes, allowing the filter to regenerate at low exhaust temperatures and/or low exhaust flow rates. Metal fiber cores tend to be more expensive than cordierite or silicon carbide cores, and generally not interchangeable with them.

(v) Disposable Paper Core Filter

Disposable paper cores are used in certain specialty applications, without a regeneration strategy. Coal mines are common users — the exhaust gas is usually first passed through a water trap to cool it, and then through the filter. Paper filters are also used when a diesel machine is to be used indoors for short periods of time, such as on a forklift being used to install equipment inside a store.

13.3.3.3 *Regeneration of Diesel Particulate Filters (DPF)*

Since the continuous flow of soot into the filter would eventually block it, it is necessary to 'regenerate' the filter by burning-off the collected particulate matter on a regular basis. Soot particulates burn-off to form water vapour and CO_2 in small quantity which is less than 0.05% of the CO_2 emitted by the engine.

Filters require more maintenance than catalytic converters. Regeneration can be done either passively (from the engine's exhaust heat in normal operation or by adding a catalyst to the filter) or actively introducing very high heat into the exhaust system. On-board active filter management can use a variety of strategies:

(i) Engine management to increase exhaust temperature through late fuel injection or injection during the exhaust stroke.

(ii) Use of a fuel borne catalyst to reduce soot burn-out temperature.

(iii) A fuel burner after the turbocharger to increase the exhaust temperature.

(iv) A catalytic oxidizer to increase the exhaust temperature, with after- injection (hydrocarbon-doser).

(v) Resistance heating coils to increase the exhaust temperature.

(vi) Microwave energy to increase the particulate temperature.

All on-board active systems use extra fuel, whether for heating the DPF, or providing extra power to the DPF's electrical system (for burning the soot), although the use of a fuel borne catalyst reduces the energy required very significantly. Typically a computer monitors one or more sensors that measure back pressure and/or temperature, and based on pre-programmed set points, the computer makes decisions on when to activate the regeneration cycle. The additional fuel can be supplied by a metering pump. Running the cycle too often while keeping the back pressure in the exhaust system low will result in high fuel consumption. Not running the regeneration cycle soon enough, increases the risk of engine damage and/or uncontrolled regeneration (thermal run away) and possible filter failure.

Diesel exhaust particulate matter burns when temperatures above 600°C are attained. This temperature can be reduced to somewhere in the range of 350 to 450°C by use of a fuel borne catalyst. The actual temperature of soot burn-out will depend on the chemistry employed. The start of combustion causes a further increase in temperature. In some cases, in the absence of a fuel borne catalyst, the combustion of the particulate matter can raise temperatures above the structural integrity threshold of the filter material, which can cause catastrophic failure of the substrate. Various strategies have been developed to limit this possibility. Note that unlike a spark-ignited engine, which typically has very low oxygen level in the exhaust gas stream before the emission control device(s), diesel engines have a very high ratio of oxygen available. While the amount of available oxygen makes fast regeneration of a filter possible, it also contributes to runaway regeneration problems.

Some applications use off-board regeneration. Off-board regeneration requires operator intervention (i.e. the machine is either plugged into a wall/floor mounted regeneration station, or the filter is removed from the machine and placed in the regeneration station). Off-board regeneration is not suitable for on-road vehicles, except in situations where the vehicles are parked in a central depot when not in use. Off-board regeneration is mainly used in industrial and mining applications. Coal mines (with the attendant explosion risk from coal damp*) use off-board regeneration if non-disposable filters are installed, with the regeneration stations sited in an area where burn off is allowed. [* coal damp—the mist, vapour, moist air, any gas other than air in coal mines].

Rhone – Poulenc Technology for Self-cleaning of Filters

This technology uses a rare earth called cerium (Lanthanide series rare earth metals, Ce, Atomic No.58, Atomic weight 140.12), which is made into a solution to blend with diesel oil easily. This additive coats the particles during their formation in combustion chamber. These particles are then trapped on the porous sides of ceramic filter fitted in the exhaust pipe. Due to catalytic action of cerium, the fusion temperature of these particles is lowered from 500–600°C to 200–250°C and therefore these particles are burnt up by the exhaust fumes which reach temperatures of 400°C when the vehicle accelerates. The ceramic filter thus gets rid of smoke particles ie it is 'self-cleaned'. This technology was first tested by a Paris transport company (RATP) on a bus.

Scrubbers for Diesel Exhaust

Water scrubbing can remove the Suspended Particulate Matter (SPM) along with soluble components of exhaust gases. This technology is very simple and effective to remove particulate matter and part of the gaseous pollutants, but is only convenient to be used for stationary diesel engines or and for diesel engines of ships.

Smoke Suppressant Additives for Diesel Fuel

Smoke suppression is possible by a number of additives. Some of the proprietary additives used for smoke reduction in CI engines are:

HYDRAX ATH (hydrated alumina); HYDRAMAX (magnesium hydroxides and hydroxy-carbonates); CHARMAX LS (low smoke); CHARMAX LS ZST and LS ZHS (zinc stannates and zinc hydroxystannates), CHARMAX AOM and MO (ammonium octamolybdate and molybdic oxide), CHARMAX ZB200 and ZB400 (zinc, magnesium, and calcium borates).

13.3.3.4 *Reduction of NO_x Emission from Diesel Engines*

NO_x Reduction based on Lowering Combustion Temperature

The emission of NO_x in engine exhaust can be controlled by the following methods which are based on reduction of combustion temperature:

(i) Exhaust Gas Recycling

A part of exhaust gas (say 15%) is recycled into the intake thus diluting the fresh charge and lowering the combustion temperature and therefore NO_x production. The exhaust gas recirculation in combination with turbo-charging is one of the important systems for diesel engines for reduction of emission. However the exhaust recycling lowers the volumetric efficiency, therefore the maximum output developed by the engine would be slightly lowered and may lead to some adverse effect on drivability of the vehicle.

(ii) Water Injection

The water or moisture injection along with fuel or combustion air reduces the combustion temperature and thus reduces the NO_x production. At richer fuel air ratio this method of reducing NO_x is more effective than at leaner fuel air ratio.

(iii) Modified Recycling with Exhaust Manifold Reactor

Installing exhaust manifold reactors along with recycling of a part of the exhaust gases can lower down the emissions. Injection of some air improves reactor performance and ensures good oxidation of carbon monoxide and hydrocarbons. This would however lead to occasional drivability problems in the vehicle.

(iv) Lean Fuel Air Ratio Operation

At the lean fuel air ratio operation, the combustion temperature is lower than when the engine is operated on normal fuel air ratio and therefore smaller amount of NO_x is generated. With such an operation of engine some drivability problem may arise during transient conditions.

Selective Catalytic Reduction of NO_x

Selective Catalytic Reduction of NO_x (generally abbreviated as SCR / $DeNO_x$) is a very good technology to reduce the NO_x emission and fuel consumption of truck and passenger car diesel engines. Euro IV regulations have been introduced by the European truck manufacturers. This technology greatly enhances the control of NO_x under the iimits imposed by the regulations.

In this technology, a 32.5% aqueous urea solution is injected upstream of the 'Catalyst'. Urea actually converts to NH_3 (ammonia) in the hot exhaust gases which reacts with NO_x to form harmless N_2 and H_2O. The urea quantity needs to be precisely dosed as a function of the engine NO_x output and the 'Catalyst' operating conditions.

Catalytic reactors in the exhaust path can cut 'NO_x emission' substantially by 'reduction of NO_x to nitrogen'. For this however presence of some CO (say1.5%) in the exhaust gases is essential. The CO acts as a reducing agent in presence of 'Catalyst' like 'copper – cobalt – alumina'.

Exhaust after treatment of exhaust gases typically consists of metal housing, ceramic monolithic substrate, and 'mounting- mats'. The 'mounting-mats' are placed between ceramic monoliths and metal housing to avoid damage to monoliths and for preventing exhaust gas-bypass. To comply with tighter emission US and European regulations, the manufacturers use SCR (Selective Catalytic Reduction) for NOx reduction on their diesel vehicles. For vehicles to be operated in very cold climate, electrically heated supply line and connector for urea/ ammonia are used.

13.3.3.5 *Reduction of CO and Hydrocarbons from Diesel Engine Exhaust*

Diesel Oxidation Catalyst (DOC) Converter

DOC is an 'Exhaust after treatment device' used for diesel engines. Palladium and platinum are used as catalyst to oxidize hydrocarbons (HC) and CO into carbon dioxide and water vapour as seen from the following reactions:

$$2CO + O_2 \rightarrow 2CO_2$$

$$[HC]^* + O_2 \rightarrow CO_2 + H_2O \text{ --------- [HC* represents hydrocarbon]}$$

Hydrocarbon Traps

Hydrocarbon trap if used in the intake air system can adsorb atmospheric hydrocarbons at vehicle shutdown and de-sorb them back into intake system upon vehicle start- up, thereby reducing evaporative emissions.

Notes: Ozone Destruction Catalyst System on Vehicle
The usual approach to pollution is to reduce emission at the tail pipe / exhaust stack. As a different approach, the 'Prem Air Catalyst system' (BASF, USA) uses a catalyst coating on vehicle radiator which converts any 'ozone' content in air to 'oxygen' as the ambient air passes over the coated surface of the radiator, thus reducing the ground level ozone concentration while the engine is working.

13.4 EMISSION STANDARDS FOR ENGINES AND TRANSPORT VEHICLES

13.4.1 Indian Standards for Emission of Smoke and Vapour etc. for Motor Vehicles

The limits imposed by Indian Environment protection Rules, 1986, as given in Schedule IV of rules (Standards for emission of smoke and vapour etc. for motor vehicles), are mentioned below:

- Idling CO (carbon monoxide) emission for four wheel petrol vehicle—3% by volume max.
- Idling CO emission limit for all petrol driven two wheeler and three wheelers—4.5 % by volume max.
- Maximum smoke density for all diesel driven vehicles at full load at speed of 60% to 70% to maximum engine speed—in (a) light absorption coefficient per meter 3.1 (b) Bosch units 5.2 (c) Hatridge units 75
- Maximum smoke density for all diesel driven vehicles at free acceleration—in (a) light absorption coefficient per meter 2.3 (b) Hatridge units 65
- For all diesel driven vehicles with effect from 1st April1992 (a) CO –14 g/kWh(max.) (b) Hydrocarbons in –3.5 g/kWh (max.). (c) NO_x –18 g/kWh (max.).

Approval Test for Vehicles:

Table 13.1b: For Petrol Light Duty Four-Wheel Vehicles Approval Type Test

Reference mass in kg	CO g / km	Hydrocarbons g / km
<1020	14.3	2.0
1020 – 1250	16.5	2.1
1250 – 1470	18.8	2.1
1470 – 1700	207	2.3
1700 – 1930	22.9	2.5
1930 – 2150	24.9	2.7
>2150	27.1	2.9

Table 13.2: For Petrol Light-Duty Four-Wheel Vehicles Conformity Production Tests

Reference mass in kg	CO g / km	Hydrocarbons g / km
<1020	17.3	2.7
1020 – 1250	19.7	2.7
1250 – 1470	22.5	2.8
1470 – 1700	24.9	3.0
1700 – 1930	27.6	3.3
1930 – 2150	29.9	3.5
>2150	32.6	3.7

(Note: Result must not exceed more than 10 %)

(a) Operating cycle for tests

Indian Environment protection Rules, 1986, Annexure II of Schedule IV gives details of the operating cycle used for the testing of vehicle comprising of idling, acceleration, steady speed, de-acceleration, etc. The values of acceleration/ de-acceleration, speed, duration, etc in the test cycle are given there.

(b) *'Reference Fuel' Characteristics for tests*

The annexure III of the same schedule gives the characteristics of 'Reference fuel' for type and production conformity test.

(c) *Limiting values of Opacity of Diesel Exhaust gases*

The limiting values of 'Opacity / absorption coefficient per meter' of 'Exhaust gases' for 'Diesel driven vehicles' for 'Engine tests' at 'Steady speed' are given in **Table 13.3** as specified in Annexure IV of Schedule IV. [The standards are becoming more and more stringent now.]

Table 13.3: Opacity Test of Diesel Vehicles Exhaust

Se. No.	Nominal flow of gas litre/s	Absorption coefficient per meter	Nominal flow of gas litre/ sec	Absorption coefficient per meter
1.	42	2.00	120	1.2
2.	45	1.91	125	1.17
3.	50	1.82	130	1.15
4.	55	1.75	135	1.31
5.	60	1.68	140	1.11
6.	65	1.61	145	1.09
7.	70	1.56	150	1.07
8.	75	1.5	155	1.05
9.	80	1.46	160	1.04
10.	85	1.41	165	1.02
11.	90	1.38	170	1.01
12.	95	1.34	175	1.00
13.	100	1.31	180	0.99
14.	105	1.27	185	0.97
15.	110	1.25	190	0.96
16.	115	1.22	195	0.95
17.			More than 200	0.93

13.4.2 Applicability of Euro Standards in India

It may be mentioned that now days in the manufacture and design of internal combustion Engines, 'Euro standards' are being mostly followed world over for automobiles. Euro standards have fixed limits for pollutants from different vehicle engines.

On October 6, 2003, the National Auto Fuel Policy was announced in India, which envisaged a phased program for introducing Euro 2 - 4 emission and fuel regulations by 2010. The implementation schedule of EU emission standards in India is summarized in **Table 13.4**

Table 13.4: Program Applicability of New Emission Standards in Indian Emission Standards (4-wheel Vehicles)

Standard	Reference	Date	Region
India 2000	Euro I	2000	Nationwide
Bharat Stage II	Euro II	2001 2003–04 2004–05	NCR*, Mumbai, Kolkata, Chennai NCR*, 10 Cities † Nationwide
Bharat Stage III	Euro III	4/2005 2004–10	NCR*, 10 Cities † Nationwide
Bharat Stage IV	Euro IV	4/2010	NCR*, 11 Cities †
		7/2015	Above plus29 cities in Haryana UP, Rajasthan, and Maharashra
		10/2015	Further, North India plus bordering districts of Rajasthan (9 states)
		4/2016	Further, Western India plus parts of South India (10 States and Union Territories)
		4/2017	Nationwide
Bharat Stage V	Euro V	-	-
Bharat Stage VI	Euro VI	4/2020	Intention to introduce nationwide from 4/2020

Note: *There is no program for introducing Bharat Stage V/Euro V.*

Abbreviations used in Tables NOx – Oxides of nitrogen, CO – Carbon monoxide, HC – Hydrocarbons, PM – Particulate matter

* National Capital Region (Delhi):

† Mumbai, Kolkata, Chennai, Bangalore, Hyderabad, Ahmedabad, Pune, Surat, Kanpur/. Secunderabad and Agra

For 2-and 3-wheelers, Bharat Stage II (Euro 2) is applicable from April 1, 2005 Bharart Stage III (Euro 3) has been introduced from April 1, 2010 (as per Table 13.4).

Bharat satage IV Standards for gasoline fueled 2-Wheelers came in force with effect from April 1st 2016. Bharat stage VI standards for SI CI run 2 and 3-Wheelers were proposed in Feb-2016.

The draft Report of CPCB dated March 2010 and up to date program of implementation of introducing Euro standards/ Bharat Stage III. IV and VI have been taken into consideration.

13.4.3 Technical Provisions under Euro Standards

It may be mentioned that certain norms of technical provisions for emission control of 'Spark Ignition (SI) engines' and 'Compression igninition (CI) engines /diesel engines' under Euro Standards have been set. These are briefly given in **Table 13.5** and **Table 13.6** respectively.

Table 13.5: Technical Provision Norms (Euro Standards)—Emission Control in SI Engines

Emission Norms	2/3 Wheelers		4 Wheelers
	2-Stroke	4-Stroke	4-Stroke
Euro I /India 2000	Intake, exhaust, Combustion optimization Catalytic converter	4-Stroke engine technology	Intake, exhaust, combustion optimization Carburetor optimization
Euro II / Bharat Stage II (B.S.II)	Secondary air injection Catalytic converter CNG / LPG (3 wheelers only)	Hot tube Secondary air injection CNG / LPG (3 wheelers only)	Fuel injection Catalytic converter Fixed EGR, Multi-valve CNG/LPG
EuroIII/ Bharat Stage III (BS.III)	Fuel injection Catalytic converter	Fuel injection Carburetor + Catalytic converter	Fuel injection +catalytic converter Variable EGR Variable valve timing Multi-valve On-board diagnostics system CNG/LPG
Euro IV / Bharat Stage IV (B.S.IV)		Lean burn Fuel injection+ catalytic converter	Direct cylinder injection Multi-Valve Catalytic converter On-board diagnostics system
Euro V B.S V	-		– Combustion improvements over Euro 4 – Faster oxygen sensors. – Catalysts improvements – oxygen storage capacity and better coatings
Euro VI B.S VI			– No changes required for port fuel-injected gasoline engines – Improvements to fuel injection timing or addition of a gasoline particle filter for gasoline direct injection (GDI) engines.

EGR—Exhaust Gas Recirculation;

Table 13.6: Technical Provision Norms (Euro Standards)—Emission Control in CI Engines

Emission Norms	Technology Options
Euro I / India 2000	(i) Retarded injection timing; (ii) Open/re-entrant bowl; (iii) Intake, exhaust and! combustion optimization; (iv) FIP~700-800 bar, low sac injectors; (v) High swirl; (vi) Naturally aspirated;

Euro II / Bharat Stage II (BS II)	(i) Turbo-charging; (ii) Injection pressure > 800 bar, moderate swirl; (iii) High pressure inline / rotary pumps, injection rate control; (iv) VO nozzles; (v) Re-entrant combustion chamber; (vi) Lube oil consumption control; (vii) Inter-cooling (optional, depends on specific power); (viii) EGR (may be required for high speed car engines); (ix) Conversion to CNG with catalytic converter;
Euro III / Bharat Stage III (BS.III)	(i) Multi valve; (ii) Low swirl – high injection pressure > 120 bar; (iii) Rotary pumps, pilot injection rate shaping; (iv) Electronic fuel injection; (v) Critical lube oil consumption control; (vi) Variable geometry turbocharger (VGT); (vii) Inter-Cooling; (viii) Oxycat and EGR(Exhaust Gas Recirculation); (ix) CNG/LPG as fuel; (x) High specific power output;
Euro IV / Bharat Stage IV (B.S.IV)	(i) Particulate trap; (ii) NO_x trap; (iii) On board Diagnostics system; (iv) Common rail injection-injection pressure >1600 bar; (v) Power by Fuel Cells; (vi) CNG/LPG as fuel;
Euro V B.S. V	– Combustion Improvement over Euro 4 – Variable fuel injection timing for DPF regeneration – Dicsel oxidation catalyst + Diesel Particulate Filter (DPF) – Some engines use Lean NO_x Traps (LNT)
Euro VI/ Bharat Stage VI (BS VI)	– Increased fuel injection pressure – Smaller and medium–size engies (< 2 liters) tend to use Diesel Oxidation Catalyst (DOC) + DPF and primarily LNT for NO_x Control – Larger engines for cars (>2L) use DOC + DPF + SCR. – Some manufacturers offer EGR-only NO_x control (with no after treatment control) and DOC + DPF on medium and larger cars.
VO- variable opening	

'Note: Government of India is planning to introduce the compliance of Bharat Stage VI/ Euro VI norms for all the vehicles to be manufactured in India. For Euro V and Euro VI norms See Appendix 13.'

13.4.4 Standard Emission Limits in India for New Internal Combustion Engines and Vehicles

The Emission standards (limits) in gm per km for light duty diesel engines are listed in **Table 13.7** The emission standards for diesel engines for trucks and buses are listed in **Table 13.8**. The emission standards for gasoline vehicles (<3500 kg) in gm/ km are given in **Table 13.9**. The emission standards for light duty diesel vehicles in g/km are given in **Table 13.10**.

Table 13.7: Emission Standards for Light Duty Diesel Engines in gm/kWh

Year	Reference	CO	HC	NOx	PM
1992	—	14.0	3.5	18.0	—
1996	—	11.20	2.40	14.4	—
2000	Euro I	4.5	1.1	8.0	0.36*
2005†	Euro II	4.0	1.1	7.0	0.15
2010	Euro III BS III	2.1-5.45	0.66-0.78	5.0	0-10
upto 4/2017	Euro IV	1.5-4.0	0.46-0.55	3.5	0.02-0.03

Table 13.8: Emission Limits for Diesel Engines for Buses & Trucks (>3500 kg) in g/kWh

Year	Reference	CO	HC	NOx	PMM
1992	—	17.3–32.6	2.7–3.7	—	—
1996	—	11.20	2.40	14.4	—
2000	Euro I	4.5	1.1	8.0	0.36*
2005†	Euro II	4.0	1.1	7.0	0.15
2010†	Euro III	2.1	0.66	5.0	0.10
upto 4/2017	Euro IV B.S IV	1.5	0.46	3.5	0.02**
2020	Euro VI BS VI	1.5	0.13	0.46	0.01**
* 0.612 for engines below 85 kW and smoke 0.8 m^{-1}; ** smoke 0.5 m^{-1}					

Table 13.9: Emission Standards for Gasoline Vehicles (<3500 kg) in g/kg

Year	Reference	CO	HC	HC+NOx
1991	—	14.3–27.1	2.0-2.9	—
1996	—	8.68–12.4	—	3.00–4.36
1998*	—	4.34–6.20	—	1.50–2.18
2000	Euro 1	2.72–6.90	—	0.97–1.70
2005	Euro 2	2.2–5.0	—	0.5–0.7
2010	Euro III B.S III	2.3	0.2	NO_x<0.15
upto 4/2017	Euro IV B.S IV	1.0	0.1	NO_x<0.08
2020	Euro VI B.S VI	1.0	0.1	NO_x0.06 PM 0.005
* *for catalytic converter fitted vehicles*				

Table 13.10: Emission Standards for light duty Diesel Vehicles in g / km

Year	Reference	CO	HC	HC+NOx	PM
1992	—	17.3–32.6	2.7–3.7	—	—
1996	—	5.0–9.0	—	2.0–4.0	—
2000	Euro 1	2.72–6.90	—	0.97–1.70	0.14–0.25
2005	Euro 2	1.0–1.5	—	0.7–1.2	0.08–0.17
2010	Euro III B.S III	0.64–0.98	NOx 0.5–0.78	0.56–0.86	0.05–0.1
upto 4/2017	Euro IV B.S IV	0.50–0.74	NO_x 0.25–0.39	0.30–0.46	0.025 – 0.06
2020	Euro VI B.SVI	0.5	NO_x 0.08	0.17	0.005

Tables 13.11 and 13.12 give the emission standards for petrol driven 3 wheelers and 2 wheelers respectively in g/km.

Table 13.11: Emission Standards for Petrol driven 3 Wheelers in g / km

Year	CO	HC	HC+ NOx
1991	12–30	8–12	—
1996	4.50	—	3.60
2000	2.00	—	2.00
2005/BS II	2.25	—	2.0
2014/BS II	1.25		1.25

Table 13.12: Emission Standards for Petrol driven 2 Wheelers in g / km

Year	CO	HC	HC+NOx
1991	12-30	8-12	—
1996	6.75	—	5.40
2000	4.00	—	2.00
2005/BS II	1.5	—	1.5
2010/BS III	1.0	—	1.0

Emission Limits for Heavy Duty Diesel Engines in Euro Standards

Table 13.13 gives the 'Emission limits' for 'Heavy Duty Diesel Engines' in g/ kWh under Euro I, II, III, IV, V, and US 2010 and Euro VI standards, while **Table 13.14** gives 'Emission limits' for 'Diesel engines for Trucks and Bus (Vehicles >3500 kg)' in g /kWh under Euro I, II, III, IV, V, VI Standards

Table 13.13: Emission Limits – Heavy Duty Diesel Engines in g /kWh (Euro and US 2010 Standards)

Pollutant	Euro I	Euro II	Euro III	Euro IV	Euro V	US 2010
NOx	8	7	<5	<3	<2	0 – 0.5
CO	4.5	4	2.1	1	1	1
HC	1.1	1.1	0.66	0.5	0.5	0.5
PM	0.36	0.15	<0.10	<0.10	0 – 0.3	0.03

Table 13.14: Emission Limits – Diesel Engines for Buses and Trucks (>3500 kg) in g/kWh (Euro Standards)

Standards	Carbon monoxide	Hydrocarbons	Oxides of nitrogen (NOx)	Particulate matter
Euro I	4.5	1.1	8.0	0.36
Euro II	4.0	1.1	7.0	0.15
Euro III	2.1	0.66	5.0	0.10
Euro IV	1.5	0.46	3.5	0.02
Euro VI	1.5	0.13	0.46	0.01

13.4.5 Emission Limits for Diesel Engines for Off-highway Vehicles/ Tractors/ Construction Equipments, etc.

13.4.5.1 *CMVR Emission limits in India for Off-road Diesel Vehicles and Equipments*

Limits as per Central Motor Vehicle Rules (CMVR) India are given in ***Table 14a***

Table 14a: Emisssion Limits as per CMVR India for Tractors/Construction Equipments (Effective from July 1999)

1. Carbon monoxide (CO) 2. Hydrocarbons (HC) 3. Oxides of nitrogen (NO_x) 4. Particulate Matter (PM)	14.0 g / kWh 3.5 g / kWh 18.0 g / kWh
With effect from April 2000, the CMVR norms for heavy duty diesel engines for off-road vehicles, tractors, construction equipments, etc. were made more stringent as mentioned below:	
(i) For Approval Tests with effect from April 2000:	
1. Carbon monoxide (CO) 2. Hydrocarbons (HC) 3. Oxides of nitrogen (NO_x) 4 Particulate Matter (PM)	4.5 g / kWh 1.1 g / kWh 8.0 g / kWh 0.36 g / kWh

(ii) For Conformity Test with effect from April 2000:	
1. Carbon monoxide (CO) 2. Hydrocarbons (HC) 3. Oxides of nitrogen (NOx 4 Particulate Matter (PM)	4.9 g / kWh 1.23 g / kWh 9.0 g / kWh 0.4 g / kWh

13.4.5.2 Off- road Vehicles — Euro and US Standards

Euro Off–road Stage 2 (COM2) Emission Levels for Diesel Engines for off- road, Tractors, Construction equipments as applicable (2011) are shown in **Table 13.15.** Emission limits in US EPA Tier 2 standards for Diesel engines capacity 19–75 kW for off-road 'Tractor, Construction Equipments' are applicable with effect from 2004.

Emission limits in US EPA Tier 3 standards for Diesel engines capacity 37–75 kW are applicable with effect from 2008 for 37–75 KW for off-road 'Tractor, Construction Equipments'. These limits for 75–130 KW diesel engines of such equipments are applicabie with effect from 2007. **Table 13.16** gives 'Tier 1, 2, 3 US standards for such Diesel engines.

Table 13.15: Euro Off Road Stage 2 (COM2) Standards for Diesel Engines

Engine power kW	Carbon monoxide (CO) g/kWh	Hydrocarbons g/kWh	Oxides of nitrogen g/kWh	Particulate Matter g/kWh
18–37	5.5	1.5	8.0	0.8
37–75	5.0	1.3	7.0	0.4
75–130	5.0	1.0	6.0	0.3
130–560	3.5	1.0	6.0	0.2

Table 13.16: US Standards for Diesel Engines (Tier 1, 2, 3) for Tractors and Off Road Equipments in g/kWh

Engine Power kW	Emission standard	Applicable wef. (year)	CO	HC	NOx	HC+ NO_x	PM
19–37	Tier 1	1999	5.5	---	---	9.5	0.80
19–37	Tier 2	2004	5.5	---	---	7.5	0.60
37–75	Tier 1	1998	--	---	9.20	---	---
37–75	Tier 2	2004	5.0	---	---	7.5	0.40
37–75	Tier 3	2008	5.0	---	---	4.7	---
75–130	Tier 1	1997	--	---	9.2	---	---
75–130	Tier 2	2003	5.0	---	---	6.6	0.30
75–130	Tier 3	2007	5.0	---	---	4.0	Being reviewed

Notes:

(1) *The Hydrocarbons taken into consideration are non-methane hydrocarbons.*

(2) *Engines of all sizes must also meet smoke standards of '20 % / 15 % / 50 % Opacity' at 'acceleration / lug / peak modes' respectively. For measuring 'Opacity'*

(a) *50 to 100 hp should be tested with 1.5" (38 mm) exhaust pipe.*

(b) *100 to 200 hp should be tested with 3.0" (76 mm) exhaust pipe.*

(c) *200 to 300 hp should be tested with 4.0" (102 mm) exhaust pipe.*

(d) *300 to 500 hp should be tested with 5.0" (127 mm) exhaust pipe.*

(e) *Engine rated over 500 hp should be tested with 6.0" (152 mm) exhaust pipe.*

Table 13.16a: Bharat Emission Standards for Non-road Diesel Engine Driven Machinery

Diesel Agricultural Machinery/Tracters					
Bharat(term) Stage III A	Power	Date	CO g/kWh	HC g/kWh	HC+NO_x g/kWh
	from < 8 to < 19 kW	4/2010	5.5	—	8.5
	19 to < 37 kW	4/2010	5.5	—	7.5
	37 to < 75 kW	4/2011	5.0	—	4.7
	75 to <130 kW	4/2011	5.0	—	4.0
	130 to < 560 kW	4/2011	3.5	—	4.0
Diesel Construction Machinery					
Bharat (CEV) Stage III	<8 kW	4/2011	8.0	—	7.5
	8 to < 19 kW	4/2011	6.6	—	7.5
	19 to <19 kW	4/2011	5.5	—	7.5
	37 to < 75 kW	4/2011	5.0	—	4.7
	75 to < 130 kW	4/2011	5.0	—	4.0
	130 to < 560 kW	4/2011	3.5	—	4.0

Note: *The above Bharat (term) stage IIIA and Bharat (CEV) Stage III Standards are now applicable in india. These are based on US Tier 2/3 requirement.*

13.4.5.3 Example of Diesel Engines for Off-highway Vehicles and Equipments Manufactured in India

The 'Diesel engines' for above mentioned purposes are being manufactured in India by various companies. M/s Kirloskar oil engines limited is one of the reputed manufacturers. They are making agricultural tractor diesel engines like Model: RV2, RV3, RB30, RB 33, 4R1040, 3R104. They are also making diesel engines for off- highway / construction equipments like Models: RB 44, RB44TC, HA494, HA694, 4R1040.

These models meet the CMVR emission limits, but only partially meet the 'Euro off stage 2 limits'. For example the emission level in actual performance of 4R1040 NA 76 hp/ 2200 rpm engine was reported as: CO—4.19; HC—1.04; NO_x—8.92; PM—0.78 (These results meet CMVR and also Euro off-stage 1 emission limits, but not that of Euro off-stage 2.

Cummins India, Caterpiliar India and Mahindra and Mahindra (India) are other manufacturers of diesel engines but new norms of emission are yet to be followed by them for off highway Vehicles, tractors and construction machinery / agricultural equipments).

13.4.6 Limits for Polycyclic Aromatic Hydrocarbons (PAH)

WHO Standard Limits /Guidelines for PAH in Environment

The Hydrocarbon (HC) exhaust gases from vehicles include Polycyclic Aromatic Hydrocarbons (PAH). The PAH has contents of 'Benzo a Pyrene' and 'Nitro- PAH' also. Most of PAHs are 'Carcinogens'. According WHO (1987), no safe level can be recommended for BaP due to its carcinogenicity. Complete removal of PAH from the environment is impossible, but they can only be controlled. The primary basis for setting up standards for PAH is the dose-effect relationship, fully demonstrated in PAH carcinogenesis in experiments on animals and the result extrapolated to man.

On the basis of such evaluation, the USSR Ministry of Health set up the maximum allowable concentration of BaP and the same has been recommended by FAO/WHO "Joint Expert Committee". The maximum allowable concentration of BaP in air in the proposal is '15 µg/100 m^3 for air of industrial working zones' and 0.1 µg/100 m^3 for 'the atmospheric air. The 'maximum allowable concentration'for the other PAHs apart from BaP is still to be decided. Integrated and long term monitoring is being carried out in the countries like Netherlands Sweden, and the United States for PAH.

The World Health Organization (WHO) has already added PAHs into the list of the priority pollutants in both air and water. France, Japan, Germany, Netherlands Sweden and Switzerland have prescribed emission standards for most of 'Hazardous Air Pollutants (HAPs)' including PAHs.

The WHO and the Netherlands (WHO, 1987 and OECD, 1995) have even prescribed ambient air quality guidelines (AQG) for the PAHs. As given in **Table 13.17a** [The guideline values for B(a)P corresponding to an excess lifetime cancer risk of 10-5 was estimated as 0.7mg/litre in Guideline for drinking water quality (WHO, 1998)].

Table 13.17a: Ambient Air Quality Standard for the PAHs (WHO and Netherlands)

Compound		Limit value	Guide value	Measuring Period
Netherlands—	PAH	5 ng/m^3	0.5 ng/m^3	Year as ng BAP/m^3
WHO-AQG	PAH		1.0 ng/m^{3-}	Year as ng BAP/m^3

Proposed Ambient Limits for PAH in India

An expert working group was constituted to propose 'Ambient Air Quality Standards for Benzene and BaP' for 2005 and 2010. The meeting of the expert working group on ambient air quality standards (Benzene and BaP) was held on March 27, 2002 at CPCB, Parivesh Bhawan, Delhi-32. The limits proposed are as given in **Table 13.17b.**

Table 13.17b: Proposed Limits for Benzene and B(a)P in India

Se.No.	Air Toxic Pollutants	Ambient Air Quality Standard upto 2005	Ambient Air Quality Standard upto 2010
1.	Benzene	10 μg/m^3*	5 μg/m^3
2.	Benzo(a)Pyrene, ie. B(a)P	5 ng/m^3**	1 ng/m^3

* Reducing by 1 μg/m^3 every year from 2005 to 2010 for Benzene.

** Reducing by 1 ng/m^3 every year from 2005 to 2010 for B(a)P.

Proposed ambient limits for PAH indicator parameter i.e. 'Benzo(a)pyrene' in India of 5 ng/m^3 (Ist January 2005) should be reduced to 1 ng/m^3 by Ist January 2010. Measurement of the concentration of Benzo (a)pyrene in ambient air and monitoring the other relevant PAH's at limited number of sites is needed. Other PAH compounds should include Benzo(a)pyrene, Benzo(b)flouranthene, Benzo(k)flouranthene, Indeno(cd)pyrene, Dibenzo(ah)anthracene, Benzo(ghi)perylene, Chrysene and Flouranthene.

New Proposed Limits of European Union for PAH in Atmosphere

Proposed new European Union ambient limits for PAH indicator parameter i.e. Benzo(a)pyrene is 6.0 ng/m^3. European Commission requires action plan to be developed to measure the concentration of Benzo(a)pyrene in ambient air and also to monitor other relevant PAH's at limited number of sites. Other PAH compounds should include Benzo (a)pyrene, Benzo(b)flouranthene, Benzo(i)flouranthene, Benzo(k)flouranthene, Indeno(cd) pyrene, Dibenzo(ah)anthracene and Flouranthene. European Commission directive also recommends to setup long term objective for containing the levels of these parameters.

13.5 CONTROL OF EMISSIONS FROM AVIATION

13.5.1 Air Pollution by Aviation

The major pollutants emitted by aircrafts are oxides of nitrogen, carbon monoxide, hydrocarbons and particulate matter/soot. Emission of noise* and fumes from aircrafts is also a big concern. [Regarding Aircraft noise and its effects see **Para 17.1.6, Chapter 17**]. The level of emissions depends on the type of aircraft. Emissions from aircraft are difficult to quantify once they are airborne and once they exceed 1000 m altitude from the ground because the effects of high altitude winds disperse the emissions. Below 1000 m, the air pollutants in exhaust gases are principally carbon dioxide, oxides of nitrogen, non-methane volatile organic compounds (hydrocarbons), and carbon monoxide. The fumes and smells consist of Volatile Organic Compounds (VOCs) which are also very prominent during fuel handling, filling and maintenance and during start up or shutting down. The small amount of sulphur in fuel would give rise to sulphur dioxide and some sulphate particles.

Because of complex chemical and photolytic reactions in the atmosphere, the pollutants in exhaust gases like oxides of nitrogen, hydrocarbons, may help formation of ozone and sometimes smog like condition in troposphere. (see **Chapters 1 & 2** for effects of Ozone and Smog).

The present day commercial aircraft fly at altitudes of 8–13 km. The high altitude aircraft releasing NO_x is a cause for protective ozone layer reduction in stratosphere. It is reported that pollutants from more than five million civil and nearly 2,800 military flights each year are also depleting 'Earth's protective ozone layer in stratosphere and the exhaust gases from air flight may be accounting for a significant amount of global greenhouse effect. (For details see **Chapter 20, Para 20.1**).

13.5.2 Measures to Reduce Emissions from Aviation

Continued progress in the areas of improvement of aircraft aerodynamics, reduction of weight of aircraft, traffic control, noise reduction, better maintenance, etc. is being carried out in the aviation sector.

Improvement of Aerodynamics

Technological developments for aerodynamic improvements include the design of winglets for wing tips (which reduce turbulence and vortex generation by the wings), laminar flow controls or systems for wing surfaces to reduce drag, improved manufacturing techniques to produce smoother surface, etc.

Weight Reduction

Development of new and improved metal alloys and composite materials* to reduce aircraft weight and to improve structural performance has been under way. Significant improvement of control systems has come about by replacing mechanical and hydraulic systems with electrical systems, which often reduce system weight while providing more precise control. See' Note on Use of Composites and Advanced Materials in Aircraft'.

Traffic Control Changes

Although airports, airlines, airplane manufactures aim to reduce carbon dioxide produced, but it might take some years. However it is reported that Governments in the Asia-Pacific region agreed to curb airline pollution through a range of measures such as reducing the number of planes circling airports as they wait to land. Australia had announced traffic control changes at its airports to minimize aviation fuel consumption.

Reduction of Noise Pollution

Noise** nuisance from aircraft can be reduced significantly by changing the way the planes come in to land. An international consortium has found that lining up with the runway as far as 70 km away and making a steady descent can more than halve the acoustic energy that reaches the ground. Newer models of commercial aircraft are quieter.

Maintenance and Pollution Prevention in Aviation

There is a guide entitled 'Aviation Unit 'Maintenance and Pollution Prevention Guide'. The pollution prevention guide ('P2'guide) has the purpose of identifying opportunities to reduce pollution generated by unit-level aviation maintenance and operations (AVUM). This guide also provides template calculations showing waste reduction estimates and economic analysis for each P2 opportunity. These calculations are based on material usages and waste generations and its disposal fees of an average light infantry aviation unit maintenance facility. Better and preventive maintenance and replacements reduce the carbon dioxide emission as well as other emissions.

**Note on Use of Composites and Advanced Materials in Aircraft:*

High-speed forward-swept-wing airplanes would not have been possible without the development of composite materials to keep their wings from bending out of shape.

Composites are the most important materials to be adapted for aviation since the use of aluminum began. Composites are materials are combinations of two or more organic or inorganic components. One material serves as a "matrix," which is the material that holds everything together, while the other material serves as reinforcement, in the form of fibres embedded in the matrix. Until recently, the most common matrix materials were "thermosetting" materials such as epoxy, bismaleimide, or polyimide. The reinforcing materials can be glass fibre, boron fibre, carbon fibre, or other more exotic mixtures. Fiberglass is the most common composite material, and consists of glass fibres embedded in a resin matrix.

A further development was a composite material like 'Boron fibre and graphite, embedded in epoxy resins'. Such materials were used in military aircrafts for some components. Composite materials can be both lightweight and strong.

Making a Structure of Composite Material

To make a composite structure, the composite material, in tape or fabric form, is laid out and put in a mould under heat and pressure. The resin matrix material flows and when the heat is removed, it solidifies. It can be formed into various shapes. In some cases, the fibres are wound tightly to increase strength. A very convenient feature of composite materials is that they can be layered, with the fibres in each layer running in a different direction which allows to design structures that behave in certain ways [such as to design a structure that will bend in one direction, but not in another].

Thermoplastic materials

Thermoplastics are a relatively new material that is replacing thermosets as the matrix material for composites and hold much promise for aviation applications. These materials components are easy to produce and are also more durable. They are tougher than thermosets, particularly for light impacts, such as when a small thing drops on it accidentally. The fall of a wrench could easily crack a thermoset material, but would bounce off a thermoplastic composite material.

Some Problems with the Use of Composite Materials

Despite the strength and low weight of composite materials, some problems in their use are:

- Composites are hard to inspect for flaws.
- Some of them absorb moisture.
- To make components of composite material is labour intensive
- Require complex and expensive fabrication machines
- Damaged components of composite material can not be repaired like that of aluminium and have to be completely replaced and therefore the maintenance is difficult and expensive.

Use of Advanced Materials in Aircraft

Because of the difficulties in maintenance, military aircraft use substantially greater percentages of composite materials than the commercial passenger aircraft primarily because of the different ways that commercial and military aircraft are maintained. Modern military aircraft use composites for at least a third of their structures, and some experts have predicted that future military aircraft will be more than two-thirds composite materials. However, modern commercial airliners are also using significant amounts of composites to achieve lighter weight (such as about ten percent of the structural weight of the Boeing 777 is composite material).

Aluminum still remains a very useful material for aircraft structures. Some better aluminum alloys have been developed. Aluminum-lithium alloy is the most successful of these alloys. It is approximately ten percent lighter than standard aluminum.

The adoption of Aluminum- lithium alloy by commercial aircraft manufacturers has been slower due to the expense of lithium and the greater difficulty of using aluminum-lithium (in particular, it requires much care during welding). But it is likely to become eventually, a widely used material for both commercial and military aircrafts.

13.5.3 Measures by Aircraft Manufacturers and Airlines for Limiting Emissions

Manufacturers are using lighter materials to build aircrafts and powering them with more fuel-efficient jet engines. Modifications of 'aerodynamics' of some types of aircrafts has been done to improve fuel efficiency. Air lines try to minimize weight on board encompassing everything from magazines to carpets.

Airlines always press air controllers and airports to shorten and simplify routes and landing approaches. The aim is not just to save the planet, but also to save jet fuel which is a major operating cost. Examples of some air lines and manufacturers which have taken steps in right direction are mentioned below:

(i) 'Airways New Zealand' has taken the following steps:

- Turn off auxiliary power units when the planes are on the ground and change over to ground power as quickly as possible.
- Engine wash program to clean blades, which can improve fuel efficiency.

(ii) 'AIR New Zealand' along with aircraft engine manufacturers has taken steps to use alternative fuels like bio fuels by introducing new technologies.

(iii) Airbus has also been working on the use of alternative fuels in aircraft.

(iv) 'Boeing' has taken the following steps:

- Use of Hydrogen fuel cells
- Investing in Algal fuels
- Making lighter planes

(v) Boeing have designed plane named Dreamliner for which, it is claimed that:

- Lighter composite materials make up nearly 50 percent of the plane,
- It can carry as many as 330 people, making it far lighter than other planes of the same size,
- 20 % more fuel-efficient producing 20% less emissions than aircrafts of the same size.

The aircraft has been produced commercially. but still requires certain improvements as per newspaper reports.

13.5.4 New Developments

New developments are taking place in the design of aircraft. A few examples are given below:

(i) *Airliner Based on Hydrogen as Fuel*

This hydrogen-burning hypersonic airliner with engine 'designated A2'can fly more than twice as fast as the fastest Concorde. Hydrogen burnt as fuel just produces water vapour and no carbon dioxide. European Union's Long-term Advanced Propulsion Concepts and Technologies project (LAPCAT) wanted an airliner that can fly from Brussels to Sydney in less than four hours. It is reported that the A2 engine was made keeping in mind the failures of its doomed supersonic predecessor, the Concorde. The A2 engine has two modes ie it is a combination of turbojet and ramjet propulsion system which would make it efficient both at lower speeds and high speeds and give it incredible speed capabilities.

HY4 became the worlds first passenger aircraft with an engine powered by a hydrogen fuel cell. Its first flight took place at stutlgart airport, Germauy on September 29th 2016, It had 9 kg hydrogen, 4 × 11 kW fuel cells and 2 × 20 kWh batteries.

(ii) *Green Jet Engine by Pratt & Whitney*

Pratt & Whitney has developed a geared turbofan engine which consumes 12 to 15% less fuel and emits lesser carbon dioxide than other engines. In other engines, the turbofan which sucks air into combustion chamber works at the same high speed as turbine, but in the engine developed by Pratt & Whitney, the turbo fan works at about one-third speed of turbine as it is a geared fan (not a direct driven fan). It is well known that fan efficiency is higher at low speeds and they make less noise, while turbines work best at higher speeds. The developed engine has higher efficiency, consumes less fuel, and emits lesser carbon dioxide and lesser noise than others. It is also claimed that it emits lesser oxides of nitrogen (NO_x) and it would cut ownership costs. Pratt & Whitney engine was thoroughly tested and more than 40,000 take offs and landings were simulated.

(iii) *Solar Powered Air planes*

World's only Swiss solar – powered single–seater aircraft, Solar Impulse (S12) set off from Abu Dhabi (Al Bateen airport) on 9th March 2015 on a bid to make the first round –world tour. The plane was powered by 17,000 solar cells built into its wings, the plane's wingspan being as wide as that of Airbus A380 super jumbo, but the weight of the plane being only 2.3 tonne. The Studies, design and construction took 12 years and the first version of the craft rolled out in 2009 broke records for heights and distances travelled by a manned solar plane. (Ref # 10)

(iv) *Solar Panels – Aircraft Noise*

'BP Solar' has installed solar PV panels on the roof of a church hall under a Heathrow flight path. These solar panels not only generate 23000 kWh of electricity per annum, but also insulate against aircraft noise. The south-facing roof, which is designed to prevent the passage of resonant sound, is covered by 174 solar panels.

(v) *New Material Technique - Quieter Planes*

Aircraft designs incorporate light-weight materials with a honeycomb like structure into their wings and cabins. It is a material that makes up the floor and ceiling of most air plane cabins. The sand- witched honeycomb structure makes it strong and light weight and therefore more fuel efficient. But the honeycomb structures are bad at blocking low frequency noise like the noise of aircraft engine. Adding insulations to limit the noise would add significant weight to the aircraft, making them less fuel efficient. Scientists of North Carolina State University and Massachusetts Institute of Technology (USA) have found a solution to this problem: 'A thin lightweight membrane (like that of a drum) covers one side of honeycomb structure. When the sound waves hit the membrane, they bounce off rather than passing through'.

Note: Aviation Fuel Characteristics:

(i) Aviation gasoline — Density 0.715 g / ml; Energy 43.71 MJ / kg or 31 MJ /L ;

(ii) Jet fuel 'Wide- cut'— Density 0.762 g / ml; Energy 43.54 MJ / kg or 33.18 MJ /L; Flash point < 0°C, Freezing point (–)72°C (for JP-4)

(iii) Jet Fuel Kerosene— Density 0.810 g / ml; Energy 43.28 MJ / kg or 35.06 MJ/L;

ASTM D 1655 for Jet fuel specifies the following:

Acidity 0.1 mg KOH /g; Aromatics 25% max Volume; Sulphur 0.3% by mass;

Mercaptans Sulphur 0.003% by mass; Naphthalene vol. 3% max.; Flash point 38°C; Freezing point (–) 40°C; kinematic Viscosity at (–) 20°C--- 8 mm^2/s ; Net heat of combustion 42.8 MJ / kg(minimum).

13.5.5 Standards of Emissions for Aircraft Engines

Key Facts Associated with Commercial Jet Aircraft NO_x Emission Standards:

Some key facts associated with commercial jet aircraft NO_x emissions standards are enumerated below:-

- International Civil Aviation Organization (ICAO) established by Chicago Treaty of 1944 regulates commercial aircraft emissions and member states adopt these regulations.
- ICAO Committee on Aircraft Environmental Protection (CAEP) adopted CAEP/1 NO_x standard in 1981 which applied to newly manufactured engines starting from 1986.
- CAEP/1 established a ceiling on NO_x emissions at 100 grams per kilo Newton (g/kN) of rated engine thrust for Engine Pressure Ratio (EPR) 30.
- EPR, or the ratio between maximum internal pressure in the engine and the ambient atmospheric pressure, is a key parameter governing NOx emissions.
- ICAO adopted CAEP/2 NO_x standard in 1993 that reduced emission limits by 20% to 80 g/kN for EPR 30. This standard applied to newly certified engines starting in 1996 and to already-certified newly manufactured engines starting in 2000.
- In 1997, U.S. EPA adopted ICAO CAEP/2 NO_x standard that harmonized US aircraft emission standards with ICAO.
- In 1999, ICAO adopted CAEP/4 NO_x standard for newly certified engines that will reduce emission limits by an additional 16% to 67 g/kN for EPR 30 and smaller percentage reductions for EPR greater than 30 beginning in 2004. This standard is the subject of the current rulemaking.
- ICAO standards are "technology following" since by the date the standards are adopted, most engines in production already meet the standard by various margins.
- A primary consideration in standard setting by ICAO is the number of in production already-certified engines that would not meet the proposed standard. The CAEP/4 standard impacted only a small fraction of the fleet. In fact, if CAEP/4 had been set at a level 10% below the adopted level, only 35 out of 124 (28%) of the in-production engines would emit at levels above the standard.
- CAEP/4 responded to the concern about the potential impact of new standards on the asset values of existing fleet aircraft by only applying the new stringency level to newly certified engines. The CAEP/4 standard therefore does not apply to newly-manufactured engines from old designs.
- Two ICAO studies of thc impact of CAEP/2 NO_x standards on the financial value of existing aircraft assets did not reveal any correlation between approval of CAEP/2 NO_x emission standards and aircraft values. ICAO was unable to definitively assess the effect CAEP/4 NO_x standards would have on fleet aircraft values.
- In 2004 ICAO applied the CAEP/4 standards to already-certified newly manufactured engines starting in either 2006, 2008, or 2012.
- A production cut-off date for already-certified engines would prevent manufacturers from offering these engines for sale unless the engines were redesigned to meet the standard and recertified.
- Nearly all (94%) already-certified engines currently meet or perform better than the CAEP/4 NOx standard.
- Only two engine families with 4 currently certified engines do not meet the CAEP/4 NOx standard and would not be required to meet the standard.
- The NO_x benefit associated with application of the proposed standard, through a production cut-off, to these 4 engines is estimated to be negligible.
- There is only one remaining newly-designed growth engine being developed by Pratt & Whitney Canada was expected to be certified in 2004 or 2005 that would need to meet the CAEP/4 NO_x standard.

US Federal Exhaust Emission Standards for Aircraft Engines

Table 13.18 gives US Federal Exhaust Emission Standards for aircraft engines according to the year of manufacture.

Table 13.18: Federal Exhaust Emission Standards for Newly Manufactured & In-Use Aircraft Engines (Table 4-39 in the document)

(Source: 40 CFR 87, Subparts A-D (July 1, 2000), and U.S. Environmental Protection Agency, Office of Air and Radiation)

Engine type[c]	Pollutant	Year of Engine Manufacture						
		1974-75	1976-77	1978-82	1983	1984-96	1997-99	2000+
Turboprop	Smoke					[g]187 $(rO)^{-0.168}$	[g]187 $(rO)^{-0.168}$	[g]187 $(rO)^{-0.168}$
Class T3 turbojet	CO (g/kN)[d]						118	118
Class T3 turbojet	HC (g/kN)[d]					19.6	19.6	19.6
Class T3 turbojet	NOx (g/kN)[d]						[e]40 + 2(rPR)	[f]32 + 1.6(rPR)
Class T3 turbojet	Smoke			25	25	[h]83.6 $(rO)^{-0.274}$	[h]83.6$(rO)^{-0.274}$	[h]83.6$(rO)^{-0.274}$
Class T8 turbojet	CO (g/kN)[d]						118	118
Class T8 turbojet	HC (g/kN)[d]					19.6	19.6	19.6
Class T8 turbojet	NOx (g/kN)[d]						[e]40 + 2 (rPR)	[f]32 + 1.6 (rPR)
Class T8 turbojet	Smoke	30	30	30	30	[h]83.6 $(rO)^{-0.274}$	[h]83.6$(rO)^{-0.274}$	[h]83.6$(rO)^{-0.274}$
Turbofan and turbojet engines other than Classes T3, T8, and TSS	CO (g/kN)[d]						118	118
Turbofan and turbojet engines other than Classes T3, T8, and TSS	HC (g/kN)[d]					19.6	19.6	19.6
Turbofan and turbojet engines other than Classes T3, T8, and TSS	NOx (g/kN)[d]						[e]40 + 2(rPR)	[f]32 + 1.6(rPR)
Turbofan and turbojet engines other than Classes T3, T8, and TSS	Smoke		[i]83.6 $(rO)^{-0.274}$	[i]83.6 $(rO)^{-0.274}$	[j]83.6$(rO)^{-0.274}$	[h]83.6$(rO)^{-0.274}$	[h]83.6$(rO)^{-0.274}$	[h]83.6$(rO)^{-0.274}$
TSS engines (supersonic aircraft engines)	HC (g/kN)					140(0.92)rPR	140(0.92)rPR	140(0.92)rPR
TSS engines (supersonic aircraft engines)	Smoke					[h]83.6$(rO)^{-0.274}$	[h]83.6$(rO)^{-0.274}$	[h]83.6$(rO)^{-0.274}$

Key to Table 13.18: Federal Exhaust Emissions Standards for Aircraft Engines

CO = carbon monoxide; g = gram; g/kN = grams of pollutant per kilo-Newton of thrust; HC = hydrocarbon, kN = kilo-Newton; kW = kilowatt; NO_x = nitrogen oxides; rO = rated output, which is the maximum power or thrust available for takeoff; rPR = rated pressure ratio

(a) Federal standards apply to all planes operating in the United States, regardless of where they were manufactured. This table primarily displays exhaust emissions standards for newly manufactured aircraft engines. Only two standards (smoke standards) have been set for in-use aircraft engines (see footnotes i and k). Therefore, unless otherwise noted, emissions in this table apply to new aircraft engines only.

(b) HC, CO, and NO_x are measured using the International Civil Aviation Organization (ICAO) Gaseous Emissions Test Procedure. Smoke is measured using the ICAO Smoke Emission Test Procedure. There is no useful life or warranty period for purposes of compliance with emissions standards.

(c) Examples of commercial aircraft that use each engine type include the following:

Class T3 turbojet-Boeing 707-320s (Class T3 engines are currently out of production, though some are still in use).

Class T8 turbojet-Boeing 727s and 737-200s, and McDonnell-Douglas MD-80s and DC-9s.

Turbofans and turbojets other than T3, T8, and TSS-Boeing 747-400s, 757s, 767-200s and 777s, and McDonnell-Douglas MD-11s; Canadair Regional Jets.

Turboprops-Used mostly in regional airliners such as ATR 72, Dornier 328, and Saab SF 340.

TSS-British Aircraft Corp./Aerospatiale Concorde (the only supersonic aircraft currently used in commercial civil aviation).

(d) Applies to engines with rO>26.7 kN.

(e) Effective as of July 7, 1997. This standard applies only to those engines of a type or model for which the date of manufacture of the first individual production model was on or before Dec. 31, 1995 and for which the date of manufacture of the individual engine was on or before Dec. 31, 1999.

(f) Effective as of July 7, 1997. This standard also applies to engines of a type or model for which the date of manufacture of the first individual production model was after Dec. 31, 1995 and for which the date of manufacture of the individual engine was after Dec. 31, 1999.

(g) Engines with rO>=1,000 kW.

(h) Engines manufactured on or after Jan. 1, 1984 and with rO>=26.7 kN. Smoke number may not exceed 50.

(i) Engines with rated output rO>=129 kN. This is also the in-use standard for all such aircraft engines.

(j) Engines with rO<26.7 kN. Smoke number may not exceed 50.

(k) Class T8 turbojet engines shall not exceed a smoke number of 30 beginning Feb. 1, 1974.

13.6 REDUCTION OF POLLUTION IN SHIPPING

13.6.1 Pollution from Shipping Operations

Ships and motor boats emit more pollutants and greenhouse gases (GHGs) than road vehicles. World-wide shipping emissions were estimated in 1999 by an initial study of emissions of oxides of nitrogen and sulphur (NO_x and SO_x) from ocean-going shipping. As estimated in 1993, ships were responsible for over 14% of total NO_x emissions and about 5% of total SO_x emissions from combustion sources. The findings from the other similar studies were:–

- About 70% of global ship emission occurs within 400 km of land whereby the commercial shipping diesel engines contribute significantly to SO_2 together with other pollutants in many coastal regions. (The concentration of SO_2 increases very much.)
- In north Atlantic regions, the projected contribution of shipping to ambient SO_2 on typical summer day may be more than 60 % and in January it may be about 90 %

13.6.2 Options for Reduction of Pollution

It is possible to reduce emissions from existing ships through a combination of measures which may include:–

- Use of cleaner fuel,
- Engine modifications,
- Add-on retrofits,
- Modifications in various systems of ship to boost efficiency for cutting energy consumption / fuel consumption,
- Use of suitable control technologies for NO_x,
- Use of Technologies for SO_x reduction/ flue gas desulphurization.
- Use of pollution monitoring techniques,
- Better operation, maintenance, overhauling, replacement and renovation of all the plants and equipments to achieve higher energy efficiency/ lower fuel consumption

It is essential that the design changes are incorporated in the new engine- supplies, instruments and process controls are modernized and better pollution control equipments / devices are provided on the ships to reduce emissions.

It may be emphasized that the measures like good operation, maintenance and overhauling of engines including replacements, improvements in engines and connected equipments, recovery of heat from engine exhaust, modernization of controls, etc. go a long way in lowering the overall emissions by increase of efficiency and lowering of fuel combustion.

Change in fuel from oil to natural gas also can lower down the pollution level. However for this, the gas engines would have to be employed instead of oil engines.

13.6.3 Use of Lower Sulphur Oil Fuel in Ships for Minimizing SO_2 Emission

The Heavy Fuel Oil (HFO) or residual oil which on the average contains 2.7 % or more of 'Sulphur', is more often used as the fuel in large marine ships, while lighter distillates which contain about 1 % or more sulphur are used for smaller coastal vessels. The reduction in sulphur content of fuel is the most appropriate action which should be taken. The reduction of sulphur from HFO from average of 2.7 % to about 0.5 %, can cut down the SO_2 emission by about 80 %. The major part of particulate matter in the smoke consists of metals oxides and metal sulphates. Therefore reduction of sulphur from fuels would reduce the formation of sulphates and therefore cut the emission of particulate matter to some extent. No modification in the engine is necessary for use of low sulphur fuels and it would also cut maintenance problems related to corrosion. These advantages would offset the cost of removal of sulphur from the fuel to a large extent.

In order to reduce the sulphur there can be three approaches:

- Obtaining and processing of lower sulphur crude oils and pay little higher cost, but this would depend on availability of such crude oils in the market.

- Desulphurize the Heavy Fuel Oil (HFO), but the option may be more costly than the first option. However this option is well within the capability of fuel refiners.
- Re blending of oil to lower sulphur content, but significant quantities of lower sulphur oil may not be possible to be obtained although the cost of the option is less.

13.6.4 Removal of Particulate Matter and Sulphur Dioxide from Exhaust Gases by Seawater Scrubbing

Seawater scrubbing of engine exhaust gases to desulphurize the gases is a very good option. The method is dealt in **Chapter 7, Para 7.4.5,** in connection with desulphurization of flue gases from boilers. For diesel engines of ships, the sea water scrubbing is based on the same principles. A flow diagram of sea water scrubbing for Diesel engines of ships is shown in **Fig. 13.1.** The sea water scrubbing washes down the exhaust gases and considerably cleans them of the particulate matter and also removes SO_2 satisfactorily. The sea water scrubbing is well established cost effective method of desulphurization and can easily be used. This technology can remove more than 90% of SO_2 and particulate matter easily. The sea water scrubbing of exhaust gases would be cheaper than reducing sulphur in the fuel for large marine vessels.

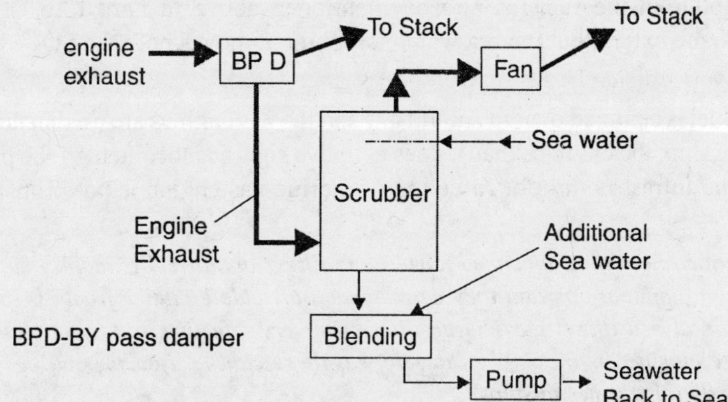

Fig. 13.1: Seawater Scrubbing for Removal of Sulphur Dioxide and Particulate Matter

However the impact of the discharge of effluents directly into sea may be kept in mind as the pH value of the effluent may be lower than the sea water. If necessary, the effluent can be neutralized/treated before discharge into sea, but mostly it may not be necessary.

13.6.5 Reduction of Emission of Oxides of Nitrogen in Exhaust Gases from Ships

The NO_x reduction technologies may include engine modification, pre-engine technologies and post combustion technologies such as mentioned below:

- Engine modifications such as engine de-rating, injection timing retard, fuel injector upgrades, etc., which may reduce NO_x by ~15-20%, but may slightly increase 'emissions of particulate matter' and 'fuel consumption'.
- Water injection to the combustion chamber would reduce combustion temperature thereby cut the NO_x generation by 30% or even more, but may result in increase of particulate matter emission in case of use of HFO as fuel.
- Instead of water injection to combustion chamber, the water vapour can be added to the combustion air (which is called 'Humid Air Motor' or 'HAM'). This measure may work out better.

- Mixing of water with fuel particularly residual oil creating stable emulsion may reduce NOx up to 40% or more and Particulate Matter (PM) by about 15% or more. It may be pointed out that the engines using residual oils are designed for about 20% water mixture.

All the above measures, however, may result in slightly lower efficiency.

SCR (Selective Catalytic Reduction)

It is a post combustion technology of reduction of NO_x of exhaust gases in a reactor where the gases are treated with ammonia (or urea) in presence of a catalyst. All the oxides of nitrogen react with ammonia in presence of catalyst and are reduced to nitrogen. This process is very effective and can remove even more than 80% of NO_x.

The SCR process has already been described in **Chapter 5, Para 5.4.1 and 6.7.1,** in connection with NO_x removal from the flue gases of boilers / waste heat boilers of combined cycle power plants. 'Compact type of SCR reactors specially meant for ships' are available, which are based on the same basic principles.

13.6.6 Measures to Reduce Particulate Matter in Ship Exhaust

The reduction of sulphur in the fuel (as already mentioned above in Para 13.6.3), can reduce the particulate matter to some extent, but the sea water scrubbing removes both the SO_2 and particulate matter very efficiently.

In case, if the oil fuel is changed over to natural gas and the gas engines are used instead of oil engines as prime mover for the ship, the engine exhaust does not have significant content of the particulate matter and SO_2. However, the infrastructure for refueling of marine gas engine is not properly developed at present.

Note: *It may be mentioned here that advanced particulate filters and diesel oxidation catalyst are used in land based diesel engine applications and they can reduce particulate matter from 20 to 50 % when low sulphur fuel is used. Catalyzed diesel particulate filters are very effective, but these technologies are not yet developed for large marine engines. When developed, these can be used for cutting down particulate emissions from marine diesel engines of ships.*

13.6.7 Inland Water ways

Inland navigation is known as environment friendly transport mode. Regarding atmospheric emissions, the greenhouse gases emitted per- ton-kilometre are much less than road transport. On an average, the CO_2 emissions of an inland vessel are only 1/3 of those of a truck per ton-kilometre (t km) which is considered outstanding.

With the introduction of emission limits for road transport since the early 1990s has led to a significant reduction of the pollutant emissions of NO_x and PM (Particulate Matter) on road, but the emission limits for inland navigation were not tightened. Consequently, the advantage in the environmental performance of inland vessels compared with trucks in this regard has become smaller, and with the introduction of EURO IV and EURO V limits (**See Table 13.13**) for road transport in 2006 and 2009, respectively, these new trucks may emit less NOx and PM per tkm than inland vessels.

It can be clearly seen that the emission standards for road transport, EURO II up to EURO V for Europe and the upcoming EPA (Environmental Protection Agency) standards for the USA, are by far stricter than the ones for inland waterway transport like the emission standards by the CCNR (Central Commission for the Navigation on the Rhine) and the ones by the European Commission, EU-stage IIIA (see **Table 13.28**), introduced in 2002 and 2005, respectively.

These emission standards are related to the energy consumption of the means of transport. As the energy consumption per (t.km) of inland vessels is on average 3 times lower than the one of trucks, the emission standards may remain 3 times higher with respect to equal environmental impact. A larger difference, however, may result in a worse environmental performance of inland water vessels than the one of trucks.

A major concern is, moreover, the longer lifetime of an inland vessel engine compared to a truck engine. Truck engines are replaced on average every 5 years. This means that five years after the introduction of a new emission limit, the average truck fleet complies with this limit. Vessel engines may remain in service for 30 years or even longer, and, therefore, it will take also much longer in order to achieve a compliance with new emission standards compared with trucks.

13.6.8 Emission Standards for Ships

Floating vessels in-land waterways and marine vessels/ boats / ships have different set of emission limits. There are three categories of marine engines ie category 1, category 2, category 3 for the purpose of Emission standards. The use, displacement per cylinder of such engines are mentioned in the **Table 13.19.**

Table 13.19: Categories of Marine Engines for Purpose of Emission Standards

Marine-engine Category	Displacement per cylinder (D); Tier 1&2	Displacement per cylinder (D); Tier 3&4	Basic engine Technology
1	D< 5dm^3. (power≥^{3}7 kW)	D<7 dm^3.	Land based Non-road diesel engine
2	5 dm^3 ≤D <30 dm^3.	7 dm^3 ≤D <30 dm^3.	Locomotive engine
3	D ≥ 30 dm^3.		Unique marine engine design

Category 1 and Category 2

The Category 1 and category 2 marine diesel engines typically range in size from about 500 to 8,000 kW (700 to 11,000 hp). These engines are used to provide propulsion power on many kinds of vessels including tugboats, push-boats, supply vessels, fishing vessels, and other commercial vessels in and around ports. They are also used as stand-alone generators for auxiliary electrical power on many types of vessels. Category 1 engines which are applicable for non road engines can be used for certain applications on ships / marine vessels.

Category 3

Category 3 engines are in the range of 2,500 to 70,000 kW (3,000 to 100,000 hp). These are very large marine diesel engines used for propulsion power on ocean-going vessels such as container ships, oil tankers, bulk carriers, and cruise ships. The EPA estimated that there is an increased engine NO_x emissions from 20 to 50% and PM from 750% to 1250% (sulphate particulates) when residual oil is used as compared to when distillate fuel is used.

Table 13.20: Category1 and 2 – Tier 2 Marine Emission Standards

Category	Displacement (D) in dm^3 per cylinder	CO g/kWh	NOx + HC g/kWh	PM g/kWh	Effective date
1	Power $\geq$ 37 kW, D < 0.9	5	7.5	0.4	2005
	$\geq$ 37 kW, 0.9 $\leq$ D < 1.2	5	7.2	0.3	2004
	$\geq$ 37 kW, 1.2 $\leq$ D < 2.5	5	7.2	0.20	2004
	$\geq$ 37 kW, 2.5 $\leq$ D < 5.0	5	7.2	0.20	2007*
2	$\geq$ 37 kW, 5.0 $\leq$ D < 15	5	7.8	0.27	2007*
	Power <3300 kW; 15 $\leq$ D < 20	5	8.7	0.50	2007*
	$\geq$ 3300 kW, 15 $\leq$ D < 20	5	9.8	0.50	2007*
	$\geq$ 3300 kW, 20 $\leq$ D < 25	5	9.8	0.50	2007*
	$\geq$ 3300 kW, 25 $\leq$ D < 30	5	11	0.50	2007*

Emission Standards—Category 1 and 2: see Table 13.20

Tier 1 standards are equivalent to the MARPOL Annexure Tier I NO_x limits *Tier 1 certification requirement starts in 2004

Abbreviations used in this Annexure: HC—Hydrocarbons; NO_x—Oxides of nitrogen; PM—Particulate Matter; CO—Carbon monoxide; D—Displacement;

NTE— US EPA's not-to-exceed (NTE) emission limits and testing requirements

IMO—International Maritime Organization

Category 3: In the 2003, EPA adopted Tier 1 'NOx emission standards' for Category 3 engines, which are equivalent to the international 'IMO MARPOL Annexure VI' (see note about it) limits. These limits range from 17 to 9.8 g/kWh depending on the engine speed, with higher limits for slower engines.

The EPA Tier 1 limits are in effect for new engines built in 2004 and later. These limits are to be achieved by engine-based controls, without the need for exhaust gas after-treatment. Emissions other than NOx are not regulated.

Category 3 Tier 2-3 Standards: According to rules adopted in 2009, EPA has introduced Tier 2 and Tier 3 emission levels standards for newly built engines of Category 3. The Tier 2 standards apply with effect from beginning of 2011 and require the use of engine-based controls, like engine timing, engine cooling, and advanced electronic controls. The Tier 2 standards result in a 15 to 25% reduction of NOx as compared with Tier 1 levels.

Tier 3 standards would be applicable with effect from beginning of 2016 and can be met with the use of high efficiency emission control technology like Selective Catalytic Reduction (SCR) which may result in 80% reduction of NO_x as compared with Tier 1 levels.

The EPA Tier 2-3 'NO_x limits' are equivalent to the respective 'IMO Tier II-III standards'. Depending on the engine speed, Tier 2 standard limits 'NOx' in the range from 14.4 to 7.7 g/kWh, while Tier 3 standard limits NO_x in the range from 3.4 to 1.96 g/kWh. In addition to the NOx limits, EPA adopted a HC emission standard of 2.0 g/kWh and a CO standard of 5.0 g/kWh for new Category 3 engines. No emission standard was adopted for PM, but manufacturers are required to measure and report PM emissions.

Blue Sky Series Program Under 1999 Regulation Program

This voluntary program (extending through 2010) permits the manufacturers for certification of their engines for more stringent emission standards shown in **Table 13.21**.

Table 13.21: Blue Sky Series Voluntary Emission Standards

Displacement (D) dm^3 per cylinder	NOx + HC g/kWh	PM g/kWh
Power $\geq$ 37 kW & D < 0.9	4.0	0.24
0.9 $\leq$ D < 1.2	4.0	0.18
1.2 $\leq$ D < 2.5	4.0	0.12
2.5 $\leq$ D < 5.0	5.0	0.12
5.0 $\leq$ D < 15	5.0	0.16
15 $\leq$ D < 20 and Power < 3300 kW	5.2	0.30
15 $\leq$ D < 20 and Power $\geq$ 3300 kW	5.9	0.30
20 $\leq$ D < 25 and Power $\geq$ 3300 kW	5.9	0.30
25 $\leq$ D < 30 and Power $\geq$ 3300 kW	6.6	0.30

Recreational Vessels (2002 Rule)

Recreational vessels standards are phased-in beginning in 2006, depending on the size of the engine as listed in **Table 13.22**. These standards are similar to the Tier 2 standards for Category 1 commercial vessels.

Recreational engines are also subject to 'NTE Limits' (see notes about it). There are no smoke requirements for recreational marine diesel engines. Similar to commercial vessels, a voluntary "Blue Sky Series" limits exist for recreational vessels, which are based on a 45% emission reduction beyond the mandatory standards.

Table 13.22: Recreactional Marine Diesel Engines Standards
(*Similar to Category 1, Tier 2 commercial vessels*)

Displacement (D) in dm^3 per cylinder	CO g/kWh	NOx + HC g/kWh	PM g/kWh	Date
0.5 $\leq$ D < 0.9	5.0	7.5	0.40	2007
0.9 $\leq$ D < 1.2	5.0	7.2	0.30	2006
1.2 $\leq$ D < 2.5	5.0	7.2	0.20	2006
D $\geq$ 2.5	5.0	7.2	0.20	2009

Tier 3-4 Standards

The standards and implementation schedules are shown in **Table 13.23** through **Table 13.24**. The engine-based Tier 3 standards were phased over 2009-2014. The after-treatment-based Tier 4 standards for commercial marine engines at or above 600 kW were phased over 2014-2017. For engines of power levels not included in the Tier 3 and Tier 4 tables, the previous tier of standards—Tier 2 or Tier 3, respectively—continues to apply.

A differentiation is made between high power density engines and standard power density engines, with a cut point between them at 35 kW/dm^3 (47 hp/dm^3).

Table 13.23: Tier 3 Emission Standards for Marine Diesel Engines Category 1 Commercial Standard Power Density ≤35 kW/ dm^3. Engines

Power (P) kW	Displacement (D) dm^3 per cylinder	(NOx + HC)* g/kWh	PM g/kWh	Date
P<19	D < 0.9	7.5	0.40	2009
19 ≤ P <75	D < 0.9	7.5 /4.7[b]	0.30/ 0.30[b]	2009/ 2014
75 ≤ P < 3700	D<0.9	5.4	0.14	2012
	0.9 ≤ D< 1.2	5.4	0.12	2013
	1.2 ≤ D< 2.5	5.6	0.11[c]	2014
	2.5 ≤ D< 3.5	5.6	0.11[c]	2013
	3.5 ≤ D< 7	5.8	0.11[c]	2012

*Tier3 standards for (NOx + HC) do not apply to 2000 to 3700 kW engines

(a) < 75 kW engines ≥ 0.9 dm^3/cylinder are subject to the corresponding 75-3700 kW standards.

(b) Option: 0.20 g/kWh PM & 5.8 g/kWh NOx + HC in 2014.

(c) This standard level drops to 0.10 g/kWh with respect to PM only for enforcing in 2018 for < 600 kW engines.

Table 13.24: Tier 3 Emission Standards for Marine Diesel Engines Category 1 Commercial Standard Power Density >35 kW/ dm^3. Engines

Power (P) kW	Displacement (D) dm^3 per cylinder	(NOx + HC)* g/kWh	PM g/kWh	Date
P<19	D < 0.9	7.5	0.40	2009
19 ≤ P <75	D < 0.9	7.5/4.7[b]	0.30/0.30[b]	2009/2014
75 ≤ P < 3700	D < 0.9	5.8	0.15	2012
	0.9 ≤ D< 1.2	5.8	0.14	2013
	1.2 ≤ D< 2.5	5.8	0.12	2014
	2.5 ≤ D< 3.5	5.8	0.12	2013
	3.5 ≤ D< 7	5.8	0.11[c]	2012

(a) < 75 kW engines ≥ 0.9 dm^3/cylinder are subject to the corresponding 75-3700 kW standards.

(b) Option: 0.20 g/kWh PM & 5.8 g/kWh NOx+HC in 2014

(c) This standard level drops to 0.10 g/kWh with respect to PM only for enforcing in 2018 for < 600 kW engines.

Table 13.25: Tier 3 Standards for Marine Diesel Engines *of Category 2

Power (P) kW	Displacement	(NOx + HC)** g/kWh	PM g/kWh	Date
P < 3700	7 ≤ D < 15	6.2	0.14	2013
	15 ≤ D < 20	7.0	0.27[a]	2014
	20 ≤ D < 25	9.8	0.27	2014
	25 ≤ D < 30	11.0	0.27	2014

* Option: Tier 3 PM/NOx+HC at 0.14/7.8 g/kWh in 2012, and Tier 4 in 2015.

** Tier 3 NO_x+HC standards do not apply to 2000-3700 kW engines

(a) 0.34 g/kWh for engines below 3300 kW.

In addition to the above NO_x+HC and PM standards, the following CO emission standards apply for all Category 1/2 engines starting with the applicable Tier 3 model year:

(i) 8.0 g/kWh for engines < 8 kW,

(ii) 6.6 g/kWh for engines ≥ 8 kW and < 19 kW,
(iii) 5.5 g/kWh for engines ≥ 19 kW and < 37 kW,
(iv) 5.0 g/kWh for engines ≥ 37 kW.

Table 13.26: Tier 4 Standards for Marine Diesel Engines of Category 1 and 2

Power (P) kW	NOx g/kWh	HC g/kWh	PM g/kWh	Date
P ≥ 3700	1.8	0.19	0.12[a]	2014+
		0.19	0.06	2016b+
2000 ≤ P < 3700	1.8	0.19	0.04	2014+
1400 ≤ P < 2000	1.8	0.19	0.04	2016+
600 ≤ P < 1400	1.8	0.19	0.04	2017c

(a). 0.25 g/kWh for engines with 15-30 dm^3/cylinder displacement.
(b). Optional compliance start dates can be used within these model years. + Option for Cat. 2: Tier 3 PM/NOx + HC at 0.14/7.8 g/kWh in 2012, and Tier 4 in 2015.
(c). The Tier 3 PM standards continue to apply for these engines in model years 2014 and 2015 only.

Table 13.27: US Emissions Standards for Commercial Marine Engines (US Medium and Large Category 2 and 3 Commercial Marine engines)*

Engine	Effective year	NMHC + NOx g /bhp hr	PM g/ bhp hr
5.0–14.9 litre/ cylinder	2007	5.8	0.20
15–19.9 litre/cylinder (<4425 hp)	2007	6.5	0.37
15–19.9 litre/cylinder (≥4425 hp)	2007	7.3	0.37
20.0–24.9 litre/cylinder	2007	7.3	0.37
25.0–29.9 litre/cylinder	2007	8.2	0.37
30 litre/cylinder and above	2004	7.35–12.75	No standards

Note: ** Commercial marine engine standards are THC + NOx (as opposed to NMHC + NOx). However, since HC emissions are low compared to NOx, it is not expected that this difference is significant for purposes of this comparison. Also, commercial marine standards are expressed in g/kWh; the conversion rate to g/bhp-hr is 0.7457 as one bhp hr = 0.7457 kWh*

Notes:-- International Maritime Organization (IMO); International Convention on the Prevention of Pollution from Ships (MARPOL); MARPOL Annexure VI

IMO is an agency of the United Nations which has been formed to promote maritime safety. It was formally established by an international conference in Geneva in 1948, and became active in 1958. When the IMO Convention entered into force, the original name was the Inter-Governmental Maritime

Consultative Organization, or IMCO, but the name was later changed in 1982 to IMO which currently groups 167 Member States and 3 Associate Members.

IMO ship pollution rules are contained in the "International Convention on the Prevention of Pollution from Ships", known as MARPOL 73/78. On 27th September 1997, the MARPOL Convention was amended by the "1997 Protocol";, which included Annex VI entitled "Regulations for the Prevention of Air Pollution from Ships". MARPOL Annex VI set limits on NOx and SOx emissions from ship exhausts, and prohibited deliberate emissions of ozone depleting substances.

The IMO emission standards are commonly referred to as Tier I, II, III Standards

Tier I standards were defined in the 1997 version of Annex VI, while the Tier II and III standards were introduced by Annex VI amendments adopted in 2008, as per the following schedule:

1997 Protocol (Tier I)—The "1997 Protocol" to MARPOL, which included Annex VI, became effective 12 months after being accepted by 15 States with not less than 50% of world merchant shipping tonnage. On 18th May 2004, Samoa deposited its ratification as the 15th State (joining Bahamas, Bangladesh, Barbados, Denmark, Germany, Greece, Liberia, Marshal Islands, Norway, Panama, Singapore, Spain, Sweden, and Vanuatu). On that date, Annex VI was ratified by States with 54.57% of world merchant shipping tonnage.

Accordingly, Annex VI entered into force on 19 May 2005. It applied retroactively to new engines greater than 130 kW installed on vessels constructed on or after January 1, 2000, or which underwent a major conversion after that date. The regulation also applied to fixed and floating rigs and to drilling platforms (except for emissions associated directly with exploration and/or handling of sea-bed minerals). In anticipation of the Annex VI ratification, most marine engine manufacturers have been building engines complying with the above standards since 2000.

2008 Amendments (Tier II and II)—Annex VI amendments adopted in October 2008 introduced (i) new fuel quality requirements beginning from July 2010, (ii) Tier II and III NOx emission standards for new engines, and (iii) Tier I NOx requirements for existing pre-2000 engines.

The revised Annex VI entered into force on 1st July 2010. By October 2008, Annex VI was ratified by 53 countries (including the Unites States), representing 81.88% of tonnage.

Not To Exceed Emission Limits (Testing)

The US EPA has introduced not-to-exceed (NTE) emission limits and testing requirements as an additional instrument to make sure that heavy-duty engine emissions are controlled over the full range of speed and load combinations commonly experienced in use. The NTE testing requirements were first introduced as part of the 1998 Consent decrees with heavy-duty engine manufacturers. NTE testing requirements were then gradually extended on other engine categories, including mobile nonroad engines and marine engines.

NTE testing does not involve a specific driving cycle of any specific length (mileage or time). Rather, it involves driving of any type that could occur within the bounds of the NTE control area, including operation under steady-state or transient conditions and under varying ambient conditions. Emissions are averaged over a minimum time of thirty seconds and then compared to the applicable NTE emission limits. (Source: Ref: US EPA)

13.6.9 Emission Standards for Inland Water Ways Vessels

The current standards for Inland Waterway Vessels in European Union are mentioned below. The Directive 97/68/EC, dubbed as 'mother' of non-road engine regulation are applicable to all types of non-road diesel engines. However, engines of inland waterway vessels weren't regulated until April 2004 under Directive 2004/26/EC

This directive defines Stage III A (**Table 13.28**) to cover engines from 19 to 560 kW including constant speed engines, railcars, locomotives and inland waterway vessels in case of 'Inland Waterway Vessels'. These Directives are applicable to diesel fuelled engines for inland waterway's vessels of rating from 19 kW to 560 kW.

Unlike the Stage I/II non-road legislation, the Stage III A standards also cover engines used in inland waterway vessels. Engines are divided into categories based on the displacement (swept volume) per cylinder and net power output. The engine categories and the standards are harmonized with the US standards for marine engines. There are no Stage III B or Stage IV standards for waterway vessels.

Table 13.28: Stage III A Standards for Inland Waterway Vessels

Category	Displacement (D) dm^3 per cylinder	Date	CO g/kWh	NOx + HC g/kWh	PM g/kWh
V1:1	$D \leq 0.9$, $P > 37$ kW		5.0	7.5	0.40
V1:2	$0.9 < D \leq 1.2$	January 2007	5.0	7.2	0.30
V1:3	$1.2 < D \leq 2.5$		5.0	7.2	0.20
V1:4	$2.5 < D \leq 5$		5.0	7.2	0.20
V2:1	$5 < D \leq 15$		5.0	7.8	0.27
V2:2	$15 < D \leq 20$, $P \leq 3300$ kW	January 2009	5.0	8.7	0.50
V2:3	$15 < D \leq 20$, $P > 3300$ kW		5.0	9.8	0.50
V2:4	$20 < D \leq 25$		5.0	9.8	0.50
V2:5	$25 < D \leq 30$		5.0	11.0	0.50

Source Retrieved from: "http://transportpolicy.net/index.php?title=EU:_Marine:_Emissions&oldid=12648" "Mother directive" for non-road emissions Directive 97/68/EC ; Standards for inland waterway vessel emissions, Directive 2004/26/EC ; Sulphur content of marine fuel, Directive 2012/33/EUDirectives on emissions from non-road mobile machinery 4.2 Notes ↑ European Parliament News: Cleaner shipping fuels to save lives - September 11, 2012

It may be mentioned that Inland waterway vessel engines are tested using the ISO 8178-4:2002 [E] and IMO MARPOL 73/78, Annex VI (NOx Code) tests. Emission durability period for inland waterway vessel engines is 10,000 hours.

Inland waterways emission standards in USA are being tightened up and the stage V limits, would be applicable to propulsion (IWP) and auxiliary (IWA) engines above 19 kW of all types.

REFERENCES

1. Aerospace-Engineering, Energy-Environment-Engineering PermaLink - Online Guide to Airplane Pollution posted by Ecacofonix : 7:29 AM 3rd march 2008 [The Ultimate Web Guide to Green Airplanes and Aircraft Pollution].
2. Asia Pacific Tech Monitor, 13(3), 1996: 5

3. Bharat Stage II, III, IV emission standards.
4. Bose et al. 2001. Transportation in Developing Countries: Greenhouse Gas Scenarios for Delhi, India. Pew Center on Global Climate Change. Arlington, Virginia; Zhou et al. 2001. Transportation in Developing Countries: Greenhouse Gas Scenarios for Shanghai, China. Pew Center on Global Climate Change, Arlington, Virginia; and Prozzi etal. 2002. Transportation in Developing Countries: Greenhouse Gas Scenarios for South Africa. Pew Center on GlobalClimate Change. Arlington, Virginia.
5. CMVR Emission limits in India for Off-road Diesel Vehicles & Equipments.
6. Dave Demerjian @ Greener Jet Engine could reduce carbon foot print – (ecoaviation 23 hmt).
7. Dhanajay Kumar Srivastav, Avinash K. Agarwal, Tarun Gupta, 'Effect of Engine Load on Size and Number Distribution of Particulate Matter Emitted from a Direct Injection Compression Ignition Engine', Aerosol and Air Quality Research, 11, 7, 915-920 (2011).
8. Ernst, Peter, 'Emission reduction', Sulzer Technical Review 4/ 2009.
9. Environment (Protection) Rules, 1986, India.
10. EURO Emission standards for vehicle.
11. Hindustan Times dated 10th March 2015, Lucknow.
12. Hindustan Times, Lucknow Dated 29th April 2015.
13. "http://transportpolicy.net/index.php?title=EU:_Marine:_Emissions&oldid=12648"; "Mother directive" for nonroad emissions Directive 97/68/EC ; Standards for inland waterway vessel emissions, Directive 2004/26/EC ; Sulphur content of marine fuel, Directive 2012/33/EU Directives on emissions from non-road mobile machinery 4.2 Notes; ↑ European Parliament News: Cleaner shipping fuels to save lives - September 11, 2012.
14. IMO (International Maritime Organization); Prevention of air pollution from ships Marine Environment protection committee MEPC/ 53/4/1 dated 7th April 2005, Back ground Paper by a coalition of NGOs.
15. International Maritime Organization (IMO); Marine Environment Protection Committee, 53rd session, Agenda item 4, MEPC 53/4/1, 7 April 2005; 'Prevention of Air Pollution from Ships', Reducing Shipping Emissions of Air Pollution Feasible and Cost-effective Options, by Clean Air Task Force Bluewater Network, European Environ mental Bureau, North Sea Foundation, Sea at Risk, European Federation for transport and environment, Swedish NGO Secretariat on Acid Rain.
16. Jitendra N. Gangwar, Tarun Gupta and Avinash K. Agarwal, 'Composition and comparative toxicity of particulate matter emitted from a diesel and biodiesel fueled CRDI engine', Atmosperic Environment 46, 472-481 (2012).
17. Off- road Vehicles — Euro and US Standards.
18. Reports from Water and Energy International Vol.68, No.9, Sept. 2011, Vol.68 No.10, Oct. 2011.
19. Sharma M., 'Combating Vehicular Pollution: Options and Approach', Enviromental pollution Control journal, Vol.1 No.3 (1997).
20. Tarun Gupta, Abhishek Kothari, Dhananjay Kumar Srivastav, Avinash k. Agarwal, 'Measurement of number and Size Distribution of Particles Emitted from a Mid-Sized Transportation Multipoint Port Fuel Injection Gasoline Engine', Fuel, 89, 2230-2233 (2010).
21. The Hindu Dated 3rd Oct 2011.
22. US EPA – EP A420-F-07 035, April 2007.
23. US EPA (1998), .Control of Emissions of Air Pollution from Nonroad Diesel Engines; Final Rule,. 63 Fed. Reg. 56968 (October 23, 1998), at p.56970; US EPA (1999), .Control of Emissions of Air Pollution from New Marine Compression-Ignition Engines at or above 37 kW; Final Rule,. 64

Fed. Reg. 73300 (December 29, 1999), at p.73307 (hereinafter .EPA 1999 C1-C2 Marine Rule.); EPA C3 Rule, 68 Fed. Reg. at 9761.

24. US Emission Standards Marine Diesel Engines (EPA 1998, 1999, 2002, 2003, 2004, 2008, 2009).
25. US EPA (2004), .Control of Emissions of Air Pollution from Nonroad Diesel Engines and Fuel; Final Rule,. 69 Fed. Reg. 38958 (June 29, 2004), at pp. 38971 and 38980 (hereafter .EPA 2004 Nonroad Rule.); and EPA1999 C1-C2 Marine Rule, 64 Fed. Reg. at 73307; EPA C3 Rule, 68 Fed. Reg. at 9761.
26. WHO Standard Limits /Guidelines for PAH in Environment.

QUESTIONS

1. List the following in increasing amounts from the exhaust of an idling automobile:----NO_x, SO_x, N_2, O_2 , CO_2 , CO, HC (unburned hydrocarbons).
2. What are the general measures or works to be undertaken to reduce pollution from inland transports?
3. Briefly enumerate emission reducing general design measures for with SI engines
4. Enumerate general design measures to reduce the emissions from diesel engines
5. What are the options for use of 'Filters' for removal of particulate matter from exhaust of diesel engines of vehicles?
6. Write short notes on (i) solar powered vehicles (ii) Hydrogen-fuel-cell-powered vehicles
7. What are the effects of emissions from aircraft and what measures are being taken by aircraft manufacturers to reduce them? Which emission standards are followed for emission from aircraft engines?
8. Enumerate the developments in the field of reduction of aircraft emissions.
9. High altitude aircrafts lead to:
   (a) Destruction of oxygen
   (b) Destruction of ozone
   (c) Formation of ozone
   (d) None of the above.
10. Write a short note on PAH emission from vehicles.
11. An automobile traveling at 50 km per hour emits 0.1% CO from the exhaust. If the exhaust rate is 80 m^3 per minute, what is the CO emission in grams per kilometer assuming temperature of exhaust to be 127°C?
12. Compare Indian standards of emission for vehicles of 1986 with Euro standards now adopted (named 'Bharat 2000, Bharat stage II, stage III, and stage IV).
13. Compare the 'CMVR (India) emission limits' for heavy duty diesel engines for off-road diesel vehicles with 'Euro off–road stage 2 (COM2) standards' for diesel vehicles.
14. What is the importance of SO_x, NO_x, and particulate matter emissions from coastal shipping and ocean going vessels respectively?
15. Describe the measures which should be taken for reducing emissions from ships.
16. Write short notes on air pollution by inland water way vessels and compare them with road transport trucks.
17. Write short note on emission standards for marine engines.

ANNEXURE 13

EUROPEAN EMISSION STANDARDS
(Source: From Wikipedia, the free encyclopedia)

1. **CO_2 emission**

   Within the European Union, road transport is responsible for about 20% of all CO_2 emissions, with passenger cars and vans contributing about 15%. The target fixed at Kyoto Protocol was an 8% reduction of emissions in all sectors of the economy compared to 1990 levels by 2008 to 2012.

   Relative CO_2 emissions from transport have risen rapidly in recent years, from 21% of the total in 1990 to 28% in 2004, but currently there are no standards for limits on CO_2 emissions from vehicles. EU transport emissions of CO_2 currently account for about 3.5% of total global CO_2 emissions.

2. **Obligatory labelling**

   The purpose of Directive 1999/94/EC of the European Parliament and the Council of 13 December 1999 relating to the availability of consumer information on fuel economy and CO_2 emissions in respect of the marketing of new passenger cars is to ensure that information relating to the fuel economy and CO_2 emissions of new passenger cars offered for sale or lease in the Community is made available to consumers in order to enable consumers to make an informed choice. EU member countries are also in the process of introducing consumer-friendly labels.

3. **Obligatory vehicle CO_2 emission limits**

   EU Regulation No 443/2009 sets an average CO_2 emissions target for new passenger cars of 130 grams per kilometre. The target has been gradually phased in between 2012 and 2015. A target of 95 grams per kilometre will apply from 2021. For light commercial vehicle, an emissions target of 175 g/km applies from 2017, and 147 g/km from 2020.

4. **Toxic emission:**

   The stages are typically referred to as Euro 1, Euro 2, Euro 3, Euro 4, Euro 5 and Euro 6 for Light Duty Vehicle standards. The legal framework consists in a series of directives, each amendments to the 1970 Directive 70/220/EEC.The following is a summary list of the standards, when they come into force, what they apply to, and which EU directives provide the definition of the standard.

   - **Euro 1 (1993):**
   - For passenger cars—91/441/EEC.
   - Also for passenger cars and light trucks—93/59/EEC.
   - Euro 2 (1996) for passenger cars—94/12/EC (& 96/69/EC)
   - For motorcycle—2002/51/EC (row A)—2006/120/EC
   - Euro 3 (2000) for any vehicle—98/69/EC
   - For motorcycle—2002/51/EC (row B)—2006/120/EC
   - Euro 4 (2005) for any vehicle—98/69/EC (& 2002/80/EC)
   - Euro 5 (2009/9) for light passenger and commercial vehicles—715/2007/EC
   - Euro 6 (2014) for light passenger and commercial vehicles—715/2007/EC

These limits supersede the original directive on emission limits 70/220/EEC.

In the area of fuels, the 2001 Biofuels Directive required that 5.75% of all transport fossil fuels (petrol and diesel) should be replaced by biofuels by 31 December 2010, with an intermediate target of 2% by the end of 2005. However, MEPs have since voted to lower this target.

5. **Emission standards for passenger cars and light commercial vehicles:**

These are summarized in **Table A13.1, Table A13.2, Table A13.3 and Table A13.4.** Since the Euro 2 stage, EU regulations introduced different emission limits for diesel and petrol vehicles. Diesels had more stringent CO standards, but were allowed higher NO_x emissions. Petrol-powered vehicles were exempted from Particulate Matter (PM) standards through to the Euro 4 stage, but vehicles with direct injection engines are subject to a limit of 0.005 g/km for Euro 5 and Euro 6. A particulate number standard (P) or (PN) has been introduced in 2011 with Euro 5b for diesel engines and in 2014 with Euro 6 for petrol engines.

Table A 13.1: European Emission Standards for Passenger Cars (Category M*), g/km

Tier	Date	CO	THC	NMHC	NOx	HC + NOx	PM	P [#/km]
Diesel								
Euro 5a	September 2009	0.50	—	—	0.180	0.230	0.005	—
Euro 5b	September 2011	0.50	—	—	0.180	0.230	0.005	6×10^{11}
Euro 6	September 2014	0.50	—	—	0.080	0.170	0.005	6×10^{11}
Petrol (Gasoline)								
Euro 5	September 2009	1.0	0.10	0.068	0.060	—	0.005**	—
Euro 6	September 2014	1.0	0.10	0.068	0.060	—	0.005**	6×10^{11}***

* Before Euro 5, passenger vehicles > 2500 kg were type approved as light commercial vehicles N1-I
** Applies only to vehicles with direct injection engines
*** 6×10^{12}/km within first three years from Euro 6 effective dates

Table A 13.2: Emission Standards for Light Commercial Vehicles ≤1305 kg (Category N1-I), g/km

Tier	Date	CO	THC	NMHC	NOx	HC + NOx	PM	P [#/km]
Diesel								
Euro 5a	September 2009	0.500	—	—	0.180	0.230	0.005	—
Euro 5b	September 2011	0.500	—	—	0.180	0.230	0.005	6×10^{11}
Euro 6	September 2014	0.500	—	—	0.080	0.170	0.005	6×10^{11}
Petrol (Gasoline)								
Euro 5	September 2009	1.000	0.1002	0.068	0.060	—	0.005*	—
Euro 6	September 2014	1.000	0.100	0.068	0.060	—	0.005*	6×10^{11}

* Applies only to vehicles with direct injection engines

Table A 13.3: European Emission Standards for Light Commercial Vehicles 1305 –1760 kg (Category N1-II), g/km

Tier	Date	CO	THC	NMHC	NOx	HC + NOx	PM	P [#/km]
Diesel								
Euro 5a	September 2010	0.630	-	-	0.235	0.295	0.005	-
Euro 5b	September 2011	0.630	-	-	0.235	0.295	0.005	6×10^{11}
Euro 6	September 2015	0.630	-	-	0.105	0.195	0.005	6×10^{11}
Petrol (Gasoline)								
Euro 5	September 2010	1.810	0.130	0.090	0.075	-	0.005*	-
Euro 6	September 2015	1.810	0.130	0.090	0.075	-	0.005*	6×10^{11}
* Applies only to vehicles with direct injection engines								
HC - Hydrocarbons, THC - Total Hydrocarbons, NMHC - Non methane Hydrocarbons.								

Table A 13.4: European Emission Standards for Light Commercial Vehicles >1760 kg, max 3500 kg. (Category N1-III & N2), g/km

Tier	Date	CO	THC	NMHC	NOx	HC + NOx	PM	P [#/km]
Diesel								
Euro 5a	September 2010	0.740	-	-	0.280	0.350	0.005	-
Euro 5b	September 2011	0.740	-	-	0.280	0.350	0.005	6×10^{11}
Euro 6	September 2015	0.740	-	-	0.125	0.215	0.005	6×10^{11}
Petrol (Gasoline)								
Euro 5	September 2010	2.270	0.160	0.108	0.082	-	0.005*	-
Euro 6	September 2015	2.270	0.160	0.108	0.082	-	0.005*	6×10^{11}
* Applies only to vehicles with direct injection engines								

5. **Emission standards for trucks and buses**

Whereas for passenger cars, the standards are defined by vehicle driving distance, g/km, for lorries (trucks) they are defined by engine energy output, g/kWh, and are therefore in no way comparable. The following table contains a summary of the emission standards and their implementation dates. Dates in the tables refer to new type approvals; the dates for all type approvals are in most cases one year later (EU type approvals are valid longer than one year). The official category name is heavy-duty diesel engines, which generally includes lorries and buses. **Table A13.5** gives the emission standards.

Table A13.5: EU Emission Standards for HD Diesel Engines, g/kWh (Smoke in m^{-1})

Tier	Date	Test cycle	CO	HC	NOx	PM	Smoke
Euro V	October 2008		1.5	0.46	2.0	0.02	0.5
Euro VI	31 December 2013[20]		1.5	0.13	0.4	0.01	
* for engines of less than 0.75 dm^3 swept volume per cylinder and a rated power speed of more than 3,000 per minute. EEV is "Enhanced environmentally friendly vehicle".							

6. **Enhanced Environmentally Friendly Vehicle (EEV)**

   EEV is a term used in the European emission standards for the definition of a "clean vehicle" > 3.5 tonne in the category M_2 and M_3. The standard lies between the levels of Euro V and Euro VI.

7. **Emission standards for non-road mobile machinery**

   European standards for non-road diesel engines harmonize with the US EPA standards, and comprise gradually stringent tiers known as Stage I–IV standards. The Stage I/II was part of the 1997 directive (Directive 97/68/EC). It was implemented in two stages with Stage I implemented in 1999 and Stage II implemented between 2001 and 2004. In 2004, the European Parliament adopted Stage III/IV standards. The Stage III standards were further divided into Stage III A and III B were phased in between 2006 and 2013. Stage IV standards are enforced from 2014.

   As of 1 January 2015, EU Member States have to ensure that ships in the Baltic, the North Sea and the English Channel are using fuels with a sulphur content of no more than 0.10%. Higher sulphur contents are still possible, but only if the appropriate exhaust cleaning systems are in place.[21]

   In 2017, the European Union will introduce testing in real-world conditions called Real Driving Emissions, using Portable Emission Measurement Systems in addition to laboratory tests. take into account different test conditions such as PEMS weight (305–533 kg in various ICCT testing), driving in the middle of the traffic, road gradient, etc.

CHAPTER 14

DAMAGE DUE TO AIR POLLUTION TO ELECTRONIC / ELECTRICAL EQUIPMENTS, AND PROTECTIVE MEASURES

14.1 CORROSION TO ELECTRICAL AND ELECTRONIC COMPONENTS BY AIR POLLUTION

14.1.1 Effect of Corrosive Atmosphere

The metallic components of various machines, equipments including mechanical, electrical, and electronic equipments in industrial environment may get corroded due to presence of small quantities of various pollutants in the air. The pollutants which cause most of the damage are:

(i) Acidic gases like sulphur dioxide (SO_2), sulphur trioxide (SO_3), oxides of nitrogen (NO_x), hydrogen sulphide gas (H_2S), hydrochloric acid gas (HCl), hydrofluoric acid gas (HF), etc. or fumes of various acids like sulphuric acid, HCl, HF, nitrous acid (HNO_2), nitric acid (HNO_3), etc.

(ii) Strong oxidant gases like chlorine (Cl_2), ozone (O_3), etc.

(iii) Other gaseous reactants like ammonia (NH_3), etc.

High humidity conditions

Changes in humidity influence the resistance of electrical insulating materials. These changes result in static electricity, particle formation and corrosion between different materials. It will also cause corrosion by:

(i) Directly chemical reaction on metals and plastics.

(ii) Corrosive compounds forming with other gases in the air, e.g. sulphuric acid/ sulphurous acid (H_2SO_4 / H_2SO_3) with sulphur dioxide (SO_2) and nitrous acid/ Nitric acid (HNO_2 /(HNO_3) with oxides of nitrogen (NO_x).

(iii) Electrochemically as an electrolyte on two different metals causing corrosion. For example —a copper plate coated with gold, if moisture condenses on it electrolysis may result causing hairline cracks.

Some insulating materials adsorb moisture at high relative humidity. This increases the electrical conductivity of insulating material and with electrical leakage causes equipment damage. Dust particles in the air, below one μm can adsorb moisture and gaseous contaminants. This may accelerate equipment corrosion.

Corrosion due to gaseous contaminants grows exponentially with an increase in relative humidity. If the air relative humidity increases above 80%, the silver used in the circuit cards may develop a "migration phenomenon" causing short circuits. The gold used to fasten chips to their beds also suffers from humidity and the chips may loosen and fail.

High humidity may not pose a problem in automation /electrical equipment room with well-designed ventilation systems, as air movement at the correct temperature removes the moisture from the space. However if the intake air temperature is low, local moisture will condense. A failure in room over-pressurization and a poor moisture barrier will cause an increase in the local relative humidity.

It may be mentioned that electronic and electrical components including connectors, wiring, instruments, and electrical protection equipments are most sensitive to corrosion and even minute quantities of the above mentioned gases in C&I rooms (Control and Instrumentation rooms) may be the cause of trouble. The corrosion of electronic control and instrumentation can cause ghost signals, misinformation, malfunction / improper control, equipment outage, and even damage to main equipment being controlled or monitored by it. Special attention is needed for the quality of environmental air during manufacture, storage, installation, operation, repairs, testing and calibration. The following types of components, specially, are the targets of corrosion attack:

- Integrated circuit chips, circuit boards, and IC plug-in sockets
- Relay switches
- Edge connectors and pin connectors
- Wire- wraps connections
- Springs and thermal overload elements of protection circuits
- Electronic interlocking systems / protection / signaling equipments for electrical substations and motor controls
- Electrical heavy current systems particularly contactors, motor starters, circuit breaker contacts.

It is important to point out that the build up of corrosion products on contacts can result in overheating and drastic reduction of life. Insulation of wires, windings, coils, etc. also deteriorates due to the above mentioned pollutants and oxidants, combined with the dust and moisture.

14.1.2 Effect of Temperature on Electrical Equipments

It may be pointed out that the temperature is also an important parameter in adverse effects on electrical and electronic equipments and therefore at least a good ventilation to drive away the heat is required for the rooms housing electrical and electronic components (cooling/ air conditioning may also be required).

For effects of temperature on electrical equipment both air and equipment temperatures should be considered. The equipment temperature depends on its electrical loading and its ability to liberate this heat to the surroundings. The convection to the surrounding air has a major influence on the rate of heat transfer.

When the equipment has a cover, the thermal conductivity of the equipment and the path to the cover is an essential factor to consider. The actual temperatures at which a fault occurs, varies with different equipment. For example, semiconductor silicon components can tolerate a range of 125 to 175°C while germanium components can only tolerate 70 to 100°C. The memory in a hard disk is damaged with a temperature in the 70°C range. The failure frequency of plant depends on the surrounding conditions, the load and the age of the equipment. The high temperature affects the electrical equipment both directly and indirectly.

(i) Direct effects of high temperature

The main *direct effects of high temperature* and typical faults in electrical equipment as per standard EN 60068-1 are as follows:

Main effects—Thermal aging, Oxidation, Cracking, Chemical reactions, Growth of mechanical stress, Softening, Melting, Sublimation, Reduction of viscosity, Evaporation and Thermal expansion.

Typical faults—Insulation faults, Structural taults, Impairment of lubricating properties, and Increase in the wearing of moving parts.

(ii) Indirect effects of high temperature:

High temperatures also damage equipment indirectly by accelerating chemical reactions, resulting in corrosion by the air contaminants. For example, when the temperature rises from 20°C to 30°C, the reaction rate doubles. Evaporation of solvents and insulating materials resulting in an acceleration of gaseous contaminants and further increases fouling of contact surfaces.

Influence of Temperature Rise on Electronic Equipment

The temperature rise has a great influence on the failure frequency of especially ***electronic equipment.*** It is said that when the temperature rises by 10°C, the frequency of failure doubles. It is assumed that the failure frequency of electronic components follows *Arrhenius' law:*

$$z = z_0 \exp(-E/KT)$$

where z = failure frequency
z_0 failure frequency in the normal conditions
K = Boltzmann constant
E = activation energy of the fault mechanism
T = component temperature in Kelvin

The Meantime Between Failures (MTBF) is the ***reverse*** *of frequency of failure.* The formula shows that temperature increase influences exponentially **the failure** frequency of components.

The temperature rise will also influence the efficiency of electronic component, besides increase of failure frequency. The frequency of failure of electronic component would increase steeply beyond a certain temperature say about 50 °C (i.e. the meantime between failures would decrease steeply at about 50 °C). If the electrical stress rate is high, thc temperature effect increases the failure rate.

Effect of Subzero Temperatures

It is not that only high temperatures affects the electrical equipment, but the low temperatures in subzero range also affects them adversely. The failure rate would be fairly constant between 0° and 20°C, but increases as temperatures fall below zero degree Celsius.

Instability of equipment can cause a change in the parameter values, such as temperature reduction, humidity, and air movement. If the intake air temperature is near room air dew point, moisture will condense on the surface of electrical equipment, with serious results.

When the temperature is below 0°C, moisture condensing on surfaces will freeze in narrow spaces causing joint failure. When the temperature drops to –40°C, the failures of different components are about 10-fold compared to normal conditions..

The main effects of low temperatures and typical faults in electrical equipment as per EN60068-1 would be as follows:

Main effects— Increase in viscosity, Ice formation, Embrittlement, Shrinking, Impairment of mechanical strength.

Typical faults—Insulation faults, Impairment of lubricating properties, Sealing faults, Cracking, Failure, Structural faults, and Increase in wearing of moving parts.

14.1.3 Environmental Classification for Electrical Equipments

The basis of the environmental classification for electrical equipment is covered in EN60721-3-0 standard. International recommendations for environmental conditions in electrical equipment rooms are covered by the European standard EN 60721-3-3.
Environmental factors covered in EN 60721-3-3 have been divided into the following groups.

(i) Climatic conditions (K)
(ii) Climatic special conditions (Z)
(iii) Biological conditions (B)
(iv) Chemically active substances (C)
(v) Mechanically active substances (S)
(vi) Mechanical conditions (M)

Electrical equipment manufacturers define the special requirements of each device according to EN 60721-3-3 in the following manner: A definition such as the following is used. **3K1/3Z1/3Z4/3B1/3C1/3S1/3M1:**

Code 3K1 - Defines climatic conditions. For the code class 3K1, the target temperature value is 20 – 25°C in the range 18–27°C. The relative humidity target value is 30–60 % in the range 20–75 % and the absolute humidity ranges from 3,5–15 g/kg of dry air.

Code 3Z1 - Z-class describes the special climatic conditions, including the surroundings thermal radiation.

Code 3Z4 - shows the permitted air velocity if different from the K-class value. (In this case it is 5 m./s.). With special condition classes, it considers how a device reacts to water.

Code 3B1 - describes biological conditions. Organisms or animals are not accepted in the 3B1 class.

Code 3C1 - A demand that covers chemically active substances. This defines the permissible concentrations of corrosive gases in the space.

Code 3S1 - Mechanically active substances. Defines the permissible concentration of particle contaminants e.g. dust, sand etc in the space.

14.2 CORROSION PROTECTION – DESIGN OF ELECTRONIC EQUIPMENT

Some methods of corrosion protection used in design and constructions of electronic equipments include the following:

(i) Coating of PCB (Printed Circuit Board) with lac / varnish (called lacquering)— is not effective as lacquering may get broken or it may entrap corrosive material / gas/ acid/ at the time of manufacture.

(ii) Covering of the conducting surface of PCB track with a very thin (4 to 8 μm) gold plating. This may also not be so effective because of porous nature of thin gold plating, it may adsorb polluting gases permitting the pollutants to attack the copper / nickel below the plating layer. The corrosion products build up and may lift off the gold plating or the corrosion product may come out from pores in plating; thus the plating may fail to provide good contact surface. The modern generations of computers are very sensitive to this type of corrosion as the voltages involved are very small and these may not be enough to break down the resistance offered by the corrosion products for flow of current. *[The surfaces of contacts e.g. edge connectors, IC sockets or contacts of relays are not coated because electric current has to pass through].*

It is therefore very essential that pollution free atmosphere is provided during manufacture, assembly, repairs, etc. of various electronic components.

14.3 PROTECTION BY PROVIDING CLEAN ATMOSPHERE

It is to be emphasized that the 'National Ambient Air Quality Standard (NAAQS)' as per Schedule VII in Environment Protection Rules, 1996, are required to be followed by all industries by effectively controlling the emissions. The emission control is the most important aspect. The air purification for inside atmosphere of a building where electrical / electronic equipment are installed is for futher improving the air quality. *[NAAQS is given in **Para .15.6, Chapter 15** for reference.]*

The practical and positive method of protecting the electronic equipments is by removal or reducing concentration of polluting gases in air i.e. having clean air / atmosphere where electronic equipment is manufactured, assembled, stored, installed, operated, repaired, or calibrated.

Although most of the water soluble pollutants and dust can be removed efficiently by wet scrubbing of air, but the increase in humidity prohibits this approach for air cleaning for the purpose being discussed. The other approach for air cleaning is dry scrubbing which may be done by filters using various media such as:

- Activated carbon or carbon impregnated with some suitable chemical
- Activated Alumina impregnated with potassium permanganate or with other chemicals
- Combination of carbon filter and alumina filter both impregnated with suitable chemicals

Air filtration involves passing the air through a packed bed of filter media. When the air passes through the bed of activated carbon, the pollutants are adsorbed on the carbon surface. The adsorption is a physical process where by gaseous contaminants are captured by media, but this is a reversible process and some amount of contaminant gas may be desorbed from the media. The removal by physical adsorption depend upon many factors such as media used, characteristics of contaminants in space around media , depth of media, contaminated air velocity through the media, resistance offered by media to air flow, etc.

The use of chemically impregnated media in the filter uses 'Chemisorption' technique. If the carbon used is impregnated with some chemical, the pollutant gases are adsorbed and also chemically altered based on the chemical with which the carbon is impregnated. The chemi-sorption is an irreversible process as the chemical present in the media can convert the contaminant gas to harmless gas/ vapour/ salts. In this case, desorption of contaminant does not take place once it is absorbed. However many of the same factors which affect the physical adsorption also affect the chemi-sorption process. The capture of contaminant is directly related to the residence time of air stream within the media bed, type and concentration of contaminants, temperature, relative humidity of air stream, etc.

Different media impregnated with different suitable chemicals may be used for different contaminants prevalent in certain industrial atmosphere. When the filter media used is a bed of 'potassium permanganate impregnated alumina', the pollutant gases are adsorbed into the pores of alumina and then chemically oxidized. Spherical alumina pellets are used here.

The 'Impregnated carbon' is effective for removal of narrow spectrum of polluting gases. If many different polluting gases are present in air, a combination of filter banks of 'Activated carbon', 'Impregnated carbon', and 'Impregnated alumina' can be used for air cleaning. It may be mentioned that dust filtration is an integral part of the air filtration unit. A typical air filtration system include a 'Pre-filter', one or two banks of dry scrubbing media contained in 'Refillable / disposable modules', a 'Blower section', and a 'Final filter'.

The use of two or more dry scrubbing media may be required for efficient removal of contaminant gases. It is preferable to use single media in each filter, but sometimes due to physical constraints, it may not be feasible. For such situations, filters with blends of media may be used.

The pressure drop in a new filter bed is generally a function of bed height, size of solids/ adsorbent material pieces, the adsorbent bulk shape, and air flow rate.

*[The subject of air cleaning is similar to dry scrubbing of gases which has been dealt with in Para **8.2.5.2, Chapter 8.** Removal of odours has been dealt with in **Chapter 18** and details about 'Activated carbon are given in note in **Para 18.2.4, of the same chapter**. This may also be referred].*

Common Adsorbents for Different Impurities

The common adsorbents for the different impurities in air are as follows:

(i) Acidic gases — Sodium hydroxide impregnated activated carbon

(ii) Alkaline gases — Commercially available acid impregnated activated carbon

(iii) Air drying — Silica gel

(iv) Organic solvents, non-methane hydrocarbons — Activated carbon but sometimes zeolites

(v) Odours — Activated carbon

(vi) Oxidation requirement such as for removal of H_2S, NO — Activated carbon impregnated with potassium permanganate/ Alumina impregnated with potassium permanganate

(vii) Indoor air cleaning in buildings for removal of smoke and other contaminants —Activated carbon and synthetic zeolites

(viii) Mercury (Vapour) removal from air — Activated carbon impregnated with sulphur or iodine.

Branded Media for Removal of Various Contaminant Gases

Branded media for air cleaning are being produced by many companies. As an example branded media are available from Purafil Inc, Doraville, Georgia (USA) for removal of various contaminants of air as mentioned below in **Table 14.1** with their break through capacities (by % weight) of 99.5 % removal of the contaminant gas as claimed by the company:

Table 14.1: Media for Removal of Contaminant Gases

Se. No.	Name of media /medias	Contaminant removal typical break through test capacity by % weight
1	Purakol AM media	NH_3 (5.83%).
2	Purakol media	Cl_2 (10.87%), NO_2 (6.61%), O_3 (21%)
3	Puracarb media	Cl_2 (9.82%), H_2S (14.7%), SO_2 (7.98%)
4	Purafil Select media	H_2S (12.82%), NO (4.28%), SO_2 (6.18%)
5	Purafill Select + Pura Carbcarb / Pura kol media	HCl (4%), HF (3.5 %)

Notes: (i) *Purakol—activated carbon;*
(ii) *PurakolAM- -activated carbon impregnated with phosphoric acid;*
(iii) *Puracarb—activated carbon + activated alumina impregnated with potassium hydroxide;*
(iv) *Purafil Select—activated carbon impregnated with potassium permanganate*

Problems Associated with Media Filters

There are some problems associated with media filters such as:

- High pressure drop / plugged media,

- Incorrect choice of media,
- Moisture/ wetness of media,
- Oil vapours in air and soiling of media with oil,
- Insufficient air flow and inappropriate air to media ratio,
- Improper air distribution over the media
- Cleaning of media/ regeneration of media not done properly
- Air leakage/ air bypass (air without passing through media goes to the clean air side).
- Abrasion of media
- Chemical attack on media by some chemical vapour/ gaseous chemical present in air.
- High temperature/ inappropriate temperature of air entering the media.
- Fire in the media (particularly media based on carbon).
- Blower malfunction.
- Dampers/ gates malfunction

The above mentioned problems have to be guarded against in design, operation and maintenance of the media filters.

14.4 VARIOUS DESIGNS OF AIR FILTRATION UNITS

(a) Positive Pressurization Units

The positive pressurization of a space requiring high standards of air quality is very important so as not to permit any outside unclean air in the space. Positive pressurization units can be employed for indoor use where gaseous pollutants in air are in low to medium concentration. Such unit would typically consist of two dry scrubbing media contained in refillable / disposable modules, a final filter and blower along with a damper for control of pressurization of air. Both particulate and chemical impregnated media are used in such units.

(b) Deep Bed Dry Scrubber Units

Deep bed scrubbing units are designed to be installed outside the protected space when the air contains comparatively higher amounts of pollutants. The unit provides clean pressurized air to the protected space. Typical deep bed scrubber unit would consist of a rain louver with bird screen, blower section, two or more beds of dry scrubbing media in series, and a final filter. The unit takes outside air, cleans it and supplies pressurized air to the protected space.

(c) Tub Scrubbers

These units like the deep bed filters can be used for filtration of contaminated gases. They employ single bed of media filter (instead of multiple bed filters) targeted for a single pollutant gas in air. They can be used as recirculation units also.

(d) Circulating Air units

When outside air is highly corrosive, such units can be installed inside the protected space (large) to purify inside air and recirculate. These units are employed where there is low to medium concentrations of gaseous contaminants to further filter and polish the room air in order to maintain very low level of air contaminants. Typical circulating air units include a pre- filter, one or two banks of dry- scrubbing media contained in refillable / disposable modules, blower section, and a final filter. To admit fresh air a small entry can be provided.

Air Filtration Units Used in Conjunction with HVAC Systems

The best way is to integrate the air filtration system with Heating Ventilation Air Conditioning (HVAC) system. However separate air filtration units can be installed as retrofits to existing HVAC System before fresh air inlet. Packed bed filter systems are a good choice for industrial applications because of their high removal efficiencies, lesser maintenance, and longer service life. Such filters are designed on the following considerations:

- Expected types, numbers, concentrations and characteristics of air contaminants
- Desired removal efficiency
- Air flow of HVAC
- Pressure drop within HVAC system.

14.5 STANDARDS OF AIR QUALITY FOR CONTROL ROOMS

The following air quality standards have been issued in connection with control rooms by various authorities:

- ANI /SA-71.04-1985 'Environmental conditions for process measurements and control system: Air borne contaminants, issued by 'Instrument Society of America'.
- IEC Standard IEC 60654-4 (1987-04) 'Operating conditions for industrial process measurements and control equipment. Part 4: Corrosive and erosive influences' issued by 'International Electrotechnical Commission'.
- Japan's Standard LEIDA-29-1990 'Standard operating conditions of industrial computer control system' issued by 'Japan Electronic Industry Development Association'.

These standards involve the 'Reactivity measurements' of specially prepared copper strips called 'Corrosion Classification Coupons (CCC)' normalized to a 30 days period of exposure to the environment under scrutiny for defining overall corrosion potential. The measurement of corrosion film layer is expressed in 'Angstroms i.e. 10^{-10} m'). Four levels of severity of corrosion have been established according to ANI/SA-S71.04-1985 as mentioned in **Table 14.2.**

Table 14.2: Copper Coupon Reactivity/Corrosion Classification Copper Coupon (CCC)

Class	Severity level	Copper reactivity	Comments
G1	Mild	<300 A	Environment is said to be well controlled and corrosion may not be considered for reliability of equipment.
G2	Moderate	<1000 A	Corrosion effects are measurable, corrosion is to be considered for reliability of equipment.
G3	Harsh	<2000 A	High probability of corrosion Harsh levels of corrosive attack should prompt further evaluation resulting in environmental controls or use of specially designed packed equipment.
GX	Severe	>2000 A	Only specially designed packaged equipment can survive in such environment.

Corrosion classification coupons are good for measurement of average corrosion, but they do not allow for the measurement of any variations that may have occurred during the test period. It may be possible that all the corrosion occurred in one day during the test period. The system does not permit measurement of short term variations, although the long term corrosion or the cumulative corrosion of electronic equipment is important for its performance. The development of real- time corrosion monitor which can electronically measure the long term as well as the incremental corrosion over a time can eliminate the difficulty associated with this classification. However the copper coupon corrosion classification is a good guide for air purity conditions

Silver Corrosion Monitoring

The 'Copper CCC' (Copper Coupon Corrosion Classification) is not sufficiently sensitive to 'Chlorine' which is very detrimental to electrical equipment. Silver reactivity monitoring is better because it is more sensitive than copper to low level of corrosive gases including chlorine. Silver corrosion is not much affected by humidity and therefore provides a clearer picture of total reactivity level of a specific environment.

Air Monitoring

Air monitoring is essential for maintenance of air quality. This can provide short term data for management and mitigation of pollutants in air. The important characteristics are sensitivity of monitoring, cost and simplicity. Electronic devices for real time gas monitoring capable of detecting pollutant levels in low ppb range are available now for a number of pollutants. Each pollutant may be monitored individually, but the cost is high. The real- time monitors are available for ammonia, formaldehyde, hydrogen chloride, hydrogen sulphide, nitrogen oxides, sulphur dioxide, ozone, etc.

14.6 REQUIREMENTS OF AIR PURITY FOR EQUIPMENTS

The requirement of air quality/ purity for different types of equipments is mentioned below. In addition, it may be noted that hot and humid weather conditions are bad for electronic equipments and therefore electronic control rooms should always be air conditioned.

- Process computer system, micro process based process control or instrumentation systems or electronic equipments should be located inside a control rooms where there is G1 air conditions.
- For discrete instrument type process control systems ie control with separate controllers, indicators and recorders, G 2 air purity conditions may suffice.
- For Motor Control Centers (MCC) and substations with programmable logic controllers (PLC's), electronic control systems, thyristor drives, and UPS's (uninterrupted power supplies) G 2 air purity condition may be sufficient.
- For MCC and substations containing heavy current switchgear require air purity condition where the following pollutant concentrations are not exceeded:
  (SO_2) – 200 ppb, (H_2S) – 30 ppb, (Cl_2) and reactive chlorine compounds – 10 ppb,
  HF-10 ppb.

14.7 APPLICATION OF AIR CLEANING IN DIFFERENT INDUSTRIAL AREAS

14.7.1 Semiconductor Industry

In 'Semiconductor industry' various reactants and toxic gases are used in various stages of production of silicon integrated circuit as clear from the 'Process' described below. The safety would therefore be required to be ensured for personnel involved in the fabrication of semiconductors (and by-products) and for the community around the facility.

Industrial Process

Oxidation: The silicon wafers are cleaned using various chemical solvents. Oxidation in a high temperature furnace is done to grow an oxide layer over the entire surface as a first step. In this, the reactants used are chlorine and hydrogen chloride.

Photo-lithography: Photo- lithography technique is used to transfer a mask pattern on to the wafer surface with the use of ultra violet light, and photo mask.

Etching: The wafer is then etched to remove the silicon dioxide inside the openings of the mask pattern. The gases used here are 'Boron tri-chloride' and 'Chlorine'. The regions free of oxide can be accessed by diffusion or ion implantation to add dopants to these areas.

Diffusion: It is a high temperature process in which certain chemicals or dopants are introduced into the surface layer of semiconductor material to change electrical characteristics. The gases used as dopants are 'Arsine', 'Boron tri-chloride', 'Di-borane', 'Phosphine', etc.

Ion-implantation: The technique as an alternative to 'Diffusion' is *'Ion-implantation'*. By this, the *dopant atoms* can be placed into selected areas of slice at room temperature (without the need for high temperatures). The technique is more precise, but the wafer can be processed one at a time only, not in large batches. The gases used are Arsine, Boron- tri-chloride, Boron tri-fluoride, Chlorine, Hydrogen sulphide, Phosphine, Silane, Silicon tetra-fluoride, Phosphorous Penta-fluoride, etc.

Annealing: When ion implantation is used, *annealing* is done to correct any structural damage to wafer. In this, there is a slow re-growing of a crystal from amorphous material through application of heat.

For most of the products, the process of *oxidation, photo-masking, etching, doping* via diffusion or *ion implantation* have to be repeated several times to get a desired structure.

The final steps are; (i) *'Passivation'* by putting inorganic coating onto the 'Wafer' to protect it from external chemicals, (ii) *'Metallization'* (ie to attach some device to which external leads could be connected) which is done by putting thin metal films onto certain areas of the chip by evaporation process, and (iii) *'Electrical testing'* to assure proper functioning of IC, etc.

Filtration of Air in Semiconductor Industry

Because of use of many toxic and harmful gases in the semi-conductor manufacture as mentioned above, it is necessary to continuously clean the internal environment. The deep bed filters may be used for venting gas cabinets, and venting the flushing of purge assembly and piping from the gas cylinders to the point of use. The outside fresh air is cleaned through filtration unit having several banks of media and the clean air goes to HVAC system for the manufacturing factory.

14.7.2 Pulp and Paper Industry

The electronic control equipments are installed in various control rooms of the industry. A typical paper and pulp mill would have more than 25 rooms where the air purification equipments may be required.

Industrial Process

The process would broadly consist of the following:

- Debarking of wood logs and making wood chips,
- Digester for wood chips—the lignin is dissolved leaving cellulose and black liquor,
- Cellulose is separated and sent for screening, washing and bleaching and then to paper making machine,

- The black liquor is thickened and sent to recovery unit—[the chemicals used in digestion are recovered and recycled. Normally the black liquor is evaporated and fired in furnace and the resulting smelt is dissolved in water and lime is added to it to precipitate calcium carbonate and the resulting liquid called white liquor containing recovered chemicals is recycled].

Pollutants and their Removal

The objectionable gases emitted to cause air pollution during the process are:

- Hydrogen sulphide
- Methyl mercaptans, and vapours or fumes of Di-methyl sulphide, and Di- methyl di-sulphide,
- Carbon monoxide
- Particulate matter
- Chlorine and chlorine dioxide during bleaching operations

A level of about 5 ppm of hydrogen sulphide may exist throughout 'Paper and pulp mill' area. About 1 ppm of chlorine or compounds of chlorine may be present in bleach plant and paper machine areas. In order to prevent corrosion in computerized process control equipments, the level of hydrogen sulphide should be brought down to 3 ppb (parts per billion) and the chlorine down to 1 ppb level.

The filtration units would be employed in various areas. These may be positive pressurized units, circulating air units, deep bed scrubbers, and tub- scrubbers according to the suitability at a particular location. Electronic equipments will be put in air conditioned rooms provided with filtration equipment to exclude or remove pollutants from the air of these rooms

14.7.3 Iron and Steel industry

Pollutants

In the coal coking plant area, the following pollutants in atmosphere may be present:

- Sulphur dioxide (SO_2)— average about 2 ppm (parts per million), while the peak value may be 300 ppm
- Oxides of nitrogen (NO_x) with peak value of about 5 ppm
- Ammonia (NH_3) with peak values may be between 20 and 50 ppm

In the areas of 'Blast furnace' and 'Steel making furnaces (open hearth / Electric arc / basic oxygen process)', the most dominant pollutant in atmosphere is SO_2 and its average concentration in air may be about 200 to 300 ppb, but the peak concentration of SO_2 may go as high as 5000 ppb. In 'Blast furnace' area, the air may contain some hydrogen sulphide (H_2S) and ammonia (NH_3) also, besides SO_2 and Suspended Particulate Matter (SPM). The SO_2 would also be present in air of the areas of 'Sinter plant' and 'Briquetting plant'. In all these areas fumes and pollutants get generated.

In 'Steel Mill' section areas where the steel cast or rolled slabs are converted to structural steel of various sections (plates, rods, sheets, etc.), the average SO_2 in air may be low say an average of about 40 ppb with peaks of 2 ppm, while the H_2S on the average may be about 15 ppb with peaks of about 300 ppb. The air in the mill areas may not contain ammonia or oxides of nitrogen (NO_x).

The atmosphere in the area of 'Galvanizing and tinning' section, may have concentration of fumes of 'Sulphuric acid (H_2SO_4)' and 'Hydrochloric acid (HCl)'. SO_2 and chlorine (Cl_2) gas may also be present in the average concentration of about 3 to 10 ppm.

Around a hydrochloric acid regeneration plant or a facility where hydrochloric acid is used, the atmosphere may have very high concentration of HCl fumes and chlorine (Cl_2). The average value of HCl / Cl_2 may be 5 ppm, while the peak value may be as high as 120 ppm.

Around Annealing furnaces, which is located in mill section may have SO_2 in air of which the peak value may be higher than 2 ppm.

Air Cleaning

The electrical and electronic equipments are located in various places/ rooms in the whole plant and various types of air cleaning / air filtration plants are provided according to requirements. Electronic equipments should be located in air conditioned rooms and air cleaning / filtration equipment would be provided to exclude / remove polluting gases from the atmosphere of these rooms. The electrical switch gear rooms should be provided with clean and pressurized air. Various types of air filtration units such as 'Positive pressurizing units', 'Deep bed scrubbers', 'Tub scrubbers', 'Circulating air units', etc. are employed according to suitability

14.7.4 Oil and Petrochemical Industry

Pollutants and removal

In the oil refinery area, polluting gases such as sulphur dioxide , hydrogen sulphide (H_2S), chlorine (Cl_2), oxides of nitrogen (NO_x), and some times ammonia (NH_3) may be present in the air. In petrochemical production areas, SO_2, H_2S, and NO_x may also be present in atmosphere.

Distillation area, catalytic cracking area, desulphurization area, petroleum coke plant area, power generation area, and oil byproduct production and handling areas, have air pollutants in the air and therefore air purification equipments would be required to protect the electrical and electronic equipments.

Air filtration equipment such as positive pressurizing units, deep bed scrubbers, tub scrubbers, circulating air units, etc are installed for various rooms housing the electrical and electronic equipments according to suitability. The electronic equipments would be located in air conditioned rooms and air cleaning / filtration equipment would be provided to exclude / remove polluting gases from the air of these rooms. The electrical switchgear rooms would be provided with clean and pressurized air.

Note: *Also* ***refer to Para. 8.9 of Chapter 8*** *regarding 'new filtering technology as protection against air-borne very fine particles ie SXC-ESP'. It works by placing a charge on air borne particles and then using an electrical field to trap the particles. It also irradiates biological air- borne pollutants to inactivate them. This device may be used to any limited indoor space- in hospitals, residences, control rooms (of factories), etc.*

REFERENCES

1. Standard: ANSI/SA S71.04-1985 'Environment conditions for process measurements and control systems': Air borne contaminants, Instrument society of America, Research Triangle Park, NC. 1985.
2. Abbott WH, The effects of operating environments on electrical and electronic equipments reliability in pulp and paper industry', IEEE conference Record, Institute of electrical and electronic engineers, Inc. New York , 1983.
3. Muller C O, ' Combination corrosion coupon needed for todays's control equipment', Pulp and paper Vol. 64, No.2, Feb 1990., pp. 165-169.
4. England WG, et al., 'Application of a real-time electronic contact corrosion monitor ', Proceedings of Advance in instrumentation and control, Vol.46, pp 929-955, Instrument society of America, Anhiem, 1991.

5. 'Break through Capacity test results(typical)at a 99.5 % efficiency- final report', Purafil Inc., Doraville, Georgia, GA, 1998.
6. Christopher O Mullar (Purafill Inc., Doraville, Georgia USA), Control of corrosive gases to avoid electrical equipment failures.

QUESTIONS

1. Enumerate the air pollutants which corrode/ damage the electrical and electronic components/ equipments. Also name the components/ equipments which are subjected to such type of damage. What type of corrosion protection is provided in the design of electronic equipments?
2. Write a short note on copper coupon corrosion classification.
3. Write short note on provision of clean atmosphere for better performance of electrical / electronic equipment especially in control rooms.
4. What is the copper coupon corrosion classification system and why the silver reactivity monitoring is considered better? Enumerate the requirements of air purity for different types of electrical and electronic equipments.
5. Enumerate the air pollutants likely to encountered even when proper emission control are provided and working:
   (i) Semiconductor manufacturing industry,
   (ii) Pulp and paper industry,
   (iii) Blast furnace and steel making furnace area,
   (iv) Oil and petrochemical industry.

CHAPTER 15

EMISSION LIMITS IMPOSED BY INDIAN REGULATIONS

15.1 INTRODUCTION—INDIAN REGULATIONS FOR EMISSIONS

The effluent discharge and emission standards of India are intended to be complied by all sectors including industrial-, power-, transport-, manufacturing-, etc. The emission limits for air pollutants are mainly given in *Environment (Protection) Rules, 1986; Bio-medical Waste (Management and Handling) rules, 1998; Municipal Solid Waste (Management and Handling) Rules, 2000; and Noise Pollution (Regulation and Control) Rules, 2000.* The 'Central Pollution Control Board (CPCB), India' and 'State Pollution Control Boards' are supposed to monitor and exercise control over the environmental pollution and they have powers to punish the entity responsible for noncompliance of these rules.

The various Indian Acts and other relevant Indian Rules (apart from those mentioned above) in connection with air pollution control are:

(i) Air (Prevention and control of Pollution) Act, 1981; (ii) Air (prevention and control of pollution) Rules, 1982; (iii) Air (prevention and control of pollution), Union Territories Rules, 1983; (iv) Environment (Protection) Act, 1986; (v) Hazardous Waste (Management and Handling) Rules, 1989; (vi) Ozone Depleting Substances (Regulation and Control) Rules, 2000; (vii) Batteries (Management and Handling) Rules 2011, (viii) Manufacture, Storage and Import of Hazardous Chemicals Rules, 1989; (ix) Indian Factories Act 1948; (x) The Rules for the Manufacture, Use, Import, Export, and Storage of Hazardous Micro-organisms/ genetically engineered organisms or cells 1989; (xi) e-waste (Management and Handling) rules 2011.

It is obvious that all the rules and acts are to be read with 'up to date amendments'.

Various Schedules contained in Environment (protection) Rules 1986 *(EP Rules 1986)* are mentioned below:

(i) *Schedule I* of the E P Rules 1986 gives the limits of emission of pollutants and discharge of effluents from various industries / processes / operations. Some examples of emission limits from this Schedule 1 are given in **Tables 15.1 to 15.6**. [The emission limits for DG sets are given in **Tables 15.2** and **15.3**, while the standards for ambient air quality at mines are given in **Tables 15.4, 15.5,** and **15.6**] *Abbreviations used in Tables are*:
NO_x – Oxides of nitrogen, CO – Carbon monoxide, HC – Hydrocarbons, SPM – Suspended Particulate matter / PM—Particulate matter, SO_x—Oxides of Sulphur, SO_2—Sulphur dioxide, BaP—Benzo(a)Pyrene.

(ii) Schedule II – has been omitted by GSR 801 (E) dated 31/12/93.

(iii) *Schedule III* – gives Ambient Air Quality Standards with respect to 'Noise'.

(iv) *Schedule IV* – gives Standards for emission of smoke and vapour etc. for motor vehicles.

(v) *Schedule V* – gives a list of authorities/agencies to be intimated in case of pollutants being emitted or discharged are in excess of prescribed limits by an establishment.

(vi) *Schedule VI* – 'General Standards for Discharge of Environmental Pollutants', - Part A for 'Liquid effluents'; Part B for 'Waste water generation standards'; Part C for 'Load based Standards for liquid effluents'; Part D-I. for 'Concentration based standards'; Part D-II for 'Equipment Based Standards', Part D-III for 'Load/ mass based standards', Part E is for 'Noise standards'

(vii) *Schedule VII* – 'National Ambient Air Quality Standards (NAAQS)'

Notes:

(i) *It may be noted that those discharges/ emissions which are not covered in Schedule I are governed by General Standards for discharge of environmental pollutant in Schedule VI of E.P. Rules 1986 (amended from time to time).*

(ii) *Not withstanding anything contained in sub-rules (1) or Sub rule (2) of EP Rules 1986 from January 1994, emission or discharge of environment pollutants from the industries, operations or processes other than those of industries, operations or processes for which standards have been specified in schedule I of E.P. Rules 1986 (examples given in Table 1) shall not exceed the relevant parameters and standards specified in schedule VI of E. P. Rules 1986. (See Para15.4 below).*

15.2 EMISSION LIMITS FOR VARIOUS INDUSTRIES AS PER SCHEDULE I OF ENVIRONMENT PROTECTION (EP) RULES, 1986, (INDIA)

The emission limits for important industries as per Schedule I of EP Rules, 1986 of India are mentioned **below in Table 15.1.** Revised emmission limits are shown in remarks column of **Table 15.1A**

Table 15.1: Emission Limits as per Schedule I of E.P. Rules 1986

Se. No.	Item Sch. I	Industry	Pollutant/ parameter	Limits	Remarks
1.	10.	Cement Plants (i) 200 t/d capacity (ii) > 200 t/d capacity (i) SPM for existing units located in critically polluted/ urban areas* (ii) SPM for new units after the notification	A. Total dust B SPM Emissions of cement plants including grinding units	400 mg/ Nm^3 250 mg/ Nm^3 (i) 100 mg/ Nm^3 (ii) 50 mg/ Nm^3	** cities with population > one lakh including 5 km from boundary Revised by Govt. notification GSR 496 (E) dated 9/5/2016, and dated 10/5/2016 shown in Table 15.1A*
2	11	Stone crushing units 3 m to 10 m from any process equipment	SPM	600 mg/ Nm^3	

3	17	Phosphatic fertilizer units (i) Phosphoric acid Manufacture.------ (ii) Grinding, granulation of rock Phosphate------ (iii) Urea—Prilling tower	(i) Fluoride (ii) SPM (iii) SPM *	(i) 25 mg/ Nm^3 (ii) 150 mg/ Nm^3 (iii) 50 mg/ Nm^3	***or 0.5 kg/t of product Revised limits: 5/2015* (i) *20mg/Nm³* (ii) *125 mg/Nm³* (iii) *no change*
4	19	Calcium chloride—kiln — Arc furnace	(i) SPM (ii) SPM	(i) 250 mg/ Nm^3 (ii) 150 mg/ Nm^3	
5	20	Carbon black	SPM	150 mg/ Nm^3	
6	21	Copper, Lead, & Zinc Smelting. (Off-gases must be utilized for acid manufacture)	(i) SPM (ii) SO_2.	(i) 150 mg/ Nm^3 (ii) SO_2.emission from stock be limited to 4 kg/ t of concentrated 100 % acid	Revised by Govt. Notification dtd 2/5/2011 GSR 354 (E) (Shown in table 15.1A)
7	22	Nitric Acid Manufacture	NO_x	3 kg /t of weak acid (before concentration) produced	
8	23	Sulphuric Acid Manufacture	SO_2	4 kg/t of (100%) concentrated acid produced	Revised by Govt. Notification dtd 7/5/2008 GSR 344 (E) as Shown in Table 15.1A
9	24 & 30	Integrated Steel plant (i) Sintering plant (ii) Normal operation of Steel making (iii) Oxygen lancing during Steel making. (iv) Rolling mill (v) Coke oven (vi) Refractory material plant	 SPM SPM SPM SPM CO SPM SPM	 (i) 150 mg/ Nm^3 (ii) 150 mg/ Nm^3 (iii) 400 mg/ Nm^3 (iv)150 mg/ Nm^3 (v) 3 kg/t coke produced 50 mg/ Nm^3 (vi) 150 mg/ Nm^3	Revised by Govt. Notification dtd 31/3/2012 GSR 277 (E) Shown in Table 15.1A
10	25	Thermal power plants capacity (i) $\geq$ 210 MW (ii) < 210 MW	SPM	 (i) 150 mg/ Nm^3. (ii) 350 mg/ Nm^3.	Revised limits shown in table 15.1A

11	27	Asbestos products manufacture (fiber 5 μm long/ dia 3 μm	(i) Asbestos fiber (ii) Total dust	0.5 fiber per ml for first one year from date of notification. 0.2 fiber per ml after above one year. (ii) 4 fiber per ml, Emission 2 mg / Nm^3	
12	28	Caustic Soda— (a)Mercury cell— hydrogen gas holder stack (b) All processes— hypo tower (c) All processes— from HCl plant	(a) Mercury (b) Chlorine (c) HCl vapour/ mist	(a) 0.2 mg/ Nm^3 (b) 15 mg/ Nm^3 (c) 35 mg/ Nm^3	
13	29	Large pulp and paper mill	(i) SPM (ii) H_2S	(i) 250 mg/Nm^3 (ii) 10 mg/ Nm^3	
14	31	Reheating reverberatory furnaces	SPM Sensitive areas SPM in other areas	150 mg/ Nm^3. 450 mg/ Nm^3.	
15	32	(i) Cupola > 3 t/hr' arc furnace / Induction furnace (ii) Cupola <3 t/hr	SPM	(i) 150 mg/ Nm^3 (ii)450 mg/Nm^3.	
16	33	Thermal power plants (i) for 500 MW and above (ii) for 200 to < 500 MW (iii) for less than 200 MW	SPM Stack height Stack height Stack height	150 mg/Nm^3 (i) 275 meters (ii) 220 meters (iii) Find by formula $H = 14Q^{0.3}$ *	**H= height in meters Q = emission rate of SO_2. in kg/h*
17	34	Small boilers > 15 t/hr capacity	SPM	150 mg/Nm^3	
18	35	Oil Refineries (i) Distillation-atmospheric./ vacuum (ii) Catalytic cracker (iii) Sulphur recovery unit	Sulphur dioxide	 (i) 0.25 kg/t of feed** (ii) 2.5 kg/t of feed** (iii) 120 kg/t of sulphur in feed**	***Feed indicates feed to the process*

19	36	Aluminum plants (a) Alumina plant (i) Raw material handling (ii) Precipitation area (b) Smelter plant (i) Green anode shop (ii)Anode bake oven (iii) Pot room (c) Standards for forge fluoride	 (a) (i) SPM (ii) SPM Calcination*-- (b) (i) SPM (ii) SPM (iii) Total fluoride Fluoride	 (a) (i)150 mg/Nm^3 (ii)250 mg/Nm^3 -----1 % CO (b) (i) 150 mg/Nm^3 (ii) 50 mg/Nm^3. (iii) 2.8 kg F/t for Soderberg technology 0.8 kg F/t for pre-baked technology 40 ppm one year average; 60 ppm 2 months average; and 80 ppm monthly average	**Find Stack height by formula against item 16/33*
20	48	(A) Glass industry Soda lime and Borosilicate and other special glass other than lead glass—Furnaces. (B) Lead Glass ------------- *The stack height is to be calculated by the formula given against item 16 / 33*	SPM Total fluoride SPM------------ Lead-----------	0.8 kg/ t of product drawn 5 mg/Nm^3. ------50 mg/Nm^3 ------20 mg/Nm^3	Particulate matter for small furnaces <60 T/ Day, SPM-20 kg/h
21	49	Lime Kiln	SPM	150 mg/Nm^3	For small kilns up to 40 T/ D limit is 500 mg/Nm^3
22	59	Agricultural residue fired including Bagasse Boiler	SPM	250 mg/Nm^3	Limit for spreader stoker boilers — 800 mg/Nm^3
23	61	Ceramic industry kilns (for Tunnel, Top Hat, Chamber,& Tank furnace	SPM Fluoride Chloride	150 mg/Nm^3 10 mg/Nm^3 100 mg/Nm^3	
24	65	Briquette industry (coal) (i) Capacity < 10 T (ii) Capacity > 10T	SPM	 (i) 350 mg/Nm^3 (ii) 150 mg/Nm^3	*Values corrected to 6% CO_2*
25	66	Soft coke industry	SPM	350 mg/Nm^3	*Values corrected to 6% CO_2*
26	70	Small boilers capacity 15t/hr and above	SPM	150 mg/Nm^3	*Values corrected to 12% CO_2*

27	74	Brick kilns capacity: (i) 30000 bricks/D or more (ii) 15000-30000 bricks/D	SPM stack height	(i) 750 mg/Nm^3 30 m (minm), (ii) 750 mg / Nm^3. 27 m (minm.)	*(iii) For capacity <30000 bricks/D Min stack height 30 m*
28	76	Cupola furnace	SO_2 (corrected to 12% CO_2.)	300 mg/Nm^3.	*[Use water scrubber & a stack height = 6x (dia .of cupola) If this not possible, use stack height by formula given in item 16/33 of sch. I thermal power plant]*
29	79	Coke oven plant with by-product recovery (i) Stack emission of coke oven (ii) SPM emission in charging (iii) SPM emission during quenching operations (iv) (BaP) in work zone air (a) Coke oven Battery area (top of battery) (b) other units in coke oven plants	 (i) SO_2, NOx SPM (ii) SPM -------- (iii) SPM -------- BaP -------- BaP --------	 800 mg/Nm^3. 500 mg/Nm^3. 50 mg/Nm^3. 25 mg/Nm^3. 50 g/t of coke produced 5 μg/ m^3. 2 μg/ m^3.	
30	81	(i) Battery manufacturing industry—Lead -acid Type In PVC section (ii) Dry cell manufacture (iii) Secondary lead smelters (Minimum height of stack 30 m)	Lead SPM SPM SPM Manganese Lead SPM	10 mg/Nm^3. 25 mg/Nm^3. 150 mg/Nm^3. 50 mg/Nm^3. 5 mg/Nm^3 as Mn 10 mg/Nm^3. 50 mg/Nm^3.	
31	82	Gas/Naphtha based thermal station units set up after 1st June 1999 for various capacities *Natural gas Based* (a) 400 MW and above (b) 100 to <400 MW (c) Less than 100 MW *For all Naphtha based power plants*	NO_x reported as vol/vol NO_x	*For natural gas* (a) 50 ppm (vol) (b) 100 ppm (vol) (c)100 ppm (vol) 100 ppm (vol)	*Stack height to be calculated by formula given against Se. no. 16, Thermal power stations(Minimum height 30 meters*
32	85	Coal Washeries emission	Fugitive SPM	150 mg/m^3.	

33.		Coal Pulverizers /Crushers [use suitable pollution control measures]	Final SPM emission from their enclosure	150 mg/Nm^3 through min stack height of 30 m	
34	87	Rayon industry for plants expanded or set up after 1st June 1999 [At least 3 air quality monitoring stations should be setup and Stack height be kept as per formula prescribed against item 87 in Sch. I]	Limits of emission of CS_2 & H_2S	CS_2.= 21 kg per tonne of fibres H_2S = 6.3 kg per tonne of fibers	WHO limits CS_2.—100 µg/ m^3. H_2S—150 µg/ m^3. For ambient air quality monitoring
35	97	Boilers using Agricultural Waste as fuel (i) Step Grate type (ii) Horseshoe/ pulsating type (iii) Spreader stoker type	SPM (at 12 % CO_2.)	(i) 250 mg/Nm^3. (ii) 500 mg/Nm^3. (iii) 500 mg/ Nm^3.	

Table 15.1A: Revised Limits as per New Notifications of Ministry of Environment, Forest, and Climate Change/Central Pollution Control Board, India

Item No. of E.P. Rules Sched. 1	Industry	Pollutant	Limits in mg/Nm^3	Notification dated
10	Cement (Rotary kiln without co-processing), standalone clinker grinding plant or blending for plants	SO_2 when pyritic sulphur in limestone is <0.25%, 0.25 to 0.5% & > 0.5%	100, 700 and 1000 respectively.	G.S.R 496(E) 9th May, 2016
	After 25/8/2014	NO_x	600	
	Before 25/8/2014 (i) for rotary kiln with in Line Calciner (ILC) technology. (ii) for rotary kiln using mixed stream of ILC, Separate Line Calciner (SLC) and suspension preheater technology or SLC technology alone or without calciner	NO_x	800 1000	
10A	Cement Plant with coprocessing of wastes (Rotary Kiln)	SPM for all plants even before 25/8/2016	30	GSR 497(E) dated 10 May 2016
		SO_2 and NO_x Limits same as for item No 10		

21	Copper, Lead or Zinc smelting plant (a) Concentrator unit	SPM for existing units	100	GS.R 354(E), 2/5/2011
		SPM for new unts	75	
	Copper, Lead or Zinc smelting plant (b) sulpher Dioxide Recovery Unit Limitting Concentration in mg/Nm^3, Plant capacity for 100% convertible concentration of Sulphuric Acid	(i) SO_2—existing units up to 300 t/D (ii) SO_2—existing units above 300 t/D (iii) SO_2—new units up to 300 t/D (iv) SO_2—new units above 300 t/D	(i) 1370 (ii) 1250 (iii) 1250 (iv) 950	
		(v) Acid mist or SO_3—existing units up to 300 t/D (vi) Acid mist or SO_3—existing units above 300 t/D (vii) Acid mist or SO_3—new units up tc 300 t/D (viii) Acid mist or SO_3—new units above 300 t/D	(v) 90 (vi) 70 (vii) 70 (viii) 50	
25	Thermal power plants installed before 31st Dec. 2003	SPM	100	SO 3305(E) Dated 7th Dec. 2015
		SO_2—for units<500 MW	600	
		SO_2—for units<500 MW and above	200	
		NO_x—for all units	600	
		Mercury—all units	0.03	
	Thermal power plants installed after 2003 up to 31st Dec. 2016	SPM	50	
		SO_2—for units<500 MW	600	
		SO_2—for units 500 MW and above	200	
		NOx—all units	300	
		Mercury-all units	0.03	
	Thermal power plants installed from I^{st} Jan 2017	SPM—all units SPM—all units NO_x—all units Mercury—all units	30 100 100 0.03	

24 and 30	Integrated steel plants Normal operation of plant	SPM	100	Notification dated 31/3/2012
	Oxygen Lancing during steel making	SPM	300	
	Blast Furnace stack New units	SPM SO_2 NO_x	30 200 150	
	Existing units	SPM SO_2 NO_x	50 250 150	
23	Sulphuric acid manufacture			Revised: 7/5/2008
	For plant up to 300 t/D for bigger plants	SO_2	2.5 kg/t of acid 2.0 kg/t of acid	

Item No. 88 of Schedule 1

Emission limits for petrol or kerosene engine driven generator sets (Revised:7/8/2013)

Displacement	CO g/kWh	$(HC + No_x)$g/kWh
Up to 99 ml	< 250	<12
19 to 225 ml	< 250	< 10
Above 255 ml	< 250	< 08

Item No. 94 of schedule I

Noise limits for DG sets up to 1000 kVA / 800 kW size

All the Diesel Engine driven Generator sets (DG sets) up to 1000 kVA or 800 kW shall be provided with acoustic enclosures for a minimum insertion loss of 25 dB(A) and a exhaust 'Muffler' with a minimum insertion loss of 25 dB(A) while the maximum permissible noise limit at 1 meter from the surface of enclosure is 75 dB(A).

Item No. 95 of Schedule I:

gives the emission limits for DG sets upto 800 kW size (see Table 15.2)

Table 15.2: Emission limits for DG sets upto 800 kW size (Revised by Notification dated 1st April 2014

Engine capacity	(NOx + HC) g/kWh	CO g/kWh	SPM g/kWh	Light absorption Coefficient/m at full load (smoke measure)
up to 19 kW	<7.5	<3.5	<0.3	<0.7
19 to 176 kW	<4.7	<3.5	<3.0	<0.7
176 to 800 kW	<4.0	<3.5	<0.2	<0.7

Item No. 96 of Schedule 1:

gives the emission limits for DG sets of capacity > 800 kW size (see Table 15.3)

Table 15.3: Emission Limits for DG Sets > 800 kW Rating for Industries/ Power Plants Commissioned on or after 01. 07.2005

Emission limits for pollution			Remarks
Oxides of nitrogen (NO_x)	Up to 150 MW capacity plant	710 ppmv	For plants commissioned before 1.7.2003—1100 ppmv, while for plants commissioned from 7/2003 to 7/2005—970 ppmv
Oxides of nitrogen (NO_x)	More than 150 MW capacity plant	360 ppmv	For plants commissioned before 1.7.2003 - 1100 ppmv, while for plants commissioned from 7/2003 to 7/2005—710 ppmv
Carbon Monoxide (CO)	For all capacity plants	150 mg/ Nm^3.	For all plants commissioned even before 1.7.2005, the limit is same ie 150 mg/Nm^3.
Particulate Matter (PM)	For all plants for fuels other than HSD or LDO	100 mg/ Nm^3.	For plants commissioned before 1.7.2003 —150 mg/Nm^3.
Particulate Matter (PM)	Up to 5 MW plants with HSD or LDO	75 mg/Nm^3.	For plants commissioned before 1/7/2005, the limit is same ie 75 mg/Nm^3
Particulate Matter (PM)	Up to 5 MW plants with HSD or LDO	75 mg/Nm^3.	For plants commissioned before 1/7/2005, the limit is same ie 75 mg/Nm^3
(NMHC) Non Methane Hydrocarbon	For all capacity plants	100 mg/ Nm^3.	For plants commissioned before 1.7.2003—150 mg/Nm^3.

Notes:

1. ***Condition for reporting emission for DG sets***
   Emission reported at 15% oxygen in exhaust gases on dry bases in; NO_x reported in ppmv (vol./ vol.) normalized at 25°C and 1.01 kPa, while carbon monoxide (CO), particulate matter (PM), and non methane hydrocarbons as carbon (NMHC as C) are reported in mg/Nm^3. for the entire plant (Total capacity).
2. ***Stack height***
   Stack height (H) in meters is determined by the formula: $H = 14\,Q^{0.3}$.
   Where Q is % of sulphur dioxide (SO_2) in emission from total plant
3. ***Fuel oil used:***
   (a) *For areas within municipal limits or within 5 km from the municipal limits of cities or towns of population more than one million : HSD and LDO are used for power plant only up to 5 MW capacity, while beyond 5 MW plant capacity other oils (fuels) are used whose sulphur content is less than 2%.*
   (b) *For areas beyond 5km from municipal limits of cities or towns of population more than one million : Sulphur content of oil fuel shall be limited to less than 4%.*
4. ***NOx monitoring***
   Continuous monitoring of NO_x shall be used for more than 50 MW capacities

5. ***European Emission Standards for Stationary Diesel Power Plants***

   In the absence of emission regulation in Europe for stationary diesel power plants, the German power plant emission standard TA-Luft (Technische Anleitung zur reinhalt der Luft) is used as a national regulation in Europe. Emissions are measured at steady state operation at max. 5% exhaust oxygen content. This standard is applied in Germany, Austria, Luxembourg, Switzerland and the Netherlands and is gaining acceptance in other European countries as well.

6. ***Model Emission Limits in German TA Luft Standards:***

   For 'Stationary diesel engines of capacity less than 3 MW, the TA Luft Emission regulations' at full (100%) load has a typical Emission limits as follows:

   CO -- 0.65; HC -- 0.15; NO_x -- 4.0; PM -- 0.13 in grams per Normal cubic meter.

Item No.90 of Schedule I --- Standards for Air Quality at Coal Mines

(i) ***Clause 1***

The concentration of Suspended Particulate Matter (SPM), respirable particulate matter (RPM), Sulphur dioxide (SO_2), Oxides of nitrogen (NO_x), in air in the downwind direction (considering predominant wind direction) at a distance of 500 m from the dust generating source shall not exceed the standards mentioned in **Tables 15.4, 15.5 and 15.6.**

The dust generating sources would include coal loading or unloading, coal transportation road, coal handling plant (CHP), coal crushing plant, railway siding, drilling site, blasting site, over-burden dumps, coke ovens, briquette industry, roads with vehicular movement, or any other source of dust generation.

Three categories of mines:

Category I: This category includes 'New Coal mines' which commenced operation after publication of this notification [G.S.R 742 (E) dated 25.9.2000. with effect from 25.9.2000.]

Category II: This category includes the existing coal fields/ mines before 25-9-2000 given below: Karanpur, Ramgargh, Giridih, Rajhara, Wardha, Nagpur Silewara, Pench Kanhan, Patharkhera, Umrer, Korba, Chirimiri, and Central India coal fields including Baikunthpur, Bisamberpur. Singrauli, Ib valley, Talchar, Godavari-valley and any other.

Category III: This category includes coal mines located in the coal fields of Jharia, Raniganj, and Bokaro.

(ii) ***Clause 2***

Frequency of sampling of air quality

Air quality monitoring at a frequency of once in a fortnight at the dust generating source as indicated in clause 1 shall be carried out. But if, as a result of monthly monitoring, it is found that the value of the pollutants are less than 50% of the specified standards for three consecutive months, the frequency of sampling may be changed to two days per 3 months.

In case, the value of any pollutant has exceeded the specified standards, the air quality sampling shall be done twice a week. If the result of four consecutive weeks indicate that the concentration of pollutants is within the specified limits, fortnightly monitoring may be reverted to.

Noise Level: The noise level should be monitored once a fortnight. The noise level would not exceed 75 dB(A) between 6 hours to 22 hours and 70 dB(A) between 22 hours to 6 hours.

Notes: *The annual arithmetic mean for the measurements taken in a year would follow the clause 2 above. The 24 hourly values shall be met 92% of the time in a year. For 8% of time the values may exceed the standards but not on two consecutive days.*

Table 15.4: Ambient Air Quality at Category I of New Coal Mines

Se. No.	Pollutant	Period of Weighted average	Concentration in ambient air in µg/m³.
1	SPM (Suspended Particulate Matter	Annual 24 hours	360 500
2	RPM (Respirable Particulate Matter size <10 µm)	Annual 24 hours	180 250
3	SO_2 (Sulphur dioxide)	Annual 24 hours	80 120
4	NO_x (Oxides of nitrogen)	Annual 24 hours	80 120

Table 15.5: Ambient Air Quality at Category II of Existing Coal Mines

Se. No	Pollutant	Period of weighted average	Concentration in ambient air in µg/m³.
1	SPM (Suspended Particulate Matter)	Annual 24 hours	430 600
2	RPM (Respirable Particulate Matter size <10 µm)	Annual 24 hours	215 300
3	SO_2 (Sulphur dioxide)	Annual 24 hours	80 120
4	NO_x (Oxides of nitrogen)	Annual 24 hours	80 120

Table 15.6: Ambient Air Quality at Category III Coal mines (Jharia, Ranigang, Bokaro mines)

Se. No	Pollutant	Period of weighted average	Concentration in ambient air in µg/m³.
1	SPM (Suspended Particulate Matter)	Annual 24 hours	500 700
2	RPM (Respirable Particulate Matter size <10 µm)	Annual 24 hours	250 300
3	SO_2 (Sulphur dioxide)	Annual 24 hours	80 120
4	NOx (Oxides of nitrogen)	Annual 24 hours	80 120

Note: *For Emission Standards for **'Vehicles including off- road or off- highway diesel engine driven vehicles, tractors and mobile equipments' refer Chapter 13.***

15.3 NOISE LEVEL LIMITS FOR DIFFERENT AREAS (EP RULES, 1986)

The noise limits in ambient air as per Schedule III, Environment Protection Rules (E. P. Rules), 1986 (India) are mentioned in **Table 15.7.** According to 'Noise Pollution (Regulations and Control) Rules 2000' Schedule 3(1) and 4(1), following limits have to be followed.

Table 15.7: Ambient Air Quality Standards with respect to Noise
[as per Schedule III of Environment Protection Rules, 1986 (India)]

Area code	Category of area	Day time noise limit [6—21 hrs]; dB(A)	Night time noise limit [21—6 hrs]; dB(A)
A	Industrial area	75	70
B	Commercial area	65	55
C	Residential area	55	45
D	Silence Zone	50	40

Notes:

1. *The silence zone is defined as areas up to 100 m around such premises as hospitals, educational institutions, courts or any premises declared as silence zone by competent authority. Use of vehicular horns, loud speakers, bursting of crackers shall be banned in such zones.*
2. *The mixed category of area should be declared as one of the category 'A, B, C, or D' by competent authority.*
3. *Day time is reckoned as between 6 a.m. to 10 p.m. Night time is taken as 10 p.m. to 6 a.m.*
4. *d(B) denotes the frequency weighting in measurement of noise and corresponds to frequecy response characteristics of the human ear.*

15.4 GENERAL STANDARDS FOR DISCHARGE OF ENVIRONMENTAL POLLUTANTS (E.P RULES, 1986)

As already mentioned in **paragraph 15.1 of this chapter**, the limits given in the general standards for emission and discharge of environmental pollutants *(Schedule VI of EP rules 1986)* would be applicable only for those processes / operations / industries which are not mentioned in schedule I. An extract of these general standards applicable regarding emission in atmosphere are mentioned in **Table 15.8, 15.9 and 15.10.**

General Emission Standards Schedule VI Part D

Table 15.8: *(Part D I. Concentration Based Standards)*

Se. No.	Parameter	Standards
1.	Particulate Matter (SPM or PM)	150 mg/Nm^3
2.	Total fluoride (F)	25 mg/Nm^3.
3.	Asbestos----Fiber -----Dust	4 fibres / cc 2 mg/Nm^3.
4.	Mercury (Hg)	0.2 mg/Nm^3.
5.	Chlorine (Cl)	15 mg/Nm^3.

6.	Hydrochloric acid vapour and mist (HCl)	35 mg/Nm^3.
7.	Sulphuric acid mist (H_2SO_4)	50 mg/Nm^3.
8.	Carbon dioxide (CO)	1 % max. by volume (v/v)
9.	Lead (Pb)	10 mg/Nm^3.

Table 15.9: *(Part D II. Equipment Based Standards Schedule)*

Se. No.	Parameter	Equipment	Standards
	SO_2 (Sulphur dioxide)	Plant capacity	Stack height H in meters
(i)	SO_2.	500 MW	275 m
(ii)	SO_2.	200 / 220 MW up to less than 500 MW	220 m
iii)	SO_2.	Power generation or steam generation Plants smaller than above	Stack height to be calculated by the formula: $H = 14 (Q)^{0.3}$. Q = Emission rate of SO_2. in kg /h

Table. 15.10: *(Part D III. Load /Mass Based Standards)*

Se. No.	Industry	Parameter	Standard Per tonne of product
1.	Fertilizer (Urea) Commissioned before 1/1/1982.	Particulate matter	2 kg/ t
2	Fertilizer (Urea) Commissioned after 1.1.1982.	Particulate matter	0.5 kg / t
3.	Copper, Lead, Zinc Smelter sulphur dioxide converter	Sulphur dioxide SO_2.	4 kg/ t of concentrated (100%) acid produced
4	Nitric acid	Oxides of nitrogen (NO_x)	3 kg/t of weak acid produced (before concentration)
5	Sulphuric acid	Sulphur dioxide SO_2.	4 kg/t of concentrated 100% acid produced
6	Coke oven	Carbon monoxide (CO)	3 kg/t of coke produced
7	Oil Refinery Atmospheric & vacuum distillation	Sulphur dioxide SO_2.	0.25 kg SO_2 /t of feed in the process
8	Catalytic cracker of oil refinery	Sulphur dioxide SO_2.	0.25 kg SO_2. /t of feed in the process
9	Sulphur Recovery unit of oil refinery	Sulphur dioxide SO_2.	120 kg SO_2. /t of sulphur (S) in the feed

10	Aluminum (Al) Plants Anode bake oven	Total Fluoride (F)	0.3 kg of F / t of Al
11	Aluminum (Al) Plants— (i) Anode Bake Oven (ii) Pot room: (a) Vertical Stud Soderberg (VSS) (b) Horizontal Stud Soderberg (HSS) (c) Pre Baked Side Work (PBSW) (d) Pre Baked Centre Work (PBCW)	Total Fluoride (F)	(i) 0.3 kg /t of Aluminum (ii) (a) 4.7 kg of F/t of Al (b) 6 kg of F / t of Al (c) 2.5 kg of F /t of Al (d) 1.0 kg of F/ t of Al
12	Glass industry furnace (i) upto product draw capacity of 60 tonne / hour (ii) upto product draw capacity of more than 60 tonne / hour	Particulate matter	(i) 2 kg / hour (ii) 0.8 kg / hour

Annexure II of Schedule VI (for purpose of part D)

This annexure specifies certain guide lines for State Pollution Control Boards for some industries as mentioned below:

(I) Industries: Emission of Particulate Matter

(a) ***Cement Plants****—Emission of particulate matter (total from all sections of plant) shall be limited to within 400 mg/Nm^3 for plants up to capacity of 200 t / day and within 250 mg/Nm^3 for plants of more than 200 t/ day.*

(b) ***Calcination Process at Aluminum Plants****—The particulate matter emission from 'Kilns' and 'Step grate Bagasse fired boilers' shall be limited to within 250 mg/Nm^3.*

(c) ***Thermal power plants****--Commissioned prior to 01. 01.1982 of capacity less than 625 MW shall limit the particulate emission to within 350 mg/Nm^3.*

(d) ***Lime kilns****—The particulate emission shall be limited to within 500 mg/Nm^3 for kiln capacity from 5 t/day to 40 t/ day.*

(e) ***Bagasse fired boilers*** *—The particulate emission shall be within 500 mg/Nm^3 for horse shoe/ pulsating grate and 800 mg/Nm^3 for spreader stoker type boiler (equalized at 12% CO_2)*

(f) ***Asbestos dust****—The asbestos dust shall not exceed 2 mg/Nm^3.*

(g) ***Urea plants commissioned after 01.01.1982, coke oven and lead glass units****—the particulate emission shall be within 50 mg/Nm^3.*

(h) ***Small boiler of capacity less than 2 t/hr and between 2 to 5 t/hr****—The particulate matter emission shall be limited to within 1600 mg/Nm^3 and 1200 mg/Nm^3 respectively.*

(i) ***Integrated steel plants****—The particulate emission during oxygen lancing shall be allowed up to 400 mg/Nm^3.*

(j) ***Stone crushing units****—The suspended particulate matter contribution value at a distance of 40 m from a controlled, isolated as well as from a unit located in the cluster should be less than 600 μg/Nm^3 (i.e. 0.6 mg/Nm^3). These units are required to adopt the following pollution control measures:*

(i) *'Dust containment'- cum 'Dust- suppression system foı equipment'*

(ii) *Construction of wind breaking walls.*

(iii) *Construction of metalled roads within the premises.*
(iv) *Regular cleaning and wetting of ground within the premises*
(v) *Growing of a green belt along the periphery*

(k) ***Ceramic industry***—*Sources of particulate matter emission include basic raw material processing operations, heat recovery dryers, and mechanical finishing operations. All preventive measures should be taken to control particulate matter emission as far as practicable.*

(2) ***Total Fluoride Emission from Glass units and Phosphatic Fertilizer Units –***

- *Total fluoride shall not exceed 5 mg/Nm^3 for glass units*
- *Total fluoride shall not exceed 25 mg/Nm^3 for phosphatic fertilizer units*

(3) ***Utilization of Gases from Smelting of Copper, Zinc, Lead***

- *The off – gases from smelting of copper, zinc, and lead shall be utilized for manufacture of sulphuric acid as far as possible.*

(4 a) ***For Cupolas of Foundries Melting Rates less than 3 tonne /hour***

- *Particulate matter emission shall be limited to within 450 mg/Nm^3.*
- *The stack should be constructed over the cupola beyond the charging door and its height should be at least six times the diameter of cupola.*

(4 b) ***For Arc furnaces and Induction furnaces***

- *Provision has to be made for collecting the fumes before discharging the emission through the stack.*

15.5 NOISE LIMITS FOR AUTOMOBILES AND SOME APPLIANCES/ EQUIPMENTS (General standards schedule VI, EP Rules, 1986):

(i) Noise Limits for Automobiles:

The noise limits with effect from 2005 for four wheel vehicles, and vehicles with carriages for passengers or carriages for goods are given in part E of Schedule VI of General Standards for Discharge of Environmental Pollutants, EP Rules 1986. These limits are for manufacturing stage at 7.5 m (free field distance). For different type of vehicles, the limits vary from 74 dB(A) to 80 dB(A)

(ii) Noise Limits for Domestic Appliances and Construction appliances:

The noise level as per part E of Schedule VI (General Standards for Discharge of Environmental Pollutants, EP Rules 1986) for certain equipments are as given below:

- (a) *Window Air Conditioner of 1 tonne to 1.5 tonne – 68 dB(A),*
- (b) *Air Coolers – 60 dB(A),*
- (c) *Refrigerators – 46 dB(A),*
- (d) *Compact Rollers, Front loaders, Concrete mixers, Movable cranes –75 dB(A)*

15.6 NATIONAL AMBIENT AIR QUALITY STANDARDS (NAAQS) [Schedule VII, EP Rules, 1986]

Period of Weighted Average

For the purpose of determining the mean concentration of pollutant in a year or 24 hours / 8 hours/ I hour the following guidelines as provided in the EP rules, are to be followed:

(i) The annual arithmetic mean of minimum 104 measurements in a year at a particular site taken twice a week 24 hourly at uniform intervals.

(ii) The 24 hourly or 8 hourly or 1 hourly monitored values, as applicable, shall be complied with 98% of the time in a year. For 2% of the time, they may exceed the limits, but not on two consecutive days of monitoring.

(iii) Whenever and wherever monitoring results on two consecutive days of monitoring exceed the limits specified above for the respective category, it shall be considered adequate reason to institute regular or continuous monitoring and further investigations.

Schedule VII of EP Rules 1986

The schedule VII of Environment protection Rules gives the national ambient air quality standards as given in **Table 15.11.**

Table 15.11: National Ambient Air Quality Standards (NAAQS)

Notes: *The revised and newly introduced standards in 2009 are given in brackets*

Pollutant	Average Period	Concentration of pollutant in ambient air per cubic meter		
		Industrial area	Residential area .	Sensitive area .
SO_2. in μg/m^3.	Annual	80 μg/m^3.	60 μg/m^3.(revised 50)	15 μg/m^3 (revised to 20)
SO_2 in μg/m^3.	24 hours	120 μg/m^3	80 μg/m^3. (not revised)	30 μg/m^3 (revised to 80)
NOx in μg/m^3.	Annual	80 μg/m^3.	60 μg/m^3 (revised 40)	15 μg/m^3 (revised to 30)
NOx in μg/m^3.	24 hours	120 μg/m^3 .	80 μg/m^3.(not revised)	30 μg/m^3.(revised to 80)
SPM in μg/m^3.	Annual 24 Hours	360 μg/m^3 500 μg/m^3.	140 μg/m^3.(not revised) 200 μg/m^3.(not revised)	70 μg/m^3.(not revised) 100 μg/m^3.(not revised)
RPM size <10 μm	Annual 24 hours	120 μg/m^3 150μg/m^3.	60 μg/m^3.(not revised) 100 μg/m^3.(not revised)	50 μg/m^3.(revised to 60) 75 μg/m^3.(revised to 100)
RPM Size <2.5 μm	Annual 24 hours	....	(40 μg/m^3. new norm) (60 μg/m^3. new norm)	(40 μg/m^3. new norm) (60 μg/m^3. new norm)
Lead (Pb) μg/m^3.	Annual	1.0 μg/m^3 .	0.75 μg/m^3(revised 0.5)	0.5 μg/m^3 (revised also 0.5 μg/m^3)
Lead (Pb) μg/m^3.	24 hours	1.5 μg/m^3 .	1.0 μg/m^3 (not revised)	0.75 μg/m^3.(revised to 1.0)
CO in mg/m^3	8 hours 1 hour	5 mg/m^3 10 mg/m^3.	2.0 mg/m^3 (not revised) 4.0 mg/m^3 (not revised)	1.0 mg/m^3 (revised to 2.0) 2.0 mg/m^3.(revised to 4.0)
Ozone (O_3) μg/m^3.	8 hours 1 hour	----- -----	(100 μg/m^3 new norms) (180 μg/m^3 new norms)	(100 μg/m^3. new norms) (180 μg/m^3. new norm)
Ammonia in μg/m^3.	Annual 24 hours	----- -----	(100 μg/m^3 new norms) (400 μg/m^3 new norms)	(100 μg/m^3. new norm) (400 μg/m^3. new norm)
Benzene μg/m^3.	Annual	------	(5.0 μg/m^3 new norms)	(5.0 μg/m^3. new norm)
Benzo(a) Pyrene ng/m^3.	Annual	-----	(1.0 ng/m^3. new norms)	(1.0 ng/m^3. new norm)
Arsenic ng/m^3	Annual	-----	(6.0 ng/m^3 new norm)	(6.0 ng/m^3. new norm)
Nickel (Ni) ng/m^3.	Annual	-----	20 ng/m^3(new norm)	(20 ng/m^3. new norm)

Methods of Measurements:

The rules provide various alternatives for measurements as given below:

- **For Suspended particulate matter**—High volume sampling with flow not less than 1.1 m^3 /min should be used.
- **For Sulphur Dioxide (SO_2), μg/m^3.**—Improved West and Gaeke Method / Ultraviolet Fluorescence
- **For Nitrogen Dioxide (NO_2), μg/m^3.**—Jacob & Hochheiser modified (NaOH-$NaAsO_2$) Method /Gas Phase Chemiluminescence
- **For Particulate Matter (Size less than 10μm) or PM10, μg/m^3.**—Gravimetric / TEOM / Beta attenuation
- **For Particulate Matter (Size less than 2.5μm) or PM 2.5, μg/m^3.**—Gravimetric / TEOM / Beta attenuation **For Ozone (O_3) μg/m3**-UV Photometric / Chemiluminescence / Chemical Method
- **For Lead (Pb) μg/m^3.**—AAS/ICP Method after sampling on EPM 2000 or equivalent filter paper / ED-XRF using Teflon filter
- **For Carbon monoxide (CO), mg/m^3.**—Non dispersive Infrared (NDIR) / Spectroscopy
- **For Ammonia (NH_3), μg/m^3.**—Chemiluminescence / Indophenol blue method
- **For Benzene (C_6H_6), μg/m^3.**—Gas Chromatography (GC) based continuous analyzer / Adsorption and desorption followed by GC analysis
- **For Benzo(a)Pyrene (BaP) Particulate phase only, ng/m^3.**—Solvent extraction followed by HPLC/GC analysis
- **For Arsenic (As), ng/m^3.**—AAS/ICP Method after sampling on EPM 2000 or equivalent filter paper
- **For Nickel (Ni), ng/m^3.**—AAS/ICP Method after sampling on EPM 2000 or equivalent filter paper.

15.7 PERMISSIBLE LEVELS OF CERTAIN CHEMICALS IN WORK ENVIRONMENT [Indian Factories Act, 1948]

The Indian Factories Act 1948, schedule II, prescribes the maximum permissible levels of certain chemicals in work environment as mentioned in **Table 15.12.**

Table 15.12: Permissible Levels of Certain Chemical Substances in Work Environment

[Schedule. II (See section 41F) of Indian Factories Act, 1948].

S. No.	SUBSTANCE	PERMISSIBLE LIMITS OF EXPOSURE			
1	2	3	4	5	6
		TWA* ppm*	TWA mg/m^3	STEL* ppm*	STEL mg/m^3
1	Acetaldehyde	100	180	150	270
2	Acetic Acid	10	25	15	37
3	Acetone	750	1780	1000	2375
4	Acrolein	0.1	0.25	0.3	0.8
5	Acrylonitrile- (Skn, SC)	2	4.5	-	-
6	Aldrin-skin	-	0.25	-	-
7	Allyl Chloride	1	3	2	6
8	Ammonia	0.25	18	35	27
9	Aniline-skin	2	10	-	-

10	Anisidine (o.p. isomers) (Skn)	0.1	0.5	-	-
11	Arsenic and Soluble compounds as 'As'	-	0.2	-	-
12	Benzene (HC)	05	1.5	25	7.5
13	Beryllium and Compounds as Be -(SC)	-	0.002	-	-
14	Boron tri-fluoride-(C)	1	3	-	-
15	Bromine	0.1	0.7	0.3	2
16	Butane	800	1900	-	-
17	2-Butanone (Methyl- Ethyl-Ketone MEK)	200	590	300	885
18	N-Butyl acetate	150	710	200	950
19	N-Butyl alcohol (Skn, C)	50	150	-	-
20	Butyl acetate	200	950	-	-
21	Butyl Mercaptan	0.5	1.5	-	-
22	Cadmium-dust & salts as 'Cd'	-	0.05	-	0.21
23	Calcium oxide	-	2	-	-
24	Carbaryl (Sevin)	-	5	-	0.10
25	Carbofuran (Furadan)	-	0.1	-	-
26	Carbon disulphide-(Skn)	10	30	-	-
27	Carbon monoxide	50	55	400	440
28	Carbon tetrachloride-(Skn, SC)	5	30	-	-
29	Chlordane-(Skn)	-	0.5	-	2
30	Chlorine	1	3	3	9
31	Chlorobenzene (mono-chlorobenzene)	75	350	-	-
32	Chloroform (SC)	10	50	-	-
33	Chloro-methyl ether (HC)	0.001	0.005	-	-
34	Chromic acid and chromates as Cr (water soluble)	-	0.05	-	-
35	Chromous Salts as Cr	-	0.5	-	-
36	Copper fume	0.2		-	-
37	Cotton dust, raw	-	0.2	-	0.06
38	Cresol, all isomers- (Skn)	5	22	-	-
39	Cyanides as CN- (Skn)	-	5	-	-
40	Cyanogen	10	20	-	-
41	DDT dichlorodiphenyl trichloroethane	-	1	-	3

42	Demeton-(Skn)	0.01	0.1	-	-
43	Diazinon-(Skn)	-	0.1	-	0.3
44	Dibutyl Phythalate	-	5	-	10
45	Dichlorous (DDVP)-(Skn)	0.1	1	0.3	3
46	Di-eldrin-(Skn)	-	0.25	-	0.75
47	Dinitrobenzene (all isomers)-(Skn)	0.15	1	0.5	3
48	Di-nitro-toluene- (Skn)	-	1.5	-	5
49	Biphenyl	0.2	1.5	-	-
50	Endosulfan (Thiodan)- (Skn)	- -	0.1	-	0.4
51	Endrin-(Skn)	-	0.1	-	0.3
52	Ethyl acetate	400	1400	-	-
53	Ethyl alcohol	1000	1900	-	-
54	Ethyl-amine	10	18	-	-
55	Fluorides as F	-	2.5	-	-
56	Fluorine	1	2	2	4
57	Formaldehyde (SC)	1.0	1.5	2	3
58	Formic Acid	5	9	-	-
59	Gasoline	300	900	500	1500
60	Hydrazine-(Skn, SC)	0.1	0.1	-	-
61	Hydrogen chloride-(C)	5	7		
62	Hydrogen cyanide (Skn, C)	10	10	-	-
63	Hydrogen fluoride as F-(C)	3	2.5	-	-
64	Hydrogen peroxide	1	1.5	-	-
65	Hydrogen sulphide	10	14	15	21
66	Iodine-(C)	0.1	1	-	-
67	Iron oxide fume (Fe_2O_3) as Fe	-	5	-	-
68	Isoamyl acetate	100	525	-	-
69	Isoamyl alcohol	100	360	125	450
70	Isobutyl alcohol	50	150	-	-
71	Lead and its compounds, dusts/ fumes as Pb	-	0.15	-	-
72	Lindane-(Skn)	-	0.5	-	-
73	Malathion-(Skn)	-	10	-	-
74	Manganese dust and its compounds as Mn- (C)	-	5	-	-

75	Manganese fume as Mn	-	1	-	3
76	Mercury as Hg –(Skn)				
	(i) Alkyle compounds	-	0.01	-	0.03
	(ii) all forms except alkyle vapour	-	0.05	-	-
	(iii) Aryle and inorganic compounds	-	0.1	-	-
77	Methyl alcohol (Skn)	200	260	250	310
78	Methyl cellosolve (2-methoxyethanol)-(Skn)	5	16	-	-
79	Methyl isobutyl ketone	50	205	75	300
80	Methyl Isocyanate-(Skn)	0.02	0.05	-	-
81	Naphthalene	10	50	15	75
82	Nickel carbonyl as Ni	0.05	0.35	-	-
83	Nitric acid	2	5	4	10
84	Nitric oxide	25	30	-	-
85	Nitrobenzene-(Skn)	1	5	-	-
86	Nitrogen dioxide	3	6	5	10
87	Oil mist mineral	-	5	-	10
88	Ozone	0.1	0.2	0.3	0.6
89	Parathion-(Skn)	-	0.1	-	-
90	Phenol- (Skn)	5	19		
91	Phorate (Thimet)-(Skn)	-	0.05	0.2	-
92	Phosgene (Carbonyl chloride)	0.1	0.4	-	-
93	Phosphine	0.3	0.4	1	1
94	Phosphoric acid	-	1	-	3
95	Phosphorus (yellow)	-	0.1	-	-
96	Phosphorus penta- chloride	0.1	1	-	-
97	Phosphorus trichloride	0.2	1.5	0.5	3
98	Picric acid-(Skn)	-	0.1	-	0.3
99	Pyridine	5	15	-	-
100	Silans (silicon tetra hydride)	5	7	-	-
101	Sodium hydroxide-(C)	-	2	-	-
102	Styrene, monomer (phenyl-ethylene)	50	215	100	425
103	Sulphur dioxide	2	5	5	10
104	Sulphur hexafluoride	1000	6000	-	-
105	Sulphuric acid	-	1	-	-
106	Tetraethyl lead as Pb – (Skn)	-	0.1	-	-

107	Toluene (Toluol)	100	375	150	560
108	O-Toluidine- (Skn, SC)	2	9	-	-
109	Tri-butyl-phosphate	0.2	2.5	-	-
110	Trichloroethylene	50	270	200	1080
111	Uranium natural as U	-	0.2	-	0.6
112	Vinyl chloride (HC)	5	10	-	-
113	Welding fumes	-	5	-	-
114	Xylene (o-m-p-isomers)	100	435	150	655
115	Zinc oxide				
	(i) Fume	-	5.0	-	10
	(ii) Dust (Total dust)	-	10.00	-	-
116	Zirconium compounds as Zr	-	5	-	10

*TWA—Time–weighted average concentration; **STEL—Short-teım exposure limit (15 min)

Notes for Table 15.12 above:

1. ***Amendments are**: Subs. by S.O. 170 (E), dated 2nd March, 1989. and Subs. by S.O. 342 (E), dated 19th April, 2001 (w.e.f. 19-4-2001).*
2. ***Dust measurement** : Lint-free dust as measured by the vertical clutricator cotton-dust sampler.*
3. **ppm is parts per million by volume at 25°C and 760 mm Hg*
4. ***Symbols used:***

Skn—denotes potential contribution to overall exposure by the cutaneous route including mucous membranes and eyes.; C—Ceiling Limit; SC—denotes suspected human carcinogens; HC—denotes confirmed human carcinogens.

REFERENCES

1. Environment (Protection) Rules, 1986.
2. Bio-medical Waste (Management and Handling) rules, 1998.
3. Municipal Solid Waste (Management and Handling) Rules, 2000.
4. Noise Pollution (Regulation and Control) Rules, 2000.
5. Indian factories act, 1948.

QUESTIONS

1. Write down the computing equations and other limiting standards which have been prescribed by CPCB for regulating minimum chimney height of an air polluting source.
2. Mention the national ambient air quality standards of important air pollutants prescribed in India under the Environment (Protection) Act, 1986, for different categories of areas.

CHAPTER 16

MEASUREMENT OF AIR POLLUTANTS

16.1 DETERMINATION OF POLLUTANTS IN AIR

16.1.1 Monitoring Sites

The ambient air quality standards which are applicable in India are mentioned in **Para 15.6 of Table 15.11** of the previous chapter. In order to monitor the ambient air quality, some monitoring sites are to be chosen according to the guidelines provided in IS 5182 (Part 14), 2000. It would involve sampling and measuring the values of various pollutants in air. The conditions for choosing the monitoring sites may be briefly stated as follows:

- Monitoring site should be away from the source of pollutant. There should be no nearby incinerator or furnace.
- The height of inlet of sampler should be more than 3 m or more (between 3 and 10 m); the distance of the sampler from the nearby obstacle should not be less than the two times the height of the obstacle (say, the inlet should be at least 20 m from trees).
- The air sample should be from free flowing and well mixed air. The air flow should be at least in three fourth quadrants.
- The elevation angle should not be more than 30 degree from inlet to top of the building.
- Collective samplers (samplers put together) should be 2 m apart.
- Site should be accessible throughout the year in all seasons.

Number of Sampling Sites

The number of sampling sites would depend on:

(i) Size of area to be covered;

(ii) Variability of concentration of pollutants over the area;

(iii) Data requirement related to monitoring and criticality of pollutant for health which may be indicated by population of the area;

(iv) Terrain of the area (for irregular terrain more stations are needed), etc.;

The number of sites may be fixed according to the guideline of IS 5182, (part 14), 2000 and the distribution of stations may be done generally in accordance with the guidelines of WHO, 1977.

(a) Special Downtown Pedestrian Exposure Monitoring Stations (Type A—CPCB*)

If the average daily travel on the street exceeds 10000 vehicles with average speed of less than 6.7 m/s in the central urban downtown street surrounded by buildings where the pedestrians walk, a monitoring station is to be located. Monitoring probe is to be set 0.5 m from the curb** at a height of 3 ± 0.5 m. *[*CPCB—Central Pollution Control Board, India; **curb – pavement edge]*

(b) Special Downtown Neighborhood Exposure Monitoring Station (Type B—CPCB)

The monitoring station is located in the central urban area, but not close to any major street. The station should be located at least 100 m away from streets with average daily travel exceeding 500 vehicles. Typical locations are parks, malls or landscaped areas having no traffic. Probe height is to be 3 ± 0.5 m above the ground.

(c) Special Residential Area Monitoring Station (Type C—CPCB)

A monitoring station can be located in the midst of a residential area or suburban area, but it should be more than 100 m away from any street having a traffic volume exceeding 500 vehicles /day and the station probe height should be 3 ± 0.5 m.

(d) Collection of Mateorological Data i.e. Mesoscale Station (Type D—CPCB)

To collect meteorological and air quality data at upper elevations locate station in the urban area at appropriate height. The purpose of this station is not to monitor human exposure but to gather trend data and meteorological data at different height. Typical locations are tall buildings and broadcast towers. The height of the probe, along with the nature of the station location should be carefully documented in each case.

(e) Non – urban Monitoring Station (Type E—CPCB)

In a remote non- urban area having no traffic and no industrial activity the purpose of this station is to monitor for trend analysis for non – degradation assessments and for large scale geographical surveys, the location or height should not be changed during the period over which trend is examined. The height of the probe should be documented in each case. A suitable height is 3 ±0.5 m.

(f) Specialized Source Survey Station like Industry (Type F—CPCB)

The Monitoring station can be located near a particular air pollution source for scrutiny or at any location where air quality is to be monitored. The purpose of the station is to determine the impact on air quality, at specified locations, of a particular emission source of interest. Station probe height should be 3 ± 0.5 m unless special considerations of the survey require non – uniform height.

16.1.2 Selection of Parameters for Monitoring and Period of Weighting Average

- The selection of 'Parameters' for monitoring should generally be done on the following criteria:
- Sensitive location--- for Sulphur dioxide(SO_2), Nitrogen dioxide(NO_2), Respirable Particulate Matter (RPM)
- Health impact stations including populated areas---- All pollutants which affect health most and are likely to be present at the location
- Traffic intersection sites--- Pollutants like Hydrocarbons, Carbon monoxide (CO), SO_2, NO_2, Suspended Particulate Matter (SPM) and Respirable Particulate Matter (RPM)
- Downtown (Accumulative values 50 m away from traffic intersection) ---- Pollutants like Hydrocarbons, CO, SO_2, NO_2, SPM, RPM and Ozone.

Determination of Weighted Average Concentration of a Pollutant in Ambient Air

For the purpose of determining the mean concentration of pollutant in a year or 24 hours / 8 hours/ I hour, the guidelines given in **Para 15.6,** of previous chapter may be followed. *[These guidelines are repeated below for ready reference:*

(i) The annual arithmetic mean of minimum 104 measurements in a year at a particular site taken twice a week 24 hourly at uniform intervals.

(ii) The 24 hourly or 8 hourly or 1 hourly monitored values, as applicable, shall be complied with 98% of the time in a year. For 2% of the time, they may exceed the limits, but not on two consecutive days of monitoring.

(iii) Whenever and wherever monitoring results on two consecutive days of monitoring exceed the limits specified above for the respective category, it shall be considered adequate reason to institute regular or continuous monitoring and further investigations].

Necessary Meteorological Measurements at the Time of Sampling

For correlating various pollution effects and measured values, certain meteorological readings (like wind velocity, wind direction, temperature, humidity, barometric pressure, cloudiness/ fog/ rain etc.) and sensory pollution effects (like odours, visibility fall, irritation in eyes, skin irritation, breathing difficulty, change in mouth taste, etc.) should be noted at the time of sampling as far as possible.

16.1.3 Basic Methods of Analysis of Air Samples

The following are the basic methods of analysis of air samples:

- Gravimetric analysis
- Volumetric analysis
- Colorimetric method
- Chromatographic method
- Optical measurement of light passing through the sample
- Emission spectrometry
- Mass spectrometric method
- Absorption spectrophotometry
- Atomic absorption spectrometry
- X-ray diffraction method
- Polarography / electrochemical analysis
- Refractometric method
- Thermal conductivity method
- Radioactivity method

The principles of gravimetric-, volumetric- and colorimetric analysis methods are not being given here. These methods are extensively used in all chemistry laboratories of schools and colleges. The broad principles of other methods are briefly mentioned below:

Chromatographic Method

Here the sample which is a mixture of various components is subjected to adsorption on adsorbing columns, or on especially sensitized paper. The elution of various substances adsorbed on various columns/ sensitized paper with selected solvents would separate the components which can be quantitatively analyzed. This method is therefore very useful for separation of substances for their quantitative estimation. The sensitivity of this method is high and is suitable for trace metals and VOCs.

Optical Measurement of Light Passing through the Sample (Opacity)

Such a method is suitable for determination of Suspended Particulate Matter (SPM) in the gas / air by noting the 'Opacity readings'. Even continuous measurement of SPM can be managed by equipment based on this principle.

An option with this method is that the precipitates obtained with specific reagents can be made into suspensions through which light is passed and the 'Opacity readings' relate to the quantity of solids present in the suspension.

Emission Spectrometry

Every element in an excited state has its own 'Emission-spectrum' which have specific characteristic lines relating to the element. The quantity of energy emitted for a spectral line of an element is proportional to the number of excited atoms and consequently to the quantity of the element in the sample. The method is useful for trace elements like sodium, potassium, calcium, lead, beryllium, cadmium, antimony, etc. in gas/air. The method is widely used to determine the composition of compound or mixture and for exploratory examination of particulate matter.

Mass Spectrometric Method

The gas sample is ionized by electric discharge and the collimated (parallel / straight) beam of positive ions produced is subjected to simultaneous action of an electric field and magnetic field at right angles to each other. In doing so the ions are deflected from their initial straight line path and are distributed along a curve. The deflection is proportional to the mass of the ions. The results may be recorded on a photographic plate to get a mass spectrograph. In the mass spectrometer apparatus, the radii of curvature of of paths of various ions can be altered by changing the speed of ions or intensity of the magnetic field. By this, the nature and quantity of various kinds of ions can be determined by measuring the current flowing in a collecting circuit, when the radius of path is changed. The method is useful for finding contents of hydrocarbons and compounds of hydrocarbons with chlorine, nitrogen, oxygen, etc. in the sample of air/ gas.

Absorption Spectrophotometry

The monochromatic light (within the visible spectrum) emerging from a source of radiation is passed through a layer of the substances to be examined. If the substance is coloured, the light absorption is proportional to the concentration when layers analyzed are of equal thickness (as per Lambert- beer law). This enables the determination of concentration of coloured substance. [*It may be noted that the absorption also varies with wave length of the incident radiation and therefore measurements are carried out with monochromatic light or light having a very limited spectrum.*]

The method can also be applied to the analysis of colourless substances by studying the absorption in the ultra- violet (UV) region of spectrum. The method is suitable for organic substances, like benzene, aromatic compounds. Infrared spectroscopy is more successful as it is applicable to solids, liquids and gaseous substances.

Certain substances show a phenomenon of fluorescence (the re-emission of light energy absorbed). By measurement of fluorescence, it is possible to analyze trace amount of certain atmospheric components.

Atomic Absorption Spectroscopy

The atoms of different elements have the characteristics of absorbing a discrete wave length of light and get raised to an excited stage. The 'excited state atom' of the element emits radiation on returning to ground state. The amount of energy absorbed and emitted is a function of the number of atoms present and can serve to determine concentration of the respective elements in the sample.

To determine a particular element, the monochromatic light of the desired discrete wave length (which is absorbed by the atoms of such an element according to its characteristics) is used. The sample to be measured is atomized by flame using conventional atomizer. The radiation from the source traverses the flame into the entrance slit of a monochromator which separates the desired resonance line from the other lines that may be present. The intensity of the resonance line is measured by conventional photo multiplier and amplifier. The background radiation from the flame is compensated by modulating the radiation from the source and feeding the signal from the detector to a tuned a.c. amplifier, which rejects the d.c. signal contributed by emission from back ground flame. *The method is applicable to metals, but not to non-metals.*

X-ray Diffraction Method

Each crystal has its own diffraction pattern for X-rays. Crystals act as diffraction gratings. By comparing the X-ray diffraction pattern of an unknown crystal with that of the crystals of known substances, the unknown crystal can be recognized. The intensity of diffraction lines can lead to the concentration of these crystals in the sample.

The sensitivity of the method is good because of great penetration power of X-rays. This method is suitable for qualitative and quantitative analysis of crystalline substances like 'Mineral dust and Quartz crystals' in air.

Polarography/ Voltametry /Electrochemical Method

This method can be used for identifying inorganic / organic compounds and is suitable for analysis of dilute solutions. In an electrochemical cell, when voltage is applied across the electrodes, a current passes, and the ratio of voltage to current may be interpreted to identify the ions of compounds. This principle is used in polarographic method of analysis.

Note: *In the electrochemical type dissolved oxygen analyzer of 'Electronic Instruments Ltd, UK', the 'Oxygen senser' is a galvanic cell comprising of 'Lead anode' and 'Silver cathode' in an alkaline electrolyte. The galvanic cell is separated from the sample by a membrane through which oxygen in the sample solution diffuses to be reduced at the silver cathode. The output of the cell is proportional to the concentration of oxygen in the sample.*

Refractometric Method

This technique is based on the measurement of refractive index of substance in sample. It can be used only when one or two known vapour or gases are to be determined, whose refractive index is sufficiently different from that of the air.

Thermal Conductivity Method

The method is based on measurement of thermal conductivity of gases and vapours. If the thermal conductivity of the gas mixture is measured before eliminating and after eliminating a component gas, the difference in the two thermal conductivity readings can be used for estimating the component.

The thermal conductivity of some gases differs widely. Thermal conductivities at 300 K of Carbon dioxide, Carbon monoxide, Nitrogen, Oxygen, Hydrogen, Helium are 16.55×10^{-3}, 25×10^{-3}, 25.9×10^{-3}, 26.8×10^{-3}, 183×10^{-3} and 152×10^{-3} W/mK respectively.

This method involves the use of a hot wire thermal conductivity gas-analyzer cell. The cell consists of two chambers each containing a wire filament. In one chamber the gas whose conductivity is to be measured is made to flow through it and the other chamber is sealed and contains a reference gas such as air. The temperature of the filament in measuring chamber rises as the thermal conductivity of the gases passing around it decreases because the heat is unable to pass out of chamber through its walls. The resistance of the filament wires is connected to a Wheatstone bridge. The method is convenient for determination of proportion of two gas mixture. Using this method for a complex gas mixture would involve passing the sample first through a chamber for conductivity measurement, then through an apparatus which removes one constituent gas and then through another chamber for conductivity measurement. The difference in two conductivity measurements would be linked to the component gas which was eliminated. Such instruments are often called *Katharometers.*

Radioactivity Method

`There are two types of techniques used in this method namely (i) Dilution technique and (ii) Activation technique. In the 'Dilution technique', the element to be detected is diluted with one of its isotopes and the specific activity of the diluted substance is measured with a detector. This method is very adversely affected by impurities of the sample.

`The 'Activation technique' uses the artificial transformation of the substances being examined into their radioactive isotopes and then the measurement of the quantity of disintegration products. The method is not affected by impurities.

16.1.4 Selecting Methods of Measurement of Pollutants in Air

`Selecting the method of measurement of air pollutant among the various options,' depends upon the resources available to sustain the measurement over a long time, detection limit of the methods, degree of skill required etc. Automatic analyzers are convenient, but need skilled manpower to operate and maintain them. *(For choice of Instruments see* ***Para 16.10 of this chapter****).* Measurement of pollutants by wet chemical methods is simpler if a good laboratory is available. The detection limit of the method should be lower than the expected concentrations in an area. Instruments used for the air quality monitoring should be easy to use and calibrate and should require minimum maintenance.

For determination of concentration of gaseous pollutant in air, it is necessary to obtain representative samples of air at a number of strategic points. The concentrations are likely to vary with time and distance from the source of pollution depending upon the nature of release, meteorological factors and local conditions such as topography and presence of buildings and vegetation. The measured concentrations are also known to depend upon the period of sampling. These aspects are required to be taken into account for sampling and interpretation of the measured values.

For oxides of nitrogen and sulphur, ozone and other pollutants, sampling is done by absorption of the pollutant in reagent solutions which react with them at atmospheric temperature and pressure when air is bubbled through the absorbing solution in the impinger.

Sampling period and rate of sampling shall vary with the type of sampling program and its purpose. Normally the sampling periods are 30 minutes, one hour, one to four hours and eight hours depending upon the expected concentration of the pollutant, its nature and the investigation patterns Based on practical experience, the air sampling rates with respect to sampling period are mentioned below.

Sampling Period	Corresponding Rate of Sampling in litre per minute (L/min)
30 minutes	2
1 hour	1
1 to 4 hours	0.5
8 hours	0.2 to 0.5
8 to 24 hours	0.1 to 0.2

Annexure 16, Table (i)/16 gives the sampler and sorption medium used for some gases/ vapours.

Interference in the Measurements

The interferences are governed by the following factors:

(i) Chemical changes such as chemical interactions among the components of the collected sample or photochemical decomposition.

(ii) Adsorption of the gases from the sample on to the walls of the container and leaks

(iii) The interval between collection and analysis of samples as short as possible, (protect the samples from light and heat).

Analysis Methods

The following methods are briefly described in this chapter:

- Sulphur dioxide – Improved West and Gaeke method
- Nitrogen dioxide – Modified Jacob and Hochheiser method
- Ozone – Chemical Buffered method
- Ammonia – Indo Phenol Blue Method
- Particulate matter of size less than 2.5 µm – Gravimetric method
- Particulate matter of size less than 10 µm – Gravimetric method
- Lead – AAS / ICP Sampling using EPM 2000 – Atomic absorption spectrometric method
- Nickel – AAS / ICP Sampling using EPM 2000 – Atomic absorption spectrometric method
- Benzo(a)pyrene (BaP) – Solvent extraction followed by HPLC/ GC Analysis

16.2 SAMPLING AND ANALYSIS FOR SULPHUR DIOXIDE IN AMBIENT AIR *(IMPROVED WEST AND GAEKE METHOD)*

Principle of the Method

This method is given in *'IS 5182 Part 2 Method of measurement of air pollution: sulphur dioxide'*. Sulphur dioxide from air is absorbed in a solution of potassium tetra-chloromercurate (TCM). On absorption in this sorbent, a di-chloro-sulphitomercurate complex, (which resists oxidation by the oxygen of the air) is formed. This complex is stable to strong oxidants such as ozone and oxides of nitrogen and therefore, the absorber solution may be stored for some time prior to analysis. The complex is made to react with pararosaniline and formaldehyde to form the intensely coloured pararosaniline methyl-sulphonic acid. The absorbance of the solution is measured by means of a suitable spectrophotometer.

Instrument/Equipment Required

The following items are necessary to perform the monitoring and analysis of sulphur dioxide in ambient air:

Analytical balance; Vacuum pump (Capable of maintaining an air pressure differential more than 0.7 atmosphere at the desired flow rate); Calibrated flow-measuring device to control the airflow from 0.2 to 1 L/min; Absorber: all glass 'Midget impinger'; Spectrophotometer (Capable of measuring absorbance at 560 nm, equipped with 1 cm path length cells); Glass wares (low actinic glassware to be used for analysis).

Reagents / Chemicals

(All the chemicals should meet specifications of 'Analytical Reagent grade'); Distilled water; Mercuric chloride; Potassium chloride / Sodium chloride; EDTA di-sodium salt; Sulphamic acid (0.6%); Formaldehyde (0.2%); Iodine solution (0.01 N); Starch indicator solution; Potassium iodate; 'Stock sodium thiosulphate solution (0.1 N); Sodium thiosulphate solution for titration (0.01N); Purified Pararosaniline stock solution—[*Take 0.5 g of Purified Pararosaniline (PRA) and dissolve it in 100 ml of distilled water and keep for two days. The solution can be preserved in refrigerator for 3 months]*;

Pararosaniline working Solution—[*10 ml of stock PRA is taken in a 250 ml volumetric flask. Add 15 ml conc. HCl and make up to volume with distilled water]*;

Absorbing Reagent, 0.04 M Potassium Tetra-chloro-mercurate (TCM)—[*Dissolve 10.86 g, mercuric chloride, 0.066 g EDTA, and 6.0 g potassium chloride or 4.68 g sodium chloride in water and bring to the mark in a one litre volumetric flask]*;
***(Caution:** It is highly poisonous; if it spills on skin, flush off with water immediately.)*

[The pH of this reagent should be approximately 4.0, but it has been shown that there is no appreciable difference in collection efficiency over the range of pH 5 to pH 3. The absorbing reagent is normally stable for six months. If a precipitate forms, discard the reagent after recovering the mercury.]

Standardized Sulphite Solution (for preparation of working sulphite-TCM solution)

Dissolve 0.30 g sodium metabisulphite ($Na_2S_2O_5$) or 0.40 g sodium sulphite (Na_2SO_3) in 500 ml of recently boiled and cooled distilled water. Sulphite solution is unstable; it is, therefore, important to use water of the highest purity to minimize this instability. This solution contains the equivalent of 320–400 µg/ml of SO_2.

The actual concentration of the sulphite solution is determined by adding excess iodine and back-titrating with standard sodium thio-sulphate solution. To back-titrate, measure, by pipette, 50 ml of the 0.01 N iodine solutions into each of two 500 ml iodine flasks A and B. To flask A (blank) add 25 ml distilled water and into flask B (sample) measure 25 ml sulphite solution by pipette. Put stopper to the flasks and allow it to react for 5 minutes. Prepare the working sulphite-TCM solution at the same time. Iodine solution is added to the flasks. By means of a burette containing standardized '0.01 N thio-sulphate', titrate each flask in turn to a pale yellow. Then add 5 ml starch solution and continue the titration until the blue colour disappears.

Concentration of sulphite solution:

$$\mathbf{C = [(V_1 - V_2) \times N \times K] / V}$$

Where, C = SO_2 concentration in µg/ ml; V_1 = Volume of thiosulphate for blank, ml;

V_2 = Volume of thiosulphate for sample, ml; N = Normality of thiosulphate

K = milli-equivalent weight of SO_2; V = Volume of standard sulphite solution, ml

Working Sulphite-TCM Solution

Measure 2 ml of the standard solution into a 100 ml volumetric flask by pipette and bring to mark with 0.04 M TCM. Calculate the concentration of sulphur dioxide in the working solution in µg/ml of SO_2.

This solution is stable for 30 days if kept in the refrigerator at 5°C. If not kept at 5°C, prepare fresh solution daily.

Sampling

Place 30 ml of absorbing solution in an impinger and sample for four hours at the flow rate of 1 L/min. After sampling, measure the volume of sample and transfer to a sample storage bottle.

Analysis

Replace any water lost by evaporation during sampling by adding distilled water up to the calibration mark on the absorber. Mix thoroughly and take out by pipette 10 ml of the 'collected sample' into a 25 ml volumetric flask. Add 1 ml 0.6% sulphamic acid and allow reaction time of 10 minutes to destroy the nitrite resulting from oxides of nitrogen. Add 2 ml of 0.2% formaldehyde solution and 2 ml pararosaniline solution and make up to 25 ml with distilled water. Prepare a blank in the same manner using 10 ml of unexposed absorbing reagent. After a 30 min.colour development interval (and before 60 minutes), measure and record the absorbance of samples and reagent blank at 560 nm wavelength of light. Use 'distilled water'; not the reagent blank, as the optical reference. The absorbance of sample minus the absorbance of blank is the true absorbance. The absorbance of sample as well as that of blank is measured against distilled water as reference.

Preparation of Standards

Transfer out by pipette 1, 2, 3, 4, 5, 6, 7, 8, ml of 'working Sulphite—TCM solution' into 25 ml volumetric flasks. Add TCM solution up to 10 ml mark in each flask. The reagents are added as per procedure given in the previous paragraph entitled 'Analysis'. A reagent blank with 10 ml absorbing solution is also prepared. The absorbance of each standard and reagent blank is read.

Standard Curve

Plot a curve 'absorbance (Y axis) versus concentration (X axis)'. Draw a line of best fit and determine the slope. The reciprocal of slope gives the calibration factor (CF).

Calculation

$$\mathbf{C\,(SO_2) = (A_s - A_b).\,CF.\,V_s\,/\,(V_a.V_t)}$$

Where, $C\,(SO_2)$ = Concentration of sulphur dioxide in µg/ m^3.;

A_s = Absorbance of sample; A_b = Absorbance of reagent blank; CF = Calibration factor; V_a = Volume of air sampled, m^3; V_s = Volume of sample, ml; V_t = Volume of liquid taken for analysis, ml

Reference: *IS 5182 Part 2 Method of Measurement of Air Pollution: Sulphur dioxide.*

16.3 SAMPLING AND ANALYSIS FOR NITROGEN DIOXIDE IN AMBIENT AIR *(MODIFIED JACOB AND HOCHHEISER METHOD)*

Principle of the Method

This method is given in IS 5182 Part 6-(Methods for Measurement of Air Pollution: Oxides of nitrogen). The ambient nitrogen dioxide (NO_2) is collected by bubbling air through a solution of sodium hydroxide and sodium arsenite. The concentration of nitrite-ion (NO_2) produced during sampling is determined colorimetrically by reacting the nitrite- ion with phosphoric acid, sulphanilamide, and N-(1-naphthyl)-

Ethylenediamine di-hydrochloride (NEDA) and measuring the absorbance of the highly coloured 'Azo-dye' at 540 nm wavelength of light.

Instrument/Equipment Required

The following items are necessary to perform the monitoring and analysis of 'Nitrogen dioxide contained in ambient air':

Analytical balance; Vacuum pump (capable of maintaining a vacuum of at least 0.6 atm. across the flow control device); Flow control device (capable of maintaining a constant flow of 200-1000 ml per minute through the sampling solution); Calibrated flow measuring device (to control the airflow from 0.2 to 1 L/min.); Absorber (a midget impinge); Spectrophotometer (capable of measuring absorbance at 540 nm equipped with 1 cm path length cells.); Glass wares (low actinic glassware to be used for analysis);

Reagents / Chemicals (All the chemicals should meet specifications of ACS Analytical Reagent grade). Distilled water; Sodium hydroxide; Sodium arsenite; Sulphanilamidl (Melting point 165 to 167°C); Hydrogen peroxide (30%); Phosphoric acid (85%); Sodium nitrite (Assay of 97% $NaNO_2$ or greater); Sodium nitrite stock solution (1000 µg NO_2 /ml); Sodium nitrite solution (10 µg NO_2 /ml.); Sodium nitrite working solution (1 µg NO_2 /ml): *Dilute with absorbing reagent, prepare fresh daily*;

Absorbing solution— *[Dissolve 4.0 g of sodium hydroxide in distilled water, add 1.0 g of sodium Arsenite, and dilute to 1,000 ml with distilled water]*;

N-(1-Naphthyl)-ethylene-di-amine Di-hydrochloride (NEDA)— *[A 1% aqueous solution should have only one absorption peak at 320 nm over the range of 260-400 nm wavelength of light. NEDA showing more than one absorption peak over this range is impure and should not be used]*;

Sulphanilamide solution— *[Dissolve 20 g of sulphanilamide in 700 ml of distilled water. Add, with mixing, 50 ml of 85% phosphoric acid and dilute to 1,000 ml. This solution is stable for one month, if refrigerated]*;

NEDA solution— *[Dissolve 0.5g of NEDA in 500 ml of distilled water. This solution is stable for one month, if refrigerated and protected from light]*;

Hydrogen peroxide solution— *[Dilute 0.2 ml of 30% Hydrogen peroxide to 250 ml with distilled water. This solution may be used for one month, if, refrigerated and protected from light]*.

Sampling

Place 30 ml of absorbing solution in an impinger and sample for four hours at the flow rate of 0.2 to 1 L/min. After sampling, measure the volume of sample and transfer to a sample storage bottle.

Analysis

Replace any water lost by evaporation during sampling by adding distilled water up to the calibration mark on the absorber, mix thoroughly. Pipette out 10 ml of the collected sample into a 50 ml volumetric flask. Pipette in 1 ml of hydrogen peroxide solution, 10 ml of sulphanilamide solution, and 1.4 ml of NEDA solution, with thorough mixing after the addition of each reagent and make up to 50 ml with distilled water. Prepare a blank in the same manner using 10 ml of unexposed absorbing reagent. After a 10 min colour development interval, measure and record the absorbance of samples and reagent blank at 540 nm wavelength of light. Use distilled water; not the reagent blank, as the optical reference.

Preparation of Standards

Transfer out by pipette 1, 2, 3, 4, 5, 6, 7, 8, 9, ml of working standard solution into 50 ml volumetric flasks. Fill to 20 ml mark with absorbing solution. A reagent blank with 10 ml absorbing solution is also prepared. Add reagents to each volumetric flask as in the procedure for analysis. Read the absorbance of each standard and reagent blank against distilled water reference.

Standard Curve:

Plot a curve absorbance (Y axis) versus concentration (X axis). Draw a line of best fit and determine the slope. The reciprocal of slope gives the calibration factor (CF).

Calculation

$$C(NO_2) = [(A_s - A_b) \cdot CF \cdot V_s] / [0.82\, V_a \cdot V_t]$$

Where, $C(NO_2)$ = Concentration of NO_2 in $\mu g/m^3$.

A_s = Absorbance of sample; A_b = Absorbance of reagent blank; CF = Calibration factor; V_a = Volume of air sampled, m^3; V_s = Volume of sample, ml; V_t = Volume of liquid taken for analysis, ml; 0.82 = Sampling efficiency

Reference: *IS 5182 Part 6 Methods for Measurement of Air Pollution: Oxides of Nitrogen*

16.4 SAMPLING AND ANALYSIS FOR PARTICULATE MATTER OF SIZE LESS THAN 10 MM (PM10) IN AMBIENT AIR

Principle of the Method (Gravimetric Method):

Air is drawn through a size-selective inlet and through a 20.3 × 25.4 cm (8 × 10 in) filter at a flow rate, which is typically 1132 L/min. Particles with aerodynamic diameter less than the cut-point of the inlet are collected, by the filter. The mass of these particles is determined by the difference in filter weights prior to and after sampling. The concentration of PM10 in the designated size range is calculated by dividing the weight gain of the filter by the volume of air sampled.

Instrument/Equipment

The following items are necessary to perform the monitoring and analysis of particulate matter of size less than 10 μm (PM10) in ambient air:

Analytical balance: Sampler (high volume sampler with size selective inlet for PM10 and automatic volumetric flow control); Calibrated flow-measuring device to control the airflow at 1132 L/min; Top loading orifice kit; Filter media (Glass fiber filter of 20.3 × 25.4 cm (8 × 10 in) size).

Sampling Method

In Field Sampling, tilt back the inlet and secure it according to manufacturer's instructions. Loosen the faceplate wing nuts and remove the faceplate. Remove the filter from its jacket and centre it on the support screen with the rough side of the filter facing upwards. Replace the faceplate and tighten the wing nuts to secure the rubber gasket against the filter edge. Gently lower the inlet. For automatically flow-controlled units, record the designated flow rate on the data sheet. Record the reading of the elapsed time meter. The specified length of sampling is commonly 8 hours or 24 hours. During this period, several hourly readings of the flow rate should be taken. After the required time of sampling, record the flow meter reading, take out the filter media from the sampler, and put it in a container or envelope.

Analysis

Filter inspection: Inspect the filter for pin holes using a light table. Loose particles should be removed with a soft brush. Apply the filter identification number or a code to the filter if it is not a numbered one. Condition the filter in an airconditioned room maintained within 20–30°C and 40-50% relative humidity or in an airtight desiccator for 24 hours. Take initial weight of the filter paper (Wi) before sampling.

Condition the filter after sampling in the airconditioned room or in an airtight desiccator for 24 hours. Take final weight of the filter paper (W_f).

Calibration

Periodical calibration of the sampler is done by 'Orifice transfer standard'.

Orifice Transfer Standard: The PM10 sampler calibration orifice consists of a 3.175 cm (1.25 in) diameter hole in the end cap of 7.62 cm diameter × 20.3 cm long (3 in × 8 in) hollow metal cylinder. This orifice is mounted tightly to the filter support in place of the inlet during calibration. A small tap on the side of the cylinder is provided to measure the pressure drop across the orifice. A flow rate of 1132 L/min through the orifice typically results in a pressure difference of several inches of water. The relationship between pressure difference and flow rate is established via a calibration curve derived from measurements against a primary standard such as a 'Roots meter' at standard temperature and pressure.

Calculation

$$\mathbf{CPM10 = (W_f - W_i) \times 10^6 / V}$$

Where, CPM10 = Concentration of PM10 in $\mu g/m^3$; W_f = final weight of filter in g; W_i = Initial weight of filter in g : V = Volume of air sampled, m^3

10^6 = Conversion of g to μg;

Reference:

Method 501, Air Sampling and Analysis, 3rd Edition, Lewis publishers IncIS 5182 Part 23 Method of Measurement of Air Pollution: Respirable Suspended Particulate Matter (PM10) cyclonic flow technique; Method IO-2.1 Sampling of Ambient Air for Total Suspended Particulate Matter (SPM) and PM10 Using High Volume (HV) Sampler.

16.5 SAMPLING AND ANALYSIS FOR PARTICULATE MATTER OF SIZE LESS THAN 2.5 MM (PM 2.5) IN AMBIENT AIR

Principle of Measurement by Gravimetric Method: The methods called 'Federal Reference Method (FRM)' and 'Reference and Equivalent Method (FEM)' are generally used for particulate matter of size less than 2.5 μm (PM2.5) and ambient air quality surveillance for particulate matter.

An electrically powered air sampler draws ambient air at a constant volumetric flow rate [16.7 litre per minute (Lpm)] maintained by a mass flow / volumetric flow controller coupled to a microprocessor into a specially designed inertial particle-size separator (i.e. cyclones or impactors) where the suspended particulate matter in the PM 2.5 size range is separated for collection on a 47 mm Poly-tetra-fluoro-ethylene (PTFE) filter over a specified sampling period. Each filter is weighed before and after sample collection to determine the net gain due to the particulate matter. The mass concentration in the ambient air is computed as the total mass of collected particles in the PM 2.5 size range divided by the actual volume of air sampled, and is expressed in $\mu g/m^3$. The microprocessor reads averages and stores five-minute averages of ambient temperature, ambient pressure, filter temperature and volumetric flow rate. The microprocessor also calculates the average temperatures and pressure, total volumetric flow for the entire sample run time and the coefficient of variation of the flow rate.

Interferences in the Determination of PM2.5 and their Control

The potential effect of body moisture or oils contacting the filters is minimized by using non-serrated forceps to handle the filters at all times. This measure also moderates the interference due to static electricity. Teflon filters accumulate a surface electrical charge, which may affect filter weight. Static electricity is controlled by treating filters with a "Static Master" i.e. static charge neutralizer prior to weighing. Placement of filters on a "Static Master" unit is required for a minimum of 30 seconds before any filter can be weighed.

Moisture content can affect filter weight. Filters must be equilibrated for a minimum of 24 hours in a controlled environment prior to pre- and post weighing. The balance room's relative humidity should be maintained at a mean value of 45 ± 5 % and its air temperature should be maintained at a mean value of 25.0 ± °C.

Airborne particulate can adversely affect accurate mass measurement of the filter. Filters undergoing conditioning should not be placed within an airflow path created by air conditioning ductwork, computer printers, or frequently opened doorways. Cleaning laboratory bench-tops and weighing areas daily, installing "sticky" floor mats at doorway entrances to the balance room and wearing clean lab coats over regular clothing can further minimize dust contamination.

Precision and Accuracy of Federal Reference Method (FRM)

The performance segment of the PM 2. 5 FRM specifies strict guidelines for controls that must be observed, as well as the range of precision and accuracy of those controls. The flow rate through the instrument is specified as 16.67 Lpm (1 m^3 /hr). This flow must be volumetrically controlled to a precision of 5% and an accuracy of 2%. The flow control must be upgraded at least every 30 seconds and recorded (logged) every five minutes. Barometric pressure, ambient temperature and filter temperature should be measured on the same schedule. Filter temperature should not exceed the ambient temperature by more than 5°C for more than 30 minutes. A fan blowing filtered ambient air through the enclosure provides the necessary cooling effect. It is necessary for the entire apparatus to provide accurate performance over a temperature range of –20 to 50°C. The supporting run-time (interval) data, which are stored in detailed 5-minute intervals in the sampler's microprocessor, as well as 24-hour integrated performance (filter) data, must be capable of being extracted at the completion of a 24-hour run. The FRM mandates the provision of an RS 232 port for this purpose. Data may be extracted to a portable computer. Mass of the filter deposit, flow rate through the filter and sampling time have typical precision of ± 0.2 mg, ± 5%, and ± 30 seconds, respectively. These uncertainties combine to yield a propagated precision of approximately ± 5 % at 10 μg/m^3 and approximately ± 2% at 100 μg/m^3.

Requirements for putting the Sampler

Samplers should be sited to meet the goals of the specific monitoring project 'For routine sampling to determine compliance with the 'National Ambient Air Quality Standards (NAAQS)'. The CPCB guidelines shall apply for locating the sampler (**see Para 16.1.1**).

Apparatus and Materials

Sampling equipment designated as FRM (Federal Reference Method) or FEM (Federal Equivalent Method); Certified 'Flow transfer standard for flow calibration'.

[Following established EPA methods and procedures, all calibration transfer standards (i.e. temperature, pressure and flow) should be certified at least once per year. Calibration of these transfer standards will be conducted by the transfer standard manufacturer.]; Certified 'Standards for pressure and temperature' (Optional); Electronic microbalance (with a minimum resolution of 0.001 mg and a precision of ±0.001 mg, supplied with a balance pan) *[The microbalance must be positioned on a vibration-damping balance support table]*; Calibration weights, utilized as 'Mass reference standards' *[should be non-corroding, range in weight from 100 mg to 200 mg, and be certified as conforming*

to NIST mass standards. The weights should be ASTM Class 1 category with a tolerance of 0.025 mg.]; Non-serrated forceps for handling filters; Non-metallic, non-serrated forceps for handling weights; Digital timer/stopwatch; 47 mm Filter: Teflon membrane, *[46.2 mm effective diameter with a polypropylene support ring or filters as recommended by FRM / FEM sampler manufacturer];* Filter support cassettes and covers; Filter equilibration racks; Relative Humidity / Temperature recorder; NIST-certified Hygrometer for calibration of relative humidity readings; NIST-certified Thermometer for calibration of temperature readings; Light box; Radioactive (alpha particle) Polonium-210 ("Static Master") antistatic strips *[for static charge neutralization however static charge gives low moderate interference in stability of reading of balance];* Antistatic, nitrate-free, phosphate-free, sulphate-free, and powder free vinyl gloves; Plastic petri-slide filter containers (Filter Cassette); Zip-lock plastic bags, 6"x 9"; Disposable laboratory wipes; Filter equilibration cabinets; Impactor (WINS); oil/grease.

Calibration

The following should be performed in accordance with the instructions of the manufacturer:

(i) Calibration and performance check of Sampler;
(ii) External leak check (following sampler repair or maintenance and at least monthly, perform a sampler external leak check according to the manufacturer's guidelines.);
(iii) Internal leak check;
(iv) Single-point ambient temperature and filter temperature;
(v) Verification check;
(vi) Ambient temperature calibration;
(vii) Filter temperature calibration;
(viii) Pressure verification check;
(ix) Pressure calibration;
(x) Single-point flow verification check;
(xi) Multi-point flow calibration according to procedure should be performed.

Filters Used

It may be noted that 'Selection of filters' is quite important. The quality of filter papers to be used should technically meet the desired specifications. It is preferable to prepare the estimate for the whole requirement and order the same in bulk with a request to supply the same batch/lot of filters to control analytical quality and blank values.

Considerations for the selection of filters are: Mechanical stability; Chemical stability; Particle or gas sampling efficiency; Flow resistance; Loading capacity; Blank values; Artefact formation; 'Compatibility' with analysis method; and 'Cost and availability'.

Required filter: diameter 47 mm, Teflon (PTFE) filter paper with Polypropylene support ring manufactured by M/s Whatman or M/s Pall Life Sciences or equivalent having 2 μm pore sizes, having very low background concentrations for ion and elements and properly inspected and readied before use.

Balance for Weighing

Gravimetric analysis of the filters needs to be performed with a microbalance. The sensitivity and reliability of the electro-balance is about + 0.001 mg or 1 μg. Though tolerances on re-weights of Teflon membrane filters are typically ± 0.010 mg, these sensitive balances require isolation from vibration and air currents. Balances placed in laminar flow hoods with filtered air minimize contamination of filters from particles and gases in laboratory air. Electrostatic effects contribute another main interference in gravimetric analysis of filters. It is established that residual charge on a filter could produce an electrostatic discharge between the filter on the pan and the metal casing of the electro balance, which induces non-gravimetric forces. This charge can be removed from most filter media by exposing the filter to a low-level radioactive source (500 Pico curies of Polonium 210) prior to and during sample weighing. For keeping good conditions for weighing in the lab, the following things are important:

Electro balance controls and calibration; Cleaning and maintenance of the sample handling room; Calibration and maintenance of balance; Cleaning; Thorough calibration of balance required once in 6 months; Stability check of balance required once in a month; Daily calibration of balance.

Field sampling

On the 'Field Data Log', fill-in the top portion of the form including the date/time of visit, the site identification, sampler identification, site name, filter ID number, sample start and stop dates and times, and field operator initials.

All necessary calibration and check procedures as above may be gone through and the necessary QA/QC checks or maintenance should be carried out. All maintenance activities should be recorded in the field log book; including time, date, and any concerns that might affect the quality of the sample.

After removing the filter to be installed from its protective filter cassette carrier, the filter should be fixed (according to the manufacturer's instructions) into place against the bottom of the 'WINS impactor'. The system clock may be checked to make sure that it is within 1 min of standard time.

For setting up the sampling 24 hour program, the operators' manual may be followed. The 'Filter setup screen' shows the start date and time and the end date and time for the next sample. To change the sampling parameters the operator's manual may be followed. Start sampling run and wait until the auto - diagnosis for all relevant parameters finishes and the sampler automatically switches over to 'SAMPLING mode'. The flow (16.7 LPM) and 'Sample volume' display on the screen should be as required.

Recovering the Sample and Data from a Completed Sample Run

From the 'Main screen', note the current sampler-operating mode (on the display). If the sampler is in 'WAIT mode' or the 'SAMP (sampling) mode', the sampler has not completed the previously scheduled sampling run. Do not disturb the sampler unless necessary.

If the sampler is in the 'DONE mode' or the 'ERR mode', press 'STOP'. This allows the sampler to write the final information into storage for the current sample run (which is to be done prior to filter exchange). The sampler mode should then indicate STOP. Open the filter exchange mechanism. Remove the filter carrier from the filter holder. Place the filter carrier in the filter cassette case.

Using the information displayed on the 'Filter data screen', complete the 'Field data log' with the following information from the completed sample run:

(a) Total sample volume; (b) Average flow rate; (c) Coefficient of variation; (d) Total run time; (e) Maximum temperature difference; (f) Minimum, average and maximum ambient temperatures; (g) Minimum, average and maximum filter temperatures; (h) Minimum, average and maximum pressures; (i) If the sampler indicated there was an error, note the error in the field log book and make any repairs as needed. Any repairs should be done prior to the next run date.

Calculation and Reporting of Mass Concentrations

The equation to calculate the mass of fine particulate matter collected on a Teflon filter is as below:

$$\mathbf{M2.5 = (M_f - M_i).10^3\ \mu g}$$

Where, M2.5 = Total mass of fine particulate collected during sampling period (μg);

M_f = Final mass of the conditioned filter after sample collection (mg); M_i = initial mass of the conditioned filter before sample collection (mg)

[10^3 = Unit conversion factor for milligrams (mg) to micrograms (μg)]

Field records of PM 2.5 samplers are required to provide measurements of the total volume of ambient air passing through the sampler (V) in m^3 at the actual temperatures and pressures measured during sampling. Use the following formula if V is not available directly from the sampler:

$$\mathbf{V = Qavg.\ t.10^{-3}\ (m^3)}$$

Where, V = total sample value (**m^3**); **Qavg** = average flow rate over the entire duration of the sampling period (L/min); t = duration of sampling period (min); [10^{-3} = unit conversion factor for liters (L) into cubic meters (**m^3**)]

The equation given below can be used to determine PM 2.5 mass concentration:

$$\mathbf{PM2.5 = M2.5 / V}$$

Where, PM2.5 = Mass concentration of PM2.5 particulates (μg/m^3); M2.5 = Total mass of fine particulate collected during sampling period (μg);

V = Total volume of air sampled (**m^3**)

References

1.0 40 CFR Parts 53 and 58 Revised Requirements for Designation of Reference and Equivalent Methods for PM2.5 and Ambient Air Quality Surveillance for Particulate Matter; Final Rule

2.0 CARB SOP MLD 055

3.0 Aerosol Science & Technology: The PM 2.5 Federal Reference Method (FRM)." 35(4):339-342

4.0 Federal Register/Vol. 72, No. 112/Tuesday, June 12, 2007/Rules and Regulations

5.0 PM2.5 Gravimetric Analysis - Revision 7, August 14, 2003, Page 2 of 24 RTI (Research Triangle Institute, US)

16.6 SAMPLING AND ANALYSIS FOR OZONE IN AMBIENT AIR

Principle of the Method (Chemical Method):

[Method 411, Air Sampling and Analysis, 3rd Edition (Determination of oxidizing substances in the atmosphere)]. Micro-amounts of ozone and the oxidants liberate iodine when absorbed in 1% solution of potassium iodide buffered at pH 6.8 $\pm$ 0.2. The iodine is determined spectro-photometrically by measuring the absorption of tri-iodide ion at 352 nm wavelength of light.
The stoichiometry is approximated by the following reaction:

$$O_3 + 3\,KI + H_2O = KI_3 + 2\,KOH + O_2$$

Instrument/Equipment

Analytical balance; Vacuum pump (Any suction pump capable of drawing the required sample flow rate of 1 to 2 L / minute); Calibrated flow measuring device (to control the air flow from 1 to 2 L/ minute); Absorber (All glass midget impinger); Spectrophotometer (capable of measuring absorbance at 352 nm wavelength of light); Glass wares (low actinic glassware to be used for analysis).

Reagents / Chemicals

(All the chemicals should meet specifications of Analytical Reagent grade)

Distilled water; Absorbing solution (1% KI in 0.1m Phosphate buffer)— [*Dissolve 13.6 g of potassium di-hydrogen phosphate ($KH_2 PO_4$), 14.2 g of di-sodium hydrogen phosphate ($Na_2 HPO_4$) or 35.8 g of the dodecahydrate salt ($Na_2 HPO_2$. 12 H_2O), and 10.0 g of potassium iodide in sequence and dilute the mixture to 1 L with water. Keep at room temperature for at least 1 day before use. Measure pH and adjust to 6.8 ± 0.2 with NaOH or $KH_2 PO_4$. This solution can be stored for several months in a glass stoppered brown bottle at room temperature without deterioration. It should not be exposed to direct sunlight*]; Stock solution 0.025 M I_2 (0.05N) – [*Dissolve 16 g of potassium iodide and 3.173 g of re-sublimed iodine successively and dilute the mixture to exactly 500 ml with water. Keep at room temperature at least 1 day before use. Standardize shortly before use, against 0.025 M $Na_2 S_2O_3$*]. The sodium thiosulphate is standardized against primary standard bi-iodate [$KH(IO_3)_2$] or potassium dichromate ($K_2 Cr_2 O_7$) ***0.001M I_2 Solution*** – Transfer by pipette exactly 4.00 ml of the '0.025 M Stock solution' into a 100 ml low actinic volumetric flask and dilute to the mark with absorbing solution. It should be protected from strong light. After use it should be discarded.

Sampling

Place 10 ml of absorbing solution in a standard impinger and sample for one hour at the flow rate of 1 L/min. Do not expose the absorbing reagent to direct sunlight. After sampling, measure the volume of sample and transfer to a sample storage bottle.

Analysis

If, appreciable evaporation of the absorbing solution occurs during sampling, add distilled water to bring the liquid volume to 10 ml. Within 30 to 60 minutes after sample collection, read the absorbance in a cuvette at 352 nm wavelength of light against a reference cuvette containing distilled water. Measure the absorbance of the unexposed reagent and subtract the value from the absorbance of the sample.

Calibration

Calibrating Iodine Solution - For calibration purposes exactly 5.11 ml of the 0.001 MI_2 solution (or equivalent volume for other molarity) is diluted with absorbing solution just before use to 100 ml (final volume) to make the final concentration equivalent to 1μl of O_3 /ml. This solution preparation accounts for the stoichiometry at standard conditions of 101.3 kPa and 25°C. Discard this solution after use.

Obtain a range of calibration points containing from 1 μl to 10 μl of ozone equivalent per 10.0 ml of solution. Prepare by individually adding 1.0, 2.0, 4.0, 6.0, 8.0 and 10.0 ml of the calibrating iodine solution to 10.0 ml volumetric flasks. Bring each to the calibration mark with absorbing reagent. Read the absorbance of each of the prepared calibration solutions at 352 nm wavelength of light against distilled water reference

Standard Curve

Plot a curve absorbance (Y axis) versus concentration (X axis). Draw a line of best fit and determine the slope. The reciprocal of slope gives the calibration factor (CF).

Calculation

$$C(O_3) = 1.962\ (A_s - A_b) \cdot CF / V_a$$

Where: $C(O_3)$ is the concentration in of O_3 in $\mu g / m^3$.

A_s = Absorbance of sample; A_b =Absorbance of reagent blank;

CF = Calibration factor; V_a = Volume of air sampled in m^3.

1.962 = Conversion factor µl to µg for ozone at 25°C

Reference: *Method 411, Air Sampling and Analysis, 3rd Edition (Determination of oxidizing substances in the atmosphere).*

16.7 SAMPLING AND ANALYSIS FOR AMMONIA (NH_3) IN AMBIENT AIR

Principle of the Method: (Indophenol - Blue Method):

Ammonia in the atmosphere is collected by bubbling a measured volume of air through a dilute solution of sulphuric acid to form ammonium sulphate. The ammonium suiphate formed in the sample is analyzed calorimetrically by reaction with phenol and alkaline sodium hypochlorite to produce indophenol. The reaction is accelerated by the addition of Sodium Nitroprusside as catalyst.

Instrument/Equipment

The following items are necessary to perform the monitoring and analysis of ammonia in ambient air: Analytical balance; Vacuum pump to maintain a flow rate up to 5 litres per minute; Calibrated flow measuring device to control the air flow from 1 to 2 litre/min;
Absorber: a midget impinger or a fritted bubbler; Spectrophotometer capable of measuring absorbance at 630 nm wavelength of light; Glass ware: low actinic glass wares to be used for analysis;

Reagents / Chemicals

[All the chemicals should meet specifications of ACS analytical reagent grade]
Distilled water; N Sulphuric acid (Absorbing solution); Sodium nitroprusside; 6.75 M sodium hydroxide; Sodium hypochlorite solution; Phenol solution 45% v/v; Sodium phosphate; Ammonium chloride or Ammonium sulphate; Hydrochloric acid; Ammonia stock solution (1 mg NH_3 /ml); Ammonia working solution (10 µg NH_3 / ml) (Prepare fresh daily):

Sampling

Place 10 ml of absorbing solution in an impinger and sample for one hour at the flow rate of 1 to 2 L/ min. After sampling, measure the volume of sample and transfer to a sample storage bottle.

Analysis

Transfer contents of the sample bottle to a 25 ml glass stopper graduated cylinder. Maintain all the solutions and sample at 25°C. Add 2 ml buffer. Add 5 ml of working phenol solution, mix, and fill to about 22 ml. Add 2.5 ml of working hypochlorite solution and rapidly mix. Dilute to 25 ml, mix and store in the dark for 30 minutes to develop colour. Measure the absorbance of the solution at 630 nm wavelength of light on a spectrophotometer using 1 cm cells. Prepare a reagent blank. Measure the absorbance of unexposed reagent and subtract the value from the absorbance of the sample.

Calibration

Pipette 0.5, 1.0, 1.5, 2.0 ml of working standard solution into 25 ml glass stoppered graduated cylinders. Fill each to 10 ml mark with absorbing solution. A reagent blank with 10 ml absorbing solution is also prepared. Add reagents to each cylinder as in the procedure for analysis. Read the absorbance of each standard against reagent blank.

Standard Curve

Plot a curve absorbance (Y axis) versus concentration (X axis). Draw a line of best fit and determine the slope. The reciprocal of slope gives the calibration factor (CF).

Calculation

$$C(NH_3) = (A_s - A_b) \cdot CF / V_a$$

Where, $C\ NH_3$ = Concentration of Ammonia in $\mu g/m^3$; A_s = Absorbance of sample

A_b = Absorbance of reagent blank; CF = Calibration factor; V_a = Volume of air sampled in m^3.

Reference: *Indophenol method (Method 401, Air Sampling and Analysis, 3rd Edition), Lewis publishers Inc..*

16.8 SAMPLING AND ANALYSIS FOR BENZO(A)PYRENE AND OTHER PAH IN AMBIENT AIR *(SOLVENT EXTRACTION AND ANALYSIS)*

Benzo (a) Pyrene written in short form as (BaP) is one of the most important constituent of PAH (Poly aromatic hydrocarbons) compounds and also one of the most potent carcinogens. This can be measured both in particulate phase and vapour phase. In the vapour phase the concentration of B(a)P is significantly less than in the particulate phase. The molecular formula of B(a)P is $C_{20}H_{12}$ having molecular weight 252.

Principle of the Method

It is based on BIS method IS 5182 (Part 12):2004 and USEPA method (TO-13). This method is designed to collect particulate phase PAHs in ambient air and fugitive emissions and to determine individual PAH compounds using capillary gas chromatograph equipped with flame ionization detector. It is a high volume (1.2 m^3/min) sampling method capable of detecting sub.ng/m^3 concentration of PAH in 24 hours sample (i.e. collected in 3 shifts of 8 hour each with 480 m^3 sampling volume of air).

Equipment/Instruments

PM10 high volume sampler, Whatman glass fibre (EPM-2000) or equivalent filter paper, Ultra sonicator (~40kHz frequency), Rotary evaporator (Buchi or equivalent type), 'Gas Chromatograph' with 'Flame ionization detector' fitted with a capillary column, Syringes (5 and 10 micro litre capacity), Variable volume micropipettes (0.5 & 1.0 ml capacity), Beakers (250 ml), Amber colour vials 3 ml and 5 ml capacity, Chromatographic column (200–250 mm x10 mm with Teflon stopcock).

Chemical/Solvents

All chemicals, calibration /reference standards of B(a)P, other PAHs, Tri-phenyl benzene (internal standard, ultra residue grade) solvents like Toluene, Cyclohexane (with minimum residue less than 0.005%) etc.) and other chemicals like Silica -Gel (60–80 mesh size) should be of highest purity and of reputed make with traceability/purity and analysis certificate.

Sampling

The sampling is done 24 hours using PM10 high volume sampler. Eight hourly samples use EPM- 2000 glass fibre or equivalent filter. The sampling is done twice a week, total of 104 days monitoring in a year. Particulate laden Benzo(a)Pyrene samples are collected on glass fibre filter (EPM 2000 or equivalent) using PM10 sampler at a flow rate of more than 1 m^3/min, at selected location(s). The sampling height may be between 3-10 m from the ground level for ambient air quality monitoring.

After sampling, filters are kept in the controlled laboratory conditions (20–25°C) in an envelope marked with necessary identification information if processed immediately; otherwise the filters are wrapped in aluminum foil and kept in a refrigerator at 4°C in dark to avoid photo oxidation of PAHs.

Sample Processing:

(i) Extraction:

Filter papers (half of all the filters papers collected in a day) are cut into strips using scissors and transferred to a 250 ml beaker. Add ~50 ml. of Toluene (GC/HPLC grade). These samples are extracted with toluene using ultrasonic bath for about 30 minutes. Repeat the procedure twice (50ml × 2 times) for complete extraction. Alternatively, sample can be extracted using 'Soxhlet extraction apparatus' for about 8 hours with Toluene and repeat it twice.

(ii) Filtration:

Filter the extracted samples with Whatman filter paper No.41 containing 2 g of anhydrous sodium sulphate (to remove moisture).

(iii) Concentration:

After filtration, the filtrate is concentrated using Rotary vacuum evaporator.

(iv) Clean-up with Silica gel:

To clean up the impurities, pass 2 ml of concentrated sample through silica gel column (pre-conditioned, 60–80 mesh, and 200–250 mm × 10 mm with Teflon stopcock). After cleaning add 5 ml of cyclo-hexane and collect the 'Elute' in 25 ml beaker. Repeat the process for at least 3 times and collect it in the same beaker. Alternatively, Solid Phase Extraction (SPE) may be used for cleaning up the impurities of the sample.

(v) Re-concentration with rotary vacuum evaporator:

The cleaned-up extract/filtrate (approximately 17 ml) is further concentrated using rotary evaporator and it is evaporated to nearly dryness with nitrogen.

(*vi*) Final Sample Volume:

The dried sample is re-dissolved in 1ml of toluene and transferred into 4 or 5 ml amber vials for final analysis.

(vii) Extracted Sample Storage:

Cover/cap the sample vials /tubes and mark with necessary identification. Keep it in a refrigerator at 4°C prior to the analysis.

Analysis/ Instrument Set-Up

GC Conditions: Injector: 300°C; FID Temp: 320°C;
Column: Ultra -2 (25m Length, 320μm diameter, 0.17μm film thickness) or equivalent; Oven: 120°C →2 min hold → 7°C/min → 300°C → 10 min hold; Run Time: 37.71 minutes; Carrier gas flow (N_2): 0.50 ml/min
Gases for FID Flame: H_2 flow: 40 ml/min; Zero grade air flow: 400 ml/min

Preparation of Standard Calibration Mixture

'Stock standard solution of PAH mix' is a standard solution of 16 Compounds including B(a)P (Dr. Ehrenstorfer, Germany made PAH mix 63) having concentration of 1000 mg/L or 1000 ng/µl in Toluene.

'Working standard solutions of PAH mix' (5, 10, 15, 20, 25 ng/µl concentrations) are prepared from stock solution by diluting 200 to 40 times the above stock solution of 1000 mg/L or 1000 ng/µl concentration with Toluene.

Internal Standard

1, 2, 3-Tri Phenyl Benzene of concentration ~1000 ng/µl is added in the working standard solution so that the final concentration of Internal standard is 10 ng/µl.

Detection Limit

The minimum detectable concentration in terms of B(a)P for a sampling period of 8 hours (with about 480 m^3 of air passed) will be 1ng/m^3 assuming 1.0 ml as the final volume of sample extract after clean-up and detectable concentration of 1ng/µl of that sample extract. High resolution capillary mass spectrometry or high pressure liquid chromatography can improve sensitivity down to 1ng/m^3.

References: *BIS Method; IS 5182 (Part 12):2004, USEPA Method TO-13.*

16.9 SAMPLING AND ANALYSIS FOR LEAD, NICKEL AND ARSENIC IN AMBIENT AIR *(ATOMIC ABSORPTION SPECTROPHOTOMETER METHOD)*

Principle of the Method

The 'Atomic Absorption Spectroscopy (AAS)' technique makes use of absorption spectrometry to assess the concentration of an analyte (material being analysed) in the sample. The method is based on active sampling using PM10 'High volume sampler' and then sample analysis is done by atomic absorption spectrophotometer.

Instrument/Equipment

The following items are necessary to perform the protocol for monitoring lead and nickel in ambient air: PM10 sampler (high volume design based); Hot plate; Microwave Digestive System; Analytical balance; Digestion chamber; Polyethylene or polypropylene bottle; Glass wares; Top loading orifice kit; FAAS (Flame atomic absorption spectrophotometer) or GFAAS (Graphite furnace atomic absorption spectrophotometer).

Reagents / Chemicals

Filter Paper: EPM 2000 or equivalent, 20.3 × 25.4 cm (8 × 10 in) Concentrated AR grade acids: Hydrochloric acid (HCl), Nitric acid (HNO_3), and Sulphuric acid (H_2SO_4); Metal standard solutions

(Certified standard); GR/ AR grade Sodium borohydride and Potassium iodide; distilled / de-ionized water.

Sampling

Tilt back the inlet and secure it according to manufacturer's instructions. Loosen the face-plate wing-nuts and remove the face plate. Remove the filter from its jacket and centre it on the support screen with the rough side of the filter facing upwards. Replace the face-plate and tighten the wing-nuts to secure the rubber gasket against the filter edge. Gently lower the inlet. For automatically flow-controlled units, record the designated flow rate on the data sheet. Record the reading of the elapsed time meter. The specified length of sampling is commonly 8 hours or 24 hours. During this period, several hourly reading of flow rate should be taken. After the required time of sampling, record the flow meter reading and take out the filter media from the sampler and put it in a container or envelope.

Sample Storage

After collecting samples, transport the filters to the laboratory, taking care to minimize contamination and loss of the sample. Glass fibre filters should be transported or shipped in a shipping envelope. Store these protective envelopes at temperature upto 30°C until analysis. The maximum sample holding times is usually 180 days. Analyze the samples within 180 days.

Extraction of Sample for Analysis

The collected sample on glass fibre filters may be extracted by either hot plate procedure or by microwave extraction (Method IO-3.1).

(i) Microwave Extraction

Cut 1" x 8" strip or half the filter from the 8" x 10" filter sample and place on its edge in a labeled centrifuge tube using vinyl gloves or plastic forceps. Using the plastic forceps, crush the filter strip down into the lower portion of the centrifuge tube to ensure acid volume will cover the entire filter. Add 10.0 ml of the extraction solution to each of the centrifuge tubes (3% conc. HNO_3 and 8% conc. HCl). Place the centrifuge tubes in a 'Teflon' vessel containing 31 ml of deionized water. Place the vessel caps with the pressure release valves on the vessels, and hand-tight and tighten using the capping tool (to a constant torque of 12 ft-lb/ 16.3 Nm). Place the vessels in the microwave carousel. Connect each sample vessel to the overflow vessel using the teflon connecting tubes. Place the carousel containing the 12 vessels on to the turntable of the microwave unit. Irradiate the sample vessels at 486 W (power output) for 23 min. Allow the pressure to dissipate, then remove the carousel containing the vessels and cool it with tap water for 10 min. Uncap the vessels using the caping tool, and remove the labeled centrifuge tube containing samples. Add 10 ml of deionized water to each centrifuge tube. Cap the centrifuge tube tightly and mix the contents thoroughly for 2–3 minutes to complete extraction. The final extraction volume is 20 ml based upon the above procedure. Filter the extracted fluid with Whatman No. 41 and make up the final volume to 100 ml, the filtered sample is now ready for analysis.

(ii) Hotplate Procedure

Cut a 1″ × 8″ strip or half the filter from the 8″ × 10″ filter using a stainless steel pizza cutter. Place the filter in a beaker using vinyl gloves or plastic forceps. Cover the filter with the extraction solution (3% HNO_3 and 8% HCl). Place beaker on the hotplate, contained in a fume hood, and reflux gently while covered with a watch glass for 30 min. Do not allow sample to dry. Remove the beakers from the hot-plate and allow it to cool. Rinse the beaker walls and wash with distilled water. Add approximately 10 ml reagent water to the remaining filter material in the beaker and allow it to stand for at least 30 min. Transfer the extraction fluid in the beaker to a 100 ml volumetric flask or other graduated vessel. Rinse the beaker and any remaining solid material with distilled water and add the rinses to the flask. Dilute to the mark with distilled water and shake. The filtered sample is now ready for analysis

Analysis of Samples

A light beam containing the corresponding wavelength of the energy required to raise the atoms of the analyte (material being analyzed) from the ground state to the excited state is directed through the flame or furnace. This wavelength is observed by a mono -chromator and a detector that measures the amount of light absorbed by the element, hence the number of atoms in the ground state in the flame or furnace. A hollow cathode lamp (for the element being determined) provides a source of that metal's particular absorption wavelength.

There are two procedures namely 'Flame Atomic Absorption (FAA) spectroscopy' and 'Graphite Furnace Atomic Absorption (GFAA) spectroscopy'.

'Atomic absorption spectrophotometer' analyzes the metals by flame. If the quantities are below detection limits, the 'Graphite Tube Absorption (GTA) spectrophotometer' would be required. Arsenic is analyzed by flame – Vapour Generation Assembly (VGA).

Flame Procedure

Set the atomic absorption spectrophotometer for the standard condition as follows:

Choose the suitable hollow cathode lamp, align the instrument, position the mono-chromator at the value recommended by the manufacturer, select the proper mono-chromator slit width, set the light source current, ignite the flame, regulate the flow of fuel and oxidant, adjust the burner for maximum absorption and stability and balance the meter. Run a series of standards of the metal of interest and construct a calibration curve. Aspirate the blanks and samples. Dilute samples that exceed the calibration range. For Lead (Pb) and Nickel (Ni), the wavelengths required for analysis are 217 nm and 232 nm respectively whereas for Arsenic (As), the VGA (Vapour Generation Assembly) should be attached with Flame and the wavelength required for analysis is 193.7 nm.

Furnace Procedure

As a general rule, samples that can be analyzed by FAA or GFAA may be more conveniently run with flame, since flame atomic absorption is faster, simpler and has fewer interference problems. Read the metal value in µg/L from the calibration curve or directly from the read-out of the instrument.

Calibration

Prepare standard solutions from the stock solutions. Select at least three standard solutions to cover linear range as recommended by the method. Aspirate the standard solutions into the flame or inject the standard solutions into the furnace and record the absorbance. Prepare the calibration graph by plotting absorbance and concentration in µg/ml.

Preparation of Standards

For each metal that is to be determined, standards of known concentration must be acquired of commercially certified standards. Standard curve is prepared by using standard solutions of known concentration.

Calculations

Sample air volume can be calculated by using the following equation:

$$V = (Q)(t)$$

Where, V = volume of air, m^3; Q = average sampling rate, m^3/min.; t = time in minutes.

Concentration can be calculated by the equation:

$$C = (M_s - M_b) . V_s . F_a / V . F_t$$

Where, C = Concentration of metal in µg/m^3; M_s = Metal concentration µg/ml; M_b = blank concentration µg /ml; V_s = Total volume of extraction in ml; F_a = total area of exposed filter in cm^2; V = Volume of air sampled in m^3;

F_t = Area of filter taken for digestion in cm^2;

References:

Method IO-2.1 Sampling of Ambient Air for Total Suspended Particulate Matter (SPM) and PM10 Using High Volume (HV) Sampler;

Method 501, Air Sampling and Analysis, 3rd Edition, Lewis publishers Inc.;

IS 5182 Part 23 Method of Measurement of Air Pollution: Respirable Suspended Particulate Matter (PM10) cyclonic flow technique;

EPA compendium method IO 3, EPA compendium method IO 3.1 and EPA compendium method IO 3.2;

Standard Method- American Public Health Association (APHA), 20th Edition, 1998.

16.10 SYSTEMS FOR CONTINUOUS EMISSION MONITORING

A Continuous Emission Monitoring System (CEMS) is defined as the equipment necessary to continuously monitor, analyze, and record the gas concentrations, particulate matter content, and/or flow rate of an emission. They include the samplers, analyzers, software, and recording devices necessary to show compliance with the environmental regulations.

A 'CEMS' can be designed to monitor a single pollutant or multiple pollutants and waste gas stream parameters. Gaseous compounds, particulate matter, opacity, and volumetric flow rate are typically monitored by continuous monitoring system.

The CEMS for exhaust gases are divided into two major categories namely, 'Extractive CEMS' and 'In-situ CEMS'. While the 'In-situ CEMS' typically have monitors and/or analyzers located directly at the duct /stack from which exhaust gases flow out. Some monitoring system designs may employ both types of systems. Proper placement of sampling ports in the waste gas stream and proper equipment selection are all critical for getting accurate and reproducible information from a CEMS.

16.10.1 Extractive Continuous Emission Monitoring System (CEMS)

In an extractive CEMS, the system extracts a sample at a specified site in the waste gas stream (duct or stack) and then transports it from the sampling probe location to a monitor in an environmentally protected area to the analyzer / monitor. This type of system protects the monitoring instrumentation from high temperatures, high velocities, high pressures, particulate matter, corrosive substances, and water vapour in the waste gas stream. In general, the sample requires some form of conditioning prior to analysis. Conditioning can include, filtering of particulate mater, removal of water vapour, cooling of the sample, and dilution of the sample.

Types of Extractive Systems

The 'Extractive systems' are generally classified based on the type of conditioning: *hot-wet, cool-dry, or dilution.*

Hot-wet systems' maintain the sample at high temperature and do not remove water vapour, while the *'Cool-dry systems'* lower the sample temperature and remove the water vapour.

The above non-dilution extractive systems are used in industries and power utilities with low emission rates.

Extractive dilution systems

Dilution extractive CEMS operate similar to the source-level CEMS, but before the sample is conditioned, it is mixed with dry, clean air in a ratio between 50:1 and 200:1. The dilution extractive method is necessary when the flue gas is too wet, too hot, or too polluted for the analyzing equipment to handle. Once diluted to the appropriate ratio, the sample is transported through a sample line to a manifold from which individual analyzers may extract a sample. Conditioning of sample may be per formed at the probe / port, sampling transfer lines, line heaters, pump, filter, or at dyer/ chiller/ condenser. Since the sample is mixed with air, this method cannot be used to measure oxygen concentration of flue gas.

Depending on the type of application, extractive CEMS sampling and conditioning equipment can include sampling probe/port, sampling transfer lines, line heaters, a pump, a filter, a condenser or dryer, and chillers. The choice of sampling system is generally application-specific.

16.10.2 In-Situ Continuous Emission Monitoring System (CEMS)

In-situ CEMS' the analyzer is physically located in the stack or duct. The effluent gas is measured in-situ as it flows through a sampling location placed in the stack or duct. Two types of in-situ measurements are possible: 'Point in-situ (in-waste stream) and 'Path in-situ (across the waste stream).

'Point in-situ' measurements take place at the precise point where the sampling cell is located, while the 'path in-situ' measurements are taken across a given path in the emissions stream. Most path measurements are taken by sending a signal across the stack/duct and reflecting it back to a detector near the source of the signal. The emissions crossing that path are then averaged over a specified period of time.

In-situ monitors require durable construction and are generally enclosed in sturdy, sealed cabinets to protect them from extreme temperatures, moisture and corrosive gases. As a result, in-situ monitors are generally more expensive than comparable extractive monitors.

Both in-situ techniques have the advantage over extractive methods of avoiding many of the sampling problems faced by the extractive methods such as absorption of sample components, moisture from condensation, and chemical reactions between the sample components. The primary advantage of in-situ monitors is the location of the monitor in close proximity to the sampling probe, which minimizes the loss of contaminate from leaks, absorption, and adsorption, and also eliminates the need for a complex and costly sampling and conditioning system. Such monitors also avoid time delay in reporting.

The service, maintenance and replacement of in-situ analyzers is more difficult than with extractive units due to their locations. Because the concentration of pollutants (especially particulate matter) in a stack is not uniform, placement of the in-situ analyzer (like placement of the extractive analyzer's probe) is a critical consideration. The sampling probe can also become contaminated or plugged.

16.10.3 Principles used in CEMS Gas Analyzers and Monitors

16.10.3.1 *General*

A monitor is a device that senses or measures a physical/chemical property of a given substance such as light absorption. The sensor or measuring device generates an electrical output signal. Strip charts and/or computer data acquisition systems record the output signal, which correlates to a pollutant concentration or other parameter (e.g. flow rate) through an equation, graph, or more complicated mathematical relationship.

CEMS convert results into units of the applicable emission standard and provide a record (typically a printed chart and/or an electronic data file). Many integrated CEMS also include a calibration system for gas analyzers that automatically perform and record the required calibrations on a periodic basis (e.g., daily). They may also provide signal to computer based equipment operation control system.

Some analyzers can control and integrate calculation data allowing them to read actual concentrations on the front of the instrument. The configurations and installation requirements vary widely between different analyzers and applications.

Critical factors in selecting the type of analyzer or monitor for a particular application include gas concentration, stack and ambient temperatures and the presence of contaminants that could damage or interfere with the sampling or analyzer systems.

CEMS analyzers for extractive systems may be based on absorption spectroscopic methods (like spectrophotometry, differential absorption, Fouier transformation infrared, etc.), luminescence methods (like Fluorescence, Chemiluminescence, or Flame photometry), electro analysis methods (like polarography, potentiometry, conductivity, etc), Para magnetic methods (like thermo magnetic, magneto pneumatic, magneto dynamic).

CEMS gas analyzers for in-situ point systems may be based on polarography, potentiometry, electro catalysis, etc., while those for path systems may be based on differential absorption, or gas filter co-relation

CEMS monitoring of particulate matter for 'Point system' may be based on back scattering of light and absorption, ion charge, or nuclear radiation, while the same for 'Path systems' may be based on attenuation or light scattering and absorption.

16.10.3.2 *Chemiluminescence (Used for NO_x)*

In this technique the Light is produced in a chemical reaction of NO with ozone (O_3) to produce NO_2. A particular wavelength band (600–900 nm) is measured by a photo-multiplier tube that produces a signal proportional to the concentration. The reaction only works with NO, so any NO_2 in the sample has to be first converted catalytically to NO. The ozone is generated inside the analyzer from dry air. The NO_2 to NO converter can usually be bypassed to allow measurement of only the existing NO. This permits the calculation of NO, NO_2, and total NO_x in the sample. Excess ozone should be provided to the reaction chamber. The sample flow as well as temperature of reaction chamber, NO to NO_2 converter, and catalyst should be monitored. Other oxides of nitrogen in NO_x and ammonia interfere with the reaction. *[Examples of instruments using this technique are: API models 252 & 200; Thermo Environmental Instruments models 42 & 10AR]. There can be dual channels in analysis in a system ie one with 'NO to NO_2 converter' and the other without converter.]*

16.10.3.3 *Electro-catalysis* (O_2 Measurement by Instrument with Zirconium oxide)

A zirconium oxide cell type instrument for measuring oxygen is based on such a principle. A solid electrolyte (commonly zirconium oxide ZrO_2) coated with platinum is maintained at approximately 850°C. Sample gas is constantly passed over one side, while reference gas of high concentration is passed over the other side. The ions move across the electrolyte in order to reach equilibrium. The voltage difference between two sides and the partial pressure of the reference gas can be used to calculate the concentration of the sample gas. The presence of a combustible gas at the temperatures attained in analyzer will give lower reading of oxygen.

The voltage output is related with absolute temperature, partial pressure of reference gas, partial pressure of sample gas, and cell constant by the following equation:

$$E = 0..0496\ T \log_{10} [P_0 / P_s] + C$$

Where, E = Voltage output in mV

T = Absolute Temperature

P_s = Partial pressure of sample gas

P_0 = Partial pressure of reference gas

C = Cell constant in mV

Note: *Zircona Oxygen Analyser Systems are microprocessor based instrument designed primarily to meet the requirements of boilers*

16.10.3.4 *Membrane Separation with Electrochemical Cell*

The instruments working on this principle require membrane separation of gas to be measured from the sample. The sample gas enters the upper chamber where the gas to be measured passes through a selectively permeable membrane. After passing through the membrane, the gas diffuses across the electrolyte liquid and reaches the sensing electrode. An oxidation-reduction reaction with the electrode material takes place and electrons flow through the electrode to the resistor where the current is measured. Once reaching the counter electrode, they again join a reaction involving the electrode, the electrolyte, and the byproducts of the first reaction. The electrolyte is to be replaced regularly for obtaining consistent results. The particulate matter and moisture must be removed by conditioning equipment before the sample enters the gas separation membrane chamber. There are portable instruments based on this principle.

16.10.3.5 *Fluorescence /UV Fluorescence (for SO_2)*

In the instruments based on fluorescence, the UV light between wavelength of 190–230 nm is focused into a chamber containing sample gas. The molecules of SO_2 absorb that energy and then give it off again at a different wavelength. The 'fluorescence' is measured by a photo-multiplier tube and related to the total UV energy input to give a concentration of SO_2. Bigger molecules of Hydrocarbons can interfere and therefore these should be removed upstream of detector. *[Examples of the instruments based on fluorescence Thermo Environmental Instruments model 43; Western Research 721AT; API model 152].*

16.10.3.6 *Fourier Transform Infrared Spectroscopy*

The Fourier transform infrared spectroscopy based continuous monitoring instruments can monitor up to six gaseous compounds like SO_2, NO_x, CO, HCl, CO_2, and O_2, various hazardous air pollutants, and volatile organic compounds simultaneously.

The Fourier Transformation Infrared spectroscopy (FTIR) detects compounds based on the absorption of infrared light at critical wavelengths. The amount of absorption is dependent on the molecular bonds present in the waste gas compounds. This absorption creates a unique 'Fingerprint', or 'Chemical signature', that can be analyzed to determine the compounds present and their concentrations.

The infrared spectrum of a sample is recorded by passing a beam of infrared light through the sample. When the frequency of the infrared light is the same as the vibrational frequency of a bond, absorption occurs. The transmitted light is examined and the amount of energy absorbed at each frequency/ wavelength is revealed. This can be achieved by scanning the wavelength range using a mono-chromator. Alternatively, the whole wavelength range is measured at once using a Fourier transform instrument and then a transmittance or absorbance spectrum is generated using a dedicated procedure. Analysis of the position, shape and intensity of peaks in this spectrum reveals details about the molecular structure of the sample.

Gaseous samples require a sample cell with a long path length to compensate for the diluteness. The path length of the sample cell depends on the concentration of the compound of interest. A simple glass tube with length of 5 to 10 cm equipped with infrared windows at the both ends of tube can be used for concentrations down to several hundred parts per million. Sample gas concentrations well below ppm can be measured with a special device called 'White's cell' in which the infrared light is guided with mirrors to travel through the gas. White's cells are available with optical pathlength starting from 0.5 m up to 100 m.

A beam of infrared light is produced, passed through an interferometer, and then split into two separate beams. One is passed through the sample, the other is passed through a reference. The beams are both reflected back towards a detector, however first they pass through a splitter, which quickly alternates which of the two beams enters the detector.

The two signals are then compared and a printout is obtained. This "two-beam" setup gives accurate spectra even if the intensity of the light source drifts over time.

To take the infrared spectrum of a sample, it is necessary to measure both the sample and a "reference" (or "control"). This is because each measurement is affected by not only the light-absorption properties of the sample, but also by the type of the instrument (for example, which light source is used, which type of 'infrared detector' is used, etc.). The reference measurement makes it possible to eliminate the instrument influence.

These systems have extractive sampling. Although FTIR instruments appear to be more expensive than other analyzers, but their ability to monitor multiple pollutants with one instrument improves its cost economy.

Due to the precision of the instrument, maintenance requirements are high such as: highly skilled maintenance staff, careful and lengthy calibration procedure, shorter frequency of maintenance, and costly spare parts. A schematic diagram of two beam absorption spectrometric measurement system is shown in **Fig 16.1**.

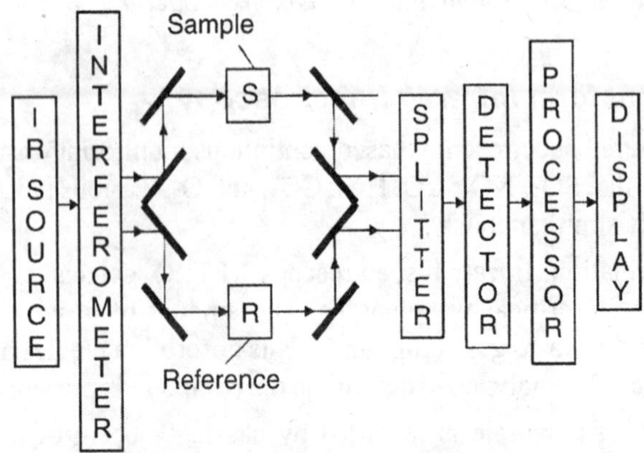

Fig. 16.1: Schematics of Two Beam Absorption Spectrometer

16.10.3.7 *Para Magnetic Analytical Technique based Instruments*

The operating principle of instruments based on 'Para-magnetism' is derived from the property of molecules of certain gases which get attracted to magnetic field. One such important gas is oxygen whose molecules when move get attracted, unlike most others, to a magnetic field. The movement of O_2 molecules through a field causes several types measurable effects utilized in several types of analyzers:

(i) In one type of analyzer, the flow of O_2 molecules over a heated coil causes a change in its electrical resistance. This property is used to measure flow of oxygen gas.

(ii) In the second type of analyzer, the changing partial pressure of oxygen in an uneven magnetic field causes changes in the field itself which swings a 'dumbbell' shaped object suspended over there. This effect is utilized to measure the flow of O_2 gas.

(iii) A third type of analyzer, measures the unbalanced flow of nitrogen (N_2) entering a chamber from two sides with O_2 in the sample gas being attracted to one side. This effect is used to measure Oxygen.

In such analyzers interference is caused by significant content of nitric oxide and nitrogen per oxide. Examples of such instruments are *Servomex '1420', Rosemount '755R', Hartmann & Braun '21113'.*

16.10.3.8 *Ion Mobility Spectrometry*

In the instruments based on ion mobility spectrometry, the sample and carrier gas is first forced through a membrane, and then ionized by a weak radiation source. It is allowed to drift through an electrical field in a tube leading to a detector. The different arrivals of the gas components are measured in 'intensity' and 'time' for making a graph. A microprocessor determines the concentration of gas. The example of this instrument is *'FP-IMS' of Environmental Technologies Group.* It may be noted that:

(i) The membrane is specially selected,
(ii) The polarity of the field can be biased (+/-),
(iii) The carrier gas can be 'doped' to suppress interference,
(iv) The drift times of compounds are very specific.

16.10.3.9 *Opacity Monitor for Particulate Matter*

Opacity monitors are in-situ path devices. These are based on the principle of transmissometry i.e. the measurement of the transmission of light through the exhaust gases or the fluid to be monitored. The absorption and scattering of light due to particulate matter in the gas stream is detected at a specified optical wavelength/ frequency that minimize absorption by other material in the stack gas. The light of known required frequency from the LED (Light emitting diode), incandescent light, or laser is passed through the gases to be monitored across the duct or stack. The opacity monitor then detects the decrease in light transmission across the gas stream due to particulate matter. Interference caused by high levels of NO_2 and water droplets can reduce accuracy.

An opacity monitor consists of a light source for generating the light, a transmisso-meter for accurately measuring the transmission of light, an internal reference system for calibration, and a data acquisition system for data collection. The example of such an analyzer is: *Dynatron '301', Datatest '900'.* A new analyzer offered *by KVB (model 'MIP LM3086EPA')* uses a low wattage laser in place of the usual light source.

16.10.3.10 *Beta Attenuation based Particulate Monitoring*

The 'Beta attenuation method' is very close to the gravimetric method already described (**Para 16.4 and 16.5 of this chapter**). In this method a strip of filter media is exposed to a known volume of the gas stream. The filter media then goes through a beta ray source and detector that measures the attenuation of the beta source by the Particulate matter on the filter. This method is subject to variation due to high beta attenuation of heavy metal in the particulate matter. A schematic diagram of the particulate matter monitoring by this method is shown in **Fig. 16.2.**

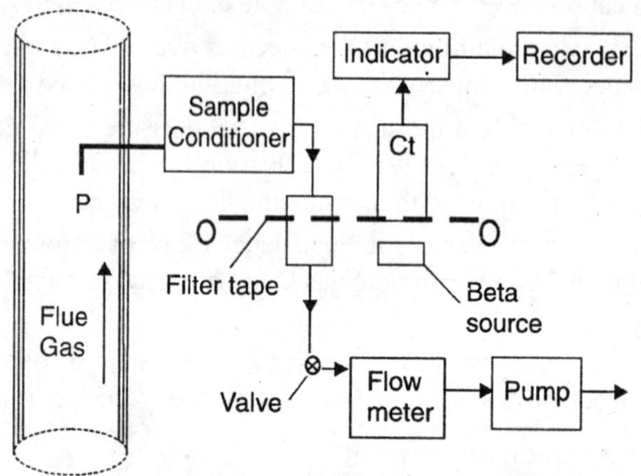

Ct–Geiger-Muller Counter; P–Probe

Fig. 16.2: Beta Attenuation Monitor for Suspended Particulate Matter

16.10.3.11 *Particulate Monitor based on 'Tribo flow'*

The 'Triboflow principle' is an impact principle of operation, where a 'Sensor' is placed in the path of emissions which are to be monitored. As the emission particles strike the sensor, a 'Triboflow' signal (electric current/voltage) is generated in relation to the particles striking the sensor. These signals are then conditioned, amplified and processed through an electronic circuitry and micro-processors and an 'Output signal' is generated which is directly proportional to the emissions flowing in the duct /stack. This principle works only for solid particles and is not affected by other fluid material present in the exhaust. It is particularly suitable for monitoring of Suspended Particulate Matter (SPM) of stack.

The sensor of the CEMS stack monitor is mounted on the stack/ chimney/ duct and the existing SPM level can be determined using conventional thimble. The CEMS stack monitor can be calibrated for direct and continuous on-line display of SPM emission level in mg/Nm^3. The milli-ampere analog output from the instrument can also be provided for recording or control purposes and for integration with the central control system.

Triboflow principle is used in the stack monitors / CEMS. These systems are internationally accepted and approved (such as by EPA of US and TUV of Germany). These instruments give a continuous digital display in mg/Nm^3 of the emissions of SPM in the stacks or ducts which are being monitored. A schematic diagram of the system is shown in **Fig. 16.3.**

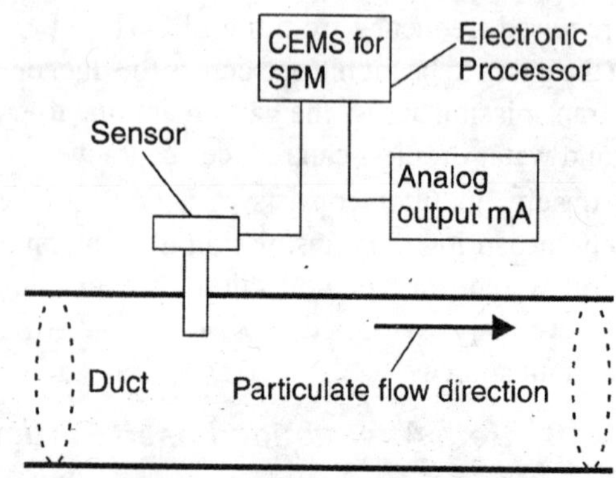

Fig. 16.3: Continuous Particulate Monitoring System Based on Tribo Flow

16.10.4 Examples of some Continuous Monitoring Systems/ Instruments

A few monitoring systems and instruments of a few manufacturers are mentioned below as examples in **Table 16.1**

Table 16.1: Certified Continuous Monitoring Systems

[Source: MCERTS Certified continuous monitoring systems CEMS — SIRA Certification Service UKAS Product Certification MCRTS]

Se. No.	Manufacturer and Model	Certified Range	Performance Standards
1	ABB Automation Model AO 2000- Multi-FID14	TOC—0 to 15 mg / m^3.	
2	ABB Automation Advance CEMAS FTIR-NT(ACF-NT)	CO—0 to 75 mg/ m^3, TOC—0 to 15 mg/ m^3, NO—0 to 200 mg/ m^3, NO—0 to 400 mg/ m^3, SO_2—0 to 75 mg/ m^3, SO_2—0 to 300 mg/ m^3, HCl—0 to 15 mg/ m^3, NH_3—0 to 15 mg/ m^3, HF—0 to 5 mg/ m^3, HF—0 to 10 mg/ m^3, O_2—0 to 25% Vol., O_2—0 to 12% Vol.,	MCERTS Performance Standard V3.1 dated July 2008 EN15267 QALI
3	ABB Automation AO 2000Limas11 UV	SO_2—0 to 75 mg/ m^3, NO—0 to 200 mg/ m^3, NO—0 to 33.5 mg/ m^3, NO_x—0 to 50 mg/ m^3, NO_x—0 to 300 mg/m^3, NO_x—0 to 125 mg/ m^3, O_2—0 to 10 % vol, O_2—0 to 25 %vol (NO_x expressed as NO_2)	MCERTS Performance Standard V3.1 dated July 2008 EN15267 QALI
4	ABB Automation AO2000 series Multigas Analyzers with SCCC-K NO/ NO_2 Converter including Uras26, Magnos206, ELsenser	CO—0 to 75 mg/m^3, and 0 to 4000 mg/ m^3, NO—0 to 100 mg/m^3, and 0 to 5000 mg/m^3, NO_x—0 to 150 mg/m^3, and 0 to 7500 mg/m^3, SO_2 0 to 75 mg/m^3 and 0 to 8000 mg/m^3, N_2O—0 to 100 mg/m^3, and 0—6700 mg/ m^3, CO_2—0 to 20% vol, O_2—0 to 10% vol and 0 to 25% vol	MCERTS Performance Standard V3.1 dated July 2008 EN15267 QALI
5	ABB Automation EL3000 Series Multigas Analyzers with SCC-K NO/NO2 converter Including: Uras26, Magnos206, El. Sensor	CO—0 to 150 mg/m^3, and 0 to 4000 mg/ m^3, NO—0 to 100 mg/m^3, and 0 to 5000 mg/m^3, NO_x—0 to 150 mg/m^3, and 0 to 7500 mg/ m^3, SO_2 0 to 300 mg/m^3 and 0 to 8000 mg/m^3, N_2O—0 to 100 mg/m^3, and 0 to 6700 mg/ m^3, CO_2—0 to 20% vol, O_2—0 to 10% vol and 0 to 25% vol	MCERTS Performance Standard V3.1 dated July 2008 EN15267 QALI

6	ABB Sp.A Dioxin Emission DECS- Continuous Sampling	Iso-kinetic Sampling Velocity 2 to 20 m/s	MCERTS Performance Standards forIsokinetic sampling V2 dated Sept 2005
7	AMETEK Process Instruments Models 909910, 919, 920 single and multigas continuous monitoring system (CEMS)	NO_2—0 to 750 ppm, and 0 to 1500 ppm, NO—0 to 500 ppm, and 0 to 1000 ppm SO_2—0 to 200 ppm, and 0 to 1000 ppm	MCERTS Performance Standard V3.2 dated Aug. 2010
8	Codel International Ltd G-CEM,4000 In-situ multigas Analyser	NO—0 to 134 mg/m^3 i.e. 0 to 100 ppm, and 0 to 1340 i.e. 0 to 1000 ppm NO_2—0 to 205 mg/m^3 i.e. 0 to 100 ppm CO—0 to 125 mg/m^3 i.e. 0 to 100 ppm and 0 to 1250 mg/m^3 i.e. 1000 ppm, SO_2—0 to 286 mg/m^3 i.e. 0 to 100 ppm, and 0 to 2860 mg/m^3 i.e. 0 to 1000 ppm, H_2O—0 to 30% vol	MCERTS Performance Standard V2.1 dated April 2003
9	Codel International Ltd G-CEM,4100 Extractive multigas Analyser	NO—0 to 134 mg/m^3 i.e. 0 to 100 ppm, and 0 to 1340 i.e. 0 to 1000 ppm, NO_2—0 to 82 mg/m^3 i.e. 0 to 40 ppm CO—0 to 125 mg/m^3 i.e. 0 to 100 ppm and 0 to 1250 mg/m^3 i.e. 1000 ppm, SO_2—0 to 286 mg/m^3 i.e. 0 to 100 ppm, and 0 to 2860 mg/m^3 i.e. 0 to 1000 ppm, H_2O—0 to 30% vol	MCERTS Performance Standard V2.1 dated April 2003
10	Dr FÖDISCH Umweltmess technik AG MCA04 Component Analyser	SO_2—0 to 75 mg/m^3 and 0 to 300 mg/m^3, NO—0 to 200 mg/m^3 and 0 to 395 mg/m^3, CO—0 to 75 mg/m^3 and 0 to 300 mg/m^3, HCl—0 to 15 mg/m^3 and 0 to 90 mg/m^3, NH_3—0 to 20 mg/m^3 and 0 to 75 mg/m^3, CO_2—0 to 20% vol, O_2—0 to 25% vol, H_2O—0 to 20% vol,	MCERTS Performance Standard V2.1 dated April 2003
11	Durag GmbH (i) D-FL100 flow monitor (ii) D-FL 200flow monitor (iii) DR290 Opacity and dust concentration monitor (iv) DR300-40 Dust concentration monitoring (v) DRx250 dust concentration and flow monitor (vi) D-R 800 Dust concentration Monitor	(i) Gas Velcity 0 to 20 m/s (ii) Gas velocity 0 to 16 m/s & 0 to 30 m/s (iii) 0 to 0.1 Extinction-opacity density (iv) Particulate concentration 0 to 3 mg/m^3 to 0 to 100 mg/m^3 (v) Particulate concentration 0 to 15 mg/m^3 and 0 to 30 mg/m^3, Velocity 0 to 30 m/s (vi) Particulate concentration 0 to 15 mg/m^3, 0 to 100% T	MCERTS Performance Standard V2.1 dated April 2003

12	Durag GmbH HM1400 TR Mercury Analyser	Total gaseous mercury 0 to 45 μg/m^3 to 0 to 90 μg/m^3,	MCERTS Performance Standard V2.1 dated April 2003
13	Emerson Process Management Manuf. GmbH & Co. OHG. UK. MLT2 Multi-gas analyzer. MLT3/4 Multi-gas analyzer	CO—0 to 75 mg/m^3, SO_2—0 to 75 mg/m^3, NO_2—0 to 50 mg/m^3, NO—0 to 200 mg/m^3, O_2—0 to 10% vol, 0 to 25% vol.	MCERTS Performance Standard V2.1 dated April 2003
14	PCME PCME QAL 181 Particulate analyzer including ICME QAL 181 ESN Senser	Particulates 0 to 15 mg/m^3, 0 to 100 mg/m^3,	MCERTS Performance Standard V3.1 dated July 2008 EN15267 QALI
15	PCME SC 620/680	Particulates 0 to 150 mg/m^3	MCERTS Performance Standard V2.1 dated April 2003
16	Environment SA CBISS Ltd (UK Agent) MIR-IS Multigas analyzer	CO—0 to 75, 0 to 500 mg/m^3, SO_2—0 to 75, 0 to 200 mg/m^3, N_2O—0 to 20; 0 to 200 mg/m^3, NO—0 to 100, 0 to 500 mg/m^3, O_2—0 to 10%, 0 to 25% vol. CH_4—0 to 10, 0 to 200 mg/m^3, CO_2—0 to 25% vol. HCl—0 to 15, 0 to 100 mg/m^3,	MCERTS Performance Standard V3.1 dated July 2008
17	Environment SA CBISS Ltd (UK Agent) MIR-FT Multigas analyzer	CO—0 to 75, 0 to 300 mg/m^3, SO_2—0 to 75, 0 to 500 mg/m^3, NO_2—0 to 200 mg/m^3, NO—0 to 200, 0 to 600 mg/m^3, O_2—0 to 25% vol CO_2—0 to 25% vol. HCl—0 to 15, 0 to 90 mg/m^3, NH_3—0 to 15 mg/m^3, N_2O—0 to 100 mg/m^3, H_2O—0 to 30% vol.;	MCERTS Performance Standard V3.1 dated July 2008

Note: *As mentioned in* ***Para 5.8 (Chapter 5)****, a number of advanced NO_x control systems based on the 'On-line optimization' of plant by means of 'Expert Systems' have been devised and are commercially available to minimize NO_x and achieving a specified target percentage carbon in ash for coal fired boilers.*

16.11 INDEX OF ATMOSPHERIC POLLUTION

16.11.1 Air Pollution Index (API)

The **Air Pollution Index (API)** is a simple and a generalized way of describing the air quality in China, Hong Kong, and Malaysia. It is calculated from several sets of air pollution data.

API Mechanics

An individual score is assigned to the level of each pollutant and the final API is the highest of those 5 scores. The pollutants can be measured quite differently. SO_2, NO_2 and PM_{10} concentration are measured as average per day. CO and O_3 are more harmful and are measured as average per hour. The final API value is calculated per day.

The scale for each pollutant is non-linear, as is the final API score. Thus an API of 100 does not mean twice the pollution of API of 50, nor does it mean twice as harmful. While an API of 50 from day 1 to 182 and API of 100 from day 183 to 365 do provide an annual average of 75, it does *not* mean the pollution is acceptable even if the benchmark of 100 is deemed safe. This is because the benchmark is a 24 hour target. The annual average must match against the annual target. It is entirely possible to have safe air every day of the year, but still fail the annual pollution benchmark.

China

China's State Environment Protection Agency (SEPA) is responsible for measuring the level of air pollution in China. SEPA monitored daily pollution level in 86 of its major cities. The API level was based on the level of 5 atmospheric pollutants, namely SO_2, NO_2, SPM less than 10 μm size (PM_{10}), CO and Ozone (O_3) measured at the monitoring stations throughout each city. The API and health implications (China) are shown in **Table 16.2.**

Table 16.2: API and Health Implications (Daily) China

API	Health Implications	Indicated by Colour
0 to 50 Pollution level: Excellent	No health implications	Green colour
51 to 100 Pollution level: Good	No health implications	Yellow colour
101 to 150 Pollution level: Slightly polluted:	Slight irritations may occur, individuals with breathing or heart problems should reduce outdoor exercise.	Orange colour
151 to 200 Pollution level: Lightly polluted	Slight irritations may occur, individuals with breathing or heart problems should reduce outdoor exercise.	Red colour
201 to 250 Pollution level Moderately polluted	Healthy people will be noticeably affected. People with breathing or heart problems will experience reduced endurance in activities. These individuals and elders should remain indoors and restrict activities.	Purple colour
251 to 300 Pollution level Heavily polluted	Healthy people will be noticeably affected. People with breathing or heart problems will experience reduced endurance in activities. These individuals and elders should remain indoors and restrict activities.	Maroon

300+ Pollution level: Severely polluted	Healthy people will experience reduced endurance in activities. There may be strong irritations and symptoms and may trigger other illnesses. Elders and the sick should remain indoors and avoid exercise. Healthy individuals should avoid out door activities.	Maroon

Note: *In mainland China, the API was replaced by an updated Air Quality Index (as described in subsequent* ***Para 16.11.2****) in early 2012.*

Hong Kong

The API has been in use in Hong Kong since June 1995. It is measured and updated hourly by the Environmental Protection Department (EPD). Moreover, the EPD makes forecast on the API for every following day.

The API is based on the level of 6 atmospheric pollutants, namely Sulphur dioxide (SO_2), Nitrogen dioxide (NO_2), Respirable Suspended Particulate, Carbon monoxide (CO), Ozone (O_3), lead (Pb), measured at all the monitoring stations throughout the territory. In Hong Kong, there are two types of API: General API and Roadside API. The EPD reports the latest APIs hourly.

The index and the air quality objectives were set in 1987; and pollutant levels are measured over varying periods, in **μg/m**^{3}. There are hourly, 24 hour and annual targets for SO_2 and NO_2 and 24 hour and annual targets for particulates. **Table 16.3** shows the official Health Implications of the respective API levels in *Hong Kong*.

Table 16.3: Health Implications of Various Reported APIs in HongKong

API	Health Implications	Air pollution level	Colour Indication
1. 0 to 25	Not expected.	Low	Green
2. 26 to 50	Not expected for the general population.	High	Pink
3. 51 to 100	Acute health effects are not expected but chronic effects may be observed if one is persistently exposed to such levels.	Very high	Purple
4. 101 to 200	People with existing heart or respiratory illnesses may notice mild aggravation of their health conditions. Generally healthy individuals may also notice some discomfort.	Very high	Purple
5. 201 to 500	People with existing heart or respiratory illnesses may experience significant aggravation of their symptoms. There may also be widespread symptoms in the healthy population (e.g. eye irritation, wheezing, coughing, phlegm and sore throats).	Severe	Maroon

Note:

In 1998, the Education Bureau recommended schools to curtail outdoor activities when the index reached 200, whereas leading healthcare advocates are urging that the level be revised to 100. The World Health Organization (WHO) revised its air quality guideline levels of SO_2, NO_2 *and ozone in 2006 in light of new scientific evidence. The WHO also introduced new measurement guidelines for very small particulates which are more dangerous to pulmonary function. At the '200' level, Hong Kong levels of* SO_2 *(800μg/m*^{3}*) and* NO_2 *(1,130 μg/m*^{3}*) are 40 times and 5½ times of WHO guidelines respectively; the equivalent for particulates (350μg/m*^{3}*) is 7 times of WHO guidelines.*

Malaysia

Similar to Hong Kong, the air quality in Malaysia is reported by API. Four pollutant components i.e. CO, 0_3, NO_2 and SO_2 of the index are reported in ppmv, but PM_{10} particulate matter is reported in $\mu g/m^3$. This scale in **Table 16.4** shows the Health classifications used by the Malaysian government.

Table 16.4: Health Classification / API in Malaysia

API	Air pollution level	Colour Indication
1. 0 to 50	Good	Green
2. 51 to 100	Moderate	Yellow
3. 101 to 200	Unhealthy	Pink
4. 201 to 300	Very unhealthy	Purple
5. 301+	Hazardous	Maroon

If the API exceeds 500, a state of emergency is declared in the reporting area. Usually, this means that non-essential government services are suspended, and all ports in the affected area are closed. There may also be a prohibition on private sector commercial and industrial activities in the reporting area excluding the food sector. *[It may be noted that Purple is very high, while maroon is severe.]*

16.11.2 Air Quality Index (AQI) and Air Quality Health Guide

The AQI is an index for reporting daily air quality. It indicates the degree of pollution in air, and the probable associated health effects to the public. The United States Environmental Protection Agency EPA (US) calculates the AQI for five major air pollutants regulated by the Clean Air Act (US). These ground level pollutants are: (i) Ozone, (ii) Particulate matter, (iii) CO, (iv) SO_2 and (v) NO_2. For each of these pollutants, EPA has established US National Air Quality Standards to protect public health.

The 'AQI yardstick' runs from 0 to 500. The higher the AQI value, greater is the level of air pollution and the greater is the health concern. For example, an AQI value of 50 represents good air quality with little potential to affect public health, while an AQI value over 300 represents hazardous air quality. An AQI value of 100 generally corresponds to the national air quality standard (US) for the pollutant, which is the level EPA has set to protect public health. AQI values below 100 are generally thought of as satisfactory. When AQI values are above 100, air quality is considered to be unhealthy-at first for certain sensitive groups of people, then for everyone as AQI values get higher. AQI ranges like 0 to 50, 51 to 100, 101 to 150, 151 to 200, 201 to 300, and above 300 have been categorized. Each category corresponds to a different level of health concern which are shown in **Table 16.5.**

Table 16.5: Air Quality Index and Health Concern

AQI	Level of Health Concern	Colour Indication
0 to 50	Good	Green
51 to 100	Moderate	Yellow
101 to 150	Unhealthy for Sensitive Groups	Orange

151 to 200	Unhealthy	Red
201 to 300	Very Unhealthy	Purple
301 to 500	Hazardous	Maroon
• *Green; 'Good' AQI: 0 to 50*: Air quality is considered satisfactory, and air pollution poses little or no risk.		
• *Yellow; 'Moderate' AQI: 51 to 100:* Air quality is acceptable; however, for some pollutants there may be a moderate health concern for a very small number of people. For example, people who are unusually sensitive to 'Ozone' may experience respiratory symptoms. • *Orange; 'Unhealthy for Sensitive Groups' AQI:101 to 150:* Although general public is not likely to be affected at this AQI range, people with lung disease, older adults and children are at a greater risk from exposure to 'Ozone', whereas persons with heart and lung disease, older adults and children are at greater risk from the presence of Particulate matter in the air.		
• Red; 'Unhealthy' AQI:151 to 200: Everyone may begin to experience some adverse health effects and members of the sensitive groups may experience more serious effects.		
• *Purple; 'Very Unhealthy' AQI: 201 to 300:* This would trigger a health alert signifying that everyone may experience more serious health effects.		
• *Maroon; 'Hazardous' AQI greater than 300:* This would trigger health warnings of emergency conditions. The entire population is more likely to be affected.		

Pollutant Concentration and AQI

The air quality index is a piecewise linear function of the pollutant concentration. At the boundary between AQI categories, there is a discontinuous jump of one AQI unit. Equation Used for Converting Concentration to AQI

$$I = \{(I_{high} - I_{low}) / (C_{high} - C_{low})\} (C - C_{low}) + I_{low}.$$

Where: $\mathbf{I}$ = Air Quality index; $\mathbf{C}$ = Pollutant concentration,

$\mathbf{C_{low}}$ = Concentration breakpoint that is $\leq$ C,

$\mathbf{C_{high}}$ = Concentration breakpoint that is $\geq$ C,

$\mathbf{I_{low}}$ = Index breakpoint corresponding to $\mathbf{C_{low}}$,

$\mathbf{I_{high}}$ = Index breakpoint corresponding to $\mathbf{C_{high}}$

EPA's Break Points

For EPA's Break Points for Particulates of ≤ 2.5 μm ($PM_{2.5}$) see **Table 16.6.**

Table 16.6: EPA's Break Points for Particulates of Size Less than 2.5 μm

C_{low}	C_{high}	I_{low}	I_{high}	**Category**
0	12.0	0	50	Good
12.1	35.4	51	100	Moderate

35.5	55.4	101	150	Unhealthy for Sensitive Groups
55.5	150.4	151	200	Unhealthy
150.5	250.4	201	300	Very unhealthy
250.5	350.4	301	400	Hazardous
250.5	350.4	301	400	Hazardous
350.5	500.4	401	500	Hazardous

Note: *For example, suppose a monitor records a 24-hour average fine particle (PM2.5) concentration of 12.0 μg/m*^{3}*. The AQI from the above equation is:* $\{(50 - 0) / (12.0 - 0)\}(12.0 - 0) + 0 = 50$
The above values are corresponding to air quality in the "Good" range. If multiple pollutants are measured at a monitoring site, then the largest or "dominant" AQI value is reported for the location. To convert an air pollutant concentration to an AQI, EPA has developed a calculator. Current ambient monitoring data and forecasts of air quality that are color-coded in terms of the air quality index are available from the web site 'AIRNow'.

United States National Ambient Air Quality Standards (NAAQS of US)

The Clean Air Act (US), which was last amended in 1990, requires EPA (US) to set National Ambient Air Quality Standards (40 CFR part 50) for pollutants considered harmful to public health and the environment. The US Clean Air Act identifies two types of national ambient air quality standards. ***Primary standards*** provide public health protection, including protecting the health of "sensitive" populations such as asthmatics, children, and the elderly. ***Secondary standards*** provide public welfare protection, including protection against decreased visibility and damage to animals, crops, vegetation, and buildings.

EPA (US) has set National Ambient Air Quality Standard for six principal pollutants, which are called "criteria" pollutants. They are listed in **Table 16.7.**

Table 16.7: EPA (US) National Ambient Air Quality Standards
[Units of measure for the standards are ppm by volume, ppb by volume, and μg/m^3 *of air]*

Pollutant	Primary / secondary	Averaging Time	Level	Form
Carbon monoxide [76FR 54297, Aug 31, 2011]	Primary	8-hour 1hour	9 ppm 35 ppm	Not to exceed more than once a year
Nitrogen dioxide [75 FR 6474, Feb 9, 2010]	Primary	1-hour	100 ppb	98 percentile averaged over 3 yrs.
[61 FR 52852, Oct 8, 1996	Primary and secondary	Annual	53 ppb	Annual Mean
Ozone	Primary and secondary	8-hour	0.075 ppm	Annual fourth-highest daily max. 8-hr concentration, averaged over 3 yrs.

Sulfur Dioxide [75 FR 35520, Jun 22, 2010] [38 FR 25678, Sept 14, 1973]	Primary	1-hour	75 ppb	99th percentile of 1-hour daily max. concentrations, averaged over 3 yrs, Not to be exceeded more than once a year.
	Secondary	3-hour	0.5 ppm	
Particle Pollution [Dec 14, 2012] $PM_{2.5}$	Primary	Annual	12 µg/m^3	annual mean, averaged over 3 yrs.
	Secondary	Annual	15 µg/m^3	annual mean, averaged over 3 yrs.
	Primary and Secondary	24-hour	35 µg/m^3	98th percentile, averaged over 3 yrs
Particle Pollution [Dec 14, 2012] PM_{10}	Primary and Secondary	24-hour	150 µg/m^3	Not to be exceeded more than once a year on average over 3 years.
Lead [73 FR 66964, Nov 12, 2008]	Primary and secondary	Rolling 3 month avg	0.15 µg/m^3	Not to be exceeded

Notes:

1. *Final rule was signed on October 15, 2008. The 1978 lead standard (1.5 µg/m^3 as a quarterly average) remains in effect until a year after the area is designated for the 2008 standard, except that in areas designated non-attainment for the 1978, the 1978 standard remains in effect until implementation plans to attain or maintain the 2008 standard are approved.*
2. *The official level of the annual NO_2 standard is 0.053 ppm = 53 ppb, which is shown here for the purpose of clearer comparison to the 1-hour standard.*
3. *Final rule signed on March 12, 2008. The 1997 ozone standard (0.08 ppm, annual fourth-highest daily maximum 8-hour concentration, averaged over 3 years) and related implementation rules remain in place. In 1997, EPA revoked the 1-hour ozone standard (0.12 ppm, not to be exceeded more than once a year) in all areas, although some areas have continued obligations under that standard ("anti-backsliding"). The 1-hour ozone standard is attained when the expected number of days per calendar year with maximum hourly average concentrations above 0.12 ppm is less than or equal to 1.*
4. *Final rule signed on June 2, 2010. The 1971 annual and 24-hour SO_2 standards were revoked in that same rule-making. However, these standards remain in effect until one year after an area is designated for the 2010 standard, except in areas designated nonattainment for the 1971 standards, where the 1971 standards remain in effect until implementation plans to attain or maintain the 2010 standard are approved.*

16.11.3 Basis of Indian Air Quality Index (IND—AQI)

Huge amount of data about ambient air quality is generated to know the quality of air environment for administering appropriate corrective actions wherever necessary, but such an endeavor would result in encyclopedic volumes of data which may not give a clear picture to a decision maker. This would also not be useful to a common man who simply wants to know how good or bad is the ambient air. The general public, usually, would not be satisfied with raw data, time series plots, statistical analysis and other complex findings pertaining to air quality. To address the above concerns, IIT Kanpur through a sponsored project from the Central Pollution Control Board, Delhi has proposed the *Indian Air Quality Index (IND - AQI)* in simple and lucid terms.

A segmented linear function is used relating the actual air pollution concentrations (of each pollutant) in a normalized number. The basis for these linear functions (for this study) is arrived after considering such functions adopted by other countries *(USEPA, 1976 and 1998; UK, 1998; Malaysia, 1997; GVAQI, 1997; Ontario, 1991; ORAQI, 1970)*.

The pollutants included for the proposed IND - AQI are: (i) *SPM, (ii) Particulate Matter of size less than 10* μm *(PM10), (iii)* SO_2*, (iv))* NO_2, (v) *CO, and (vi) Ozone.*

Since all the pollutants are not measured under the Indian National Ambient Air Quality Monitoring Program, it is necessary that at least three pollutant concentrations must be available to calculate and report the index. The over all AQI is based on maximum operator system where the maximum value of sub-index becomes the AQI. To reflect the attainment of NAAQS, the AQI is referred to as Good between the ranges 0 - 100. For the second break point (at the standard of USEPA), the AQI takes the value of 200 and referred to as Moderate. In absence of 'any other pollutant health criteria' in India, rest of categorization of Index is based on the USEPA Federal Episode criteria and Significant Harm Level.

Note: Features of Website Developed for Nation-wide Air Quality Index Display (India)

The display system is implemented using HTML, JavaScript, Java and CGI (Common Gateway Interface)/ Perl script. As the air quality data become available, online calculation of AQI is carried out and displayed as an Air Quality Meter showing index value (with a pointer) with animation on the screen. The website is comprehensively designed to indicate the pollutant responsible for index and the pollutants exceeding the standards. They developed website facilitates for intra city comparison of AQI using multiple windows for online as well as historical data of air quality. The general public can access the information through internet and other media agencies like newspaper, TV, radio, can also download the information and disseminate the information. Although this web site does not forecast the AQI for coming days, it however, does include the possible health effects so that the people can take precautions; as in all likelihood the AQI is not going to change significantly on next day. It may be mentioned that Dr Mukesh Sharma Prof, Dept of Civil Engineering, IIT Kanpur has proposed the Air Quality Index.

Table 16.8 presents the summary of the break point concentrations and AQI values for *India* (proposed) for all pollutants.

Table 16.8: Indian Air Quality Index for Different Pollutants

Index	**Category**	**SO_2 24 hr avg. μg/m^3**	**NO_2 1 hr avg μg/m^3**	**SPM (24-hr avg.) μg/m^3**	**CO (1-hr avg.) μg/m^3**	**CO (8-hr avg.) μg/m^3**	**O_3 (1-hr avg.) μg/m^3**	**NH_3 (24-hr avg.) μg/m^3**	**$PM_{2.5}$ (24-hr avg.) μg/m^3**	**Lead (24 hrs. avg.) μg/m^3**	**PM_{10} (24 hrs avg.) μg/m^3**
1	2	3	4	5	6	7	8	9	10	11	12
0-100	Good	0-80	0-80	0-200	0 to 4	0 to 2	0 to 180	0 to 60	0 to 400	0 to 1	0 to 100
101 to 200	Moderate	81 to 367	81 to 180	201 to 260	4.1 to 25	2.1 to 12	180 to 225	61 to 90	400 to 550	1 to 15	101 to 150
201 to 300	Poor	368 to 786	181 to 564	261 to 400	25.1 to 35	12.1 to 17	225 to 300	91 to 210	550 to 700	15 to 225	151 to 350

301 to 400	Very poor	787 to 1572	565 to 1272	401 to 800	35.1 to 75	17.1 to 35	301 to 800	211 to 250	700 to 900	225 to 325	351 to 420
401 to 500	Severe	>1572	>1272	>800	>75	>35	>800	>250	>900	>325	>420

The final AQI has been launched by *Indian Prime Minister on April 6th, 2015,* with slight changes in above as regards the air quality numbers and indicative colours,as shown in the **Table 16.9.**

Table 16.9: Indicative Colours of Air Quality Index for Air Pollution

AQI Numbers	Division of Quality	Pollution indicated by colours
0 to 50	Good	Dark Green
51 to 100	Satisfactory	Less Green
101 to 200	Moderate	Yellow
201 to 300	Poor	Orange
301 to 400	Very Poor	Red
401 and above	Severe	Dark Red

REFERENCES

1. Avinash Kumar Agarwal, Akhilendra Pratap Singh, Jithin Lukose, Tarun Gupta, 'Characterization of Exhuaust Particulates from Diesel Fueled Homogenous Change Compression Ignition Combustion Engine,' Journal of Aerosol Science, 58, 71-85, (2013).
2. Dharmender Kumar Singh, Lakshay and Tarun Gupta, 'Field performance evaluation during fog-dominated wintertime of a newly developed denuder-equipped PM1 sampler', Environmental Science and Pollution Research, DOI: 10.1007/s113556-013-2371-9 (2013).
3. GVAQI (1997). Greater Vancouver Regional District Air Qulaity and Source Control Department, Burnaby, BC, Canada.
4. Kamal Kant Ujinwal and Tarun Gupta, 'Development of a Multipurpose Sampler to Measure Respirable Airborne Particulate Matter to Improve Understanding of Air Quality', International Congress of Environmental Research, Bhopal, (Dec. 2007).
5. Malaysia (1997). A Guide to Air Pollutant Index in Malaysia. Department of Environment, Kuala Lumpur, Malaysia.
6. Ontario (1991). A Guideline to the Ontario Air Quality Index System. ISBN 0-7729-8230-9 Air Resource Branch, Ontario Ministry of the Environment, Toronto, Ont. Canada.
7. ORAQI (1970). Oak Ridge Air Quality Index. In Environmental Indices Theory and Practice by Wayne R. Ott Ann Arbor Science Publisher Inc. Mich., USA.
8. Philip Demokritou, Tarun Gupta and Petros Koutrakis, 'A High Volume Apparatus for the Condensation Growth of Ultrafine Particles for inhalation and Toxicological Studies', Aerosol Science and Technology, 36,1061-1072 (2002).
9. Philip Demokritou, Tarun Gupta, Stephen Ferguson and Petros Koutrakis, Development of a High Volume Concentrated Ambient Particles System (CAPS) for Human and Animal Inhalation Toxicological Studies', Inhalation Toxicology, 15 111-129 (2003).

10. Sharma Mukesh, Maheshwari, M., Sengupta, B., and Shukla, B.P., 'Design of a Website for Dissemination of Air Quality Index', Journal of Environmental Software and Modeling. 18 (2003) 405-411.
11. Sharma Mukesh, Panday Reenu, Maheshwari, M., Sengupta, B., Shukla, B.P., Gupta, N.K. and Johari, S., Interpretation of Air Quality Data using Air Quality index for City of Kanpur, India', Journal of Environmental Engineering and Science Vol. 2(6): pp 453-462 (2003).
12. Sharma Mukesh, Panday Reenu, Maheshwari Mayank, Sengupta, B., Mishra Sengupta, B., Mishra, Anuradha, Shukla, B.P., 'Air Quality index and Its Interpretation for City of Delhi,' Clean Air International Journal on Energy for a Clean Environment Volume 4, number 3, 269-283, (2002).
13. Sharma M., Bhasker S, and Sengupta B., 'Instituting Ambient Air Monitoring Protocol: Issues for Discussion', Indian Journal of Air Pollution Control, April 2002.
14. Sharma, M., Aggrawal, S, Bose P., 'Meteorology-based Forecasting of Air Quality Index using Neutral Network', International Conference ICONIP'02-SEAL'02-FSKD on Neural Network, Singapore, November, 2002.
15. Sharma M. and Maloo Shaily., 'Assessment and Characterization of Ambeint Air PM10 and PM 2.5 in the City of Kanpur, India', Atmospheric Environment 39 (2005) 6015-6026.
16. Tarun Gupta, Philip Demokritou and Petros Koutrakis, 'Development and Evaluation of an Ultrafine Particle Concentrator for inhalation Toxicological Studies', Inhation Toxicology, 16((13), 851-862 (2004).
17. Tarun Gupta, Abishek Chakrabortry and Kamal Kant Ujinwal, 'Development and Performance Evaluation of an Indigenously Developed Air Sampler Designed to Collect Submicron Aerosol,' Annals of the Indian National Academy of Engineering (INAE), Vol, VII, 189-193 (2010).
18. Tarun Gupta, Jaiprakash and Shefali Dubey, 'Field Performance Evaluation of a newly Developed PM 2.5 Sampler at IIT Kanpur,' Science of the Total Environment (doi: 10.1016/j. scitotenv.2011.05020), 409, 3500-3507 (2011).
19. U.K. (1998). Air Pollution - what it means for your health. Department of the Environment, Transport and the Regions. PO BOX No 236, Wetherby LS23 7NB, U.K.
20. USEPA (1976). Federal Register, Vol. 41 No 174 - Tuesday September 7, 1976.
21. USEPA (1998). Federal Register, Vol. 63 No. 236 / Wednesday, December 9, 1998.
22. htcitykanpur@hindustantimes.com

QUESTIONS

1. Write short notes on selection of sites for pollution monitoring.
2. Enumerate the basic methods of measurement / analysis of various pollutants.
3. Describe the method of sampling of the following mentioned pollutants in ambient air for analysing them in a laboratory:
   (a) Particulate matter of size less than 10 μm, (b) Particulate matter of size less than 2.5 μm.
4. Describe the method of sampling of (a) Sulphur dioxide, (b) Ozone, (c) Nitrogen dioxide, (d) Ammonia, in ambient air for analysing them in a laboratory.
5. Describe the method of sampling of Benzo (a) pyrine in ambient air for analysing them in a laboratory.

6. Describe the method of sampling lead, nickel, and arsenic in ambient air for analysis in a laboratory.
7. Describe briefly (a) Extractive system of continuous monitoring (b) In-situ continuous monitoring system.
8. Write short notes on: (a) Chemiluminescence, (b) Membrane separation with electrochemical cell, (c) Ion mobility spectrometry, (d) Fluorescence / UV Fluorescence.
9. Describe principle of Fourier transform infra red spectroscopy.
10. What is para magnetic property of a gas? How is it employed for analysis of gases?
11. Describe principles of continuous monitoring particulate matter in flue gases from a boiler.
12. Write short notes on: (a) Air pollution index, (b) Air quality index, (c) Air quality health guide.

ANNEXURE 16

Table (i) / 16: Collection of Gases by Absorption

Se. No.	Gas	Sampler used	Sorption medium
1	Hydrogen sulphide	Midget impinger	5% Cadmium sulphate
2	Formaldehyde	Fritted bubbler	1% sodium bisulphide
3	Benzene	Fritted bubbler glass bead column	0.1 N sulphuric acid or nitric acid
4	Carbon dioxide	Fritted Bubbler	0.1 N Barium hydroxide
5	Hydrochloric acid	Fritted bubbler	0.005 N sodium hydroxide
6	Ethyl Benzene	Fritted bubbler or Midget impinger	Spectro- grade isooctane
7	Nickel carbonyl	Midget impinger	3% hydrochloric acid
8	Mercury, diethyl or dimethyl	Midget impinger	0.1 M Iodine monochloride in 0.3N hydrochloric acid
9	Phosphine	Fritted Bubbler	0.5% Silver diethyl dithiocarbamate in pyridine
10	Lead, tetraethyl or tetramethyl	Dreschel type scrubber	0.1 M Iodine monochloride in 0.3N hydrochloric acid
11	Styrene	Fritted midget impringer	Spectro- grade isooctane
12	Toluene di-isocyanate	Midget impinger	Marcalisolution
13	Vinyl acetate	Fritted midget impringer and simple midget impinger in series	Toluene

Source: *Pagnotto, L.D., and Keenan, R.G.,* ***,*** *Sampling and analysis of gases and vapours, in The Industrial Environment-Its Evaluation and Control, Us Department of health, education and welfare, 1773*

CHAPTER 17

CONTROL OF NOISE POLLUTION

17.1 NOISE

17.1.1 General

Noise is defined as the unwanted sound consisting of a vast number of superimposed sound waves in which there is no predetermined relationship between frequencies and amplitude of waves.

It is well known that excessive noise adversely affects productivity and also the health of human beings. The adverse effects of excessive noise include, hearing loss, personal discomfort, annoyance, sleeplessness, high blood pressure, lack of concentration, mental disorientation, mental and body fatigue, stress and tension leading to major health problems, loss of working efficiency, increase in industrial accidents, interference with communications and signaling, etc. Even in wild life, it may create several effects such as inactiveness/ dullness, change of habitat thus creating environmental imbalances. The persistent noise exposure such as aircraft noise in the vicinity of airport can cause risk of 'Heart disease'. If the intensity of sound increases beyond a certain limit, pain is produced in the ear.

The sensitivity of the average human ear varies over wide ranges of intensity and frequency. The minimum intensity of sound to be audible is called threshold intensity. For being audible, there is a relationship between intensity and frequency. For human listner, detectable sound frequencies are 2 to 20,000 Hertz (vibration or cycles per second), however with the age, the hearing faculty declines. *The most sensitive range of human ear is 500 to 5000 Hz.*

The greater the frequency, the greater is the pitch or shrillness of sound. The greater the amplitude, greater is the loudness.

The velocity of sound in air is given by:

V= n λ [Where V is velocity, n is frequency, and λ is the wave length].

The intensity of sound is the energy received by the listener from the source of sound and is inversely proportional to the distance from the source of sound. The intensity of sound at a point is proportional to the square of amplitude of vibration at the point.

The intensity of sound is measured in watts per square meter. Human ear is sensitive to perceive a vast range of sound intensities from around 1×10^{-12} W/m^2 to around 1000 W/m^2. The loudness of sound is the degree of sensation of sound produced on ear of the listener. The greater the density of a medium, greater is the loudness. Human ear perceives sound that have the same intensity, but different frequencies as different loudness since the ear is a nonuniform transductor whose sensitivity varies from ear to ear and therefore, has a personal factor that cannot be easily measured. As the 'intensity' is measurable, so in order to assess loudness, we actually measure sound intensity only.

The loudness and sound intensity is greater in the direction of motion of medium (say air). The resonant body present near the source of sound would intensify the sound. The difference in loudness or technically the sound intensity level is measured in decibels written as dB which is the common unit of measurement of sound.

The Noise or the sound intensity level L is given by the formula:

$$L = 10 \log_{10} (I / I_0)$$

where I = Intensity of sound; I_0 = some standard reference intensity arbitrarily taken as 10^{-12} W/m^{-2}

One dB corresponds to I / I_0 of 1.26 approximately i.e. a 26 % change in intensity increases Noise level by 1 decibel.

There are some practical difficulties in measuring sound intensity, but it is relatively easier to measure the consequence of energy passage, i.e. the pressure variation. Since the intensity is directly proportional to the 'square of the pressure/ sound pressure level (SPL) in decibels' it is defined as:

SPL = 20 log $_{10}$ (P / P_0) decibel pressure level where P_0 is the reference pressure level usually taken as 2×10^{-5} Nm^{-2}.

The unit of weighted average sound pressure level is written as dB(A), where 'A' denotes weighted average. In acoustics where the SPL is a logarithmic number, the weighting factors are normally also logarithmic. Noise is measured generally in units of sound pressure levels called decibels, named after Alexander Graham Bell. The weighted sound levels closely match the perception of loudness by the human ear.

The dB scale begins with '0 decibel' which represents the faintest sound being audible to human ear. In this scale if there is 10 times increase in sound intensity from the faintest sound, it would be 10 dB, if sound intensity is 100 times, the scale would denote 20 dB, if the increase in sound intensity is 1000 times, the scale would denote 30 dB, if the sound intensity is 10,000 times, the scale would denote 40 dB, if the sound intensity is 100,000 times, the scale would denote 50 dB, and so on.

It may be noted that threshold limit of pain in human ear is 10 W/m $^{-2}$ which is equivalent to 130 decibels intensity, while a sound intensity of 10,000 W/m $^{-2}$ (equivalent to 160 dB) would cause instantaneous perforation of an ear drum.

Weighted Noise Level (dB(A)

For measuring environmental noise (the sounds heard by the person concerned), the measuring equipment can use a weighting function which filters the frequency of sound to mimic the characteristics of human hearing. This produces readings expressed as dB(A) and the scale remains logarithmic. [This means that two machines emitting exactly the same noise level of 80dB(A) produce a total noise of 83dB(A), not 160dB(A). A 10 dB(A) increase in sound level represents a doubling of loudness.

Between 1kHz and about 7kHz, the weighting makes very little difference, so the level in dB(A) is very similar to the level in dB. Outside this range the difference between the two measurements increases rapidly.]

Measuring noise (SPL)

Noise emissions are measured using sound-level meters (SLM), which detect and record changes in the Sound Pressure Level (SPL). As the level is constantly changing, integrating SLMs are used to obtain various "averages". These values can be used to describe the data which may have been recorded over several minutes or hours, and so is likely to be varied and complex. The descriptors always have the time interval defined (T_in the examples above), over which that description applies. With environmental monitoring, it is usually 1 hour.

Multiple Noise Sources

In most instances, a receptor will be subject to noise from a number of different sources. If this is the case, then the total SPL at that location can be obtained from the SPLs of each source by the following equation.

$$SPL = 10*\log_{10}[10^{SPL1/10} + 10^{SPL2/10} ... + 10^{SPLn/10}]$$

An example was given earlier where the result of adding two sources of noise of 80dB(A) gave a resulting SPL of 83dB(A). Where two or more sources with very different noise levels are combined, the combined level will be just slightly higher than the highest value. For example, two sources giving 68dB(A) and 56dB(A) at a location would give a combined value of:

$$SPL = 10*\text{Log}_{10}(10^{(68/10)} + 10^{(56/10)}) = 68.3 \text{ dB(A)}$$

17.1.2 Sound Power

The sound power is related to sound intensity by the following relation:

Sound power **(W) = I . (4πr²)**

where, I = intensity, r = the distance from the source

If the source is not located in free space and the ideal reflecting surfaces direct the sound in a particular direction, the sound power would be given by the following equation:

$$W = [I . (4\pi r^2)] / Q$$

where Q is the directivity factor; which is defined as the ratio of sound power of 'a small unidirectional hypothetical source' to the sound power of 'an actual source that produces the same sound- pressure level at a point of measurement.

If the sound is bounded by acoustically reflecting surfaces, the value of Q is:

Q = 1 for sound source in free space

Q = 2 for sound source located in an infinite reflecting plane

Q = 8 for sound source located at the intersection of three mutually perpendicular intersecting planes.

17.1.3 Speed of Sound in various Mediums

Speed of sound travelling in a gaseous medium which is approximately a perfect gas, is given by

$$V = [(k R T) / (M)]^{1/2}$$

Where, V = speed of sound; T = absolute temperature; R universal gas constant

M = Molecular weight of gaseous medium; k = ratio of C_p (specific heat at constant pressure) to C_v (the specific heat at constant volume). The speed of sound in air at standard temperature and pressure is 344.5 m/s. The speeds of sound in dense mediums like liquids and solids are much higher.

17.1.4 Categories of Noise

The noise can be categorized as:

(i) Broadband noise – In this the acoustical energy is distributed over a broad range of wave length having no distinguishing tone such as noise produced by a crowd of people.

(ii) Narrowband noise – In this the acoustical energy is distributed over a dominant tone such as circular saws, jet engines, etc.

(iii) Impact or impulse noise – In this acoustical energy is distributed over a short period of time (say less than 1 second duration) and generated at a frequency less than 200 periods per minute. (Example pneumatic jack hammer, punch press, etc.)

(iv) White noise – Broad band noise in which the power density (acoustical power at a particular frequency) varies with time, such as highway traffic noise, is termed as *white noise.*

(v) Pink noise – In this type of broadband noise, the power densities are equal. Pink Noise is oflenfound in nature such as waves lapping on beach leaves rustling in the trees, or a steady ranfall.

17.1.5 Typical Sources of Machine Noise

Most often the machine noises may be due to:

- Unbalancing of rotors—(Static/ dynamic)
- Unbalanced magnetic forces, irregular magnetic fields
- Spark discharges
- Worn out bearing (excessive clearance)
- Pulsation because of tapping
- Sticking of slipping surfaces (Screeching and squealing)
- Sliding on rough surface
- Cutting, Milling, Sawing, Turning, and Grinding operations
- Electric motors
- Reciprocating engines /machines
- Rotating machineries like engines, turbines, generators, pumps, etc.
- Impact Riveting operation
- Eccentric press working
- Jet plane starting / flying
- Gear changing
- Pneumatic tools operations
- Noises because of fluid flow or material flow such as:
  - Pulsating flow; Turbulent flow; Unstable flow
  - Cavitation on surfaces
  - Material loading on conveyor belts and unloading from the same
  - Material flow through pipes

The sound from machine propagates as waves in all kinds of elastic mediums like solids, fluids, gases, etc., but the characteristics of sound changes at the interface between different media. It may be noted that the acoustic energy from its original source of generation propagates predominantly as 'Structure- borne sound' inside the structure. This structure- borne sound generates vibrations on the surface which radiates the sound in the air. The sound from there propagates as air pressure wave. However there may be several parallel transmission paths of acoustic energy.

The structure- borne sound should be investigated by analyzing its transfer path. The properties pertaining to acoustic energy radiation for the entire surface of the structure should be accounted for better analysis. This can be done by acoustic intensity measurements or by microphone array (some times termed as acoustic cameras).

17.1.6 Noise from Aircrafts

A moving aircraft causes compression and rarefaction of the air, producing motion of air molecules, propagating air as pressure waves which are strong enough and are within the audible frequency range.

Different types of aircrafts have different noise levels and frequencies. The noise originates from three main sources:

(i) Aerodynamic noise

(ii) Engine and other mechanical noise,

(iii) Noise from aircraft systems

(i) Aerodynamic Noise

Aerodynamic noise arises from the airflow around the aircraft body (fuselage) and control surfaces. Aerodynamic noise increases with aircraft speed and density of air medium. (At low altitudes the density of the air is more, so the noise increases as the altitude decreases).

Jet-powered aircraft creates intense aerodynamic noise. Military aircraft when flying at high speed at low altitude creates intense aerodynamic noise. The shape of the nose, windshield or canopy of an aircraft affects the sound produced.

In propeller aircrafts, air flow around blades of impeller creates a lot of aerodynamic noise. Aerodynamic noise is also produced at the main and tail rotors of helicopter. Most of this aerodynamic noise is of low frequency and is closely related to the speed of rotor.

The air flow hitting the landing gear, wings or any other part of airframe of plane creates aerodynamic noise. Broadly noise generated at the airframe may be of two types as mentioned below:

Bluff Body Noise—Low pressure is created by the alternating vortex shedding from sides of bluff body at the core of the shed vortex. This gives rise to pressure waves (sound waves). The separated air flow around the bluff body being unstable rolls up into ring vortices breaking down later in turbulence. (Bluff body—*a body steep or upright in front which when moves through a fluid / air leaves behind it a large disorderly wake and experiences a large drag).*

Edge Noise—The fluctuation in pressure associated with the passing of turbulent flow at the end of an object or any gap in the structure (like high lift clearance gap) manifests itself as sound propagation from the edge of the object radially downward.

(ii) Engine Noise and other Mechanical Noise

As already mentioned, the major cause of noise by a propeller aircraft is of aerodynamic nature. Aerodynamically induced noise in Helicopter emanates from main and tail rotors while mechanical noise emanates from gear box and various power transmission chains. Sound of narrow band high intensity peaks related to rotational speed and movement of components is created by the mechanical sources.

In the gas turbine (Jet engine), very loud noise is created during take-off and climbs. Intense noise of a Buzz-saw (circular saw) type is created when tips of the fan reaches the supersonic speed. But in some modern jet engines, the airframe is typically noisier during landing. Major contributor of engine noise of this type of aircraft is the jet noise. There is an inherent shear layer in the high velocity jet leaving the back of the engine. This shear layer rolls up in vortices which later break down into turbulence. The sound pressure level associated with the engine noise is closely related to jet velocity and a small reduction in jet velocity would make large reduction in noise.

(iii) Noise from Aircraft Auxiliary Systems

Air conditioning and air pressurization system of cock pit and cabin is a big contributor to noise inside the aircraft during flight. Other noise contributors may be the use of auxiliary power unit, on-board generator, air compressor, and some types of electronic equipments.

Health Effects of Aircraft Noise

The exposure to aircraft noise for a long time such as by the residents near the airport may create various health problems like hearing impairment, hypertension, ischemic heart disease*, annoyance, sleep disturbance, increase workplace accident rates, stimulation for aggression, etc.

**** Noise Related Health Study in Germany – Residents Near Air Port***

A large-scale statistical analysis of the health effects of aircraft noise was done in the late 2000s by 'Umweltbundesamt', the Germany's central environmental office. The health data of over one million residents around the Cologne airport were analysed for health effects correlating with aircraft noise. The results were then corrected for other noise influences in the residential areas, and for socioeconomic factors, etc.

The German study concluded that aircraft noise clearly and significantly impairs health:- For example, a day-time average sound pressure level of 60 dB(A) increased coronary heart disease by 61% in men and 80% in women. A night-time average sound pressure level of 55 dB(A) increased the risk of heart attacks by 66% in men and 139% in women.

17.1.7 Noise Measurements

Systematic analysis of noise sources can be done by sound pressure level measurements, sound intensity measurements, acceleration measurements, microphone array. There is IEC international standard 651 'Sound level meters' for measurement of sound level. These are divided in four types: Type 0—for laboratory precision use; Type 1—for general; use, Type 2—low price instruments; and Type 3—have very wide tolerance and therfore not used in practice.

Sound Level Meters:- Sound level meters are designed to give constant and objective measurements of sound levels. These meters convert the variations in air pressure to variations in voltage which are amplified and displayed on an electrical meter calibrated in decibels. A network that is generally fitted in our sound measuring meters of day to day use, severely filters out sound of low frequency ranging between 10 to 200 Hz.

Typical sound level meters take RMS (root mean square) value of the signals. The RMS value is found to be more relevant than peak values. The meter may have 'fast' and 'slow' response settings, but may not be fast enough to accurately record an impulse sound, such as a gunshot.

'A' Scale of sound level measurement:- **Usually the noise is measured with** a single measurement by allowing the sound level meter listen to all the frequencies at one time, in the same way as the ear. An electronic circuit in the meter then emphasises the middle frequencies, just as the hearing system does. The results are usually quoted as dB(A) and the 'A' is an important tag showing the way the measurement was made.

Measurements in dB(A) are widespread and convenient, but it may be noted that they are a compromise and tell us nothing about the frequency content of a noise. The ear can tell difference between a grinder and a telephone buzzer because their sounds have a very different frequency structure, yet these sounds may give identical readings in dB(A). However the human hearing changes sensitivity with frequency, so if the noise is measured at different frequencies, different values would be obtained. The most complete measurement of a particular noise is given by separate readings of the sound level (in decibels) at standard bands of frequency and showing the results in a list or on a chart.

Many types of noise measuring equipment are commercially available such as: Sound level meter, Octave band analyzer, Magnetic tape recorder, Cassette recorder, Pen recorder, Noise average meter, Noise analyzer, Acoustic cameras / microphone array, etc.

Note: Dosimeters for noise

Noise dosimeters measure and store Sound Pressure Levels (SPL) and, by integrating these measurements over time, provide a cumulative noise-exposure reading for a given period of time, such as an 8-hour workday. Dosimeters can function as personal or area noise monitors.

Area monitoring can be used to estimate noise exposure when the noise levels are relatively constant and employees are not mobile. In workplaces where employees move about in different areas or where the noise intensity tends to fluctuate over time, noise exposure is generally more accurately estimated by the personal monitoring approach.

In occupational settings, personal noise dosimeters are often worn on the body of a worker with the microphone mounted on the middle-top of the person's most exposed shoulder. Current dosimeters are designed to provide the user with parameters such as noise dose, time-weighted average, sound exposure level, as well as peak, maximum, and minimum sound pressure levels. Most dosimeters also generate statistical and graphical representations of the collected data. ANSI S1.25 specifies that dosimeters should at least provide the following parameters:

17.1.8 Noise Levels of different Sources

Table 17.1 gives an idea of noise level (approx.) generated by different sources.

Table 17.1: Noise Generated by Different Sources

Se. No.	Sources of Sound /Noise	Decibels dB
1	Average whispering	30 dB
2	Average living room	40 dB
3	Normal conversation	50 to 60 dB
4	Two wheelers, three wheelers, etc.	75 to 80 dB
5	Heavy trucks	77 to 80 dB
6	Noisy office	70 dB
7	Average factory	80 dB
8	Oxygen torch, Construction machinery, Rock-n-roll band	120 dB
9	Farm machinery	90 dB
10	Jet aircraft at take off point	150 dB
11	Jet aircraft at 1000 feet height	110 dB
12	Rocket engine peak	180 to 195 dB
13	Fire arms, fire works peak	132 to 167 dB
14	Noise of large water falls	90 dB
15	Train at 100 meters at 100 km/h	85 dB
16	Grinding	80 to 110 dB
17	Air cooled electric motor (50 kW)	89 dB
18	Eccentric Press	95 dB
19	Spray painting	105 dB
20	Impact riveting	115 dB
21	Jet plane starting up	145 dB

Approximate Relationship between sound pressure level in decibels (dB) and sound power is given in **Table 17.2.**

Table 17.2: Approximate Relation Between Sound Pressure Level and Sound Power

Sound pressure level in decibels (dB)	Sound power in Watts
140	100
120	1.0
100	10^{-2}
80	10^{-4}
60	10^{-6}
40	10^{-8}
20	10^{-10}

17.2 NOISE REGULATIONS

17.2.1 Noise Limits in Indian Regulations

The Environment protection Rules, 1986, have prescribed certain Noise limits which are mentioned in **Para 15.3 and 15.5 of Chapter 15**. The Indian regulations are to be followed in projects in India, but where Indian regulations are silent, OSHA regulations are generally followed. (More Stringent regulation should generally be followed in the interest of environment).

17.2.2 OSHA Regulations for Noise

The Occupational Safety and Health Administration USA, (OSHA) sets legal limits on noise exposure in the workplace. The limits imposed by these regulations are also observed in the projects in India. These limits are based on a worker's time weighted average noise exposure over an 8 hour day. With noise, OSHA's permissible exposure limit (PEL) is 90 dB(A) for all workers for an 8 hour day. The OSH(A) standard uses a 5 dB(A) exchange rate. This means that when the noise level is increased by 5 dB(A), the amount of time a person can be exposed to a certain noise level to receive the same dose is cut in half.

Acceptable Noise Level

OSHA prescribes that an average of 85 decibels is the highest acceptable limit of noise in the workplace. If employees experience levels higher than this, the employer is required to take steps to reduce the noise levels. If noise reduction is simply not feasible, protective equipment must be provided by the employer and used by employees to bring noise levels under the limits advised by OSHA.

Permissible Exposures according to OSHA are as given below	
85 dB	8 hours
90 dB	4 hours
95 dB	2 hours
100 dB	1 hour
105 dB	½ hour
110 dB	¼ hour

Combined Effect of Noise

When the daily noise exposure is composed of two or more periods of noise exposure of different levels, their combined effect should be considered rather than individual effect. It can be calculated as follows:

Suppose the duration of exposure at a specific level are denoted by C_1, C_2, C_3, C_4, say at noise level of 80 dB, 85 dB, 90 dB, 95 dB levels and permissible time duration at these noise levels are denoted by T_1, T_2, T_3, T_4. In order to find whether the average noise is within permissible limits or not, the combined fraction is calculated as follows:

Combined fraction = $C_1/T_1 + C_2/T_2 + C_3/T_3 + C_4/T_4$ -----

If the combine fraction is more than one, the noise is not within permissible limits

NIOSH Recommendations

The National Institute for Occupational Safety and Health, USA, (NIOSH) has recommended that exposure of all workers to noise should be controlled below a level equivalent to 85 dB(A) for eight hours to minimize occupational noise induced hearing loss. *NIOSH* has found that significant noise-induced hearing loss occurs at the exposure levels equivalent to the OSHA permissible exposure limit (PEL) based on updated information obtained from literature reviews. *NIOSH* also recommends a 3 dB(A) exchange rate so that every increase of 3 dB(A) halves the recommended amount of exposure time.

As an example: OSHA allows 8 hours of exposure to 90 dB(A), but only 2 hours of exposure to 100 dB(A) sound levels. NIOSH would recommend limiting the 8 hour exposure to less than 85 dB(A). At 100 dB(A), NIOSH recommends less than 15 minutes of exposure per day.

Monitoring

Where there is evidence that noise levels may exceed the 85 dB limit, employers are required to implement a monitoring strategy. This strategy is designed to determine how high the noise levels are and which employees are being subjected to high noise levels. Employees experiencing high noise levels must be included in a hearing conservation strategy and provided with hearing protection. Monitoring should be conducted whenever there is a substantial change in the way the work is conducted which may increase noise levels and put employees at risk.

Note:

> *In 1981, OSHA implemented new requirements to protect all workers (in USA) in general industry (e.g. the manufacturing and the service sectors) for employers to implement a 'Hearing conservation program' where* ***workers are exposed to a time weighted average noise level of 85 dB(A)*** *or higher over an 8 hour work shift.*
>
> *The 'Hearing conservation programs' require employers to measure noise levels, provide free annual hearing examinations and free hearing protection, provide training, and conduct evaluations of the adequacy of the hearing protectors in use unless changes to tools, equipment and schedules are made so that they are less noisy and worker exposure to noise is less than the 85 dB(A).*

17.3 NOISE CONTROL

Noise abatement measures can tackle noise both at source of noise and at various places along the path of acoustic energy transmission. It is very important to locate the dominant source of noise and tackle it to abate the noise there itself. Unless the major sources of noise and most important acoustic path are not well tackled to minimize the noise, the measures taken at minor sources and unimportant paths will achieve only negligible results. Fundamentally, an attempt should first be made to first reduce the sound at the source of its generation, then hinder the noise transmission, and finally, minimize sound radiation.

It may be noted that the noise control at the source reduces the stress and strain on the machines, lowers the fatigue damage and wear, and also increases the life of machines.
There are generally three types of approaches for noise control namely control at source, restricting the noise transmission, and control at the receiver. In some cases all the three types of approaches may be necessary. Noise and vibration monitoring is an integral part of the noise control.

17.3.1 Control of Sound at Source

The techniques involved in control of noise at source may include the following:

- Balancing of masses in rotating machines ie static balancing
- Balancing of magnetic forces
- Dynamic Balancing of rotating machines and dynamic compensation
- Reduction of impacts / impulsive forces
- Reduction of frictional forces
- Isolation of Vibrations
- Shock absorbing contact points/ providing shock absorbers
- Examine if the vibration is due to resonance and if so, remove the cause of resonance say by alteration of the natural frequency of resonant element
- Increasing energy dissipation
- Changes in design of equipments
- Changes in layout of the plant
- Changes in foundations of equipments

17.3.2 Restriction in the Path of Noise

The restriction in the path of sound can be done by:

- Absorption of noise by padding with suitable noise absorbing materials
- Deflection of noise away from the receiver
- Providing enclosures to noise making equipment, acoustic hoods, sound proof walls, noise barriers, etc.

Measures influencing noise propagation

- Structure borne sound insulation by impedance discontinuities (such as blocking masses)— At an impedance change , a part of the structure- borne sound wave is reflected so that less sound propagates further.
- Structure- borne sound deadening— This can be done by deadening foil or matting. A coating by viscous material can provide additional damping effect (and also additional mass).
- Application of sound absorbing material.
- Reduction of air borne noise radiation with discontinuity of surfaces such as by perforated plates.
- Sound encapsulation as by acoustic enclosures.

Sound Absorption

Absorption of sound has a large role in reducing noise. All materials have some sound absorbing properties. When sound strikes a surface, some energy is absorbed, some portion is reflected and balance is transmitted or dissipated. A material's sound absorbing properties can be described by a sound absorption coefficient in a particular frequency range. The coefficient can be viewed as a percentage of sound being absorbed, where 1.00 is complete absorption (100%) and 0.01 is minimal (1%).

Most of the good sound reflectors prevent sound transmission by forming a solid, impervious barrier. Conversely, most good sound absorbers are not good reflectors of sound. Sound reflectors tend to be impervious and massive, while sound absorbers are generally porous, lightweight material.

There are three basic categories of sound absorbers:

- Porous materials commonly formed of matted or spun fibers;
- Panel (membrane) absorbers having an impervious surface mounted over an airspace;
- Resonators created by holes/ slots connected to an enclosed volume of trapped air.

In some cases, the absorptivity of each type of sound absorber is dramatically influenced by the mounting method employed.

(i) Porous absorbers:

Common porous absorbers include carpet, draperies, spray-applied cellulose, aerated plaster, fibrous mineral wool / slag wool / glass wool / glass fiber, open-cell foam, and felted or cast porous ceiling tile. Generally, all of these materials allow air to flow into a cellular structure where sound energy is converted to heat.

Porous absorbers are the most commonly used sound absorbing materials. Higher the thickness of the absorbing material, higher is the absorption. Thicker materials generally provide more bass* sound absorption or damping. Fabric applied directly to a hard, massive substrate such as plaster or gypsum board does not make an efficient sound absorber due to very thin layer of fiber.

**Bass—low or grave part in music/ low frequency sound.*

(ii) Panel Absorbers:

Panel absorbers are non-rigid, non-porous materials which are placed over an airspace that vibrates in a flexural mode in response to sound pressure exerted by adjacent air molecules. Common panel (membrane) absorbers include thin wood paneling over framing, lightweight impervious ceilings and floors, glazing and other large surfaces capable of resonating in response to sound.

For absorbing low frequencies, panel absorbers are usually very efficient as seen from Orchestra Platforms where wood paneling traps most of the bass sound.

(iii) *Resonators:*

The action of 'Resonators' is to absorb sound in a narrow frequency range. Resonators include some materials that have openings (holes and slots) and some perforated materials. The classic example of a resonator is the 'Helmholtz resonator', which has the shape of a bottle. The resonant frequency is governed by the size of the opening, the length of the neck and the volume of air trapped in the chamber. Typically, perforated materials only absorb the mid-frequency range unless special care is taken in designing the facing to be as acoustically transparent as possible. As the 'Slots' usually have a similar acoustic response, long narrow slots can be used to absorb low frequencies.

However, for acoustic music production rooms, low frequency energy is very valuable and its absorption is not desirable and therefore long narrow air distribution slots are not desirable.

17.3.3 Control of Sound at the 'Receiver's end'

In case after using the above methods the noise level is still undesirable, personal sound protection equipments / devices should be used by the persons working or present in the area. The period of exposure to a person to high decibel sound should be reduced in accordance with regulations already referred.

REFERENCES

1. E P Rules, 1986, India.
2. Occupational Safety and Health Administration, USA— Occupational noise exposure.
3. David R. Schwind 'Room Acoustics: Basic principles of reflection, diffusion, and absorption'.
4. http://en wikipedia.org/wiki/aircraftnoise.

QUESTIONS

1. Noise is measured in units of
   (a) Hertz; (b) Decibel; (c) Dobson; (d) Becquerel.
2. Which of the following is not an effect of noise on wild life?
   (a) Inactiveness/dullness; (b) Change of habitat;
   (c) Environmental imbalances (d) None of the above.
3. 10 decibel (dB) increase in sound level will increase the loudness of sound by
   (a) 2 times; (b) 5 times;
   (c) 10 times; (d) None of these.
4. The unit for measuring the frequency of sound is
   (a) Decibel (dB); (b) Hertz (Hz);
   (c) Dobson Unit (DU); (d) None of the above.
5. Although human ear can hear sound of all frequencies varying from 20 Hz to 20,000 Hz, but it is most sensitive to frequencies in the range of
   (a) 100 to 500 Hz.; (b) 500 to 5000 Hz.;
   (c) 5000 to 10,000 Hz.; (d) 10,000 to 20,000 Hz.
6. A network that is generally fitted in our sound measuring meters of day to day use, severely filters out sound of: (a) Low frequency ranging between 10 to 200 Hz; (b) Medium frequency ranging between 200 to 2000 Hz; (c) High frequency ranging between 2000 to 10,000 Hz; (d) None of the above.
7. Define (i) Noise, (ii) Sound velocity, frequency, and wave length, (iii) Sound intensity, (iv) Sound intensity level, (v) Sound pressure level and (vi) Sound power.
8. What are the different categories of noise? Enumerate various sources of noise.
9. Write a short note on measurement of sound / noise level.
10. Which regulations govern the noise pollution? What are the ill effects of noise pollution on human health?
11. Briefly describe noise control methods.

Answers: **1.** (b) **2.** (d) **3.** (c) **4.** (a) **5.** (b) **6.** (a)

CHAPTER 18

CONTROL OF ODOURS

18.1 ODOURS

An odour is the physiological response to a particular air borne molecular species by the olfactory nerve cells (olfactory neurons) located at the top of nose, just above the bridge of the nose. Olfaction is a sensitive physiological system that responds to contaminants at very low concentrations sometimes even at less than part per billion. The olfaction is capable of discriminating one odour from a background of different odours. However olfaction system exhibits fatigue making the individuals insensitive to an odour unless the concentration changes.

There is also a correlation between the sense of smell and taste. *[Example: Food flavours are generally associated with aromas. The aromas are airborne mixtures of some molecules which stimulate the sensory organs].*

The odour is caused by various substances even in tiny quantity and is best detected by human or animal nose, but sometimes the very odour disagreeable (foul) to one may be agreeable (good) to the other person. The 'foul odour' causing compounds / chemicals include: Hydrogen sulphide, Mercaptans, Carbon disulphide, Cresols, Sulphur compounds from crude oil, certain petroleum products, Ammonia, Chlorine, Sulphur dioxide, Acetic acid, Animal fats, certain aromatic compounds like Phenols, certain pharmaceutical products of animal origin, certain solvents, paints, pesticides, compounds in tobacco and its smoke, compounds in smoke of burning rubber, trash/ biomedical wastes, hazardous materials, etc'. The odour causing compound may be gaseous, mist, fumes, or micro particles discharged into environment by industrial, municipal, or commercial operations. Often the odour is also caused by decay of organic matter like garbage, sewage, vegetable matter, food items, leather, animal flesh, eggs, fish, etc. Some natural materials, animals, vegetable matter, algae, fish, animal and human wastes, etc. have foul smell.

Industries which emit foul odours include: tanneries, pulp and paper mills, sugar factories, fertilizer factories, oil refineries, oil processing units, poultries, slaughter houses, food processing including meat processing, chemical and pharmaceutical factories, rubber and rubber product factories, foundries, welding shops, handling and disposal of municipal and biomedical waste, etc.

Various terms are used to describe the type of odour such as: repulsive, nauseating, pungent/ irritating, rotten egg odour, rubber like, sharp, fishy, musty, vinegary, garlic, leather like, and also pleasant, fragrant, sweet, fruity, etc. While dealing with odour control, it is to be kept in mind that:

(i) Odour travels in the direction of wind.

(ii) Nature and strength of odour may change with the dilution of odorous material and distance from the source.

(iii) Constant exposure to the same intensity of odour to a person makes him unaware of sensation of odour and he would only notice it when the intensity varies.

(iv) If a person is exposed to an odour for a long time, perception fatigue occurs and he may not be able to perceive similar odour of other material, however he would perceive dissimilar odour.

(v) An unfamiliar odour may invite more complaints than a familiar odour.

(vi) Two or more different odours when mix, they may become un- recognisable or may cancel each other or be dominated by the more strong odour. Weak odour may not be perceived in presence of strong odour.

(vii) Some substances can have similar odour even when they are dissimilar.

(viii) The like and dislike of a certain odour may be different for individuals and may be influenced by his life experiences associated with the particular odour.

(ix) Some animals have a better perception of smell and can differentiate the odours better than human beings.

Odour intensity

To quantify odours at least three independent properties are needed:

- Quality / character—a description of the odour using familiar functional groupings
- Acceptability/ hedonic tone—sensual pleasure, annoyance
- Intensity or strength of a response

The intensity of odour would generally increase with the concentration of odour causing substance in the air being inhaled. The relation between intensity of odour and the odour causing substance i.e. odorant concentration in air may be expressed by ***Weber-Fechner law*** as:

Odour intensity = Constant x log (odorant concentration)

The threshold concentration against a background, is the minimum identification level of odour causing substance in air being inhaled. However detection threshold may be different for different backgrounds. A widely used scale of odour intensity is from 0 to 4.

0—No odour

1—Threshold level, i.e. barely perceptible

2—Definite odour

3—Strong odour

4—Overpowering odour

In order to determine the odour intensity, a panel of experienced persons (6 to12) may be chosen who are deputed to sniff the air at a given location at the same time to report the nature and intensity of odour individually. The odour intensity stated by each observer is noted and then average value is arrived at.

*Odour Index**

An additional way to categorize odours is the odour index which is calculated as follows:

Odour index = (actual vapour pressure) / (threshold level partial pressure)

*Dilution Factor**

The dilution factor is defined as the number of volumes to which one volume of air saturated with the odorant vapour must be diluted to reach the odour threshold.

*[*As the odour index and dilution factor are large, these are sometimes reported in terms of logarithms].*

18.2 METHODS OF ODOUR CONTROL

Various methods of odour control are available which are discussed below. It may be mentioned that different odour control measures/ methods may be combined to achieve high efficiency of odour removal. The odour control technologies being used will depend on the composition, characteristics, and quantity of odorous gases being generated or emitted. The options available for control of odour are:

- Dilution
- Confinement of Stream of Gases or Material Emitting Odour
- Masking the Foul Odour by Stronger Scent
- Adsorption of Foul Odour Emitting Component
- Scrubbing of Odorous Gas by Liquid Solvent or Reagent
- Biological Method of Removing Odours
- High Temperature Oxidation of Odorous Substances
- Catalytic Oxidation/ Incineration
- Oxidation of Odorous Substance by Powerful Oxidants
- Irradiation
- Odour Control by Patented Formulations
- Process Modifications

18.2.1 Dilution

As the odour intensity is dependent on concentration of the odour emitting substance, the odour in a closed space can be minimized by ventilation of the space. The ground level odour can be considerably reduced by release and dispersal of odorous gases from a high stack.

18.2.2 Confinement of Stream of Gases or Material Emitting Odour

The stream of odorous material / liquid / gases should be confined to a limited space and any leakages or spills should be strictly avoided. Confinement methods may also be followed during handling of odorous material.

18.2.3 Masking the Foul Odour by Stronger Scent

As a strong odour would mask a weak odour, the foul odour in a limited space can be masked by use of strong pleasant odorant in that space. Various odorants are available in the market with varieties of good smells. Some odorants have anti-bacterial properties also and can be used when required.

Many masking patented products are available for the above mentioned purpose commercially which are mentioned in ***Para. 18.2.11, of this chapter.***

18.2.4 Adsorption of Foul Odour Emitting Component

When the odorous stream is passed through a bed of adsorbent material like activated carbon, the odour emitting component of the substance can be adsorbed by the adsorbent. Adsorbents commonly used are: 'Activated carbon*' or 'Alumina pellets impregnated with permanganate'. These materials are highly porous and have a large surface area upon which adsorption of odorous substance may occur. Activated carbon exhibits rapid adsorption kinetics or the highest volume of adsorbing porosity of any material known to mankind. The forces which attract and attach the molecules to the surface of the pores are known as 'Van der Waals forces'.

Activated carbon is generally used for adsorption of some organic gases and vapours, some inorganic gases and a few metallic vapours. The activated carbon is good for adsorbing compounds with a boiling point over 0°C. Before odorous stream is passed through activated carbon, it should be free from dust and moisture and its temperature should be near the ambient and therefore the odorous stream should be cooled, dehumidified and dedusted. Activated carbon needs to be replaced before it is saturated. The used activated carbon can be regenerated or disposed.

Activated carbon can be used for various types of odours including ammonia, Most of the 'Adsorber vessels' containing adsorbent are pressure vessels constructed of carbon steel, stainless steel or plastic. Large systems for drinking water are often constructed in concrete. In some cases, a moving or pulsed bed adsorber is employed to optimize the use of the granular activated carbon. High performance 'Impregnated activated carbon pellets'* are used for the removal of hydrogen sulphide, organic mercaptans (thiols) and certain organic sulphides from air.

Note on Activated Carbon

Activated carbon is a form of carbon that has been processed to make it extremely porous and thus to have a very large surface area available for adsorption or chemical reactions. Many varieties of activated carbon products are commercially available. They may exhibit different characteristics depending upon the raw material and activation technique used in their production.

'Activation process' for carbon increases the adsorption properties of carbon. Activated carbon can be produced from various 'carbonic source materials' like nutshells, wood and coal. The raw material used in the production has large influence on the characteristics and performance of activated carbon product. In selecting an activated carbon, it is important to develop a clear understanding of both the adsorptive and physical characteristics of the material in order to optimize the performance capabilities. The activated carbon can be granulated or pelleted. The 'Granular activated carbon' has irregular shaped particles and may generally have size ranging from 0.2 to 5 mm. This type of activated carbon is preferred for both liquid and gas phase applications. The 'Pelleted activated carbon' is extruded cylindrical shaped pellets. This type of activated carbon is mainly used for gas phase applications because of their low pressure drop, high mechanical strength and low dust content.

The use of treated activated carbon is determined in relation to the nature of the odorous compound present. Activated carbon can be bought in the chemically impregnated form for enhanced loading capacity and removal efficiency for special applications.

18.2.5 Scrubbing of Odorous Gas by Liquid Solvent or Reagent

Compared with incineration technique and adsorption on activated carbon, scrubbing by a liquid may be economical when volume flow of odorous stream is high say when the volume of odorous gas to be treated is more than 1.4 m^3/s (about 5000 m^3/h). It is necessary to cool and remove the dust and moisture from the hot moist odorous gas stream before treating it in a scrubbing tower. Indirect condenser is preferably used to remove the moisture from the stream.

The odorous gases are brought into intimate contact with the scrubbing liquid in a well designed scrubbing tower or other type of equipment. The gas absorption equipments include packed towers, plate or tray towers, spray towers, venturi and fluidized-bed scrubbers. The most frequently used absorbing solutions are as mentioned below:

- Sodium carbonate solution – to absorb hydrogen sulphide and mercaptans.
- Sodium hydroxide solution – for absorbing, sulphur dioxide, hydrogen chloride, nitric acid fumes or nitric oxide, hydrogen cyanide, ozone, chlorine, ammonia and mercaptans,
- Amine – used to trap hydrogen sulfide, mercaptans
- Calcium hydroxide slurry – for removal of chlorine, hydrogen chloride gas
- Sodium hypochlorite, potassium permanganate, or hydrogen peroxide solution – for absorbing unsaturated organic compounds

- Dilute sulphuric acid – for absorbing ammonia
- Water – for absorbing water soluble gases like ammonia, sulphur dioxide, chlorine, hydrogen chloride gas, it can also be used even for removal of dioxins
- Ammonium sulphite solution or Zinc oxide slurry – for absorbing Sulphur dioxide. [Refer to **Para.8.3, Chapter 8** for removal of various gases].

Control of Odour caused by Organic Compounds

The removal of organic odours is more difficult than inorganic odour. The removal of organic odour is normally done in two stages. The odour or odourous material is first absorbed in some solvent and then it is oxidized in a less malodorous material. Examples are given in **Table 18.1.**

Table 18.1: Typical Organic Odour Control System

Industry	Odour causing material	Absorbent	Oxidant
Pulp and paper mill	Organic sulphides	Sulphuric acid	Chlorine dioxide (ClO_2) .
Food processing	Vegetable oil	Sulphuric acid	Potassium permanganate ($KMnO_4$)
Silver plating	Cyanides	NaOH solution	Sodium hypochloride (NaOCl)
Iron foundry	Amines	$NaHSO_4$.	.Potassium permanganate ($KMnO_4$)
Brass foundry	Phenol formaldehyde	Na_2CO_3 solution	.Potassium permanganate ($KMnO_4$)
Varnish plant	Resin	Na_2CO_3 solution	Potassium permanganate ($KMnO_4$)
Hardboard	Linseed oil	Na_2CO_3 solution	.Potassium permanganate ($KMnO_4$)
Rendering	Hem fat	Na_2CO_3 solution	Sodium hypochloride (NaOCl)

Odour Control in Sewage Handling Plants

Odours emitted by sewage treatment are typically an indication of an anaerobic or 'Septic' condition. Early stages of processing will tend to produce foul smelling substances like hydrogen sulphide and mercaptans. In large plants in cities often treat the odours with carbon reactors, a contact media with bio-slimes, small doses of chlorine or circulating fluids to biologically capture and metabolize the noxious gases. Other methods of odor control include addition of iron salts, hydrogen peroxide, calcium nitrate, etc. to reduce hydrogen sulphide levels. For reducing odour, the sludge handling should be done through hermetically closed pipe-work with high density sludge handling pumps.

18.2.6 Biological Method of Removing Odours

There are two biological methods of removing odour, namely 'Biofiltration' and 'Bio-scrubbing'. The odour causing substance should be biodegradable and should not be a biocide itself, otherwise these methods would not be successful. The outgoing gases from the bio-filter / bio-scrubbing may carry the bacteria with it and therefore should be disinfected before being released in the environment.

In bio-filtration, the odour is removed by action of bacteria. The bacterial mold or culture is held on inert supports in a filter and the odorous stream is passed through it, bringing the stream in intimate contact with the bacteria which consume the odorous substance for their growth or convert it to other non- odorous substance. The strain of bacteria is to be specially chosen for a particular odorous substance. The intimate contact between the bacteria and the odorous stream is to be ensured by proper design of the biofilters and recirculation. Two or three biofilters may be put in series for ensuring removal of odour. The process is self- sustaining, but the bed may have to be replaced on regular basis otherwise mechanical failures may occur.

Bio-scrubbing uses both the liquid scrubbing and the bio-filtration for removal of odorant from the gaseous stream. Bio-scrubbers can only be used successfully if the contaminants can be removed from waste gases by absorption in water, or activated sludge mixed with water.

18.2.7 High Temperature Oxidation of Odorous Substances

The odorous stream is passed through an incinerator kiln or furnace the temperature of which is maintained between 650 and 815°C by burning a fuel. The odorous component is oxidized in the kiln by oxygen of air which is supplied to the furnace (for burning the fuel and oxidizing the odorous substance). The odorous stream is preheated by the outlet gases from the kiln / furnace in a heat exchanger. The outlet gases on exit from the heat exchanger are still hot and can be further used to heat water or raise low pressure steam for other uses. The optimum temperature for oxidation may be different for different odorous streams according to their constituents which, therefore, should be found out by trials. The length of furnace/kiln and flow through the furnace is to be adjusted to give sufficient residence time for oxidation of odorous substance. *[Refer **Para. 8.3.9.1, chapter 8,*** for removal of 'Dioxin' by combustion].

18.2.8 *Catalytic Oxidation/ Incineration*

The high temperature oxidation entails high cost of fuel consumption which can come down if the oxidation is possible at lower temperature. The catalytic oxidation of most of the substances can take place at lower temperatures and can save fuel. The catalyst can be chosen suitable for the particular odorous substance, but Platinum, and platinum-rhodium alloy are good catalysts. Some other metals or metal oxides are also used as catalysts. Catalytic oxidation of many organic odorous substances can take place at lower temperatures (say, around 260 °C). The catalysts should have high specific surface area and should be widely distributed and supported securely in the furnace / kiln. The odorous gases admitted in the furnace must be free from materials (such as mercury, lead, zinc, particulate matter) which are poison for the catalysts. They may react with them or cover them to make them in-effective. There are many commercially available catalysts which are being used for catalytic oxidation of odorous stream from varnish / paint manufacture, refuse/ biomedical waste incinerators, coffee roasting, chemical plants, ovens for baking of enamel, etc.

18.2.9 *Oxidation of Odorous Substance by Powerful Oxidants*

The powerful oxidants like ozone and chlorine can destroy / oxidize the odorous substance quite effectively. These oxidants are biocides and destroy the odour producing bacteria, fungi, and other microorganisms. Chlorine is used in the oxidation of paper mill effluents. The partial oxidation of highly odorous mercaptans may still yield odorous products (though less odorant), hence it is better to completely oxidize the odorant.

The ozone reacts with organic matter, but on partial oxidation, may give out products like aldehyde, acetones and organic acids and therefore it is better to fully oxidize the objectionable gases. The oxidants should not be used in excess as these themselves are odorous.

18.2.10 Irradiation

The odour problem caused by micro-organisms can be removed by irradiation by ultra violet rays. Ultra violet method is used for removing bacterial pollution from drinking water. If the odour is due to micro-organisms, it gets removed. [*The other odorous gases are removed by prior filtration through activated carbon].*

18.2.11 Odour Control by Patented Formulations

Many manufacturers have patented formulations for odour control. Two such examples are mentioned below.

Bio- World Odour Neutralizer

As an example the Bio-World Odor Neutralizer (BON) is claimed to be an unique liquid formulation that eliminates offensive odours at the source. These odour control products are said to neutralize unpleasant odours and make them undetectable to our sense of smell, while many other commercial products only mask odour by strong scents. These products are said to have been developed for the control of major industrial type odours and are used in factories, food processing facilities, warehouses, large office complexes, shopping malls, wastewater treatment facilities, restaurants, hospitals and many other places. For application, these formulations are to be diluted and fogged / sprayed to get rid of odour problem

Epoleon Unique Formulations (patented)

Epoleon's unique formulations allow the chemicals to react with a variety of odour causing gases both acidic and alkaline. For example the Epoleon N-100 and N-7 (patented) can control odour. This is clear from the following reactions:

NH_3 + $-CH_2COOH \rightarrow -CH_2COONH_4$

[Ammonia odour] [No undesirable odour]

$(CH_3)_3$ N + $-CH_2COOH \rightarrow -CH_2COONH(CH_3)_3$

[Trimethylamine odour] [No undesirable odour]

H_2 S + $-CH_2COO$ Na $\rightarrow$ NaHS + $-CH_2COOH$

[Hydrogen Sulphide odour] [No undesirable odour]

$CH_3SH + -CH_2COONa \rightarrow CH_3SNa + -CH_2COOH$

[Methyl Mercaptan odour] [No undesirable odour]

Some industries use Sodium Hypo-chloride (NaClO) to reduce H_2S levels, but this chemical does not reduce H_2S level below 5 ppm, still leaving it detectable by human nose. Further reduction of odour causing H_2S below the human threshold of recognition can be done by above chemicals of Epoleon. ***(Ref.:*** *Epoleon Corporation, Torrance, California 90503., USA).*

18.2.12 Process Modifications

The process modification for odour control may include:

- Replacement of the raw material and /or fuel to control the generation of less odorous emissions
- Advance purification of materials used in process
- Substitution of highly odorous solvents by low odour solvents
- Change of the production process / technology to reduce the generation of odorous compounds

- Adjustment of process variables including temperature, pressure, reaction time, ratios of substances used, mixing, turbulence, and flow rates;
- Recovery of odour creating compounds

Besides the above mentioned odour control measures, some other technologies may be used to reduce the odour generation in some cases. (As for example: Diet optimization for animals to limit odour produced from livestock).

18.3. MEASURES TO MITIGATE SPECIFIC ODOROUS GASES

For control of various types of polluting odorous gases, reference may also be made to the following paragraphs of *chapter 8:*

Para. 8.3.3 — for sulphur dioxide;

Para. 8.3.4 — for Hydrogen chloride;

Para 8.3.5 — for Chlorine;

Para.8.3.7 — for Hydrogen sulphide, Mercaptans, Carbonyl sulphide, Carbon disulphide;

Para. 8.3.8 — for Ammonia;

Para.8.3.9 — for Dioxins;

Para. 8.5 — for VOC

It may be pointed out that the control measures in the above referred paragraphs sometimes may not be able to completely eliminate the odour and the help of methods of odour control discussed in the preceding ***Para.18.2*** may still be required.

QUESTIONS

1. Write short note on: odour perception, differentiation and intensity, and odour index.
2. Enumerate various options available for odour control.
3. What is activated carbon and what is its role in the fiela of odour control?
4. Elaborate how the chemical scrubbing can control various odours.
5. Write short notes on: (a) Process modifications for control of odour, (b) Oxidation of odorous substance for control of odour, (c) Biological control of odours.

CHAPTER 19

ENVIRONMENTAL IMPACT ASSESSMENT (EIA) FOR PROJECTS

19.1 IMPORTANCE OF EIA

Before setting up any industry or before construction of a factory in any area, it is necessary to assess the environmental impact of the project in advance to know the projected estimated level of pollutants in the environment when the proposed industry comes into being. This is necessary in order to decide what measures are required to be incorporated in the project / industry to keep the level of pollutants within the permissible limits. This would also indicate if the industry is environmentally sustainable and feasible at the proposed location even after providing all required control equipments and measures. No new project of setting up of an industry, factory or power station can be started anywhere without its clearance from environmental angle by the concerned authorities.

The actual ground level pollution in air would depend on meteorological factors, the number and position of pollution sources present in the area, number and position of new sources of pollutants proposed to be introduced by the project, the background values of various pollutants present in ambient air (at the time when the proposed project comes into being), and the assessed values of various pollutants which would get introduced in the environment by the project. It is necessary to look into the pollution of not only air, but also ground water, surface water and land. The effect on agriculture, plants, forestry, human health and animal health, land utilization, land pollution and social factors are also to be looked into.

If the proposed facilities for mitigation of pollution can keep the level of pollutants in the ambient air, water, and land still within the 'Standards applicable', and if there is no perceived major environmental disadvantage, the project may be feasible from environmental angle. All this can be judged only after environmental assessment is properly done. This is essential for sustainable development which was defined for the first time by the World Commission on Environment and Development (i.e. the Brundtland Commission) in the year 1987.

An 'Environmental Impact Assessment (EIA) study' is made for an approved length of time covering all seasons (preferably a year so that all seasons are covered well) and then the EIA report is submitted to *'Competent authorities'* for project clearance. The competent authorities after going into the details of EIA report may approve the project if the projected pollution levels are found within limits permitted by Environment Protection Rules of the country.

The EIA study would involve finding out existing environmental parameters (i.e. the parameters before setting up of project) called 'Background Environment' which requires the study of climate, terrain, natural or manmade drainage, geology and soil, hydrology, hydro-geology, ground water and surface water, air quality, noise level, ecology including flora and fauna, existing specific health problems/ diseases, land use including forest cover and agriculture, existing industries, mines, historical sites, archeological sites, other current or proposed developmental projects already cleared, etc. for the area. The project should have 'Inbuilt design and implementation measures' for 'Pollution mitigation and control' so that the already existing conditions of environment are not aggravated.

The EIA study would assess the level of pollution and environmental degradation created by the project. It may be pointed out that to assess the impact of operating a plant under the project being set up, the mathematical modeling for dispersion of air pollutants is an important part of the exercise. The subject of air pollution dispersion modelling has been briefly discussed in ***Chapter 3.*** The fundamentals of Gaussian air pollution models are discussed in detail in ***Para. 19.3** of this chapter.*

The net level of pollution i.e. 'the level of projected pollution created by the proposed project superimposed on existing level of pollution' should still be within the *limits imposed by environment protection rules* both during the construction of the project and during operation and maintenance of the facility, industry, or power station.

19.2 CONTENTS OF EIA REPORT

Broadly an EIA Report for a project would normally contain:

(a) Impact assessment,

(b) Environment management plan and

(c) Environment monitoring plan.

The contents of report are given below:

19.2.1 Environmental Impact Assessment:

1. Project Outline

2. Study Area for EIA

3. Project Justification

3.1. Demand for the commodity to be manufactured / produced by setting up of the project.

3.2. Justification of proposed site selection.

3.3. Various options for the production of the desired commodity (to be produced in the facility) including options for meeting the demand of commodity if the project is not sited in this region.

3.4. Details of site area selection.

3,5. Other Industries in the area, their location on area plan and their brief details about emissions and waste disposal.

3.6. Project Developer's conclusive views on the above matters.

4. Project Overview:

4.1. Physical location, and geographical, and historical features of the project site.

4.2. Process design of industry detailing raw materials used, intermediate products, final products, waste products including solids, liquids and gases and their ultimate disposal.

4.3. Plant layout

4.4. Important design features of plant

4.5. Raw material requirements, supply, storage, and handling details.

4.6. Fuel requirement, fuel characteristics, supply, storage and handling system.

4.7. Electrical energy requirement and details of its supply or other options.

4.8. Water systems
   (a) *Water requirement and supply system*
   (b) *Required water treatments systems for the various plants or systems.*
   (c) *Water conservation measures*

4.9. Details of chemical recovery system incorporated in the plant design and characteristics of waste products.

4.10. Air pollution mitigation systems for:
   (a) *Particulate matter, respirable particulate matter, micro particles of lead, mercury or other toxic metal, material or compounds*
   (b) *Oxides of sulphur*
   (c) *Oxides of nitrogen*
   (d) *Halogens, halogen compounds, and ozone depleting compounds*
   (e) *Hydrogen sulphide and mercaptans*
   (f) *Ammonia*
   (g) *Carbon Sulphide*
   (h) *Hydrocarbons including Polycyclic Aromatic Hydrocarbons (PAH) and Polychlorinated Biphenyls (PCB), Nitro PAH, and Dioxins*

4.11. Liquid waste management, treatment system, utilization, recycling' and disposal system for:
   (a) *Chemical effluents*
   (b) *Waste water*
   (c) *Wash water*
   (d) *Surface drainage*
   (e) *Oil contaminated water*
   (f) *Sewage*

4.12. Management system for biological waste, and protection against biological pollutants, allergens, pathogens and virus.

4.13. Control of pollution by radioactive wastes, their treatment and ultimate disposal.

4.14. Solid waste management and disposal for:
   (a) *Electronic Waste*
   (b) *Municipal solid waste*
   (c) *Metallic scrap*
   (d) *Ash of fuel*
   (e) *Other solid wastes*

4.15. Hazardous materials/ intermediate products / final products / waste products, if any, involved in the process, and their treatment / mitigation/ storage / disposal system

4.16. Noise pollution and its mitigation for different equipments

4.17. Fire prevention, fire detection, fire protection systems for various plants

4.18. Details of stack /stacks from which emission gases go out to atmosphere

4.19. Special measures provided for control of greenhouses cases

4.20. Special measures for material conservation

4.21. Rehabilitation of population displaced and their employment

4.22. Landscaping and architecture

5. Existing Environment – Physical

5.1. Climate, relief and natural drainage

5.2. Seismic activity in the area

5.3. Geology and soil

5.4. Existing hydrology and hydro-geology of area

5.4.1. *Ground water level (GWL) / groundwater abstraction rate by adjoining industries and their impact in terms of sinking of groundwater level / availability*

5.4.2. *Surface water availability and the impact of adjoining industries / projects / on availability*

5.5. Existing air quality i.e. background air quality

5.5.1 *Available air quality data of the area from other independent establishments*

5.5.2 *Individual emitters in the area and tentative details of emission*

5.5.3 *Radiation emission if any —Details to be incorporated.*

5.5.4 *Base line air quality studies*

(a) *Setting up of high volume sampling stations (8hourly values) and their results*

(b) *Results of NO_2 diffusion tube survey (Technology developed at Warren Spring Lab. and Harwell Lab) or results of other types of air quality surveys*

5.5.5 *Existing noise level*

6. Environmental Impacts and Mitigation Measures during Construction

6.1. Construction overview

6.2. Waste disposal during construction

6.3. Transport during construction

6.4. Construction machinery use

6.5. Noise

6.6. Air quality impact during construction

6.7. Water supplies for construction / Drainage

6.8. Impact on water quality during construction

6.9. Safety during construction

6.10. Socio- economic issues for construction

6.11. Impacts of construction on Flora

6.12. Impacts of Construction on Fauna

7. Water Abstraction, Discharge during operation of project and Water Quality Studies

7.1. Projected water requirement of cooling water, water for various processes

7.2. Projected consumptive water

7.3. Water quality in existing surface water reservoirs / ponds / lakes / rivers as per results of monitoring stations set up in connection with the project

7.4. Groundwater quality results from monitoring wells and monitoring tube wells set up in connection with the project

7.5. Existing water quality data of groundwater and surface water and trend of water quality deterioration over the years as per information from the adjoining establishments.

7.6. Proposed quantities of water abstraction from various sources

7.7. Projected quantities of effluents from outlets of various plants and their treatment,utilization / recycling within the plant

7.8. Projected quantities and qualities of final water discharges from the premises

7.9. Projected impact on quality and future availability

7.10. Possibility of sea water intrusion if any

7.11. Details of Radiation by radioactive material used if any

8. *Quantities, Types, Qualities of Solid Waste and Liquid Waste Discharged during Operation of Plant, and Options for Disposal of these Wastes.*

9. *Control of Impacts on Air Quality during plant Operation*

9.1. Emission control and monitoring measures adopted

9.2. Projected rate of emissions of various pollutants discharged to atmosphere.

9.3. Proposed details of control over radiation by radioactive materials used in the plant / project, if any

10. *Air Pollution Modeling for Prediction of Air Quality during Plant Operation*

10.1. Air pollution model selection and its application to the particular case

10.2. Predicted short term average concentrations of air pollutants

10.3. Predicted long term annual concentrations of air pollutants.

10.4. Sample details of air pollution model

11. *Noise during Plant Operation*

11.1 Noise sources in the plant

11.2 Predicted noise levels

12. *Predicted Impacts on ecological systems during Plant Operation*

Note on Air dispersion models: *For determination of the air dispersion of pollutants, various models are used as described in* **Chapter 3..** *The projected level of maximum ground level pollutants are to be found out and given in the EIA report.*

19.2.2 Environment Management Plan:

1. Scope of plan
2. Works-plans and schedules for various materials including raw material, fuels (solid / liquid / gaseous), bulk chemicals, hazardous materials or radioactive materials (if any).
3. Water management and plan for raw water, cooling water, potable water, drainage, liquid waste.
4. 'Biological waste' management plan and 'Solid waste' management plan including ash management.
5. Emission management plan
6. Radiation management plan
7. Maintenance plan concerning environment

8. Fire Protection plan
9. Disaster management plan for protection of plant and personnel
10. Compensatory Tree Planting Plan
11. Any other contingency plan
12. Resource, implementation, and training

19.2.2 Environmental Monitoring Plan:

1. Atmospheric emission monitoring
2. Water / effluent discharge quality monitoring
3. Radiation monitoring
4. Ground level air quality Monitoring
5. Monitoring of discharge in environment of toxic materials in solid wastes or in particulate matter.
6. Monitoring of leachate quality percolating in ground water
7. Monitoring quality of raw materials, water, fuels, chemicals used
8. Noise monitoring
9. Monitoring of occupational safety implementation

Notes:

1. **EIA in India**

The Ministry of Environment and Forest (MoEF) of India has been making constant efforts in Environmental Impact Assessment (EIA) in India. The main laws in action are the Water Act (1974), the Indian Wildlife (Protection) Act (1972), the Air (Prevention and Control of Pollution) Act (1981) and the Environment (Protection) Act (1986). Biological Diversity Act(2002). The responsible body for this is the Central Pollution Control Board. EIA studies need a significant amount of primary and secondary environmental data. Primary data are those collected in the field to define the status of the environment (like air quality data, water quality data etc.). Secondary data are those collected over the years that can be used to understand the existing environmental scenario of the study area. The EIA studies are conducted over a short period of time and therefore the understanding of the environmental trends based on a few months of primary data, has limitations. Ideally, the primary data must be considered along with the secondary data for complete understanding of the existing environmental status of the area. In many EIA studies, the secondary data needs could be as high as 80% of the total data requirement. Environmental Information Centre (EIC) is the repository of one stop secondary data source for EIA in India.

The Environmental Impact Assessment (EIA) experience in India indicates that the lack of timely availability of reliable and authentic environmental data has been a major bottleneck in achieving the full benefits of EIA. The environments being a multi-disciplinary subject, a multitude of agencies are involved in collection of environmental data. However, no single organization in India tracks available data from these agencies and makes it available at one place in a form required by EIA practitioners. Further, environmental data is not available in enhanced forms that improve the quality of the EIA. This makes it harder and more time-consuming to generate EIA's and receive timely environmental clearances from regulators. With this background, the EIC has been set up to serve as a professionally managed clearing house of environmental information that can be used by MoEF, project proponents, consultants, NGOs and other stakeholders involved in the process of EIA in India. EIC caters to the need of creating and disseminating the organized environmental data for various developmental initiatives all over the country.

EIC stores data in GIS format and makes it available to all EIA studies and to EIA stakeholders in a cost effective and timely manner. Section 34A, Environmental Quality Act. 1974 requires developments that have significant impact on the environment are required to conduct the Environmental impact assessment.

2. EIA *MANUAL (Ministry of Environment and Forests Government of India)*

A Manual of EIA has been issued by Ministry of Environment and Forest India in January 2001, the preface of which is reproduced below for the benefit of readers.

***Preface:** 'Environmental Protection and Sustainable Development have been the cornerstones of the policies and procedures governing the industrial and other developmental activities in India. Ministry of Environment & Forests has taken several policy initiatives and enacted environmental and pollution control legislations to prevent indiscriminate exploitation of natural resources and to promote integration of environmental concerns in developmental projects. One such initiative is the Notification on EIA of developmental projects issued on 27.1.1994 under the provisions of Environment (Protection) Act, 1986 making EIA mandatory for 29 categories of developmental projects. One more item was added to the list in January, 2000.*

EIA is a planning tool that is now generally accepted as an integral component of sound decision-making. The objective of EIA is to foresee and address potential environmental problems/concerns at an early stage of project planning and design. EIA/EMP should assist planners and government authorities in the decision making process by identifying the key impacts/issues and formulating mitigation measures. The Ministry had issued sectoral guidelines some time ago.

A compendium of the procedures and questionnaires entitled 'Application Form and Questionnaire for Environmental Clearance' was published in September, 1999 in association with the Confederation of Indian Industry.

As part of the continued efforts to ensure transparency in the procedures of environmental clearance and to assist the project authorities in improving the quality of EIA documents, this Manual is now being brought out by the Ministry. The Manual has been designed to cover the whole gamut of issues like regulatory requirements, the EIA methodology including baseline studies, identification of key issues and consideration of alternatives, impact analysis and remedial measures in a systematic way. It also delineates the process of reviewing the adequacy of EIA and EMP reports and post-project monitoring. To make the Manual comprehensive and self-contained, information pertaining to legislative regime, base line data generation and monitoring, thumb rules for pollution control measures etc. have been annexed to the main text.

It has been our experience that EIA documents are often voluminous but much of the base-line information included in these is not fully utilized in the impact analysis and prediction. Some of the impacts of the proposed development are of little significance to the decision making process. Hence a brief reference has been made to 'Scoping'. A section on Risk Assessment and Hazard analysis also has been included. It is hoped that project proponents, EIA consultants and regulatory authorities will find this EIA Manual useful.'

Contents of the Manual:

(Environmental Impact Assessment, Impact Assessment Division,
Ministry of Environment and Forests (MOEF) Government of India, January, 2001)

PREFACE

Chapter 1- Introduction

Chapter 2- Identification of key issues
Chapter 3- Reviewing the EIA Report
Chapter 4- Review of EMP monitoring

REVIEW CHECK-LIST

Section-A	Guidance Notes for Pre-Appraisal
	Preliminary Project Appraisal
Section-B	Guidance for Appraisal
1	Description of the Project
2	Project and Process Alternatives
3	Description of the Environment
4	Description of Environmental Impacts
5	Mitigation Measures
6	Difficulties in Compiling Information
7	General Presentation
8	Non -Technical Summary

ANNEXURES

i	EIA Notification
ii	List of Ecologically Sensitive Areas
iii	International Agreements and Commitments to Conventions
iv	Methods of Monitoring And Analysis
v	Models for Impact Predictions
vi	Checklist for Ecological Impact Assessment
vii	Guidance for Relevant issues for Different Project Types
viii	Good Practices for Prediction
ix	Occupational Health Impacts
x	Risk Assessment
xi	Impact Mitigation Measures

QUESTIONS

1. What is EIA and what is its importance?
2. Which items have to be covered in the study of background environment.
3. Environmental Impact Assessment is prepared for:
   (a) Every specified polluting developmental project;
   (b) Every natural disaster to pre-assess its adverse impacts on humanity and environment;
   (c) Both (a) and (b);
   (d) none of the above.

4. An Environmental Impact Assessment is intended to identify the environmental, social and economic impacts of a proposed development:
   (a) Prior to the decision to sanction a project is taken;
   (b) During the execution of a project;
   (c) After the execution of a project a assess its beneficiaries;
   (d) None of the above.
5. An Environmental Impact Statement should include:
   (a) Description of the project with location, design, size, etc.;
   (b) Description of its significant effects on environment and society;
   (c) Mitigating measures to reduce the possible adverse impacts;
   (d) All the above.
6. The type of projects, which among the following, will require public consultation under EIA process to be carried out under Indian EIA notification 2006, is:
   (a) Building/construction/area development projects;
   (b) Expansion of roads and highways which do not involve any further land acquisition;
   (c) Projects concerning national defence and security;
   (d) Nuclear power projects.

Answers: **3.** (a) **4.** (a) **5.** (d) **6.** (d)

CHAPTER 20

MISCELLANEOUS TOPICS

20.1 STRATOSPHERIC OZONE

20.1.1 Ultraviolet Radiation

The ultraviolet radiations are electromagnetic radiations coming to earth from sun like other radiations including that of visible light. The frequency and wave lengths and photon energy of these radiations for 'Extreme ultra violet (EUV)' radiations are:300 to 30 PHz, 10 to 100 nm, and 1.24 keV to 124 eV respectively, while the same for 'Near ultra violet (NUV)' are 30 to 3 PHz, 100 nm to 1μm, and 124 to 12.4 eV respectively. These radiations cause excitation of molecular/ atomic valence electrons and eject electrons thereby damaging molecules. The extreme ultra violet radiations are ionizing radiations (produce ions and free radicals in materials including the living tissue). The ultra violet radiations threaten life on earth as they can cause much more damage to life than simple heating.

Although the concentration of ozone in the ozone layer in stratosphere is very small (less than 10 ppm), it is vitally important to life because it absorbs biologically harmful ultraviolet (UV) radiation coming from the sun. The average ozone content of air at the surface of earth is only about 0.3 ppm.

Extremely short wavelength UV (10 to 100 nm) is screened out by nitrogen. UV radiation capable of penetrating nitrogen is divided into three wavelength categories and these are referred to as UV-A (400 to 315 nm), UV-B (315 to 280 nm), and UV-C (280–100 nm).

UV-C is very harmful to all living things, but is entirely screened out by a combination of oxygen molecule (< 200 nm wavelength by oxygen) and ozone (> about 200 nm wavelength by ozone) by around 35 km altitude.

UV-B radiation can be harmful to the skin and is the main cause of sunburn. Also, the excessive exposure to it can also cause cataracts, immune system depression, and genetic damage, and even skin cancer. The UV-B radiation also affects the plants adversely on its growth, physiology and yield of crops. The ozone layer is very effective at screening out UV-B.

The ozone layer absorbs UV radiations from about 200 nm to 310 nm (maximum around 250 nm wavelengths). The intensity of radiation with a wavelength of 290 nm, at the top of the atmosphere is 350 million times stronger than at the Earth's surface. Nevertheless, some UV-B, particularly at its longest wavelengths, reaches the surface.

Ozone does not absorb UV-A so most of this longer wavelength UV radiation reaches the earth's surface. Most of the UV reaching the Earth is UV-A which is significantly less harmful to DNA, although it may cause physical damage like premature aging of the skin, and on long exposure, indirect genetic damage and skin cancer.

World Ozone Day

Due to environmental problems created by ozone, United Nations decided in 1994 that World Ozone Day be observed on 16th Sept every year.

20.1.2 Ozone Shield

A 20 km thick ozone layer between 15 km and 35 km height (in stratospheric zone) over the earth is present as a shield which the nature has provided against ultra violet radiation coming to earth. This layer of ozone is in dynamic equilibrium with ultra violet radiations. Some ozone / oxygen molecules dissociate due to ultra violet rays to form atomic oxygen. The atomic oxygen in turn reacts with other oxygen molecules to form ozone

$$O_2 = O + O; O_3 = O_2 + O;$$

$$\text{and then } O_2 + O = O_3$$

This natural ozone shield in stratosphere is under real threat of depletion by chlorofluorocarbons (as mentioned in Para. 20.1.4 of this chapter) and chemical catalysts like oxides of nitrogen (NOx), etc. The chlorofluorocarbons (CFCs) are released into the environment as they are largely used in refrigeration, air conditioning, and cleaning solvents industries; they are also used sometimes in aerosol sprays and as blowing agents to create certain kinds of plastic foams.

The nitric oxide (NO) which may reach stratosphere reacts with atomic oxygen (formed in the above mentioned reaction) and forms nitrogen dioxide (NO_2)

$$NO + O = NO_2.$$

Thus the nitric oxide absorbs some atomic oxygen thereby slowing the reaction forming ozone in stratosphere and hence depleting the ozone layer.

Similarly, the nitrous oxide whose life in atmosphere is high about 120 years reaches stratosphere and absorbs the atomic oxygen therefore slows down the production of ozone. This therefore also depletes ozone shield.

The Chlorofluorocarbons (CFC) are very stable compounds and reach stratospheric zone. There they dissociate due to photolysis or through reaction with atomic oxygen. Free chlorine is a strong catalyst for dissociation of ozone. Thus chlorofluorocarbons are considered very strong ozone depleting compounds. The chlorine may eventually get converted to hydrochloric acid by reaction with hydrogen ion present there descend in troposphere and get washed down by rain lowering its pH value (one of the causes of acid rain).

20.1.3 Ozone Hole

Atmospheric models predict thinning of stratospheric ozone layer across all latitudes, but because of complexity of measurement operations and cost, this has not been spectrally observed except in 'Polar regions'. In 1987 March to October, a multinational expedition in Antarctica found that:

(a) The average natural concentration of stratospheric ozone over Antarctica dropped by 50% during the period 15th August 1987 to 7th October 1987.

(b) In some patches of observed area of Antarctica, the Stratospheric ozone disappeared. These ozone-less patches are now called 'Ozone hole'. The Ozone hole was found to be 19.3 million square kilometre in Sept of that year.

The Antarctica 'Ozone Hole' reached its Peak on 12th September 2011, while the ozone reading over the 'South pole' dropped to 102 Dobson units*. The ozone reading is usually between 250 to 350 DU

Some readings of ozone hole over Antarctica as per 'NASA- Ozone Watch' are given in **Tables 20.1 and 20.2.**

Table 20.1: Readings of Ozone Hole Over Antarctica as per 'NASA- Ozone Watch'

Year	Ozone hole area detected over Antarctica from 7th Sept to 13th Sept
1984	10.1 million square km
1990	19.2 million square km
1994	23.6 million square km
2000	24.8 million square km
2002	12 million square km
2006	26.6 million square km
2011	24.7 million square km

Table 20.2: Readings of Ozone Hole Over Antarctica as per 'NASA- Ozone Watch'

Year	Ozone layer detected over Antarctica from 21st Sept to 16th Oct in DU
1984	163.6 DU
1990	124.2 DU
1994	92.3 DU
2000	98.7 DU
2002	157.4 DU
2006	98.2 DU
2011	106.2 DU

Probable Reason for Fast Depletion of Stratospheric Ozone in Polar Regions

Because of angle of earth's rotation, the jet stream in northern hemisphere and in southern hemisphere move in opposite direction and these two general circulation patterns collide over the poles and create polar vortex. The pollutants like nitrous oxide, nitric oxide, Chlorofluorocarbons, etc. transported in stratosphere accumulate over the poles and destroy the ozone layer rapidly creating ozone hole observed in the polar region.

Notes:

1. ***Dobson unit of Ozone:*** *The Dobson Unit (DU) is the unit of measure for total ozone. If all the ozone is taken in a column of air stretching from the surface of the earth to space, and that ozone is brought to standard temperature of 0 °C and pressure of 1013.25 millibars, or one atmosphere (atm), the column would be about 0.3 cm thick. A column of 300Du of atmospheric ozone would form a 3 mm layer of pure ozone gas at the surface of the carth of Standard Temperature and Pressure (STP) Thus, the total ozone would be 0.3 atm-cm. To make the units easier to work with, the "Dobson Unit" is defined to be 0.001 atm-cm. (0.3 atm-cm would be 300 DU.) This unit is called a Dobson Unit (D.U.), after G. M. B. Dobson, one of the first investigators of atmospheric ozone. A Dobson unit is the amount of ozone in a column which if unmixed with the rest of the atmosphere, would be 10 μm thick at standard temperature and pressure.*
2. ***Effect of Increase of Greenhouse Gas at Stratospheric Temperature***

   When greenhouse gas/ CO_2 centent is high, most of the outgoung infrared energy is absorbed in troposphere and lesser amount of energy is left out to be absorbed in Stratosphere. At these altitudes, most of the pho:on energy is absorbed by water vapour and ozone. Thus increase of carbon dioxide or greenhouse gases contribute to cooling of stratosphere.

20.1.4 Ozone Depleting Substances, Greenhouse gases and Global warming [A]

The main cause of ozone layer depletion is the presence of pollutants like Chlorofluorocarbons (CFCs), Hydrochlorofluorocarbons(HCFCs), Halon compounds [such as Halon 1211, i.e. Bromo- chloro-difluoro-methane, Halon 1301, i.e. bromo- trifluoro-methane, Halon2402 ie Dibromo-tri-fluoroethane], Bromo fluorocarbons, Hydro bromo- fluorocarbons, Hydro-fluorocarbons, Per- fluorocarbons, Sulphur-hexa-fluoride, Carbon-tetra-chloride, Chloroform, Methyl bromide, etc. in atmosphere. The oxides of nitrogen (NO_x) and nitrous oxide, etc. in atmosphere aid the depletion of ozone as already mentioned above.

There is extensive industrial importance of Chlorofluorocarbons (CFCs), but due to their harmful effects their replacement by lesser harmful compounds is now being done. The Hydro chlorofluorocarbons (HCFC) i.e. 'CFC with additional hydrogen atoms' break down faster in the lower atmosphere itself and therefore their global warming potential as well as ozone depleting potential are much lower than those of CFCs. The Hydro-fluorocarbons (HFCs) contain no chlorine atom at all and have industrial properties similar to those of the HCFCs having a reduced potential of depleting ozone. The replacements of CFCs are more expensive. Halon compounds such as Halon 1211 and Halon1301 which were used in fire extinguishing services have also been put under restrictions as these are also potentially ozone depleting compounds.

In India, 'The Ozone Depleting Substances Regulations and control Rules, 2000' gives a list of such substances in its Schedule I besides putting restrictions on production and consumption of such substances.

The **Montreal protocol** signed in September 1987 by 35 countries is the first international agreement to limit the production and consumption of ozone depleting substances. The Montreal Protocol is the UN-sponsored international ozone treaty. Since signing in 1987 this protocol has triggered the reduction of production and consumption of 96 ozone-depleting chemicals by 97%. The CFCs* and HCFCs** are regulated under this agreement.

The **United Nation Framework Convention on Climate Change (UNFCCC)** is an international body which deals with global climate change and the **1997 Kyoto Protocol** is an agreement that resulted from negotiations under the UNFCCC.

The Fluoronated gases (F-gases) such as CFCs, HCFCs, HFCs, etc. are thousands of times more potent as greenhouse gases than carbon dioxide. Their major applications are in refrigeration and air-conditioning, foams, aerosols, fire extinguishers and solvents. It is the same heat-trapping properties that make most F-gases such good refrigerants which also make them extremely powerful greenhouse gases as clear from **Table 20.3.**

The Hydrofluorocarbons (HFCs)*** were advocated in place of CFCs and HCFCs as their ozone depleting effect is not significant. The HFCs were introduced to save the ozone layer in the upper atmosphere from destruction, but scientists have found that they are beginning to pose a serious threat to the global climate because of their very high greenhouse effect. The HFCs are already the refrigerants of choice in most industrialized countries. But if HFCs are used as substitutes for all the ozone-destroying chemicals they were designed to replace and end up in the atmosphere—they will have a devastating impact on the climate. These chemicals could be one of the most dangerous and yet most avoidable chapters in our environmental history. 'Greenpeace' is therefore calling on governments and corporations in industrialized countries to rapidly phase out the use of HFCs and to immediately move to environmentally safer alternatives in the interest of avoiding damage to climate.

Table 20.3: Greenhouse Gas Concentrations (Updated December 2011) **[as per T.J. Blasing DOI: 10.3334/CDIAC/atg.032]**

GAS	Recent Tropospheric Concentrations in parts per trillion (ppt)	GWP (100-yr time horizon) taking GWP of CO_2 as unity	Atmospheric lifetime (years)
CFC-11 (tri-chloro-fluoro-methane) (CCl_3F)	241 to 239	4,750	45
CFC-12 (CCl_2F_2)	534 to 532	10,900	100
CF-113(CCl_2FFClF_2)	75	6,130	85
HCFC-22($CHClF_2$)	218 to 194	1,810	12
HCFC-141b(CH_3CCl_2F)	22 to 19	725	9.3
HCFC-142b(CH_3CClF_2)	22 to 19	2,310	17.9
Halon 1211 ($CBrClF_2$)	4.3 to 4.1	1,890	16
Halon 1301 ($CBrClF_3$)	3.33 to 2	7,140	65
HFC-134a(CH_2FCF_3)	62 to 52	1,430	14
Carbon tetrachloride (CCl_4)	87 to 86	1,400	26
Sulfur hexafluoride (SF_6)	7.12 to 6.73	22,800	3200
Other Halocarbons	Varies by substance		
Carbon dioxide	388.5**	1	<100
Methane (CH_4)	1870 to 1748	25	12
Nitrous oxide (N_2O)	323 to 322	298	114
Tropospheric ozone (O_3)	34	NA	Hours-days

Notes:

1. **CFCs (Chlorofluorocarbons), the first and most well-known class of F-gases (Fluorinated greenhouse gases) are both ozone-depleting and very powerful greenhouse gases. From 2010, CFC production or consumption anywhere in the world is illegal. They are regulated by the Montreal Protocol.*
2. ***HCFCs (Hdro-chloro-fluoro-carbons), the second generation of F-gases, introduced as a temporary replacement for CFCs are less ozone-depleting and not contributing to global warming as much as CFCs, but they are still extremely potent. They will continue to be used until 2020 in the developed countries and 2030 in developing countries. They are also regulated by the Montreal Protocol.*
3. **** HFC (Hydro-fluoro-carbons), the third generation of F-gases, yet another 'temporary' replacement for CFCs, although do not deplete the ozone layer, but they are extremely potent greenhouse gases, some even more powerful than CFCs. There is currently no international agreement to phase them out. They are included in the United Nations Framework Convention on Climate Change (UN-FCCC) basket of controlled greenhouse gases.*
4. *HFO (Hydro-fluoro-olefins), the fourth generation of F-gases, are actually just HFCs marketed under a different name to avoid the negative connotations of HFCs. According to the chemical industry that is manufacturing them, they do not damage the ozone layer and they have a minimal impact on the climate. However, there are other environmental and health risks of these chemicals that make their use unacceptably dangerous. [Financial Times (US)- July 15, 2009 'HFCs: A growing threat---'].*
5. [A] ***Note on greenhouse gases and global warming:***
   The Solar energy in the form of heat is retained in atmosphere mainly in some gases present in atmos-

phere which are called 'Greenhouse gases' and the effect of heat retention by such gases is named as green house effect. The rise in content of greenhouse gases in atmosphere is raising the average atmospheric temperatures which is called 'Global warming'.

It may be mentioned that Carbon dioxide, Methane, Nitrous Oxide, Ozone in Troposphere, CFCs, HCFC, and HFCs are important greenhouse gases which are causing global warming. As water vapour present in atmosphere also retains a lot of heat, it is also included in the category of greenhouse gases, but the amount of water vapour in atmosphere varies from season to season and is not cumulative while other gases have a cumulative effect. (Man has no control over water vapour).

The rise in contents of other greenhouse gases is due to various anthropogenic activities. Some natural causes are also there in generation of greenhouse gases (except CFCs, HCFCs, HFCs) but there are natural sinks also, due to which these gases are absorbed/ adsorbed/ converted to other compounds/ destroyed. The rise of greenhouse gases in atmosphere is largely on account of anthropogenic activities and therefore its control would also have to be done by human beings. The Control of greenhouse gases can be done mainly by conservation of resources, stopping destruction of forests, carbon sequestration in soil, trees, etc., decreasing the consumption of fuel by adopting various technologies, increasing process efficiencies, cutting out wastes, using materials and technologies which do not create greenhouse effects or pollution.

20.2 ACIDIC GASES AND ACID RAIN

20.2.1 Background

Acidic gases like sulphur dioxide, nitrogen dioxide, hydrogen chloride or any halogen chloride, etc. are very harmful to humans, animals and plants. Because of their corrosive nature, these are harmful to metals and all materials (particularly which have chemically basic nature).

'Acid rain' is a rain or any other form of precipitation that is unusually acidic i.e. having a low pH value and possessing elevated levels of hydrogen ions concentration. It can have harmful effects on plants, aquatic animals, and infrastructure. 'Acid rain' is a popular term referring to the deposition of rain, snow, sleet, fog, and dew and dry acidic components (acidifying particles and gases). [Distilled water on removal of carbon dioxide, has a pH value of ' 7' ie neutral. Liquids with a pH less than 7 are acidic, and those with a pH greater than 7 are alkaline. However the rain whose pH value is less than 5.6 is generally termed as 'Acid Rain'].

The acidic rain was discovered in 1853, but it was not until the late 1960s that scientists began widely observing and studying the phenomenon. Harold Harvey, a Canadian, was among the first to research a 'Dead lake' i.e. lake devoid of life. Public awareness of acid rain in the U.S increased in the 1970s after The New York Times published reports from the Hubbard Brook Experimental Forest in New Hampshire of the myriad deleterious environmental effects shown to result from it.

20.2.2 Causes of Acid Rains

Occasional very low pH values of rain and fog have been reported in industrialized areas. Acid rain is caused mainly by emissions of sulphur dioxide and nitrogen oxide from thermal stations, factories, and motor vehicles. These gases react with the water molecules in the atmosphere to produce acids. Some other emission of acidic gases like hydrochloric acid gas, hydrofluoric acid gas may also be involved.

Electrical power complexes and factories utilizing coal/ lignite /fuel oil are among the greatest contributors to gaseous pollutants responsible for acidic rain. The gases can be carried hundreds of km in the atmosphere (particularly from high/ tall stacks) before they are converted to acids and deposited. *[In the past, thermal stations and factories had short stacks to let out smoke, but this caused many problems locally: so now taller smoke stacks have been adopted].*

Some natural phenomena contribute towards the production of acidic- gaseous air pollutants which are soluble in water. These gases (like SO_x, HCl, NO_x, etc.) would therefore produce acids. The principal natural phenomenon is emissions of acidic gases from volcanoes. Acid-producing gasses are also created by biological processes that occur on the land, in wetlands, and in the oceans. The major 'Biological source' of sulphur containing compounds is 'Di-methyl-sulphide'. Some amounts of oxides of nitrogen are also produced by electrical activity in the atmosphere such as lightning and photo activity. [It may be noted that small amount of nitric acid and nitrous acid (HNO_3 and HNO_2) in rainwater is an important source for nitrogen fixing in plant life].

20.2.3 Ill-effects of Acid Rains

Acid rains have adverse impacts on forests, fresh-waters and soils, killing insect and aquatic life-forms as well as causing damage to buildings and monuments (specially those using marble/ lime stone), corrosion of exposed metallic surfaces.

Effect on Aquatic life

Both the lower pH and higher aluminum concentrations in surface water that occur as a result of acid rain can cause damage to fish and other aquatic animals. At pH values lower than '5' most fish eggs do not hatch and lower pH values may kill adult fish. The biodiversity goes on reducing, as lakes and rivers become more and more acidic. In some lakes and streams, the acid rain has eliminated insect life and some fish species, and other types of aquatic animals. 'Coral reefs' are damaged by acidification of sea water due to various reasons including acid rain.

Effect on Soils

Soil biology and soil chemistry can be seriously damaged by acid rain. Some microbes are unable to tolerate changes to low pH values and are killed. The enzymes of these microbes are denatured by acid so they no longer function. The hydrogen ions of acid rain also mobilize (increase) toxins such as 'Aluminum', and leach away essential nutrients and minerals such as 'Magnesium'.

$$2\,H^+\,(aq) + Mg^{2+}\,(clay) \leftrightarrow 2\,H^+\,(clay) + Mg^{2+}\,(aq)$$

The chemistry of soil may be dramatically changed when base cations, like calcium and magnesium, get leached away by acid rain thereby affecting sensitive species, such as sugar maple (Acer saccharum).

Effect on Forests and other Vegetation

Adverse effects on forest and vegetation may be indirectly related to acid rain, like the acid's effects on soil (see above) or high concentration of gaseous precursors to acid rain. The 'High altitude forests' are especially vulnerable as they are often surrounded by clouds and fog which are more acidic than rain.

Many plants may be damaged by acid rain, but the effect on food crops can be minimized by the application of lime and fertilizers to replace lost nutrients. In cultivated areas, limestone dust may also be added to increase the ability of the soil to keep the pH value stable, but this tactic is largely unusable in the case of forests, plantations and natural vegetation on lands. Leaching of calcium from the needles of red spruce *(a coniferous tree)*, makes these trees less cold tolerant; they exhibit winter injury and even die.

Effect on Human health

Acid rain does not directly affect human health as the acid in the rainwater is too dilute to have direct adverse effects. However, the gases responsible for acid rain (sulphur dioxide and nitrogen oxides) do have an adverse effect. Increased amounts of fine particulate matter in the air do contribute to heart and lung problems including asthma and bronchitis.

Dry Deposition

Acid deposition also occurs via dry deposition in the absence of precipitation. This can be responsible for as much as 20 to 60% of total acid deposition. This occurs when acidic gases and particulate matter stick to the plants, ground or other surfaces.

20.2.4 Control of Acid Rains

The control of acid rain is mainly the control of the emission of acid gases like oxides of sulphur, oxides of nitrogen, etc. These control of emissions are amply dealt with earlier (***Chapter 5, 6, 7, 8, 13, etc.***).

International Treaties Regarding Acid Rain

There are a number of international treaties on the long range transport of atmospheric pollutants, such as 'Sulphur Emissions Reduction Protocol under the Convention on Long-Range Trans- boundary Air Pollution'. Canada and the US signed the Air Quality Agreement in 1991. Most European countries have also signed such treaties/ protocols

20.3 POLLUTION BY SPONTANEOUS COMBUSTION IN COAL OR LIGNITE STACKS/ STORAGE

As already mentioned in **Para 8.8 notes of chapter 8**, the spontaneous combustion is phenomenon associated with coal and lignite. When such a solid fuel in 'fuel stacks' is slowly oxidized by the atmospheric oxygen present in air pockets of the fuel stack, the temperature slowly builds up as the natural cooling at the particular point may be inadequate. When the temperature goes up to ignition point, the fire would appear spontaneously. The phenomenon is seen in coal or lignite stacks at mines, thermal power stations or industries (burning coal or lignite). The higher the volatile content of solid fuel, the higher is the chance of spontaneous combustion. The high volatile coals / medium volatile coals and lignite (which normally have higher volatile content) exhibit this phenomenon more often.

The spontaneous combustion would result in emission of smoke along with greenhouse gases like nitrous oxide (N_2O), carbon dioxide (CO_2), and methane (CH_4) as well as pollutants like nitric oxide (NO), nitrogen dioxide (NO_2), sulphur dioxide (SO_2), carbon monoxide (CO), and other hydrocarbons. It is not feasible to eliminate them once these gases are out in atmosphere.

Control of Spontaneous Combustion

The phenomena of spontaneous combustion can be minimized by:

- Effective Monitoring of coal stack temperatures,
- Consolidation of fuel stacks while stacking the fuel layer by layer avoiding air pockets which are the potential source of oxygen for oxidation of fuel,
- Avoiding long storage of fuel at a place,
- Adequate arrangement of water sprinkling for cooling and fire fighting.

20.4 UTILIZATION OF SOLAR ENERGY AND WIND ENERGY AS AN ANTI-POLLUTION MEASURE

As the utilization of carbon based fuels like coal, lignite, oils, etc. for industries, domestic consumption, or for generation of electric power causes wide spread air pollution, it is necessary that alternative non polluting sources of energy are tapped in a big way. The '*Hydroelectric resources*' are already utilized to the maximum as far as feasible world over. The solar energy and wind energy are the two non- polluting resources provided by nature abundantly which should be harnessed to substitute for thermal energy of carbon based fuels.

The solar energy can be directly utilized for heating or it can be harnessed to generate electric power. Solar energy can be used for irradiation of some products.

Solar Heating

Solar heating of water can meet most of the hot water requirement except on non-sunny days / cloudy days. In tropical and subtropical regions, requirement of hot water in homes is only in winters, but in cold countries the hot water requirement is all the year round. Solar thermal can be combined with any other source of energy to cater for non-suny days.

'Solar still' for brackish water distillation is very advantageous. The device is very simple. The brackish water is evaporated by solar heat. The whole system is thoroughly insulated against escape of heat. The vapour is condensed to obtain distilled water.

Many other domestic heat utilization systems such as solar cooking, solar drying, solar air conditioning of buildings, solar space cooling for storage of vegetables like potato, and fruits, solar drying / moisture removal, etc. are being used, but initial cost is high for which the governments declare subsidies time to time. Solar heating is already quite popular both for domestic purposes, agricultural purposes and for certain purposes in industries.

Solar Electric Power

`The solar energy can be used for generation of electric power either by *'Solar thermal energy concentrating systems for electricity generation' or via 'Solar photovoltaic (PV) electricity generation systems'*.

Solar thermal Power Generation— Solar energy concentration is done for use in solar electric power generation. The working fluid in the engine for thermal power generation can be hydrogen, steam, or any other gas. The solar concentrator may be in the form of parabolic dishes, parabolic troughs, heliostats, helio-mirrors, stretched membrane and bubble heliostats. The prime mover for electric generator can be an engine working on thermodynamic cycles like *Rankine cycle* working on steam raised in the tubes of heat receiving surface, or *Sterling cycle* working on air or hydrogen heated by the solar energy concentrating system. The heat receiver systems can be a central receiver system where all field reflectors / concentrators focus the heat to a central receiver. Alternatively, the heat receiver system can be a distributed receiver system where the working fluid is heated and piped to a central engine coupled with generator facility. The auto sun tracking system can ensure the maximum collection of solar energy during the sun shine hours. Surplus power generated during bright sunshine can be stored in a battery* (with inverter) and used via inverter when required. A back up supply can be provided by grid / diesel unit. A configuration of solar thermal power generation system is shown in **Fig. 20.1.**

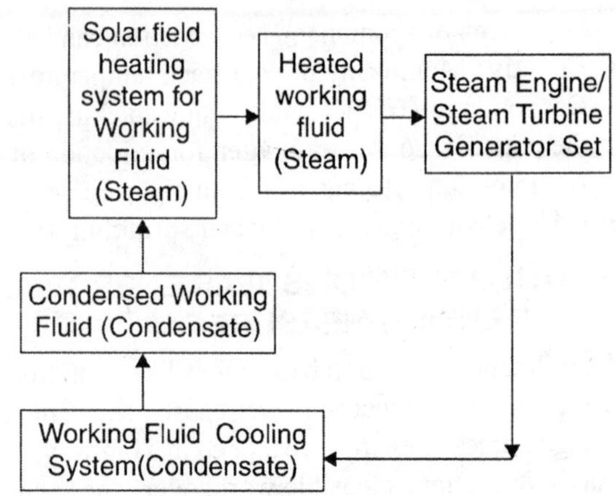

Fig. 20.1: Configuration of Solar Thermal power Generation System

* *'Ion-Lithium Batteries' have been developed by Tesla in USA as announced by them in May, 2015. These batteries are claimed to be highly efficient and have five times more capacity than conventional lead acid batteries. [Hindustan times, Lucknow, dated 10th may 2015].*

Solar Photovoltaic (PV) Power generation — Many solar electric devices are being used nowadays such as solar lighting, solar lantern, solar pumping specially in remote areas, solar power computers / calculators / mobile chargers / toys / other instruments, etc. The solar PV cell is central to conversion of light energy to electricity which is used in all such devices or for power generation in solar power station.

The technology of solar power generation is developing fast with the aim to make it more efficient and reduce its installation cost. The cost of solar PV cells is, however, coming down very fast because of new researches in cell material technology, large scale manufacturing, and competition. Solar electric generation systems can be combined with any other sources of energy to cater during the time when there is no sun shine. The solar PV Electric Power Generation System is shown in **Fig. 20.2.**

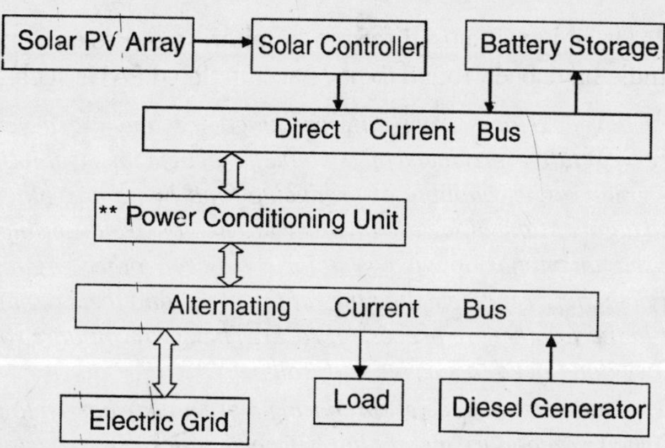

** *POWER CONDITIONING UNIT—has features of Bi-directional inverter, Load voltage regulation, Battery management / control, Diesel generator management / control, Monitoring*

Fig. 20.2: Solar PV Electric Power Generation System

Wind Energy

Utilization of wind energy in industry is age old practice. Wind driven pumps, wind driven mills for grinding of corns, wind driven air blowers are some of the examples. Wind energy can be utilized to drive turbines which can drive a number of processing machines in small industries. However due to non- continuous nature of wind energy direct utilization in industries is not being generally made nowadays. The wind energy is being used to generate electricity which is stored in batteries. The wind turbine can drive air compressors and the compressed air can be stored which can be used in various ways in industry. The wind turbine driven pumps can pump water in over head storage tanks which can be used in various ways in industries. Wind electricity generation can be combined with other generating systems or it can be used for production of non-polluting fuel like hydrogen from water. A configuration of wind electric power generation system with battery storage is shown in **Fig. 20.3.**

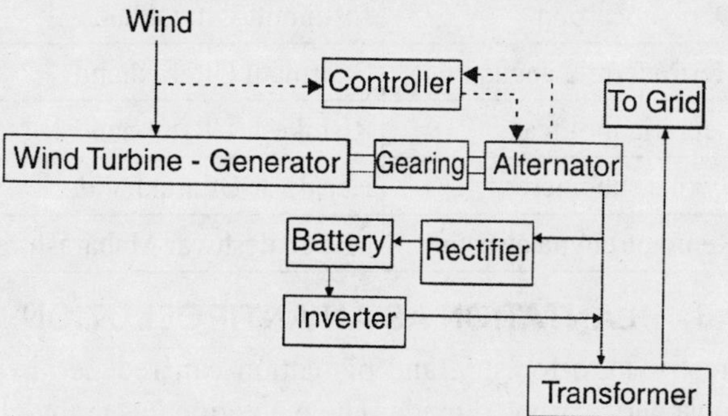

Fig. 20.3: Configuration of Wind Electric Power Generation System with Battery Storage

20.5 LICHENS* AS BIO-MONITORS FOR POLLUTION

Lichens* have no roots and take all nutrition (including water) from air. In the process they absorb/ adsorb various substances from atmosphere including air pollutants. Lichen diversity and distribution pattern is an important biometric tool to assess the prevailing environmental condition in the area. There is a distinct increase in the number of species and their density when we move away from air -polluted area.

Lichen Genus 'Usnea' is suitable organism for air pollution monitoring for a term in the world. The density and abundance of Usnea species can be correlated with changes in the forest ecosystem. The species have a preference for temperate regions and therefore it helps to evaluate the shifts due to climate changes.

Lichens accumulate PAHs and can be used for bio-monitoring of PAHs. Some lichens sampled out from different regions of India have been found to have accumulated PAHs such as given in **Table 20.4.**

Note: on Lichens: *Lichens have partners from different kingdoms and are a very diverse ecologically interesting group. They are vegetative products of a mutually beneficial symbiotic relationship between fungus and cyanobacteria/ green algae, capable of producing food by photosynthesis. Some 'Lichens' are symbiosis between fungi and algae. Many algae are involved in symbiotic relationships with other organisms. For example, corals and sea anemones involve symbiosis between other organisms and single-celled dinoflagellate algae. Coral bleaching represents the loss of the algae and their pigments.*

Lichens are also important in the weathering process of rocks. They grow on bare rock and release acids as part of their metabolism. They don't need to survive with soil.

Lichens are common primary successional species, preparing a previously barren landscape for other plants and organisms to exploit. Lichen's secondary metabolites (in particular the weak acids), begin the breakdown of sedimentary and igneous rocks, providing a loose soil matrix in which other nonvascular and vascular plants can take root. Lichens are integral part of all ecosystem and they are often responsible for fixing or capturing essential nutrients from air. The potential of discovering and utilising useful metabolites in lichens has opened new pathways for research.

Table 20.4: Total PAHs concentration in Lichen samples from various regions in India
(Source: *Vertica Shukla and DK Uprati Environews 2011)*

Lichen Species	Locality	Total concentration of PAHs in μg /g
Acarospora bullata	Mana Uttarakhand	30.07
A. praeradiosa	Mana Uttarakhand	22.98
Dermatocarpon	Joshimutt Uttarakhand	33.72
Hertrodermia angustiloba	Badrinath Uttarakhand	32.98
Lepraria lobificans	Rishikesh Uttarakhand	43.1
Pyxine subcinerea	Haridwar Uttarakhand	187.3
Remototrchyna awasthii	Mahabaleshwar Maharashtra	54.78

20.6 PLANTATION AS AN ANTIPOLLUTION MEASURE

Forestry, agro-forestry, and plantation can reduce air pollution, noise, etc., apart from sequestering atmospheric carbon dioxide. The plantation has many other advantages as mentioned below:

1. Increase in percolation of rain water by the roots of trees improves the ground water level as well as improves the quality of ground water;

2. Reducing heavy metals from soil and promoting soil improving bacteria;
3. Providing timber, fuel wood, fruits, and other useful products like medicines, gum, etc.;
4. Plantation softens the extremes of temperatures, radiation and humidity, and has positive effects on climate and ecology;
5. Many plants perform function of bio-indicators for pollution of air, water and land;
6. Many water plants remove heavy metals from water and purify it.
7. Plants provide shade and shelter to humans and animals and habitat for avifauna?
8. Serve as barriers/ fence, improve landscape;
9. Trees check soil erosion and retain soil on slopes.
10. Trees check desertification.

Measures Required for Boosting Plantations

In view of many advantages of growing plants, the following measures should be taken to boost the plantations:

- *Large Scale Planting of Trees*— Large scale plantations on all available land (which is not being used for agriculture or for other purposes) should be done and proper care and maintenance of plantations should be undertaken so that the plants do not die out. It is necessary to encourage the planting of trees alongside of roads, railway lines, rivers and streams and canals, and on other un-utilized lands under State/corporate, institutional or private ownership. Green belts should be raised in urban/industrial areas as well as in arid tracts. Agro-forestry should be practiced to fulfil the need of rural communities for fuel wood and timber, and for secondary benefits of having income through fruits, seeds and other products.
- *Control on Felling of Trees*— Unauthorised tree felling and avoidable tree felling in the name of development must be eliminated. In developmental projects, like industrial, power and irrigation projects, and housing projects only such trees should be cut which are really necessary to be removed and others be left as such. Compensatory forestation (as per rules for number of trees cut) should be done.
- *Utilization of Biomass for making manure/ Compost.*
- *Carrying out Soil improvement.*
- *Planting Leguminous Varieties of Plants (which fix atmospheric nitrogen for soil enrichment.*
- **Utilization of indigenous and acclimatized species of trees according to soil condition, climate, irrigation facility, need of community.*
- *Utilization of 'Salt affected land' for plantation* – Soil improvement measures should be taken in the salt affected areas, which then can be utilized for carbon sequestration both in soil and plants. It is possible to develop green cover on such lands with diverse plant communities consisting of herbs, shrubs, and trees.
- ***Restoration of 'Mined out areas'.*
- *Utilization of Mined out Areas for Plantations*— The trees planted in mined out areas should be tolerant for minerals present in the soil/ water at the location.
- *Practicing 'Social Forestry'.*
- *Better care against fires*

***Notes on suitable varieties of plants for various conditions**

(1) Trees suitable for Saline Lands

(a) ***For high saline land*** *– Date palm, Zizyphus jujube (ber), Tamarindus indica (Tamarind / Imli), Cordia dichotama (Lasoda), Butea monosperm (Palas / Dhak), Diospyros melanoxylon / D.tomentosa (Tendu), Tamarax aphylla (Farash), etc. (Some Herbs like Latjira, Shail kanta, Moringa, Bhat-kataiya, etc. can grow on highly saline soil.)*

(b) ***For moderately saline land*** *– In addition to plants mentioned above in (a), the other varieties of suitable trees are: Azadirachta indica (Neem), Madhuca latifolia (mahua), Coconut, Syzygium cuminii / S.jambos (Jaman), Fig, Carrissa carandas (Karonda), Prosopis juliflora, etc. (Some fruits like Cherry, Phalsa, Pomegranate etc can also be grown.)*

(c) ***For low saline condition*** *– In addition to all the plants mentioned above in (b), other varieties of suitable trees are: Emblica officinalis (Amla /Aonla), Pisidium (Guava), Mango, Aegle marmelos (Bael), etc. (Grapes can also be grown.)*

(d) ***Plants for estuarine areas/ saline areas***

Halophytes—Salicornia brachiata*

(2) Suitable Plants for Dusty Areas

- ***Areas near Cement Factories (High Cement Dust Pollution):*** *Trees like Palas / Dhak (Butea monosperm), Gulmohar, Madhuca indica (Mahua), Terminalia arjuna (Arjun), Tamarindus Indus (Imli), shrubs like Bougain- villea, Behaya, and red / yellow Kaner, etc. survive in high cement dust pollution around cement factories.*
- ***Suitable Plants for Road-side Plantation:*** *The plants suitable for road-side plantation include: Neem, Satwin, Satavar, Peepal, Sheesham, Tamarind / Imli, Kadamb, Jangal Jalebi, Pakkar, Goolar, Mango, Jaman, Palas / Dhak (Butea monosperm), Gulmohar, Mahua, Arjun, Ashok, Shahtoot, Bamboo, Ber, Khajoor, Babool, etc.*
- ***Low Height Plants for Dividers of Roads:*** *Plant suitable for dividers should normally have low height. These may include: Kaner, Bottle- brush, China rose, Chandni, Bougainvillea, Harshingar, Satavar, Rubber plant, Shahtoot, Ashok, etc.*
- ***Suitable Plants for Growing on Soil with Ash:*** *(Redundant ash ponds or already filled ponds with ash, over laid by soil and pits filled with good soil on the ash pond) — Babul, Subabul, Arjun, Ber, Lasoda, Behya, etc.*

(3) Household plants for Healthy Indoor Environment

NASA has published a list of 50 suitable household plants that can remove chemical vapours, resist insects and are easy to maintain. [Areca palm (Areca lutescens) is a natural 'Humidifier'; Spathiphyllum Wallis/ Peace Lily which is good for removing Benzene vapours; Gerbera Daisy flowers absorb 'Benzene and Formaldehyde', and 'Trichloroethylene/ vapours'; Dracaena absorbs 'Formaldehyde vapours'; Bamboo palm is a good 'Air purifier',etc.].

*****Note on Land Restoration of Mined out Areas***

The plantation in the degraded area would naturally add to the efforts of abatement of pollution and global warming. The following procedure may be adopted for restoration of mined out areas:

(i) *The mining authorities should be instructed to store the top soil of the land being mined (say up to 100 cm depth is the most fertile) during mining, while the stones and rocks out of overburden should be stacked separately for use as material for construction of roads and buildings.*

(ii) *Carry out 'Ecological survey' of major vegetative association and natural succession ie identifying pioneering indigenous species of grasses, herbs, shrubs, crops and trees.*

(iii) *Carry out 'Soil Characterisation' of mine spoils/ overburden/ dumps. Carry out soil testing with a view to find limiting factors for plant growth. Soil and mine spoils may contain some metallic compounds / ingredients* and therefore the plant species tolerant to those ingredients would be required to be selected.*

(iv) *While doing land restoration plan, carry out consolidation, grading, drainage, and restoration of top soil. Stabilize the slopes against land sliding and soil erosion by proper stage wise consolidation. For higher dumps, use two or three stair type formations. Use mechanical means for stabilization. Back fill the pits where necessary. Spread top soil if stacked separately. Apply the required soil amendments according to soil characteristics including addition of biofertilizers /manures and ploughing, mulching, etc. If top soil was not stacked in previous mining, carry out dressing by any available soil mixed with bio-manure, compost, and soil amendments.*

(v) *Select plant species according to (a) the result of ecological survey, (b) the actual soil characteristics, (c) trials in nurseries set up at site to eslabtish -the capability of plants for colonising in degraded area, (d) capabilities of plants in fixing atmospheric nitrogen (leguminosae varieties), conserving soil, and tolerating soil ingredients present. Give preference to the grasses having fodder value, plants of economic value, medicinal plants, fruit trees, plants required by community having faster growth which can survive under the prevailing condition on the degraded land.*

(vi) *Apply required moisture /water, and to plant grasses, shrubs and trees as required.*

(vii) *Make provision for 'Water harvesting and water conservation' in order to provide water for survival of plants.*

(viii) *Make provision for protection of plants against destruction by animals, pests, etc.*

(ix) *Make provision for collection of plant litter (without burning the same) for making compost and application of the same to the soil regularly.*

(x) *Test the soil year after year to monitor the changes in its characteristics so that when the soil becomes ready for more economical plantations, the new varieties can be introduced slowly in due course of time.*

REFERENCES

1. Vartika Gupta et al.' Air pollution monitoring in around Faizabad city utilizing Lichen distribution pattern'. Vth International conference on plants and environmental pollution; ICPEP-5, Feb 2015 NBRI, Lucknow.
2. DK Upretti,et al.' Species of lichen genus usnea; most suitable organism for long term pollution monitoring.; Vth International conference on plants and environmental pollution; ICPEP-5, Feb 2015 NBRI, Lucknow.
3. Climate Change 2007: The Physical Science Basis, Contribution of Working Group I to the Fourth Assessment Report of IPCC (Intergovernmental Panel on Climate Change).
4. "Trends in Atmospheric CO_2 by Dr Pieter Trans NOAA/ESRLwwwesrl.gov/gmd/ccgg/trends & Dr Ralph Keeling Scripps Institution of Oceanography Scripps CO_2 ucsd.edu ,Mauna Loa, Hawaii, as in Dec 2011 is 391.8ppmv.
5. T.J. Blasing DOI: 10.3334/CDIAC/atg.032, the green house gas concentration in troposphere, GWP for 100 years time horizon, atmospheric time in years, and increased radiative forcing W/m^2.
6. Environment Protection Rules, India, 1986.
7. Vertica Shukla and DK Uprati Environews 2011.

QUESTIONS

1. The ozone depletion in the stratosphere (outer atmosphere) is chiefly caused by the release of
   (a) Chlorofluorocarbons (CFCs); (b) Carbon dioxide;
   (c) Methane; (d) None of them.
2. The chlorofluorocarbons (CFCs) are released into the environment by humans, as they are largely used
   (a) In refrigeration and air conditioning (b) As cleaning solvents in factories
   (c) In aerosol sprays
   (d) As blowing agents to create certain kinds of plastic foams
   (e) All of the above.
3. Depletion of ozone layer in the outer atmosphere is likely to increase the incidence of
   (a) Skin cancer; (b) Lung cancer;
   (c) Bronchitis; (d) None of them.
4. Pollution by depletion of ozone layer in stratosphere, threatening the environment, is caused due to reaction of ozone with
   (a) Carbon monoxide; (b) Chlorine;
   (c) Sulphur dioxide; (d) Nitrous oxide.
5. Harmful U.V. radiations emanating from the sun are prevented from reaching the Earth by the presence of ozone in the
   (a) Thermosphere; (b) Mesosphere;
   (c) Stratosphere; (d) Troposphere.
6. The ozone layer thickness is measured in (a) Decibels (db); (b) Dobson Units (DU); (c) Becquerel (Bq); (d) None of them.
7. One Doboson Unit (DU) is equal to
   (a) 1 mm; (b) 0.1 mm;
   (c) 0.01 mm; (d) 0.001 mm.;
   (e) None of them.
8. The depletion of stratospheric ozone thickness to as low a value as about 150 DU (1.5 mm) as against the usual value of 350 DU (3.5 mm) to 250 DU (2.5 mm), creates ozone hole, the area of which in the southern hemisphere has been observed to be maximum in Sept. 2000 and Sept. 2006, equaling about
   (a) 26 million sq. km; (b) 19 million sq. km;
   (c) 9 million sq. km; (d) None of them.
9. In order to control depletion of ozone in the stratosphere, it is necessary to reduce CFC's and other such substances which release chorine atoms, which, in turn, react with ozone to convert it into ClO and, thereby reducing ozone. To achieve this objective, an international treaty was signed in 1987, imposing bans on the use of substances which release CFC's. This treaty is known as
   (a) Geneva Convention (b) London Amendment
   (c) Kyoto Protocol (d) Montreal Protocol.

10. The ozone strongly absorbs ultraviolet light in the region having wavelength in the range of
    (a) 360–460 nm; (b) 220–330 nm;
    (c) 135–200 nm; (d) None of the above.
11. The major green house gas which is responsible for causing about 60% of the green house effect on Earth, among the following is:
    (a) Carbon monoxide; (b) Methane; (c) Nitrous oxide; (d) Carbon dioxide.
12. An ordinary rain will generally be:
    (a) Slightly acidic; (b) Slightly alkaline; (c) Neither acidic nor alkaline.
13. A rainfall is generally classified as acidic, if this pH is less than equal to:
    (a) 5.6; (b) 6.5; (c) 7.0; (d) 7.5.
14. The fossil fuels responsible for causing acid rains are used in
    (a) Industries; (b) Domestic cooking and heating;
    (c) Automobiles; (d) All of the above.
15. One third of acid rains are caused by nitrogen oxides which are chiefly produced by
    (a) Auto and industrial emission; (b) Nitrogenous fertilizers;
    (c) Forests and grass lands on fire; (d) All of the above.
16. Which out of the following statements is incorrect in relation to checking and controlling acid rains?
    (a) Expensive scrubbers are installed in industries and thermal power stations.
    (b) Catalytic converters are installed in cars.
    (c) Use of petrol cars is promoted by discouraging the use of diesel cars.
    (d) None of the above.
17. What is the importance of having ozone in the stratosphere?
18. What is the effect of chlorofluorocarbons entering the stratosphere?
19. Describe the roll of Lichens as the biomonitors of pollution.

Answers: **1.** (a) **2.** (e) **3.** (a) **4.** (b) **5.** (c) **6.** (b) **7.** (c)
8. (a) **9.** (d) **10.** (b) **11.** (d) **12.** (a) **13.** (a) **14.** (d)
15. (d) **16.** (d)

Additional References on Air Pollution and Control

1. "Acidic Precipitation, "Vols. 1-5. Advances in Environmental Sciences, Springer-Verlag, London, 1989-90.
2. Agrawal, S.B. and Agrawal, M. (eds) (2000) *Environmental Pollution and Plant Responses,* CRC Press, Boca Raton, Florida.
3. Agarwal, Avinash Kumar, Tarun Gupta and Abhishek Kothari, 'Toxic Potential Evaluation of Particulate Matter Emitted from a Constant Speed Compression Ignition Engine: A Comparison between Straight Vegetable Oil and Mineral Diesel', **Aerosol Science and Technology, 44 (9)**, 724-733 (2010).
4. Agarwal, Avinash Kumar, Tarun Gupta and Abhishek Kothari, 'Particulate Emissions from Biodiesel Vs. Diesel Fuelled Compression Ignition Engine', **Renewable and Sustainable Energy Review** (DOI:10.1016/j.rser.2011.04.002), 15(6), 3278-3300 (2011).
5. Ahmad, R., Cartwright, M. and Taylor, F. (eds) (2001) *Analytical Methods for Environmental Monitoring*, Pearson Education, Harlow.
6. American Meteorological Society, **Bull. Am. Meteorol. Soc. 72**, 57-59 (1991).
7. Anjaneyulu, Y., Air Pollution and Control Technologies. Allied Publishers, Bangalore, 2002.
8. Arya, S.P. (1999), *Air Pollution Meteorology and Dispersion,* Oxford University Press, New York.
9. Ashby, E., and Anderson, M., "The Politics of Clean Air. " Oxford Univ. Press (Clarendon), New York, 1981.
10. Barrett, E. C., and Curtis, L.F., "Introduction to Environmental Remote Sensing," 3rd ed. Chapman & hall, London, 1992.
11. Benedick, R.E., "Ozone Diplomacy: New Directions in Safeguarding the Planet." Harvard University Press, Cambridge, MA,1991.
12. Bergin, M.H., S.N. Tripathi et al., 'The discoloration of the Taj Mahal due to particulate carbon and dust deposition', **Environmental Science & Technology, 49(2)**, 808-812, 2015.
13. Berthouex, P.M., Kurki-Suonio, I., Noll, K.E. and Ouano, E.A.R. 1995. Air Pollution Control, UNESCO Series of Learning Materials in Engineering Sciences.
14. Books, B.O., and Davis, W.F., "Understanding Indoor Air Quality. "CRC Press, Boca Raton, FL, 1992.
15. Brasseur, P. Orlando, J.L. and Tyndall, G.S. (eds) (1999) *Atmospheric Chemistry and Global Change*, Oxford University Press, New York.
16. Bridgman, H.A., "Global Air Pollution: Problems for the 1990's." Belhaven,London,1990.
17. Cadle, R.D., " The Measurement of Airborne Particles." Academic Press, New York, 1976.
18. Colvile, R.N., Hutchinson, E.J., Mindell, J.S. and Warren, R.F. (2001) 'The transport sector as a source of air pollution', **Atmospheric Environment 35:** 1537-1565.

19. Council on Environmental Quality, Annual Reports. Washington, DC,1984-1992.
20. De Nevers, N. 1995. Air Pollution Control Engineering. McGraw Hill & Co., New York.
21. Dwivedi D, Agarwal A.K. and Sharma M., 'Particulate Emission Characterization of a Biodiesel Vs. Diesel Fuelled Compression Ignition Transport Engine: A Comparative Study', **Atmospheric Environment Vol. 40**, 5586:5595, 2006.
22. Evans, G.F., Lumpkin, T.A., Smith, D.L., and Somerville, M.C., **J. Air Waste Manage. Assoc. 42**, 1319-1323, (1992).
23. Faith, W.L., and Atkission, A.A., Air Pollution, 2nd ed. Wiley, New York, 1972. 'U.S. Interagency Team Proposes Program to Quantify Effects of Kuwait Oil Fires', **J. Air Waste Management Assoc. 41,(6)**, June 1991.
24. Fenger, J., Hertel, O. and Palmgern, F. (eds) (1998) Urban Air Pollution: European Aspects, Kluwer Academic Publishers, Dordrecht, The Netherlands.
25. Garbut, J.H. (2000) *Environmental Law: A Practical Handbook,* 3rd edn, Bembridge Palladian.
26. Georgii, H.W., " Atmospheric Pollutants in Forest Areas: Their Deposition and Interception.", Kluwer Academic Publishers, Boston, 1986.
27. Graedel, T.F., and Crutzen, P.J., "Atmospheric Change: An Earth System Perspective." W.H. Freeman, New York, 1993.
28. Hansen, D.L. and David, L. (1999) *Indoor Air quality Issues,* Taylor & Francis, London.
29. Harrison, R.M. and Perry, R. (eds) (1986) *Handbook of Air Pollution Analysis*, Chapman and Hall, London, UK.
30. Harrison, R.M., and Young, R.J. (eds.), "Handbook of Air Pollution Analysis," 2nd Ed. Chapman & Hall, London, 1986.
31. Heck, R.M. and Farrauto, R.J. (1995) *Catalytic Air Pollution Control: commercial Technology,* Wiley, New York.
32. Heumann, W.L. (1997) Industrial Air Pollution Control Systems, McGraw-Hill, New York, USA.
33. Hinds, W.C. (1999) Aerosol Technology: Properties, Behaviour and Measurement of Airborne Particles, Wiley, New York.
34. Holgate, S.T., Samet, J.M., Koren, H.S. and Maynard, R.L. (Eds.), 1999, Air Pollution and Health, Academic Press, London, UK.
35. Holland, D.M., Principe, P.P. and Sickles II, J.E. (1999) ' Trends in atmospheric sulphur and nitrogen species in the eastern United States for 1989-1995', **Atmospheric Environment 33:** 37-49.
36. Hughes, D. (1998) *Air Pollution: Law and Legislation,* Parpworth, Joan Upson.
37. Hunt, G.T., and Maisel, B.E., **J. Air Waste Mange. Assoc. 42,** 672-680 (1992).
38. Jacob, D.J. (1999) Introduction to Atmospheric Chemistry, Princeton University Press, Princeton, New Jersey.
39. Jitendra Gangwar, Tarun Gupta, Sudhir Gupta and Avinash Kumar Agarval, 'Emission from Diesel Vs. Blodiesel Fuel used in a CRDI SUV Engine: PM Mass and Chemical Composition', Inhalation Taxicology, (DOI:10.3109/08958378.2011.582189), 23(8), 449-458 (2011).
40 Jitendra N. Gangwar, Tarun Gupta and Avinash K. Agarval, 'Composition and comparative toxicity of particulate matter emitted from a diesel and biodiesel fueled CRDI engine', Atmospheric Environment 46, 472-481 (2012).
41. Jones, A.P. (1999) ' Indoor air quality and health', **Atmospheric Environment 33:** 4535-4564.

42. Jones, K.C. and de Voogt, P. (1999) 'Persistent organic pollutants (POPs): state of the science', **Environmental Pollution 100:** 209-221.
43. Kamal K. Ujinwal and Tarun Gupta, 'Development of a Multipurpose Sampler to Measure Human Exposure to Respirable Airborne Particulate Matter', Indian Aerosol Science and Technology Association Conference, New-Delhi, (14th - 16th Nov, 2007).
44. Kirpa Ram, S. Singh, M.M. Sarin, A.K. Srivastava and S.N. Tripathi, 'Variability in aerosol optical properties over an urban site, Kanpur, in the Indo-Gangetic Plain: A case study of haze and dust events', **Atmospheric Research, 174-175**, 52-61, 2016.
45. Lal, S., S. Venkataramani, S.N. Tripathi et al., 'Loss of crop yields in India due to surface ozone: An estimation based on a network of observation, **Environmental Science and Pollution Research'**, DOI 10.1007/s11356-017-9729-3, 2017.
46. Lal, R., A. Nagpure, L. Luo, S.N. Tripathi et al., 'Municipal solid waste and dung cake burning: Discoloring the Taj Mahal and human health impacts in Agra', **Environmental Research Letters, 11(10)**, doi:1088/1748-9326/11/10/104009, 2016.
47. Lippmann, M. (ed.) (2000) *Environmental Toxicants,* Wiley, New York.
48. Lloyd, A.C. and Cackette. T.A. (2001) 'Diesel engines: environmental impact and control', **Journal of Air and Waste Management Association 51:** 809-847.
49. Lundgren, D.A., Harris, F.S., Jr., Marlow, W.H., Lippmann, M., Clark, W.E., and Durham, U.D. (eds.), " Aerosol Measurement." University Presses of Florida, Gainesville, 1979.
50. Masters, G.M. (1998) Environmental Engineering and Science, Prentice Hall, New Jersey, USA.
51. McMurry, P.H. (2000) ' A review of atmospheric aerosol measurements;, **Atmospheric Environment 34:** 1959-1999.
52. Mahajan, S.P. 2001. Pollution Control in Process Industries, 14th Reprint. Tata McGraw Hill & Co., New Delhi.
53. Manish Kumar, Mueksh Sharma and Tarun Gupta, 'Particulate Mass and Number Emission from CNG and Diesel-powered Three-Wheelar: Insight into OC-EC Characteristics', 6th International Symposium on Fuels and Lubricants, New Delhi. (9th -12th Mar, 2008).
54. McCabe, W.L., Smith, J.C., and Harriot, P. 1993. Unit Operations of Chemical Engineering. McGraw Hill, Inc. New York, 5th Edition.
55. Mellanby, K. (ed.), " Air Pollution, Acid Rain and the Environment." Elsevier Science Publishers Essex, England, 1988.
56 Mishra, S.K., N. Saha, S.N. Tripathi et al., 'Morphology, mineralogy and mixing of individual atmospheric particles over Kanpur (IGP): Relevance of homogeneous equivalent sphere approximation in radiative models', **Journal of Metrology Society of India (MAPAN), 32(3)**, 229-241, 2017.
57. Monteith, J.L. and Unsworth, M.H. (1990) Principles of Environmental Physics, Edward Arnold, London, UK.
58. Nagy, G.Z., The odor impact model. **J. Air Waste Manage. Assoc. 42**, 1567 (1992).
59. National Research Council, "Human exposure assessment for airborne pollutants: advances and opportunities. " National Academy Press, Washington, DC, 1991.
60. National Society for Clean Air and Environmental Protection (annual) NSCA Pollution Handbook, NSCA, Brighton, UK.

61. Newman, L. (ed.), "Measurement Challenges in Atmospheric Chemistry." American Chemical Society, Washington, DC, 1993.
62. Noll, K.E., and Miller, T.L., "Air Monitoring Survey Design." Ann Arbor Science Publishers, Ann Arbor, MI, 1977.
63. Pablo Ruiz, Joy Lawrence, Mike Wolfsone, Stephen Ferguson, Tarun Gupta, Choong-Min Kang and Petros Koutrakis, 'Development and Evaluation of a Photochemical Chamber to Examine the Toxicity of Coal - Fired Power Plant Emission', Inhalation **Toxicology, 19 (8)**, 597-606 (2007).
64. Pablo A. Ruiz, J.E. Lawrence, J.M. Wolfson, S.T. Ferguson, Tarun Gupta, C.-M. Kang and P. Koutrakis, 'Development of a Photochemical Chamber for the Toxicological Evalution of Coal Combustion Emissions', Annual Conference of American Association for Aerosol Research (AAAR), Austin, TX, USA (17th -21st Oct, 2005).
65. Patange, O., N. Ramanathan, S.N. Tripathi et al., 'Reductions in Indoor black carbon concentration from improved biomass stoves in rural India', **Environmental Science & Technology, 49(7),** 4749-4756, 2015.
66. Philip Demokritou, Tarun Gupta, Stephen Ferguson and Petros Koutrakis, 'Development and laboratory Characterization of a Prototype Coarse Paticle Concentrator for Inhalation Toxicological Studies', **Journal of Aerosol Science, 33**, 1111-1123 (2002).
67. Philip Demokritou, Tarun Gupta, Petros Koutrakis, 'A high Volume Apparatus for the Condensation Growth of Ultarfine Particles for inhalation and Toxicological Studies', **Aerosol Science and Technology, 36,** 1061-1072 (2002).
68. Phili Demokritou, Tatun Gupta, Stephen Ferguson and Petros Koutrakis, 'Development of a High Volume Concentrated Ambient Particle System (CAPS) for Human and Animal Inhalation Toxicological Studies', **Inhalation Toxicology, 15**, 111-129 (2003).
69. Peirce, J.J., Weiner, R.F. and Vesilind, P.A. (1998), *Environmental Pollution and Control*, Butterworth-Heinemann, Boston, USA.
70. Salthammer, T. (1999) *Organic Indoor Air Pollutants: Occurrence-Measurement- Evaluation,* Wiley-VCH, Weinheim.
71. Samet, J.M., and Spengler, J.D. (eds), "Indoor Air Pollution: A Health Perpective." Johns Hopkins University Press, Baltimore, 1991.
72. Sarish, R., P.M. Shamjad, Thamban Navaneeth, S.N. Tripathi et al., 'Temporal characteristics of brown carbon over the central Indo-Gangetic Plain', **Environmental Science & Technology, 51(12)**, 6765–6772, 2017.
73. Seinfeld, J.H., " Atmospheric Chemistry and Physics of Air Pollution," Wiley, New York, 1986.
74. Seinfeld, J.H. and Pandis, S.N. (2000) Atmospheric Chemistry and Physics, Wiley, New York.
75. Sen, I.S., Michel Bizimis, S.N. Tripathi and Devajyoti Paul, 'Lead isotopic fingerprinting of aerosol to characterize the source of atmospheric lead in an industrial city of India', **Atmospheric Environment, 129**, 27–33, 2016.
76. Sickles, J.E., II, **Adv. Environ. Sci. Technol. 25**, 51-128 (1992).
77. Singh, S.N. (ed.) (2000) *Trace Gas Emissions and Plants,* Kluwer Academic, Dordrecht, 2000.
78. Sloane, C.S., and Tesche, T.W., "Atmospheric Chemistry: Models and Predictions for Climate and Air Quality." Lewis Publishers, Chelsea, MI, 1991.
79. Snider, G., S.N. Tripathi, A. Misra, 'Variation in global chemical composition of PM2.5: Emerging results from SPARTAN,' **Atmospheric Chemistry and Physics, 16**, 9629–9653, 2016.

80. Stanley-Wood N.G., and Lines, R.W. (eds.), "Particle Size Analysis." Royal Society of Chemistry, Cambridge, UK, 1992.
81. Summerhays, B.E., Procedures for estimating emissions from the cleanup of Superfund sites, **J. Air Waste Management Assoc. 40(1)**, January 1990.
82. Tarun Gupta, Philip Demokritou and Petros Kourtrakis, 'Development and Evaluation of an Ultrafine Particle Concentrator for Inhalation Toxicological Studies', **inhalation Toxicology, 16 (13)**, 851-862 (2004).
83. Thamban, N., S.N. Tripathi et al., 'Internally mixed black carbon in the Indo-Gogetic Plain and its effect on absorption enhancement', **Atmospheric Research, 197**, 211-223, 2017.
84. Tiwari, Surash, Philip K. Hopke, Devraj Thimmaiah, S.N. Tripathi, el al., 'Nature and sources of ionic species in precipitation across the indo–Gangetic Plains, India', **Areosol and Air Quality Research, 16(4)**, 943-957, 2016.
85. Tomatis, L. (ed.), "Air Pollution and Human Cancer." Springer-Verlag, New York, 1990.
86. Treshow, M. and Anderson, F.K. (1989) *Plant Stress from Air Pollution,* Wiley, Chichester, UK.
87. United Kingdom Building Effects Review Group (1989) *The Effects of Acid Deposition on Buildings and Building Materials,* DoE/HMSO, London, UK.
88. VanAardenne, J.A., Carmichael, G.R., Levy II, H., Streets D. and Hordijk, L. (1999) 'Anthropogenic *NOx* emissions in Asia in the period 1990-2020', **Atmospheric Environment 33:** 633-646.
89. Van Loon, G.W. and Duffy, S.J. (2000) Environmental Chemistry-a Global Perspective, Oxford University Press, Oxford.
90. Villalobos, A.M., S.N. Tripathi et al, 'Source apportionment of carbonaceous fine particulate matter (PM2.5) in two contrasting cities across the Indo-Gangetic Plain', **Atmospheric Pollution Research, 6(3)**, 398-405, 2015.
91. Wanta, R.C., and Lowry, W.P., The meteorological setting for dispersal of air pollutants, in "Air Pollution, " 3rd ed., Vol. I, "Air Pollutants, Their Transformation and Transport" (A.C. Stern, ed.). Academic Press, New York, 1976.
92. Warneck, P. (1988) Chemistry of the Natural Atmosphere, Academic Press, San Diego, USA.
93. Watkins, L.H. (1991) *Air Pollution from Road Vehicles,* HMSO, London, UK.
94. Wellburn, A. (1994) Air Pollution and Climate Change, Longman, Harlow, UK.
95. Wells, N. (1997) The Atmosphere and Ocean, Wiley, Chichester.
96. Willeke, K. and Baron, P.A. (eds) (1993) *Aerosol Measurement: Principles, Techniques and Applications*, van Nostrand Reinhold, New York, USA.
97. Winegar, E.D., and Keith, L.H., "Sapling and Analysis of Airborne Pollutants," Lewis Publishers, Boca Raton, FL, 1993
98. World Health Organization (1999), *Monitoring Ambient Air Quality for Health Impact Assessment,* WHO Regional Office for Europe, Copenhagen.
99. Yocom, J.E and McCarthy, S.M. (1991) *Measuring Indoor Air Quality,* Wiley, Chichester.
100. Yunus, M. and Iqbal, M. (1997), *Plant Response to Air Pollution,* John Wiley & Sons, Chichester.

Air Pollution and Control

ER. RAJNI KANT & DR. KESHAV KANT

FEEDBACK FORM

1. Which topics from your examination point of view are inadequately or not covered in this book?

____

____

2. Is there any misprints/ mistakes/ factual inaccuracy in this book? If so, please specify.

____

____

3. What is your assessment of this book as regards the presentation of the subject-matter, expression, precision and price in relation to the other recommended books available on this subject?

____

____

4. Which competing books you regard as better than this? Please mention their author and publishers.

____

____

5. Any further suggestions/comments you would like to make for the improvement of the book?

____

____

PS: If needed, please attach a separate sheet.

Name : ____________________

College : ____________________

Address : ____________________

Email : ____________________

Mobile : ____________________

Please mail the feedback form at:

KHANNA BOOK PUBLISHING CO. (P) LTD.

4C/4344, Ansari Road, Daryaganj, New Delhi-110002

You can also email this feedback form at contact@khannabooks.com or Whatsapp at +91 – 99109 09320

BEST FEEDBACK WILL BE REWARDED SUITABLY